GEOTECHNICAL EARTHQUAKE ENGINEERING HANDBOOK

Robert W. Day

McGRAW-HILL

New York   Chicago   San Francisco   Lisbon   London   Madrid
Mexico City   Milan   New Delhi   San Juan   Seoul
Singapore   Sydney   Toronto

Cataloging-in-Publication Data is on file with the Library of Congress

McGraw-Hill

A Division of The **McGraw·Hill** Companies

1 2 3 4 5 6 7 8 9 0   DOC/DOC   0 7 6 5 4 3 2 1

ISBN 0-07-137782-4

*The sponsoring editor for this book was Larry S. Hager, the editing supervisor
was Penny Linskey, and the production supervisor was Sherri Souffrance. It was
set in Times Roman following the HB1A design by Deirdre Sheean of McGraw-Hill
Professional's Hightstown, N.J., composition unit.*

Printed and bound by R. R. Donnelley & Sons Company.

This book was printed on recycled, acid-free paper containing
a minimum of 50% recycled, de-inked fiber.

Dedicated with love to my wife, Deborah

Chapter 4. Earthquake Structural Damage   4.1

Part 2 Geotechnical Earthquake Engineering Analyses

Chapter 5. Site Investigation for Geotechnical Earthquake Engineering   5.3

Chapter 6. Liquefaction   6.1

CONTENTS

Chapter 7. Earthquake-Induced Settlement   7.1

Chapter 8. Bearing Capacity Analyses for Earthquakes   8.1

Chapter 9. Slope Stability Analyses for Earthquakes   9.1

Chapter 10. Retaining Wall Analyses for Earthquakes   10.1

Chapter 11. Other Geotechnical Earthquake Engineering Analyses   11.1

Part 3   Site Improvement Methods to Mitigate Earthquake Effects

Chapter 12. Grading and Other Soil Improvement Methods   12.3

Chapter 13. Foundation Alternatives to Mitigate Earthquake Effects   13.1

Part 4   Building Codes

Chapter 14. Earthquake Provisions in Building Codes   14.3

Appendix A. Glossaries   A.1

Appendix B. EQSEARCH, EQFAULT, and FRISKSP Computer Programs   *B.1*

Appendix C. Conversion Factors   C.1

Appendix D. Example of a Geotechnical Report Dealing with
Earthquake Engineering   D.1

Appendix E. Solution to Problems   E.1

Appendix F. References   F.1

PREFACE

The purpose of this book is to present the practical aspects of geotechnical earthquake engineering. Because of the assumptions and uncertainties associated with geotechnical engineering, it is often described as an "art" rather than exact science. Geotechnical earthquake engineering is even more challenging because of the inherent unknowns associated with earthquakes. Because of these uncertainties in earthquake engineering, simple analyses are prominent in this book, with complex and theoretical evaluations kept to an essential minimum.

The book is divided into four separate parts. Part 1 (Chaps. 2 to 4) provides a discussion of basic earthquake principles, common earthquake effects, and typical structural damage caused by the seismic shaking. Part 2 (Chaps. 5 to 11) deals with earthquake computations for conditions commonly encountered by the design engineer, such as liquefaction, settlement, bearing capacity, and slope stability. Part 3 (Chaps. 12 and 13) discusses site improvement methods that can be used to mitigate the effects of the earthquake on the structure. Part 4 (Chap. 14) is a concluding chapter dealing with building codes.

The book contains practical analyses for geotechnical earthquake engineering. There may be local building code, government regulations, or other special project requirements that are more rigorous than the procedures outlined in this book. The analyses presented here should not replace experience and professional judgment. Every project is different, and the engineering analyses described in this book may not be applicable for all circumstances.

Robert W. Day

ACKNOWLEDGMENTS

I am grateful for the contributions of the many people who helped make this book. Special thanks are due to the International Conference of Building Officials, who sponsored the author's work. I would also like to thank the following individuals for their contributions:

- Professor Nelson, who provided Fig. 3.46 and additional data concerning the Turnagain Heights landslide.
- Thomas Blake, who provided assistance in the use and understanding of his EQSEARCH, EQFAULT, and FRISKSP computer programs.
- Professor Robert Ratay, who reviewed the proposed content of the book and provided many helpful suggestions during its initial preparation.
- Professor Charles C. Ladd, Massachusetts Institute of Technology, who reviewed the draft of the author's book titled *Geotechnical and Foundation Engineering: Design and Construction*, portions of which have been reproduced in this book.

I am also indebted to Gregory Axten, president of American Geotechnical, for his support and encouragement during the development of the book. I would also like to thank Tom Marsh for his invaluable assistance during the preparation of the book, Rick Walsh for his help with the engineering analyses, and Eric Lind for supplying some of the deep foundation photographs.

Tables and figures taken from other sources are acknowledged where they occur in the text. Finally, I wish to thank Larry Hager, Penny Linskey, Sherri Souffrance, and others on the McGraw-Hill editorial staff, who made this book possible and refined my rough draft into this finished product.

CHAPTER 1

INTRODUCTION

1.1 GEOTECHNICAL EARTHQUAKE ENGINEERING

Geotechnical earthquake engineering can be defined as that subspecialty within the field of geotechnical engineering which deals with the design and construction of projects in order to resist the effects of earthquakes. Geotechnical earthquake engineering requires an understanding of basic geotechnical principles as well as geology, seismology, and earthquake engineering. In a broad sense, *seismology* can be defined as the study of earthquakes. This would include the internal behavior of the earth and the nature of seismic waves generated by the earthquake.

The first step in geotechnical earthquake engineering is often to determine the dynamic loading from the anticipated earthquake (the anticipated earthquake is also known as the *design earthquake*). For the analysis of earthquakes, the types of activities that may need to be performed by the geotechnical engineer include the following:

- Investigating the possibility of liquefaction at the site (Chap. 6). Liquefaction can cause a complete loss of the soil's shear strength, which could result in a bearing capacity failure, excessive settlement, or slope movement.

- Calculating the settlement of the structure caused by the anticipated earthquake (Chap. 7).

- Checking the design parameters for the foundation, such as the bearing capacity and allowable soil bearing pressures, to make sure that the foundation does not suffer a bearing capacity failure during the anticipated earthquake (Chap. 8).

- Investigating the stability of slopes for the additional forces imposed during the design earthquake. In addition, the lateral deformation of the slope during the anticipated earthquake may need to be calculated (Chap. 9).

- Evaluating the effect of the design earthquake on the stability of retaining walls (Chap. 10).

- Analyzing other possible earthquake effects, such as surface faulting and resonance of the structure (Chap. 11).

- Developing site improvement techniques to mitigate the effects of the anticipated earthquake. Examples include ground stabilization and groundwater control (Chap. 12).

- Determining the type of foundation, such as a shallow or deep foundation, that is best suited for resisting the effects of the design earthquake (Chap. 13).

- Assisting the structural engineer by investigating the effects of ground movement due to seismic forces on the structure and by providing design parameters or suitable structural systems to accommodate the anticipated displacement (Chap. 13).

In many cases, the tasks listed above may be required by the building code or other regulatory specifications (Chap. 14). For example, the *Uniform Building Code* (1997), which is the building code required for construction in California, states (code provision submitted by the author, adopted in May 1994):

> The potential for soil liquefaction and soil strength loss during earthquakes shall be evaluated during the geotechnical investigation. The geotechnical report shall assess potential consequences of any liquefaction and soil strength loss, including estimation of differential settlement, lateral movement or reduction in foundation soil-bearing capacity, and discuss mitigating measures. Such measures shall be given consideration in the design of the building and may include, but are not limited to, ground stabilization, selection of appropriate foundation type and depths, selection of appropriate structural systems to accommodate anticipated displacement or any combination of these measures.

The intent of this building code requirement is to obtain an estimate of the foundation displacement caused by the earthquake-induced soil movement. In terms of accuracy of the calculations used to determine the earthquake-induced soil movement, Tokimatsu and Seed (1984) conclude:

> It should be recognized that, even under static loading conditions, the error associated with the estimation of settlement is on the order of ± 25 to 50%. It is therefore reasonable to expect less accuracy in predicting settlements for the more complicated conditions associated with earthquake loading....In the application of the methods, it is essential to check that the final results are reasonable in light of available experience.

1.2 ENGINEERING GEOLOGY

An *engineering geologist* is an individual who applies geologic data, principles, and interpretation so that geologic factors affecting the planning, design, construction, and maintenance of civil engineering works are properly recognized and utilized (*Geologist and Geophysicist Act* 1986). In some areas of the United States, there may be minimal involvement of engineering geologists except for projects involving such items as rock slopes or earthquake fault studies. In other areas of the country, such as California, the geotechnical investigations are usually performed jointly by the geotechnical engineer and the engineering geologist. The majority of geotechnical reports include both engineering and geologic aspects of the project, and the report is signed by both the geotechnical engineer and the engineering geologist.

The primary duty of the engineering geologist is to determine the location of faults, investigate the faults in terms of being either active or inactive, and evaluate the historical records of earthquakes and their impact on the site. These studies by the engineering geologist will help to define the design earthquake parameters, such as the peak ground acceleration and magnitude of the anticipated earthquake. The primary duty of the geotechnical engineer is to determine the response of soil and rock materials for the design earthquake and to provide recommendations for the seismic design of the structure.

1.3 GEOTECHNICAL ENGINEERING TERMS

Like most fields, geotechnical engineering has its own unique terms and definitions. Appendix A presents a glossary, which has been divided into five different parts, as follows:

Glossary 1: Field Testing Terminology

Glossary 2: Laboratory Testing Terminology

Glossary 3: Terminology for Engineering Analysis and Computations

Glossary 4: Compaction, Grading, and Construction Terminology

Glossary 5: Earthquake Terminology

1.4 SYMBOLS AND UNITS

A list of symbols is provided at the beginning of most chapters. An attempt has been made to select those symbols most frequently listed in standard textbooks and used in practice. Units that are used is this book consist of the following:

1. International System of Units (SI).
2. Inch-pound units (I-P units), which is also frequently referred to as the U.S. Customary System (USCS) units. Appendix C presents factors that can be used to convert USCS values into SI units.

In some cases, figures have been reproduced that use the old metric system (e.g., stress in kilograms per centimeter squared). These figures have not been revised to reflect SI units.

1.5 BOOK OUTLINE

Part 1 of the book, which consists of Chaps. 2 through 4, presents a brief discussion of basic earthquake principles, common earthquake effects, and structural damage caused by earthquakes. Numerous photographs are used in these three chapters in order to show the common types of earthquake effects and damage.

Part 2 of the book deals with the essential geotechnical earthquake engineering analyses, as follows:

- Field exploration (Chap. 5)
- Liquefaction (Chap. 6)
- Settlement of structures (Chap. 7)
- Bearing capacity (Chap. 8)
- Slope stability (Chap. 9)
- Retaining walls (Chap. 10)
- Other earthquake effects (Chap. 11)

Part 3 of the book (Chaps. 12 and 13) presents commonly used site improvement methods and foundation alternatives. Part 4 (Chap. 14) presents a brief introduction to building codes as they pertain to geotechnical earthquake engineering.

As mentioned in Sec. 1.3, a glossary in included in App. A. Other items are presented in the appendices:

- Data from the EQSEARCH, EQFAULT, and FRISKSP computer programs (App. B)
- Conversion factors (App. C)

- Example of the portion of the geotechnical report dealing with earthquake engineering (App. D)
- Solution to problems (App. E)
- References (App. F)

P · A · R · T · 1

INTRODUCTION TO EARTHQUAKES

CHAPTER 2
BASIC EARTHQUAKE PRINCIPLES

The following notation is used in this chapter:

SYMBOL DEFINITION

a_{max} Maximum horizontal acceleration at the ground surface (also known as the peak ground acceleration)

A Maximum trace amplitude recorded by a Wood-Anderson seismograph

A' Maximum ground displacement in micrometers

A_f Area of the fault plane

A_0 Maximum trace amplitude for the smallest recorded earthquake ($A_0 = 0.001$ mm)

D Average displacement of the ruptured segment of the fault

g Acceleration of gravity

M_L Local magnitude of the earthquake

M_0 Seismic moment of the earthquake

M_s Surface wave magnitude of the earthquake

M_w Moment magnitude of the earthquake

Δ Epicentral distance to the seismograph, measured in degrees

μ Shear modulus of the material along the fault plane

2.1 PLATE TECTONICS

The theory of plate tectonics in the 1960s has helped immeasurably in the understanding of earthquakes. According to the plate tectonic theory, the earth's surface contains tectonic plates, also known as lithosphere plates, with each plate consisting of the crust and the more rigid part of the upper mantle. Figure 2.1 shows the locations of the major tectonic plates, and the arrows indicate the relative directions of plate movement. Figure 2.2 shows the locations of the epicenters of major earthquakes. In comparing Figs. 2.1 and 2.2, it is evident that the locations of the great majority of earthquakes correspond to the boundaries between plates. Depending on the direction of movement of the plates, there are three types of plate boundaries: divergent boundary, convergent boundary, and transform boundary.

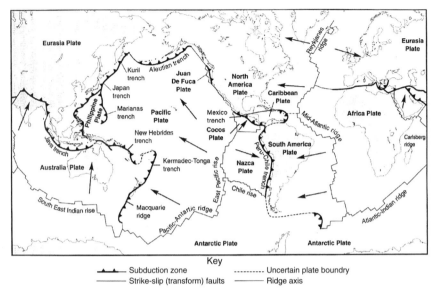

FIGURE 2.1 The major tectonic plates, mid-oceanic ridges, trenches, and transform faults of the earth. Arrows indicate directions of plate movement. (*Developed by Fowler 1990, reproduced from Kramer 1996.*)

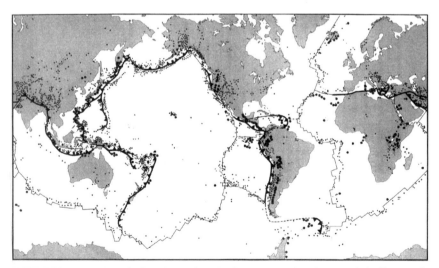

FIGURE 2.2 Worldwide seismic activity, where the dots represent the epicenters of significant earthquakes. In comparing Figs. 2.1 and 2.2, the great majority of the earthquakes are located at the boundaries between plates. (*Developed by Bolt 1988, reproduced from Kramer 1996.*)

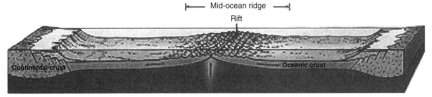

FIGURE 2.3 Illustration of a divergent boundary (seafloor spreading). (*From USGS.*)

Divergent Boundary. This occurs when the relative movement of two plates is away from each other. The upwelling of hot magma that cools and solidifies as the tectonic plates move away from each other forms spreading ridges. Figure 2.3 illustrates seafloor spreading and the development of a mid-ocean ridge. An example of a spreading ridge is the mid-Atlantic ridge (see Fig. 2.1). Earthquakes on spreading ridges are limited to the ridge crest, where new crust is being formed. These earthquakes tend to be relatively small and occur at shallow depths (Yeats et al. 1997).

When a divergent boundary occurs within a continent, it is called *rifting*. Molten rock from the asthenosphere rises to the surface, forcing the continent to break and separate. Figure 2.4 illustrates the formation of a continental rift valley. With enough movement, the rift valley may fill with water and eventually form a mid-ocean ridge.

Convergent Boundary. This occurs when the relative movement of the two plates is toward each other. The amount of crust on the earth's surface remains relatively constant, and therefore when a divergent boundary occurs in one area, a convergent boundary must occur in another area. There are three types of convergent boundaries: oceanic-continental subduction zone, oceanic-oceanic subduction zone, and continent-continent collision zone.

1. *Oceanic-continental subduction zone:* In this case, one tectonic plate is forced beneath the other. For an oceanic subduction zone, it is usually the denser oceanic plate that will subduct beneath the less dense continental plate, such as illustrated in Fig. 2.5. A deep-sea trench forms at the location where one plate is forced beneath the other. Once the subducting oceanic crust reaches a depth of about 60 mi (100 km), the crust begins to melt and some of this magma is pushed to the surface, resulting in volcanic eruptions (see Fig. 2.5). An example of an oceanic-continental subduction zone is seen at the Peru-Chile trench (see Fig. 2.1).

2. *Oceanic-oceanic subduction zone:* An oceanic-oceanic subduction zone often results in the formation of an island arc system, such as illustrated in Fig. 2.6. As the subducting oceanic crust meets with the asthenosphere, the newly created magma rises to the surface and forms volcanoes. The volcanoes may eventually grow tall enough to form a chain of islands. An eexample of an oceanic-oceanic subjection zone is the Aleutian Island chain (see Fig. 2.1).

The earthquakes related to subduction zones have been attributed to four different conditions (Christensen and Ruff 1988):

- Shallow interplate thrust events caused by failure of the interface between the down-going plate and the overriding plate.
- Shallow earthquakes caused by deformation within the upper plate.
- Earthquakes at depths from 25 to 430 mi (40 to 700 km) within the down-going plate.

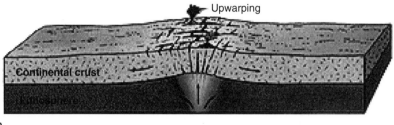

A.

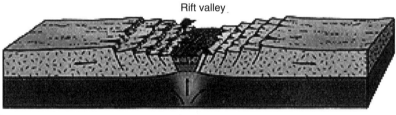

B.

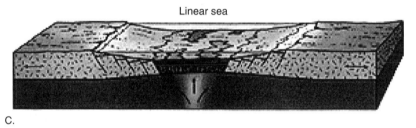

C.

FIGURE 2.4   Illustration of a divergent boundary (rift valley). (A) The upwarping of the ground surface, (B) the rift valley development, and (C) flooding to form a linear sea. (*From USGS.*)

• Earthquakes that are seaward of the trench, caused mainly by the flexing of the down-going plate, but also by compression of the plate.

In terms of the seismic energy released at subduction zones, it has been determined that the largest earthquakes and the majority of the total seismic energy released during the past century have occurred as shallow earthquakes at subduction zone–plate boundaries (Pacheco and Sykes 1992).

3. *Continent-continent collision zone:*   The third type of convergent boundary is the continent-continent collision zone, which is illustrated in Fig. 2.7. This condition occurs when two continental plates collide with each other, causing the two masses to squeeze, fold, deform, and thrust upward. According to Yeats et al. (1997), the Himalaya Mountains

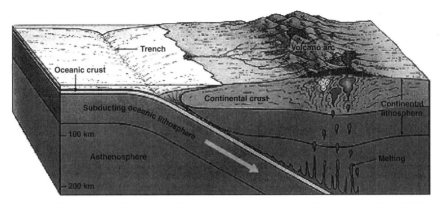

FIGURE 2.5 Illustration of a convergent boundary (oceanic-continental subduction zone). (*From USGS.*)

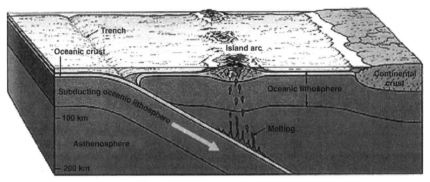

FIGURE 2.6 Illustration of a convergent boundary (oceanic-oceanic subduction zone). (*From USGS.*)

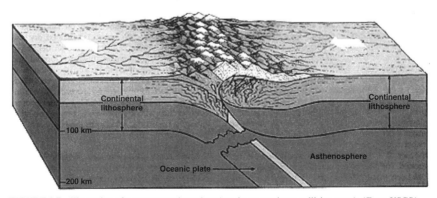

FIGURE 2.7 Illustration of a convergent boundary (continent-continent collision zone). (*From USGS.*)

mark the largest active continent-continent collision zone on earth. They indicate that the collision between the Indian subcontinent and the Eurasia plate began in early tertiary time, when the northern edge of the Indian plate was thrust back onto itself, with the subsequent uplifting of the Himalaya Mountains.

Transform Boundary.   A transform boundary, or transform fault, involves the plates sliding past each other, without the construction or destruction of the earth's crust. When the relative movement of two plates is parallel to each other, strike-slip fault zones can develop at the plate boundaries. *Strike-slip faults* are defined as faults on which the movement is parallel to the strike of the fault; or in other words, there is horizontal movement that is parallel to the direction of the fault.

California has numerous strike-slip faults, with the most prominent being the San Andreas fault. Figure 2.8 shows that large earthquakes have occurred on or near the San Andreas fault, and Fig. 2.9 presents an example of the horizontal movement along this fault (1906 San Francisco earthquake). Since a boundary between two plates occurs in California, it has numerous earthquakes and the highest seismic hazard rating in the continental United States (see Fig. 2.10).

The theory of plate tectonics is summarized in Table 2.1. This theory helps to explain the location and nature of earthquakes. Once a fault has formed at a plate boundary, the shearing resistance for continued movement of the fault is less than the shearing resistance

FIGURE 2.8   Epicenters of historic earthquakes (1812–1996). The map does not show all the epicenters of earthquakes with magnitude greater than 4.5, but rather is meant as an overview of large and destructive, fairly recent, or unusual earthquakes. Also shown are the traces of major faults. The magnitudes indicated are generally moment magnitude M_w for earthquakes above magnitude 6 and local magnitude M_L for earthquakes below magnitude 6 and for earthquakes which occurred before 1933. (*Source: USGS.*)

FIGURE 2.9   San Francisco earthquake, 1906. The fence has been offset 8.5 ft by the San Andreas fault displacement. The location is 0.5 mi northwest of Woodville, Marin County, California. (*Photograph courtesy of USGS.*)

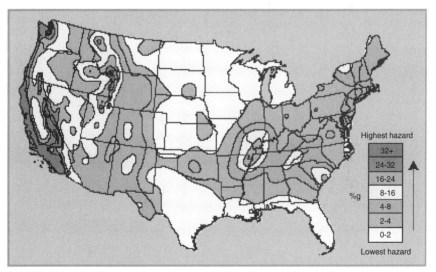

FIGURE 2.10   Seismic hazard map for the continental United States. The map indicates the lowest versus highest seismic hazard areas. (*Source: USGS.*)

TABLE 2.1   Summary of Plate Tectonics Theory

Plate boundary type	Type of plate movement	Categories	Types of earthquakes	Examples
Divergent boundary	Relative movement of the two plates is away from each other.	Seafloor spreading ridge (Fig. 2.3)	Earthquakes on spreading ridges are limited to the ridge crest, where new crust is being formed. These earthquakes tend to be relatively small and occur at shallow depths.	Mid-Atlantic ridge
		Continental rift valley (Fig. 2.4)	Earthquakes generated along normal faults in the rift valley.	East African rift
Convergent boundary	Relative movement of the two plates is toward each other.	Oceanic-continental subduction zone (Fig. 2.5)	1. Shallow interplate thrust events caused by failure of the interface between the down-going plate and the overriding plate. 2. Shallow earthquakes caused by deformation within the upper plate.	Peru-Chile trench
		Oceanic-oceanic subduction zone (Fig. 2.6)	3. Earthquakes at depths from 25 to 430 mi (40 to 700 km) within the down-going plate. 4. Earthquakes that are seaward of the trench, caused mainly by the flexing of the down-going plate, but also by compression of the plate.	Aleutian Island chain
		Continent-continent collision zone (Fig. 2.7)	Earthquakes generated at the collision zone, such as at reverse faults and thrust faults.	Himalaya Mountains
Transform boundary	Plates slide past each other, without the construction or destruction of the earth's crust.	Strike-slip fault zones (Fig. 2.9)	Earthquakes often generated on strike-slip faults.	San Andreas fault

required to fracture new intact rock. Thus faults at the plate boundaries that have generated earthquakes in the recent past are likely to produce earthquakes in the future. This principle is the basis for the development of seismic hazard maps, such as shown in Fig. 2.10.

The theory of plate tectonics also helps explain such geologic features as the islands of Hawaii. The islands are essentially large volcanoes that have risen from the ocean floor. The volcanoes are believed to be the result of a thermal plume or "hot spot" within the mantle, which forces magma to the surface and creates the islands. The thermal plume is believed to be relatively stationary with respect to the center of the earth, but the Pacific plate is moving to the northwest. Thus the islands of the Hawaiian chain to the northwest are progressively older and contain dormant volcanoes that have weathered away. Yeats et al. (1997) use an analogy of the former locations of the Pacific plate with respect to the plume as being much like a piece of paper passed over the flame of a stationary candle, which shows a linear pattern of scorch marks.

2.1.1 Types of Faults

A fault is defined as a fracture or a zone of fractures in rock along which displacement has occurred. The fault length can be defined as the total length of the fault or fault zone. The fault length could also be associated with a specific earthquake, in which case it would be defined as the actual rupture length along a fault or fault zone. The rupture length could be determined as the distance of observed surface rupture.

In order to understand the terminology associated with faults, the terms "strike" and "dip" must be defined. The "strike" of a fault plane is the azimuth of a horizontal line drawn on the fault plane. The dip is measured in a direction perpendicular to the strike and is the angle between the inclined fault plane and a horizontal plane. The strike and dip provide a description of the orientation of the fault plane in space. For example, a fault plane defined as N70W 50NE would indicate a strike of N70W (North 70° West) and a dip of 50NE (50° to the Northeast).

Typical terms used to describe different types of faults are as follows:

- Strike-Slip Fault: During the discussion of the transform boundary in Section 2.1, a strike-slip fault was defined as a fault on which the movement is parallel to the strike of the fault. A strike-slip fault is illustrated in Fig. 2.11.

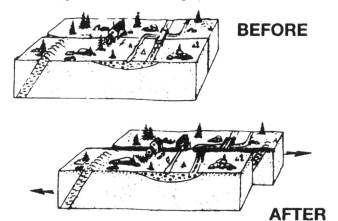

FIGURE 2.11 Illustration of a strike-slip fault. (*From Namson and Davis 1988.*)

- Transform Fault: A fault that is located at a transform boundary (see Section 2.1). Yeats et al. (1997) define a transform fault as a strike-slip fault of plate-boundary dimensions that transforms into another plate-boundary structure at its terminus.

- Normal Fault: Figure 2.12 illustrates a normal fault. The "hangingwall" is defined as the overlying side of a nonvertical fault. Thus, in Figure 2.12, the "hangingwall" block is that part of the ground on the right side of the fault and the "footwall" block is that part of the ground on the left side of the fault. A normal fault would be defined as a fault where the hangingwall block has moved downward with respect to the footwall block.

- Reverse Fault: Figure 2.13 illustrates a reverse fault. A reverse fault would be defined as a fault where the hangingwall block has moved upward with respect to the footwall block.

- Thrust Fault: A thrust fault is defined as a reverse fault where the dip is less than or equal to 45°.

- Blind Fault: A blind fault is defined as a fault that has never extended upward to the ground surface. Blind faults often terminate in the upward region of an anticline.

- Blind Thrust Fault: A blind reverse fault where the dip is less than or equal to 45°.

- Longitudinal Step Fault: A series of parallel faults. These parallel faults develop when the main fault branches upward into several subsidiary faults.

- Dip-Slip Fault: A fault which experiences slip only in the direction of its dip, or in other words, the movement is perpendicular to the strike. Thus a fault could be described as a "dip-slip normal fault," which would indicate that it is a normal fault (see Fig. 2.12) with the slip only in the direction of its dip.

- Oblique-Slip Fault: A fault which experiences components of slip in both its strike and dip directions. A fault could be described as a "oblique-slip normal fault," which would indicate that it is a normal fault (see Fig. 2.12) with components of slip in both the strike and dip directions.

- Fault Scarp: This generally only refers to a portion of the fault that has been exposed at

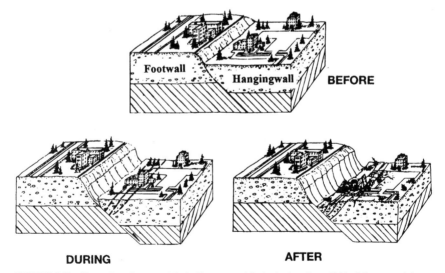

FIGURE 2.12   Illustration of a normal fault. For a normal fault, the hangingwall block has moved downward with respect to the footwall block. (*Adapted from Namson and Davis 1988.*)

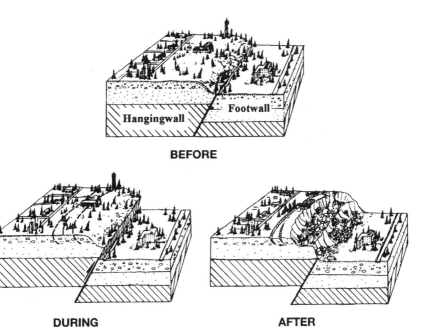

BEFORE

DURING   **AFTER**

FIGURE 2.13   Illustration of a reverse fault. For a reverse fault, the hangingwall block has moved upward with respect to the footwall block. (*Adapted from Namson and Davis 1988.*)

ground surface due to ground surface fault rupture. The exposed portion of the fault often consists of a thin layer of "fault gouge," which is a clayey seam that has formed during the slipping or shearing of the fault and often contains numerous slickensides.

2.2 SEISMOGRAPH

Most earthquakes are caused by the release of energy due to sudden displacements on faults. This is not to imply that all ground movement of a fault will produce an earthquake. For example, there can be fault creep, where the ground movement is unaccompanied by an earthquake. The major earthquake is characterized by the buildup of stress and then the sudden release of this stress as the fault ruptures.

A *seismograph* is an instrument that records, as a function of time, the motion of the earth's surface due to the seismic waves generated by the earthquake. The actual record of ground shaking from the seismograph, known as a *seismogram,* can provide information about the nature of the earthquake.

The simplest seismographs can consist of a pendulum or a mass attached to a spring, and they are used to record the horizontal movement of the ground surface. For the pendulum-type seismograph, a pen is attached to the bottom of the pendulum, and the pen is in contact with a chart that is firmly anchored to the ground. When the ground shakes during an earthquake, the chart moves, but the pendulum and its attached pen tend to remain more or less stationary because of the effects of inertia. The pen then traces the horizontal movement between the relatively stationary pendulum and the moving chart. After the ground shaking has ceased, the pendulum will tend to return to a stable position, and thus could indicate false ground movement. Therefore a pendulum damping system is required so that the ground displacements recorded on the chart will produce a record that is closer to the actual ground movement.

Much more sophisticated seismographs are presently in use. For example, the engineer is often most interested in the peak ground acceleration a_{max} during the earthquake. An *accelerograph* is defined as a low-magnification seismograph that is specially designed to record the ground acceleration during the earthquake. Most modern accelerographs use an electronic transducer that produces an output voltage which is proportional to the acceleration. This output voltage is recorded and then converted to acceleration and plotted versus time, such as shown in Fig. 2.14. Note that the velocity and displacement plots in Fig. 2.14 were produced by integrating the acceleration.

The data in Fig. 2.14 were recorded during the February 9, 1971, San Fernando earthquake. The three plots indicate the following:

1. *Acceleration versus time:*   The acceleration was measured in the horizontal direction. In Fig. 2.14, the maximum value of the horizontal acceleration a_{max}, which is also commonly referred to as the *peak ground acceleration,* is equal to 250 cm/s^2 (8.2 ft/s^2). The peak ground acceleration for this earthquake occurs at a time of about 13 s after the start of the record.

Since the acceleration due to earth's gravity g is 981 cm/s^2, the peak ground acceleration can be converted to a fraction of earth's gravity. This calculation is performed by dividing 250 cm/s^2 by 981 cm/s^2; or the peak ground acceleration a_{max} is equal to 0.255g.

2. *Velocity versus time:*   By integrating the horizontal acceleration, the horizontal velocity versus time was obtained. In Fig. 2.14, the maximum horizontal velocity at ground surface v_{max} is equal to 30 cm/s (1.0 ft/s). The maximum velocity at ground surface for this earthquake occurs at a time of about 10 s after the start of the record.

3. *Displacement versus time:*   The third plot in Fig. 2.14 shows the horizontal displacement at ground surface versus time. This plot was obtained by integrating the horizontal velocity data. In Fig. 2.14, the maximum horizontal displacement at ground surface is 14.9 cm (5.9 in). The maximum displacement at ground surface for this earthquake occurs at a time of about 10 s after the start of the record.

2.3 SEISMIC WAVES

The acceleration of the ground surface, such as indicated by the plot shown in Fig. 2.14, is due to various seismic waves generated by the fault rupture. There are two basic types of seismic waves: body waves and surface waves. P and S waves are both called body waves because they can pass through the interior of the earth. Surface waves are only observed close to the surface of the earth, and they are subdivided into Love waves and Rayleigh waves. Surface waves result from the interaction between body waves and the surficial earth materials. The four types of seismic waves are further discussed below:

1. *P wave (body wave):*   The P wave is also known as the primary wave, compressional wave, or longitudinal wave. It is a seismic wave that causes a series of compressions and dilations of the materials through which it travels. The P wave is the fastest wave and is the first to arrive at a site. Being a compression-dilation type of wave, P waves can travel through both solids and liquids. Because soil and rock are relatively resistant to compression-dilation effects, the P wave usually has the least impact on ground surface movements.

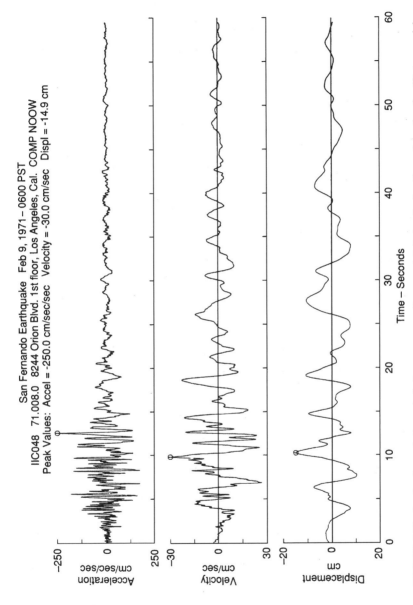

FIGURE 2.14 Acceleration, velocity, and displacement versus time recorded during the San Fernando earthquake. (*Data record from California Institute of Technology 1971, reproduced from Krinitzsky et al. 1993.*)

2. *S wave (body wave):* The S wave is also known as the secondary wave, shear wave, or transverse wave. The S wave causes shearing deformations of the materials through which it travels. Because liquids have no shear resistance, S waves can only travel through solids. The shear resistance of soil and rock is usually less than the compression-dilation resistance, and thus an S wave travels more slowly through the ground than a P wave. Soil is weak in terms of its shear resistance, and S waves typically have the greatest impact on ground surface movements.

3. *Love wave (surface wave):* Love waves are analogous to S waves in that they are transverse shear waves that travel close to the ground surface (Yeats et al. 1997).

4. *Rayleigh wave (surface wave):* Rayleigh waves have been described as being similar to the surface ripples produced by a rock thrown into a pond. These seismic waves produce both vertical and horizontal displacement of the ground as the surface waves propagate outward.

Generally, there is no need for the engineer to distinguish between the different types of seismic waves that could impact the site. Instead, the combined effect of the waves in terms of producing a peak ground acceleration a_{max} is of primary interest. However, it is important to recognize that the peak ground acceleration will be most influenced by the S waves and, in some cases, by surface waves. For example, Kramer (1996) states that at distances greater than about twice the thickness of the earth's crust, surface waves, rather than body waves, will produce peak ground motions.

2.4 MAGNITUDE OF AN EARTHQUAKE

There are two basic ways to measure the strength of an earthquake: (1) based on the earthquake magnitude and (2) based on the intensity of damage. *Magnitude* measures the amount of energy released from the earthquake, and *intensity* is based on the damage to buildings and reactions of people. This section discusses earthquake magnitude, and Sec. 2.5 discusses the intensity of the earthquake.

There are many different earthquake magnitude scales used by seismologists. This section discusses three of the more commonly used magnitude scales.

2.4.1 Local Magnitude Scale M_L

In 1935, Professor Charles Richter, from the California Institute of Technology, developed an earthquake magnitude scale for shallow and local earthquakes in southern California. This magnitude scale has often been referred to as the *Richter magnitude scale*. Because this magnitude scale was developed for shallow and local earthquakes, it is also known as the *local magnitude scale* M_L. This magnitude scale is the best known and most commonly used magnitude scale. The magnitude is calculated as follows (Richter 1935, 1958):

$$M_L = \log A - \log A_0 = \log A/A_0 \qquad (2.1)$$

where M_L = local magnitude (also often referred to as Richter magnitude scale)
A = maximum trace amplitude, mm, as recorded by a standard Wood-Anderson seismograph that has a natural period of 0.8 s, a damping factor of 80%, and a static magnification of 2800. The maximum trace amplitude must be that amplitude that would be recorded if a Wood-Anderson seismograph were

located on firm ground at a distance of exactly 100 km (62 mi) from the epicenter of the earthquake. Charts and tables are available to adjust the maximum trace amplitude for the usual case where the seismograph is not located exactly 100 km (62 mi) from the epicenter.

A_o = 0.001 mm. The zero of the local magnitude scale was arbitrarily fixed as an amplitude of 0.001 mm, which corresponded to the smallest earthquakes then being recorded.

As indicated above, Richter (1935) designed the magnitude scale so that a magnitude of 0 corresponds to approximately the smallest earthquakes then being recorded. There is no upper limit to the Richter magnitude scale, although earthquakes over an M_L of 8 are rare. Often the data from Wood-Anderson siesmographs located at different distances from the epicenter provide different values of the Richter magnitude. This is to be expected because of the different soil and rock conditions that the seismic waves travel through and because the fault rupture will not release the same amount of energy in all directions.

Since the Richter magnitude scale is based on the logarithm of the maximum trace amplitude, there is a 10-times increase in the amplitude for an increase in 1 unit of magnitude. In terms of the energy released during the earthquake, Yeats et al. (1997) indicate that the increase in energy for an increase of 1 unit of magnitude is roughly 30-fold and is different for different magnitude intervals.

For the case of small earthquakes (that is, $M_L < 6$), the center of energy release and the point where the fault rupture begins are not far apart. But in the case of large earthquakes, these points may be very far apart. For example, the Chilean earthquake of 1960 had a fault rupture length of about 600 mi (970 km), and the epicenter was at the northern end of the ruptured zone which was about 300 mi (480 km) from the center of the energy release (Housner 1963, 1970). This increased release of energy over a longer rupture distance resulted in both a higher peak ground acceleration a_{max} and a longer duration of shaking. For example, Table 2.2 presents approximate correlations between the local magnitude M_L and the peak ground acceleration a_{max}, duration of shaking, and modified Mercalli intensity level (discussed in Sec. 2.5) near the vicinity of the fault rupture. At distances farther from the epicenter or location of fault rupture, the intensity will decrease but the duration of ground shaking will increase.

TABLE 2.2 Approximate Correlations between Local Magnitude M_L and Peak Ground Acceleration a_{max}, Duration of Shaking, and Modified Mercalli Level of Damage near Vicinity of Fault Rupture

Local magnitude M_L	Typical peak ground acceleration a_{max} near the vicinity of the fault rupture	Typical duration of ground shaking near the vicinity of the fault rupture	Modified Mercalli intensity level near the vicinity of the fault rupture (see Table 2.3)
≤2	—	—	I–II
3	—	—	III
4	—	—	IV–V
5	0.09g	2 s	VI–VII
6	0.22g	12 s	VII–VIII
7	0.37g	24 s	IX–X
≥8	≥0.50g	≥34 s	XI–XII

Sources: Yeats et al. 1997, Gere and Shah 1984, and Housner 1970.

2.4.2 Surface Wave Magnitude Scale M_s

The surface wave magnitude scale is based on the amplitude of surface waves having a period of about 20 s. The surface wave magnitude scale M_s is defined as follows (Gutenberg and Richter 1956):

$$M_s = \log A' + 1.66 \log \Delta + 2.0 \tag{2.2}$$

where M_s = surface wave magnitude scale
   A' = maximum ground displacement, μm
   Δ = epicentral distance to seismograph measured in degrees (360° corresponds to circumference of earth)

The surface wave magnitude scale has an advantage over the local magnitude scale in that it uses the maximum ground displacement, rather than the maximum trace amplitude from a standard Wood-Anderson seismograph. Thus any type of seismograph can be used to obtain the surface wave magnitude. This magnitude scale is typically used for moderate to large earthquakes, having a shallow focal depth, and the seismograph should be at least 1000 km (622 mi) from the epicenter.

2.4.3 Moment Magnitude Scale M_w

The moment magnitude scale has become the more commonly used method for determining the magnitude of large earthquakes. This is because it tends to take into account the entire size of the earthquake. The first step in the calculation of the moment magnitude is to calculate the seismic moment M_0. The seismic moment can be determined from a seismogram using very long-period waves for which even a fault with a very large rupture area appears as a point source (Yeats et al. 1997). The seismic moment can also be estimated from the fault displacement as follows (Idriss 1985):

$$M_0 = \mu A_f D \tag{2.3}$$

where M_0 = seismic moment, N · m
   μ = shear modulus of material along fault plane, N/m^2. The shear modulus is often assumed to be 3×10^{10} N/m^2 for surface crust and 7×10^{12} N/m^2 for mantle.
   A_f = area of fault plane undergoing slip, m^2. This can be estimated as the length of surface rupture times the depth of the aftershocks.
   D = average displacement of ruptured segment of fault, m. Determining the seismic moment works best for strike-slip faults where the lateral displacement on one side of fault relative to the other side can be readily measured.

In essence, to determine the seismic moment requires taking the entire area of the fault rupture surface A_f times the shear modulus μ in order to calculate the seismic force (in newtons). This force is converted to a moment by multiplying the seismic force (in newtons) by the average slip (in meters), in order to calculate the seismic moment (in newton-meters).

Engineers may have a hard time visualizing the seismic moment. The reason is because the seismic force and the moment arm are in the same direction. In engineering, a moment is calculated as the force times the moment arm, and the moment arm is always perpendicular (not parallel) to the force. Setting aside the problems with the moment arm, the seismic moment does consider the energy radiated from the entire fault, rather than the energy from an assumed point source. Thus the seismic moment is a more useful measure of the strength of an earthquake.

Kanamori (1977) and Hanks and Kanamori (1979) introduced the moment magnitude M_w scale, in which the magnitude is calculated from the seismic moment by using the following equation:

$$M_w = -6.0 + 0.67 \log M_0 \qquad (2.4)$$

where M_w = moment magnitude of earthquake
M_0 = seismic moment of earthquake, N · m. The seismic moment is calculated from Eq. (2.3).

2.4.4 Comparison of Magnitude Scales

Figure 2.15 shows the approximate relationships between several different earthquake magnitude scales. When we view the data shown in Fig. 2.15, it would appear that there is an exact relationship between the moment magnitude M_w and the other various magnitude

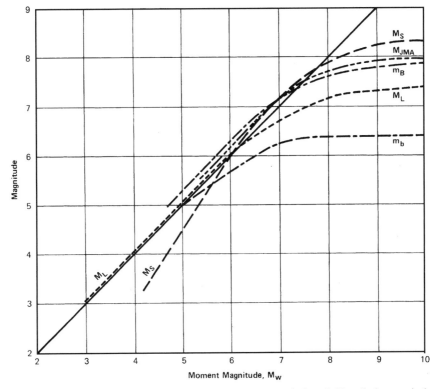

FIGURE 2.15 Approximate relationships between the moment magnitude scale M_w and other magnitude scales. Shown are the short-period body wave magnitude scale m_b, the local magnitude scale M_L, the long-period body wave magnitude scale m_B, the Japan Meteorological Agency magnitude scale M_{JMA} and the surface-wave magnitude scale M_S. (*Developed by Heaton et al. 1982, reproduced from Idriss 1985.*)

scales. But in comparing Eqs. (2.1) and (2.4), it is evident that these two equations cannot be equated. Therefore, there is not an exact and unique relationship between the maximum trace amplitude from a standard Wood-Anderson seismograph [Eq. (2.1)] and the seismic moment [Eq. (2.4)]. The lines drawn in Fig. 2.15 should only be considered as approximate relationships, representing a possible wide range in values.

Given the limitations of Fig. 2.15, it could still be concluded that the local magnitude M_L, the surface wave magnitude scale M_s, and moment magnitude M_w scales are reasonably close to one another below a value of about 7. At high magnitude values, the moment magnitude M_w tends to significantly deviate from these other two magnitude scales.

Note in Fig. 2.15 that the various relationships tend to flatten out at high moment magnitude values. Yeats et al. (1997) state that these magnitude scales are "saturated" for large earthquakes because they cannot distinguish the size of earthquakes based simply on the maximum trace amplitude recorded on the seismogram. Saturation appears to occur when the ruptured fault dimension becomes much larger than the wavelength of seismic waves that are used in measuring the magnitude (Idriss 1985). As indicated in Fig. 2.15, the local magnitude scale becomes saturated at an M_L of about 7.3.

2.4.5 Summary

In summary, seismologists use a number of different magnitude scales. While any one of these magnitude scales may be utilized, an earthquake's magnitude M is often reported without reference to a specific magnitude scale. This could be due to many different reasons, such as these:

1. *Closeness of the scales:* As discussed in Sec. 2.4.4, the local magnitude M_L, the surface wave magnitude M_s, and moment magnitude M_w scales are reasonably close to one another below a value of about 7. Thus as a practical matter, there is no need to identify the specific magnitude scale.

2. *Average value:* An earthquake's magnitude may be computed in more than one way at each seismic station that records the event. These different estimates often vary by as much as one-half a magnitude unit, and the final magnitude M that is reported can be the average of many estimates.

3. *Preseismograph event:* For earthquakes before the advent of the seismograph, the magnitude M of the earthquake is a rough estimate based on historical accounts of damage. In these cases, it would be impractical to try to determine the magnitude for each of the different magnitude scales.

4. *Lack of seismograph data:* Even after the advent of the seismograph, there may still be limited data available for many parts of the world. For example, Hudson (1970) states that not a single ground acceleration measurement was obtained for the earthquakes in Mexico (1957), Chile (1960), Agadir (1960), Iran (1962), Skopje (1963), and Alaska (1964). With only limited data, the earthquake magnitude is often an estimate based on such factors as type of damage, extent of damage, and observations concerning any surface fault rupture.

At high magnitude values, it is often desirable to determine or estimate the earthquake magnitude based on the moment magnitude M_w scale. This is because M_w tends to significantly deviate from the other magnitude scales at high magnitude values and M_w appears to better represent the total energy released by very large earthquakes. Thus for very large earthquakes, the moment magnitude scale M_w would seem to be the most appropriate magnitude scale. In terms of moment magnitude M_w, the top five largest earthquakes in the world for the past century are as follows (USGS 2000a):

Ranking	Location	Year	Moment magnitude M_w
1	Chile	1960	9.5
2	Alaska	1964	9.2
3	Russia	1952	9.0
4	Ecuador	1906	8.8
5	Japan	1958	8.7

2.5 INTENSITY OF AN EARTHQUAKE

The intensity of an earthquake is based on the observations of damaged structures and the presence of secondary effects, such as earthquake-induced landslides, liquefaction, and ground cracking. The intensity of an earthquake is also based on the degree to which the earthquake was felt by individuals, which is determined through interviews.

The intensity of the earthquake may be easy to determine in an urban area where there is a considerable amount of damage, but could be very difficult to evaluate in rural areas. The most commonly used scale for the determination of earthquake intensity is the *modified Mercalli intensity scale,* which is presented in Table 2.3. As indicated in Table 2.3, the intensity ranges from an earthquake that is not felt (I) up to an earthquake that results in total destruction (XII). In general, the larger the magnitude of the earthquake, the greater the area affected by the earthquake and the higher the intensity level. Figures 2.16 to 2.18 present the locations of U.S. earthquakes causing VI to XII levels of damage according to the modified Mercalli intensity scale. Table 2.2 presents an approximate correlation between the local magnitude M_L and the modified Mercalli intensity scale.

A map can be developed that contains contours of equal intensity (called *isoseisms*). Such a map is titled an intensity map or an isoseismal map, and an example is presented in Fig. 2.19. The intensity will usually be highest in the general vicinity of the epicenter or at the location of maximum fault rupture, and the intensity progressively decreases as the distance from the epicenter or maximum fault rupture increases. There can be numerous exceptions to this rule. For example, the epicenter of the 1985 Michoacan earthquake was about 350 km (220 mi) from Mexico City, yet there were buildings that collapsed at the Lake Zone district. This was due to the underlying thick deposit of soft clay that increased the peak ground acceleration and the site period, resulting in resonance for the taller buildings. This effect of local soil and geologic conditions on the earthquake intensity is further discussed in Sec. 5.6.

The modified Mercalli intensity scale can also be used to illustrate the anticipated damage at a site due to a future earthquake. For example, Fig. 2.20 shows the estimated intensity map for San Francisco and the surrounding areas, assuming there is a repeat of the 1906 earthquake. It is predicted that there will be extreme damage along the San Andreas fault as well as in those areas underlain by the San Francisco Bay mud.

2.6 PROBLEMS

The problems have been divided into basic categories as indicated below:

Identification of Faults

2.1 The engineering geologist has determined that a fault plane is oriented 5NW 34W. The engineering geologist also discovered a fault scarp, and based on a trench excavated

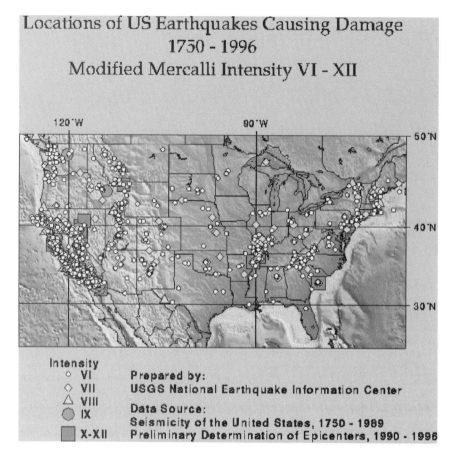

FIGURE 2.16 Locations of continental U.S. earthquakes causing damage from 1750 to 1996 and having a modified Mercalli intensity of VI to XII. (*Prepared by USGS National Earthquake Information Center.*)

across the scarp, the hangingwall block has moved upward with respect to the footwall block. In addition, the surface faulting appears to have occurred solely in the dip direction. Based on this data, determine the type of fault. *Answer:* dip-slip thrust fault.

2.2 Figure 2.21 shows the displacement of rock strata caused by the Carmel Valley Fault, located at Torry Pines, California. Based on the displacement of the hangingwall as compared to the footwall, what type of fault is shown in Figure 2.21. *Answer:* normal fault.

Earthquake Magnitude

2.3 Assume that the displacement data shown in Fig. 2.14 represents the trace data from a standard Wood-Anderson seismograph and that the instrument is exactly 100 km from the epicenter. Based on these assumptions, determine the Richter magnitude. *Answer:* $M_L = 5.2$.

Locations of Alaskan Earthquakes Causing Damage

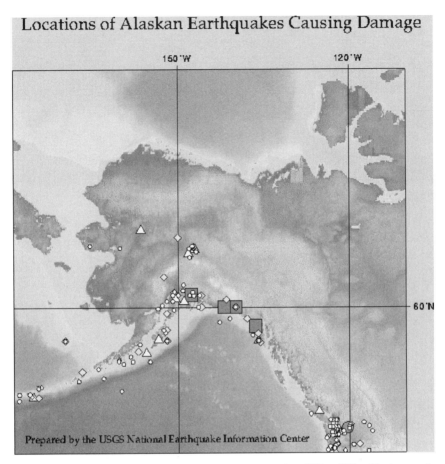

FIGURE 2.17 Locations of Alaskan earthquakes causing damage from 1750 to 1996 and having a modified Mercalli intensity of VI to XII. See Fig. 2.16 for intensity legend. (*Prepared by USGS National Earthquake Information Center.*)

2.4 Assume that a seismograph, located 1200 km from the epicenter of an earthquake, records a maximum ground displacement of 15.6 mm for surface waves having a period of 20 seconds. Based on these assumptions, determine the surface wave magnitude. *Answer:* $M_s = 7.9$.

2.5 Assume that during a major earthquake, the depth of fault rupture is estimated to be 15 km, the length of surface faulting is determined to be 600 km, and the average slip along the fault is 2.5 m. Based on these assumptions, determine the moment magnitude. Use a shear modulus equal to 3×10^{10} N/m^2. *Answer:* $M_w = 8.0$.

Earthquake Intensity

2.6 Suppose that you are considering buying an house located in Half Moon Bay, California. The house can be classified as a well-designed frame structure. For a repeat of

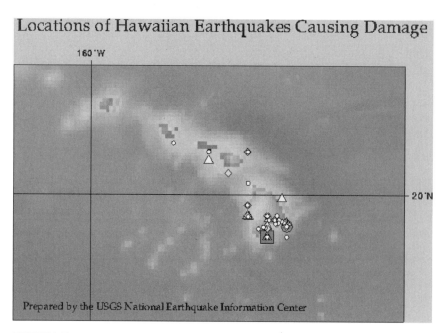

FIGURE 2.18 Locations of Hawaiian earthquakes causing damage from 1750 to 1996 and having a modified Mercalli intensity of VI to XII. See Fig. 2.16 for intensity legend. (*Prepared by USGS National Earthquake Information Center.*)

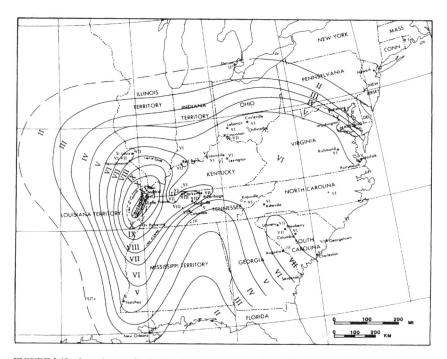

FIGURE 2.19 Intensity map for the New Madrid earthquake of December 16, 1811. (*Developed by Stearns and Wilson 1972, reproduced from Krinitzsky et al. 1993.*)

the 1906 San Francisco earthquake, what type of damage would be expected for this house? *Answer:* Based on Fig. 2.20, you should expect a modified Mercalli level of damage of IX, which corresponds to heavy damage. Per Table 2.3, at a level of IX, well-designed frame structures are thrown out of plumb.

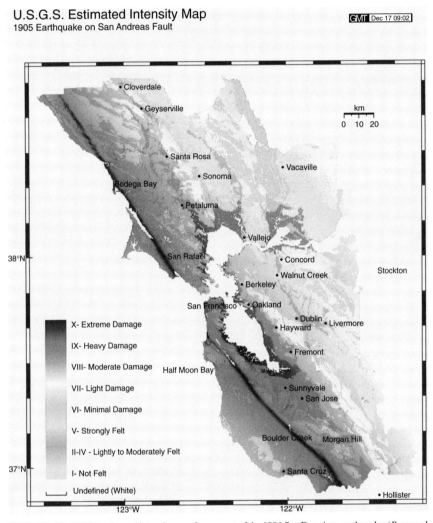

FIGURE 2.20   USGS estimated intensity map for a repeat of the 1906 San Francisco earthquake. (*Prepared by USGS.*)

FIGURE 2.21   Figure for Prob. 2.2.

TABLE 2.3   Modified Mercalli Intensity Scale

Intensity level	Reaction of observers and types of damage
I	Reactions: Not felt except by a very few people under especially favorable circumstances. Damage: No damage.
II	Reactions: Felt only by a few persons at rest, especially on upper floors of buildings. Many people do not recognize it as an earthquake. Damage: No damage. Delicately suspended objects may swing.
III	Reactions: Felt quite noticeably indoors, especially on upper floors of buildings. The vibration is like the passing of a truck, and the duration of the earthquake may be estimated. However, many people do not recognize it as an earthquake. Damage: No damage. Standing motor cars may rock slightly.
IV	Reactions: During the day, felt indoors by many, outdoors by a few. At night, some people are awakened. The sensation is like a heavy truck striking the building. Damage: Dishes, windows, and doors are disturbed. Walls make a creaking sound. Standing motor cars rock noticeably.
V	Reactions: Felt by nearly everyone, many awakened.

TABLE 2.3 Modified Mercalli Intensity Scale (*Continued*)

Intensity level	Reaction of observers and types of damage
	Damage: Some dishes, windows, etc., broken. A few instances of cracked plaster and unstable objects overturned. Disturbances of trees, poles, and other tall objects sometimes noticed. Pendulum clocks may stop.
VI	Reactions: Felt by everyone. Many people are frightened and run outdoors. Damage: There is slight structural damage. Some heavy furniture is moved, and there are a few instances of fallen plaster or damaged chimneys.
VII	Reactions: Everyone runs outdoors. Noticed by persons driving motor cars. Damage: Negligible damage in buildings of good design and construction, slight to moderate damage in well-built ordinary structures, and considerable damage in poorly built or badly designed structures. Some chimneys are broken.
VIII	Reactions: Persons driving motor cars are disturbed. Damage: Slight damage in specially designed structures. Considerable damage in ordinary substantial buildings, with partial collapse. Great damage in poorly built structures. Panel walls are thrown out of frame structures. There is the fall of chimneys, factory stacks, columns, monuments, and walls. Heavy furniture is overturned. Sand and mud are ejected in small amounts, and there are changes in well-water levels.
IX	Damage: Considerable damage in specially designed structures. Well-designed frame structures are thrown out of plumb. There is great damage in substantial buildings with partial collapse. Buildings are shifted off of their foundations. The ground is conspicuously cracked, and underground pipes are broken.
X	Damage: Some well-built wooden structures are destroyed. Most masonry and frame structures are destroyed, including the foundations. The ground is badly cracked. There are bent train rails, a considerable number of landslides at river banks and steep slopes, shifted sand and mud, and water is splashed over their banks.
XI	Damage: Few, if any, masonry structures remain standing. Bridges are destroyed, and train rails are greatly bent. There are broad fissures in the ground, and underground pipelines are completely out of service. There are earth slumps and land slips in soft ground.
XII	Reactions: Waves are seen on the ground surface. The lines of sight and level are distorted. Damage: Total damage with practically all works of construction greatly damaged or destroyed. Objects are thrown upward into the air.

CHAPTER 3
COMMON EARTHQUAKE EFFECTS

3.1 INTRODUCTION

This chapter deals with common earthquake damage due to tectonic surface processes and secondary effects. Section 3.2 deals with ground surface fault rupture, which is also referred to as *surface rupture*. Section 3.3 discusses regional subsidence, which often occurs at a rift valley, subduction zone, or an area of crust extension. Surface faulting and regional subsidence are known as tectonic surface processes.

Secondary effects are defined as nontectonic surface processes that are directly related to earthquake shaking (Yeats et al. 1997). Examples of secondary effects are liquefaction, earthquake-induced slope failures and landslides, tsunamis, and seiches. These secondary effects are discussed in Secs. 3.4 to 3.6.

3.2 SURFACE RUPTURE

3.2.1 Description

Most earthquakes will not create ground surface fault rupture. For example, there is typically an absence of surface rupture for small earthquakes, earthquakes generated at great depths at subduction zones, and earthquakes generated on blind faults. Krinitzsky et al. (1993) state that fault ruptures commonly occur in the deep subsurface with no ground breakage at the surface. They further state that such behavior is widespread, accounting for all earthquakes in the central and eastern United States.

On the other hand, large earthquakes at transform boundaries will usually be accompanied by ground surface fault rupture on strike-slip faults. An example of ground surface fault rupture of the San Andreas fault is shown in Fig. 2.9. Figures 2.11 to 2.13 also illustrate typical types of damage directly associated with the ground surface fault rupture. Two other examples of surface fault rupture are shown in Figs. 3.1 and 3.2.

Fault displacement is defined as the relative movement of the two sides of a fault, measured in a specific direction (Bonilla 1970). Examples of very large surface fault rupture are the 11 m (35 ft) of vertical displacement in the Assam earthquake of 1897 (Oldham 1899) and the 9 m (29 ft) of horizontal movement during the Gobi-Altai earthquake of 1957

FIGURE 3.1   Surface fault rupture associated with the El Asnam (Algeria) earthquake on October 10, 1980. (*Photograph from the Godden Collection, EERC, University of California, Berkeley.*)

(Florensov and Solonenko 1965). The length of the fault rupture can be quite significant. For example, the estimated length of surface faulting in the 1964 Alaskan earthquake varied from 600 to 720 km (Savage and Hastie 1966, Housner 1970).

3.2.2 Damage Caused by Surface Rupture

Surface fault rupture associated with earthquakes is important because it has caused severe damage to buildings, bridges, dams, tunnels, canals, and underground utilities (Lawson et al. 1908, Ambraseys 1960, Duke 1960, California Department of Water Resources 1967, Bonilla 1970, Steinbrugge 1970).

There were spectacular examples of surface fault rupture associated with the Chi-chi (Taiwan) earthquake on September 21, 1999. According to seismologists at the U.S. Geological Survey, National Earthquake Information Center, Golden, Colorado, the tectonic environment near Taiwan is unusually complicated. They state (USGS 2000a):

> Tectonically, most of Taiwan is a collision zone between the Philippine Sea and Eurasian plates. This collision zone is bridged at the north by northwards subduction of the Philippine Sea plate beneath the Ryuku arc and, at the south, an eastwards thrusting at the Manila trench. The northern transition from plate collision to subduction is near the coastal city of Hualien, located at about 24 degrees north, whereas the southern transition is 30–50 kilometers south of Taiwan.

With a magnitude of 7.6, the earthquake was the strongest to hit Taiwan in decades and was about the same strength as the devastating tremor that killed more than 17,000 people

FIGURE 3.2 Surface fault rupture associated with the Izmit (Turkey) earthquake on August 17, 1999. (*Photograph by Tom Fumal, USGS.*)

in Turkey a month before. The earthquake also triggered at least five aftershocks near or above magnitude 6. The epicenter of the earthquake was in a small country town of Chi-chi (located about 90 mi south of Taipei). Surface fault rupture associated with this Taiwan earthquake caused severe damage to civil engineering structures, as discussed below:

- *Dam failure:* Figures 3.3 and 3.4 show two views of the failure of a dam located northeast of Tai-Chung, Taiwan. This dam was reportedly used to supply drinking water for the surrounding communities. The surface fault rupture runs through the dam and caused the southern end to displace upward about 9 to 10 m (30 to 33 ft) as compared to the northern end. This ground fault displacement is shown in the close-up view in Fig. 3.4. Note in this figure that the entire length of fence on the top of the dam was initially at the same elevation prior to the earthquake.

- *Kuang Fu Elementary School:* Figures 3.5 and 3.6 show damage to the Kuang Fu Elementary School, located northeast of Tai-Chung, Taiwan. The Kuang Fu Elementary School was traversed by a large fault rupture that in some locations caused a ground displacement of as much as 3 m (10 ft), as shown in Fig. 3.5.

FIGURE 3.3   Overview of a dam damaged by surface fault rupture associated with the Chi-chi (Taiwan) earthquake on September 21, 1999. (*Photograph from the Taiwan Collection, EERC, University of California, Berkeley.*)

FIGURE 3.4   Close-up view of the location of the dam damaged by surface fault rupture associated with the Chi-chi (Taiwan) earthquake on September 21, 1999. Note in this figure that the entire length of fence on the top of the dam was initially at the same elevation prior to the earthquake. (*Photograph from the Taiwan Collection, EERC, University of California, Berkeley.*)

FIGURE 3.5 Overview of damage to the Kuang Fu Elementary School by surface fault rupture associated with the Chi-chi (Taiwan) earthquake on September 21, 1999. (*Photograph from the Taiwan Collection, EERC, University of California, Berkeley.*)

FIGURE 3.6 Portion of a building that remained standing at the Kuang Fu Elementary School. This portion of the building was directly adjacent to the surface fault rupture associated with the Chi-chi (Taiwan) Earthquake on September 21, 1999. Note in this figure that the ground was actually compressed together adjacent to the footwall side of the fault rupture. (*Photograph from the Taiwan Collection, EERC, University of California, Berkeley.*)

Figure 3.6 shows a building at the Kuang Fu Elementary School that partially collapsed. The portion of the building that remained standing is shown in Fig. 3.6. This portion of the building is immediately adjacent to the surface fault rupture and is located on the footwall side of the fault. Note in Fig. 3.6 that the span between the columns was actually reduced by the fault rupture. In essence, the ground was compressed together adjacent to the footwall side of the fault rupture.

- *Wu-His (U-Shi) Bridge:* Figure 3.7 shows damage to the second bridge pier south of the abutment of the new Wu-His (U-Shi) Bridge in Taiwan. At this site, surface fault rupturing was observed adjacent to the bridge abutment. Note in Fig. 3.7 that the bridge pier was literally sheared in half.

- *Retaining wall north of Chung-Hsing (Jung Shing) in Taiwan:* Figure 3.8 shows damage to a retaining wall and adjacent building. At this site, the surface fault rupture caused both vertical and horizontal displacement of the retaining wall.

- *Collapsed bridge north of Fengyuen:* Figures 3.9 to 3.11 show three photographs of the collapse of a bridge just north of Fengyuen, Taiwan. The bridge generally runs in a north-south direction, with the collapse occurring at the southern portion of the bridge.

  The bridge was originally straight and level. The surface fault rupture passes underneath the bridge and apparently caused the bridge to shorten such that the southern spans were shoved off their supports. In addition, the fault rupture developed beneath one of the piers, resulting in its collapse. Note in Fig. 3.11 that there is a waterfall to the east of the bridge. The fault rupture that runs underneath the bridge caused this displacement and development of the waterfall. The waterfall is estimated to be about 9 to 10 m (30 to 33 ft) in height.

  Figure 3.12 shows a close-up view of the new waterfall created by the surface fault rupture. This photograph shows the area to the east of the bridge. Apparently the dark

FIGURE 3.7   Close-up view of bridge pier (Wu-His Bridge) damaged by surface fault rupture associated with the Chi-chi (Taiwan) earthquake on September 21, 1999. (*Photograph from the Taiwan Collection, EERC, University of California, Berkeley.*)

FIGURE 3.8 Retaining wall located north of Chung-Hsing (Jung Shing). At this site, the surface fault rupture associated with the Chi-chi (Taiwan) earthquake on September 21, 1999, has caused both vertical and horizontal displacement of the retaining wall. (*Photograph from the Taiwan Collection, EERC, University of California, Berkeley.*)

FIGURE 3.9 Collapsed bridge north of Fengyuen caused by surface fault rupture associated with the Chi-chi (Taiwan) earthquake on September 21, 1999. (*Photograph from the USGS Earthquake Hazards Program, NEIC, Denver.*)

FIGURE 3.10 Another view of the collapsed bridge north of Fengyuen caused by surface fault rupture associated with the Chi-chi (Taiwan) earthquake on September 21, 1999. (*Photograph from the USGS Earthquake Hazards Program, NEIC, Denver.*)

rocks located in front of the waterfall are from the crumpling of the leading edge of the thrust fault movement.

- *Roadway damage:* The final photograph of surface fault rupture from the Chi-chi (Taiwan) earthquake is shown in Fig. 3.13. In addition to the roadway damage, such surface faulting would shear apart any utilities that happened to be buried beneath the roadway.

In addition to surface fault rupture, such as described above, there can be ground rupture away from the main trace of the fault. These ground cracks could be caused by many different factors, such as movement of subsidiary faults, auxiliary movement that branches off from the main fault trace, or ground rupture caused by the differential or lateral movement of underlying soil deposits.

As indicated by the photographs in this section, structures are unable to resist the shear movement associated with surface faulting. One design approach is to simply restrict construction in the active fault shear zone. This is discussed further in Sec. 11.2.

3.3 REGIONAL SUBSIDENCE

In addition to the surface fault rupture, another tectonic effect associated with the earthquake could be uplifting or regional subsidence. For example, at continent-continent

FIGURE 3.11 Another view of the collapsed bridge north of Fengyuen caused by surface fault rupture associated with the Chi-chi (Taiwan) earthquake on September 21, 1999. Note that the surface faulting has created the waterfall on the right side of the bridge. (*Photograph from the USGS Earthquake Hazards Program, NEIC, Denver.*)

collision zones (Fig. 2.7), the plates collide into one another, causing the ground surface to squeeze, fold, deform, and thrust upward.

Besides uplifting, there could also be regional subsidence associated with the earthquake. There was extensive damage due to regional subsidence during the August 17, 1999, Izmit earthquake in Turkey. Concerning this earthquake, the USGS (2000a) states:

> The Mw 7.4 [moment magnitude] earthquake that struck western Turkey on August 17, 1999 occurred on one of the world's longest and best studied strike-slip faults: the east-west trending North Anatolian fault. This fault is very similar to the San Andreas Fault in California. Turkey has had a long history of large earthquakes that often occur in progressive adjacent earthquakes. Starting in 1939, the North Anatolian fault produced a sequence of major earthquakes, of which the 1999 event is the 11th with a magnitude greater than or equal to 6.7. Starting with the 1939 event in western Turkey, the earthquake locations have moved both eastward and westward. The westward migration was particularly active and ruptured 600 km of contiguous fault between 1939 and 1944. This westward propagation of earthquakes then slowed and ruptured an additional adjacent 100 km of fault in events in 1957 and 1967, with separated activity further west during 1963 and 1964. The August 17, 1999 event fills in a 100 to 150 km long gap between the 1967 event and the 1963 and 1964 events.

The USGS also indicated that the earthquake originated at a depth of 17 km (10.5 mi) and caused right-lateral strike-slip movement on the fault. Preliminary field studies found that the earthquake produced at least 60 km (37 mi) of surface rupture and right-lateral offsets as large as 2.7 m (9 ft).

FIGURE 3.12   Close-up view of the waterfall shown in Fig. 3.11. The waterfall was created by the surface fault rupture associated with the Chi-chi (Taiwan) earthquake on September 21, 1999, and has an estimated height of 9 to 10 m. (*Photograph from the USGS Earthquake Hazards Program, NEIC, Denver.*)

   As described above, the North Anatolian fault is predominantly a strike-slip fault due to the Anatolian plate shearing past the Eurasian plate. But to the west of Izmit, there is a localized extension zone where the crust is being stretched apart and has formed the Gulf of Izmit. An extension zone is similar to a rift valley. It occurs when a portion of the earth's crust is stretched apart and a graben develops. A *graben* is defined as a crustal block that has dropped down relative to adjacent rocks along bounding faults. The down-dropping block is usually much longer than its width, creating a long and narrow valley.
   The city of Golcuk is located on the south shore of the Gulf of Izmit. It has been reported that during the earthquake, 2 mi (3 km) of land along the Gulf of Izmit subsided at least 3

FIGURE 3.13   Surface fault rupture and roadway damage associated with the Chi-chi (Taiwan) earthquake on September 21, 1999. (*Photograph from the USGS Earthquake Hazards Program, NEIC, Denver.*)

m (10 ft). Water from the Gulf of Izmit flooded inland, and several thousand people drowned or were crushed as buildings collapsed in Golcuk. Figures 3.14 to 3.18 show several examples of the flooded condition associated with the regional subsidence along the extension zone.

It is usually the responsibility of the engineering geologist to evaluate the possibility of regional subsidence associated with extension zones and rift valleys. For such areas, special foundation designs, such as mat slabs, may make the structures more resistant to the regional tectonic movement.

FIGURE 3.14 Flooding caused by regional subsidence associated with the Izmit (Turkey) earthquake on August 17, 1999. (*Photograph from the Izmit Collection, EERC, University of California, Berkeley.*)

FIGURE 3.15 Flooding caused by regional subsidence associated with the Izmit (Turkey) earthquake on August 17, 1999. (*Photograph from the Izmit Collection, EERC, University of California, Berkeley.*)

FIGURE 3.16   Flooding caused by regional subsidence associated with the Izmit (Turkey) earthquake on August 17, 1999. (*Photograph from the Izmit Collection, EERC, University of California, Berkeley.*)

FIGURE 3.17   Flooding caused by regional subsidence associated with the Izmit (Turkey) earthquake on August 17, 1999. (*Photograph from the Izmit Collection, EERC, University of California, Berkeley.*)

FIGURE 3.18   Flooding caused by regional subsidence associated with the Izmit (Turkey) earthquake on August 17, 1999. (*Photograph from the Izmit Collection, EERC, University of California, Berkeley.*)

3.4 LIQUEFACTION

3.4.1 Introduction

The final three sections of this chapter deal with secondary effects, which are defined as nontectonic surface processes that are directly related to earthquake shaking. Examples of secondary effects are liquefaction, earthquake-induced slope failures and landslides, tsunamis, and seiches.

This section deals with liquefaction. The typical subsurface soil condition that is susceptible to liquefaction is loose sand, which has been newly deposited or placed, with a groundwater table near ground surface. During an earthquake, the propagation of shear waves causes the loose sand to contract, resulting in an increase in pore water pressure. Because the seismic shaking occurs so quickly, the cohesionless soil is subjected to an undrained loading. The increase in pore water pressure causes an upward flow of water to the ground surface, where it emerges in the form of mud spouts or sand boils (see Fig. 3.19). The development of high pore water pressures due to the ground shaking and the upward flow of water may turn the sand into a liquefied condition, which has been termed liquefaction. For this state of liquefaction, the effective stress is zero and the individual soil particles are released from any confinement, as if the soil particles were floating in water (Ishihara 1985).

Because liquefaction typically occurs in soil with a high groundwater table, its effects are most commonly observed in low-lying areas or adjacent rivers, lakes, bays, and oceans. The following sections describe the different types of damage caused by liquefaction. The engineering analysis used to determine whether a site is susceptible to liquefaction is presented in Chap. 6.

FIGURE 3.19 Sand boil in Niigata caused by liquefaction during the Niigata (Japan) earthquake of June 16, 1964. (*Photograph from the Steinbrugge Collection, EERC, University of California, Berkeley.*)

3.4.2 Settlement and Bearing Capacity Failures

When liquefaction occurs, the soil can become a liquid, and thus the shear strength of the soil can be decreased to essentially zero. Without any shear strength, the liquefied soil will be unable to support the foundations for buildings and bridges. For near surface liquefaction, buried tanks will float to the surface and buildings will sink or fall over.

Some of the most spectacular examples of settlement and bearing capacity failures due to liquefaction occurred during the Niigata earthquake in 1964. The Niigata earthquake of June 16, 1964, had a magnitude of 7.5 and caused severe damage to many structures in Niigata. The destruction was observed to be largely limited to buildings that were founded on top of loose, saturated soil deposits. Even though numerous houses were totally destroyed, only 28 lives were lost (Johansson 2000).

Concerning the 1964 Niigata earthquake, the National Information Service for Earthquake Engineering (2000) states:

The Niigata Earthquake resulted in dramatic damage due to liquefaction of the sand deposits in the low-lying areas of Niigata City. In and around this city, the soils consist of recently reclaimed land and young sedimentary deposits having low density and shallow ground water table. At the time of this earthquake, there were approximately 1500 reinforced concrete buildings in Niigata City. About 310 of these buildings were damaged, of which approximately 200 settled or tilted rigidly without appreciable damage to the superstructure. It should be noted that the damaged concrete buildings were built on very shallow foundations or friction piles in loose soil. Similar concrete buildings founded on piles bearing on firm strata at a depth of 20 meters [66 ft] did not suffer damage.

Civil engineering structures, which were damaged by the Niigata Earthquake, included port and harbor facilities, water supply systems, railroads, roads, bridges, airport, power facilities, and agricultural facilities. The main reason for these failures was ground failure, particularly the liquefaction of the ground in Niigata City, which was below sea level as a result of ground subsidence.

Figure 3.19 shows a sand boil created by liquefaction during the 1964 Niigata earthquake. Some examples of structural damage caused by liquefaction during the 1964 Niigata earthquake are as follows:

- *Bearing capacity failures:* Figure 3.20 shows dramatic liquefaction-induced bearing capacity failures of Kawagishi-cho apartment buildings located at Niigata, Japan. Figure 3.21 shows a view of the bottom of one of the buildings that suffered a bearing capacity failure. Despite the extreme tilting of the buildings, there was remarkably little structural damage because the buildings remained intact during the failure.

- *Building settlement:* Figures 3.22 and 3.23 show two more examples of liquefaction-induced settlement at Niigata, Japan. Similar to the buildings shown in Figs. 3.20 and 3.21, the buildings remained intact as they settled and tilted. It was reported that there was essentially no interior structural damage and that the doors and windows still functioned. Apparently, the failure took a considerable period of time to develop, which could indicate that the liquefaction started at depth and then slowly progressed toward the ground surface.

- *Other damage:* It was not just the relatively heavy buildings that suffered liquefaction-induced settlement and bearing capacity failures. For example, Fig. 3.24 shows liquefaction-induced settlement and tilting of relatively light buildings. There was also damage to surface paving materials.

Because riverbeds often contain loose sand deposits, liquefaction also frequently causes damage to bridges that cross rivers or other bodies of water. Bridges are usually designated

FIGURE 3.20   Kawagishi-cho apartment buildings located in Niigata, Japan. The buildings suffered lique-faction-induced bearing capacity failures during the Niigata earthquake on June 16, 1964. (*Photograph from the Godden Collection, EERC, University of California, Berkeley.*)

FIGURE 3.21 View of the bottom of a Kawagishi-cho apartment building located in Niigata, Japan. The building suffered a liquefaction-induced bearing capacity failure during the Niigata earthquake on June 16, 1964. (*Photograph from the Steinbrugge Collection, EERC, University of California, Berkeley.*)

as essential facilities, because they provide necessary transportation routes for emergency response and rescue operations. A bridge failure will also impede the transport of emergency supplies and can cause significant economic loss for businesses along the transportation corridor. There are several different ways that bridges can be impacted by liquefaction. For example, liquefaction beneath a bridge pier could cause collapse of a portion of the bridge. Likewise, liquefaction also reduces the lateral bearing, also known as the passive resistance. With a reduced lateral bearing capacity, the bridge piers will be able to rock back and forth and allow for the collapse of the bridge superstructure. A final effect of liquefaction could be induced down-drag loads upon the bridge piers as the pore water pressures from the liquefied soil dissipate and the soil settles.

Figure 3.25 shows the collapse of the superstructure of the Showa Bridge caused by the 1964 Niigata earthquake. The soil liquefaction apparently allowed the bridge piers to move laterally to the point where the simply supported bridge spans lost support and collapsed.

3.4.3 Waterfront Structures

Port and wharf facilities are often located in areas susceptible to liquefaction. Many of these facilities have been damaged by earthquake-induced liquefaction. The ports and wharves often contain major retaining structures, such as seawalls, anchored bulkheads, gravity and cantilever walls, and sheet-pile cofferdams, that allow large ships to moor adjacent to the retaining walls and then load or unload their cargo. There are often three different types of liquefaction effects that can damage the retaining wall:

FIGURE 3.22   Settlement and tilting of an apartment building located in Niigata, Japan. The building suffered liquefaction-induced settlement and tilting during the Niigata earthquake on June 16, 1964. (*Photograph from the Steinbrugge Collection, EERC, University of California, Berkeley.*)

1. The first is liquefaction of soil in front of the retaining wall. In this case, the passive pressure in front of the retaining wall is reduced.

2. In the second case, the soil behind the retaining wall liquefies, and the pressure exerted on the wall is greatly increased. Cases 1 and 2 can act individually or together, and they can initiate an overturning failure of the retaining wall or cause the wall to slide outward or tilt toward the water. Another possibility is that the increased pressure exerted on the wall could exceed the strength of the wall, resulting in a structural failure of the wall.

Liquefaction of the soil behind the retaining wall can also affect tieback anchors. For example, the increased pressure due to liquefaction of the soil behind the wall could break the tieback anchors or reduce their passive resistance.

3. The third case is liquefaction below the bottom of the wall. In this case, the bearing capacity or slide resistance of the wall is reduced, resulting in a bearing capacity failure or promoting rotational movement of the wall.

FIGURE 3.23 Settlement and tilting of a building located in Niigata, Japan. The building suffered liquefaction-induced settlement and tilting during the Niigata earthquake on June 16, 1964. (*Photograph from the Steinbrugge Collection, EERC, University of California, Berkeley.*)

Some spectacular examples of damage to waterfront structures due to liquefaction occurred during the Kobe earthquake on January 17, 1995. Particular details concerning the Kobe earthquake are as follows (EQE Summary Report 1995, EERC 1995):

- The Kobe earthquake, also known as the Hyogo-ken Nanbu earthquake, had a moment magnitude M_w of 6.9.
- The earthquake occurred in a region with a complex system of previously mapped active faults.
- The focus of the earthquake was at a depth of approximately 15 to 20 km (9 to 12 mi). The focal mechanism of the earthquake indicated right-lateral strike-slip faulting on a nearly vertical fault that runs from Awaji Island through the city of Kobe.
- Ground rupture due to the right-lateral strike-slip faulting was observed on Awaji Island, which is located to the southwest of the epicenter. In addition, the Akashi Kaikyo Bridge,

FIGURE 3.24   Settlement and tilting of relatively light buildings located in Niigata, Japan. The buildings suffered liquefaction-induced settlement and tilting during the Niigata earthquake on June 16, 1964. (*Photograph from the Steinbrugge Collection, EERC, University of California, Berkeley.*)

which was under construction at the time of the earthquake, suffered vertical and lateral displacement between the north and south towers. This is the first time that a structure of this size was offset by a fault rupture.

- Peak ground accelerations as large as 0.8g were recorded in the near-fault region on alluvial sites in Kobe.

- In terms of regional tectonics, Kobe is located on the southeastern margin of the Eurasian plate, where the Philippine Sea plate is being subducted beneath the Eurasian plate (see Fig. 2.1).

- More than 5000 people perished, more than 26,000 people were injured, and about $200 billion in damage were attributed to this earthquake.

Damage was especially severe at the relatively new Port of Kobe. In terms of damage to the port, the EERC (1995) stated:

FIGURE 3.25 Collapse of the Showa Bridge during the Niigata earthquake on June 16, 1964. (*Photograph from the Godden Collection, EERC, University of California, Berkeley.*)

The main port facilities in Kobe harbor are located primarily on reclaimed land along the coast and on two man-made islands, Port Island and Rokko Island, which are joined by bridges to the mainland. The liquefaction and lateral spread-induced damage to harbor structures on the islands disrupted nearly all of the container loading piers, and effectively shut down the Port of Kobe to international shipping. All but 6 of about 187 berths were severely damaged.

Concerning the damage caused by liquefaction, the EERC (1995) concluded:

Extensive liquefaction of natural and artificial fill deposits occurred along much of the shoreline on the north side of the Osaka Bay. Probably the most notable were the liquefaction failures of relatively modern fills on the Rokko and Port islands. On the Kobe mainland, evidence of liquefaction extended along the entire length of the waterfront, east and west of Kobe, for a distance of about 20 km [12 mi]. Overall, liquefaction was a principal factor in the extensive damage experienced by the port facilities in the affected region.

Most of the liquefied fills were constructed of poorly compacted decomposed granite soil. This material was transported to the fill sites and loosely dumped in water. Compaction was generally only applied to materials placed above water level. As a result, liquefaction occurred within the underwater segments of these poorly compacted fills.

Typically, liquefaction led to pervasive eruption of sand boils and, on the islands, to ground settlements on the order of as much as 0.5 m [see Fig. 3.26]. The ground settlement caused surprisingly little damage to high- and low-rise buildings, bridges, tanks, and other structures supported on deep foundations. These foundations, including piles and shafts, performed very well in supporting superstructures where ground settlement was the principal effect of liquefaction. Where liquefaction generated lateral ground displacements, such as near island edges and in other waterfront areas, foundation performance was typically poor. Lateral displacements fractured piles and displaced pile caps, causing structural distress to several bridges. In a few instances, such as the Port Island Ferry Terminal, strong foundations withstood the lateral ground displacement with little damage to the foundation or the superstructure.

There were several factors that apparently contributed to the damage at the Port of Kobe, as follows (EQE Summary Report 1995, EERC 1995):

1. *Design criteria:*   The area had been previously considered to have a relatively low seismic risk, hence the earthquake design criteria were less stringent than in other areas of Japan.

2. *Earthquake shaking:*   There was rupture of the strike-slip fault directly in downtown Kobe. Hence the release of energy along the earthquake fault was close to the port. In addition, the port is located on the shores of a large embayment, which has a substantial thickness of soft and liquefiable sediments. This thick deposit of soft soil caused an amplification of the peak ground acceleration and an increase in the duration of shaking.

FIGURE 3.26   The interiors of the Rokko and Port islands settled as much as 1 m with an average of about 0.5 m due to liquefaction caused by the Kobe earthquake on January 17, 1995. This liquefaction-induced settlement was accompanied by the eruption of large sand boils that flooded many areas and covered much of the island with sand boil deposits as thick as 0.5 m. In this photograph, a stockpile has been created out of some of this sand. (*Photograph from the Kobe Geotechnical Collection, EERC, University of California, Berkeley.*)

3. *Construction of the port:* The area of the port was built almost entirely on fill and reclaimed land. As previously mentioned, the fill and reclaimed land material often consisted of decomposed granite soils that were loosely dumped into the water. The principal factor in the damage at the Port of Kobe was attributed to liquefaction, which caused lateral deformation (also known as lateral spreading) of the retaining walls. Figures 3.27 to 3.30 show examples of damage to the port area.

4. *Artificial islands:* On Rokko and Port Islands, retaining walls were constructed by using caissons, which consisted of concrete box structures, up to 15 m wide and 20 m deep, with two or more interior cells (Fig. 3.31). The first step was to prepare the seabed by installing a sand layer. Then the caissons were towed to the site, submerged in position to form the retaining wall, and the interior cells were backfilled with sand. Once in place, the area behind the caisson retaining walls was filled in with soil in order to create the artificial islands.

During the Kobe earthquake, a large number of these caisson retaining walls rotated and slid outward (lateral spreading). Figures 3.32 and 3.33 show examples of damage caused by the outward movement of the retaining walls. This outward movement of the retaining walls by as much as 3 m (10 ft) caused lateral displacement and failure of the loading dock cranes, such as shown in Fig. 3.34.

5. *Buildings on deep foundations:* In some cases, the buildings adjacent to the retaining walls had deep foundations consisting of piles or piers. Large differential movement occurred between the relatively stable buildings having piles or piers and the port retaining walls, which settled and deformed outward. An example of this condition is shown in Fig. 3.35.

FIGURE 3.27 Ground cracks caused by lateral retaining wall movement due to liquefaction during the Kobe earthquake on January 17, 1995. The site is near Nishinomiya Port and consists of reclaimed land. (*Photograph from the Great Hanshin Bridge Collection, EERC, University of California, Berkeley.*)

FIGURE 3.28 Lateral retaining wall movement due to liquefaction during the Kobe earthquake on January 17, 1995. As the retaining wall has moved outward, the ground surface has dropped. The site is adjacent to the east end of the Higashi Kobe Bridge. (*Photograph from the Great Hanshin Bridge Collection, EERC, University of California, Berkeley.*)

FIGURE 3.29 Settlement and lateral spreading damage due to retaining wall movement caused by the Kobe earthquake on January 17, 1995. The location is near the northwest corner of the Port Island Bridge. (*Photograph from the Great Hanshin Bridge Collection, EERC, University of California, Berkeley.*)

FIGURE 3.30 Retaining wall damage caused by the Kobe earthquake on January 17, 1995. The site is the east bank of the Maya-ohashi Bridge (Harbor Expressway). Liquefaction caused significant lateral displacement of the retaining wall, which in turn created a depression behind the wall and dropped a large truck halfway into the water. (*Photograph from the Kobe Geotechnical Collection, EERC, University of California, Berkeley.*)

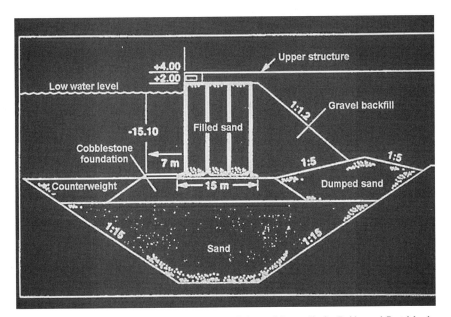

FIGURE 3.31 Diagram depicting the construction of the retaining walls for Rokko and Port islands. (*Photograph from the Kobe Geotechnical Collection, EERC, University of California, Berkeley.*)

FIGURE 3.32   Retaining wall that moved outward about 2 to 3 m (7 to 10 ft), creating a depression behind the wall that was about 3 m (10 ft) deep. The cause of the retaining wall movement was liquefaction during the Kobe earthquake on January 17, 1995. (*Photograph from the Kobe Geotechnical Collection, EERC, University of California, Berkeley.*)

There was also apparently liquefaction-induced retaining wall movement that resulted in a bridge failure. For example, the EERC (1995) stated:

> Most bridge failures on the Kobe mainland during the Hyogo-ken Nanbu Earthquake in Japan on January 17, 1995 were a result of structural design rather than a result of liquefactions. However, the photograph [Fig. 3.36] illustrates an example where the Nishihomiya Bridge may have failed due to liquefaction and lateral spreading. The bridge collapsed because of the separation of the two supporting piers, which the lateral ground displacements may have caused [Fig. 3.37].

A discussion of the design and construction of retaining walls for waterfront structures is presented in Sec. 10.3.

FIGURE 3.33 Damage on Port Island caused by retaining wall movement during the Kobe earthquake on January 17, 1995. (*Photograph from the Kobe Geotechnical Collection, EERC, University of California, Berkeley.*)

FIGURE 3.34 Collapse of a crane due to about 2 m of lateral movement of the retaining wall on Rokko Island. There was also about 1 to 2 m of settlement behind the retaining wall when it moved outward during the Kobe earthquake on January 17, 1995. (*Photograph from the Kobe Geotechnical Collection, EERC, University of California, Berkeley.*)

FIGURE 3.35 Settlement caused by lateral movement of a retaining wall during the Kobe earthquake on January 17, 1995. The industrial building is supported by a pile foundation. The Higashi Kobe cable-stayed bridge is visible in the upper right corner of the photograph. (*Photograph from the Great Hanshin Bridge Collection, EERC, University of California, Berkeley.*)

3.4.4 Flow Slides

Liquefaction can also cause lateral movement of slopes and create flow slides (Ishihara 1993). Seed (1970) states:

> If liquefaction occurs in or under a sloping soil mass, the entire mass will flow or translate laterally to the unsupported side in a phenomenon termed a flow slide. Such slides also develop in loose, saturated, cohesionless materials during earthquakes and are reported at Chile (1960), Alaska (1964), and Niigata (1964).

A classic example of a flow slide was the failure of the Lower San Fernando Dam caused by the San Fernando earthquake, also known as the Sylmar earthquake. Particulars concerning this earthquake are as follows (Southern California Earthquake Data Center 2000):

- Date of earthquake: February 9, 1971
- Moment magnitude M_w of 6.6
- Depth: 8.4 km
- Type of faulting: Thrust fault
- Faults involved: Primarily the San Fernando fault zone
- Surface rupture: A zone of thrust faulting broke the ground surface in the Sylmar–San Fernando area (northeast of Los Angeles, California). The total surface rupture was roughly 19 km (12 mi) long. The maximum slip was up to 2 m (6 ft).

FIGURE 3.36 The liquefaction-induced retaining wall movement was caused by the Kobe Earthquake on January 17, 1995. Both lateral spreading fissures and sporadic sand boils were observed behind the retaining wall. The site is near the pier of Nishihomiya Bridge, which previously supported the collapsed expressway section. (*Photograph from the Kobe Geotechnical Collection, EERC, University of California, Berkeley.*)

- Deaths and damage estimate: The earthquake caused more than $500 million in property damage and 65 deaths. Most of the deaths occurred when the Veteran's Administration Hospital collapsed.

- Earthquake response: In response to this earthquake, building codes were strengthened and the Alquist Priolo Special Studies Zone Act was passed in 1972. The purpose of this act is to prohibit the location of most structures for human occupancy across the traces of active faults and to mitigate thereby the hazard of fault rupture.

As mentioned above, the Lower San Fernando Dam was damaged by a flow failure due to the 1971 San Fernando earthquake. Seismographs located on the abutment and on the crest of the dam recorded peak ground accelerations a_{max} of about 0.5 to 0.55g. These high peak ground accelerations caused the liquefaction of a zone of hydraulic sand fill near the base of the upstream shell. Figure 3.38 shows a cross section through the earthen dam and

FIGURE 3.37   Collapse of a span of the Nishihomiya Bridge, apparently due to lateral retaining wall movement during the Kobe earthquake on January 17, 1995. (*Photograph from the Kobe Geotechnical Collection, EERC, University of California, Berkeley.*)

the location of the zone of material that was believed to have liquefied during the earthquake. Once liquefied, the upstream portion of the dam was subjected to a flow slide. The upper part of Fig. 3.38 indicates the portion of the dam and the slip surface along which the flow slide is believed to have initially developed. The lower part of Fig. 3.38 depicts the final condition of the dam after the flow slide. The flow slide caused the upstream toe of the dam to move about 150 ft (46 m) into the reservoir.

Figures 3.39 and 3.40 show two views of the damage to the Lower San Fernando Dam. A description of the damage is presented below:

Figure 3.39 is a view to the east and shows the condition of the dam after the earthquake. While nearly the entire length of the upstream portion of the earthen dam slumped downward, the main flow failure is located at the eastern end of the dam. This is the location of the cross section shown in Fig. 3.38.

Note in Fig. 3.39 that the water in the reservoir almost breached the top of the failed portion of the dam. If the water had breached the top of the dam, it would have quickly cut through the earthen dam and the subsequent torrent of water would have caused thousands of deaths in the residential area immediately below the dam.

Figure 3.40 is a view to the west and shows the condition of the dam after the reservoir has been partly emptied. The flow failure is clearly visible in this photograph. The concrete liner shown in Fig. 3.40 was constructed on the upstream dam face in order to protect against wave-induced erosion. Although initially linear and at about the same elevation across the entire length of the upstream face of the dam, the concrete liner detached from the dam and moved out into the reservoir along with the flowing ground.

As indicated in the upper part of Fig. 3.38, flow failures develop when the driving forces exceed the shear strength along the slip surface and the factor of safety is 1.0 or less. The engineering analyses used to determine whether a site is susceptible to liquefaction and a subsequent flow slide failure are presented in Chaps. 6 and 9.

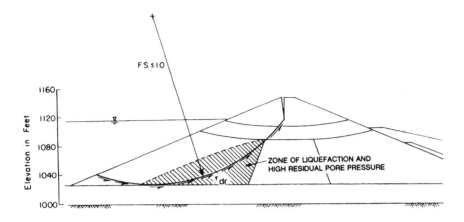

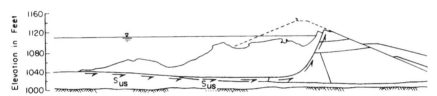

FIGURE 3.38 Cross section through the Lower San Fernando Dam. The upper diagram shows the condition immediately prior to the flow slide caused by the San Fernando earthquake on February 9, 1971. The lower diagram shows the configuration after the flow slide of the upstream slope and crest of the dam. (*From Castro et al. 1992. Reproduced with permission of the American Society of Civil Engineers.*)

3.4.5 Lateral Spreading

Lateral spreading was introduced in Sec. 3.4.3 where, as shown in Figs. 3.27 to 3.30, the principal factor in the damage at the Port of Kobe was attributed to liquefaction, which caused lateral deformation (also known as lateral spreading) of the retaining walls. This liquefaction-induced lateral spreading was usually restricted to the ground surface behind the retaining walls, and thus it would be termed *localized lateral spreading.*

If the liquefaction-induced lateral spreading causes lateral movement of the ground surface over an extensive distance, then the effect is known as *large-scale lateral spreading.* Such lateral spreads often form adjacent waterways on gently sloping or even flat ground surfaces that liquefy during the earthquake. The concept of cyclic mobility is used to describe large-scale lateral spreading. Because the ground is gently sloping or flat, the static driving forces do not exceed the resistance of the soil along the slip surface, and thus the ground is not subjected to a flow slide. Instead, the driving forces only exceed the resisting forces during those portions of the earthquake that impart net inertial forces in the downslope direction. Each cycle of net inertial forces in the downslope direction causes the driving forces to exceed the resisting forces along the slip surface, resulting in progressive and incremental lateral movement. Often the lateral movement and ground surface cracks first develop at the unconfined toe, and then the slope movement and ground cracks progressively move upslope.

FIGURE 3.39 Flow slide of the Lower San Fernando Dam caused by the San Fernando earthquake on February 9, 1971. The photograph is a view to the east, and the main flow slide is at the eastern end of the dam. (*Photograph from the Steinbrugge Collection, EERC, University of California, Berkeley.*)

FIGURE 3.40 Flow slide of the Lower San Fernando Dam caused by the San Fernando earthquake on February 9, 1971. The photograph is a view to the west and shows the dam failure after the reservoir has been partly emptied. (*Photograph from the Steinbrugge Collection, EERC, University of California, Berkeley.*)

Figure 3.41 shows an example of large-scale lateral spreading caused by liquefaction during the Loma Prieta earthquake on October 17, 1989. As shown in Fig. 3.41, as the displaced ground breaks up internally, it causes fissures, scarps, and depressions to form at ground surface. Notice in Fig. 3.41 that the main ground surface cracks tend to develop parallel to each other. Some of the cracks have filled with water from the adjacent waterway. As the ground moves laterally, the blocks of soil between the main cracks tend to settle and break up into even smaller pieces.

Large-scale lateral spreads can damage all types of structures built on top of the lateral spreading soil. Lateral spreads can pull apart foundations of buildings built in the failure area, they can sever sewer pipelines and other utilities in the failure mass, and they can cause compression or buckling of structures, such as bridges, founded at the toe of the failure mass. Figure 3.42 shows lateral spreading caused by liquefaction during the Prince William Sound earthquake in Alaska on March 27, 1964, that has damaged a paved parking area.

Lateral spreading is discussed further in Sec. 9.5.

3.5 SLOPE MOVEMENT

3.5.1 Types of Earthquake-Induced Slope Movement

Another secondary effect of earthquakes is slope movement. As indicated in Tables 3.1 and 3.2, there can be many different types of earthquake-induced slope movement. For rock slopes (Table 3.1), the earthquake-induced slope movement is often divided into falls and slides. Falls are distinguished by the relatively free-falling nature of the rock or rocks,

FIGURE 3.41 Lateral spreading caused by the Loma Prieta, California, earthquake on October 17, 1989. (*Photograph from the Loma Prieta Collection, EERC, University of California, Berkeley.*)

FIGURE 3.42 Lateral spreading caused by the Prince William Sound earthquake in Alaska on March 27, 1964. (*Photograph from the Steinbrugge Collection, EERC, University of California, Berkeley.*)

where the earthquake-induced ground shaking causes the rocks to detach themselves from a cliff, steep slope, cave, arch, or tunnel (Stokes and Varnes 1955). Slides are different from falls in that there is shear displacement along a distinct failure (or slip) surface.

For soil slopes, there can also be earthquake-induced falls and slides (Table 3.2). In addition, the slope can be subjected to a flow slide or lateral spreading, as discussed in Secs. 3.4.4 and 3.4.5.

The minimum slope angle listed in column 4 of Tables 3.1 and 3.2 refers to the minimum slope inclination that is usually required to initiate a specific type of earthquake-induced slope movement. Note that for an earthquake-induced rock fall, the slope inclination typically must be 40° or greater, while for liquefaction-induced lateral spreading (Sec. 3.4.5) the earthquake-induced movement can occur on essentially a flat surface (i.e., minimum angle of inclination is 0.3°).

3.52 Examples of Earthquake-Induced Slope Movement

Three examples of deadly earthquake-induced slope movements are described below.

December 16, 1920, Haiyuan Earthquake in Northern China ($M_w = 8.7$). This earthquake triggered hundreds of slope failures and landslides that killed more than 100,000 people and affected an area of more than 4000 km² (1500 mi²) (Close and McCormick 1922). The landslides blocked roads and buried farmlands and villages. In one area that had a hilly topography with layers of loess ranging from 20 to 50 m (65 to 160 ft) in thickness, there were about 650 loess landslides (Zhang and Lanmin 1995).

TABLE 3.1 Types of Earthquake-Induced Slope Movement in Rock

Main type of slope movement	Subdivisions	Material type	Minimum slope inclination	Comments
Falls	Rockfalls	Rocks weakly cemented, intensely fractured, or weathered; contain conspicuous planes of weakness dipping out of slope or contain boulders in a weak matrix.	40° (1.2 : 1)	Particularly common near ridge crests and on spurs, ledges, artificially cut slopes, and slopes undercut by active erosion.
Slides	Rock slides	Rocks weakly cemented, intensely fractured, or weathered; contain conspicuous planes of weakness dipping out of slope or contain boulders in a weak matrix.	35° (1.4 : 1)	Particularly common in hillside flutes and channels, on artificially cut slopes, and on slopes undercut by active erosion. Occasionally reactivate preexisting rock slide deposits.
	Rock avalanches	Rocks intensely fractured and exhibiting one of the following properties: significant weathering, planes of weakness dipping out of slope, weak cementation, or evidence of previous landsliding.	25° (2.1 : 1)	Usually restricted to slopes of greater than 500 ft (150 m) relief that have been undercut by erosion. May be accompanied by a blast of air that can knock down trees and structures beyond the limits of the deposited debris.
	Rock slumps	Intensely fractured rocks, preexisting rock slump deposits, shale, and other rocks containing layers of weakly cemented or intensely weathered material.	15° (3.7 : 1)	Often circular or curved slip surface as compared to a planar slip surface for block slides.
	Rock block slides	Rocks having conspicuous bedding planes or similar planes of weakness dipping out of slopes.	15° (3.7 : 1)	Similar to rock slides.

Sources: Keefer (1984) and Division of Mines and Geology (1997).

3.35

TABLE 3.2 Types of Earthquake-Induced Slope Movement in Soil

Main type of slope movement	Subdivisions	Material type	Minimum slope inclination	Comments
Falls	Soil falls	Granular soils that are slightly cemented or contain clay binder.	40° (1.2 : 1)	Particularly common on stream banks, terrace faces, coastal bluffs, and artificially cut slopes.
Slides	Soil avalanches	Loose, unsaturated sands.	25° (2.1 : 1)	Occasionally reactivation of preexisting soil avalanche deposits.
	Disrupted soil slides	Loose, unsaturated sands.	15° (3.7 : 1)	Often described as *running soil* or *running ground.*
	Soil slumps	Loose, partly to completely saturated sand or silt; uncompacted or poorly compacted artificial fill composed of sand, silt, or clay; preexisting soil slump deposits.	10° (5.7 : 1)	Particularly common on embankments built on soft, saturated foundation materials, in hillside cut-and-fill areas, and on river and coastal flood plains.
	Soil block slides	Loose, partly to completely saturated sand or silt; uncompacted or slightly compacted artificial fill composed of sand or silt, bluffs containing horizontal or subhorizontal layers of loose, saturated sand or silt.	5° (11 : 1)	Particularly common in areas of preexisting landslides along river and coastal floodplains, and on embankments built of soft, saturated foundation materials.
	Slow earth flows	Stiff, partly to completely saturated clay, and preexisting earth flow deposits.	10° (5.7 : 1)	An example would be sensitive clay.
Flow slides and lateral spreading	Flow slides	Saturated, uncompacted or slightly compacted artificial fill composed of sand or sandy silt (including hydraulic fill earth dams and tailings dams); loose, saturated granular soils.	2.3° (25 : 1)	Includes debris flows that typically originate in hollows at heads of streams and adjacent hillsides; typically travel at tens of miles per hour or more and may cause damage miles from the source area.
	Subaqueous flows	Loose, saturated granular soils.	0.5° (110 : 1)	Particularly common on delta margins.
	Lateral spreading	Loose, partly or completely saturated silt or sand, uncompacted or slightly compacted artificial fill composed of sand.	0.3° (190 : 1)	Particularly common on river and coastal floodplains, embankments built on soft, saturated foundation materials, delta margins, sand spits, alluvial fans, lake shores, and beaches.

Sources: Keefer (1984) and Division of Mines and Geology (1997).

3.36

Loess is a deposit of wind-blown silt that commonly has calcareous cement which binds the soil particles together (Terzaghi and Peck 1967). Usually the loess is only weakly cemented which makes it susceptible to cracking and to brittle slope failure during earthquakes.

May 31, 1970, Peru Earthquake (M$_w$ = 7.9). This earthquake occurred offshore of central Peru and triggered a large rock slide in the Andes. The mountains are composed of granitic rocks, and most of the initial rock slide consisted chiefly of such rocks. The mountains were heavily glaciated and oversteepened by glacial undercutting.

The earthquake-induced rock slide mass accelerated rapidly as it fell over glacial ice below the failure zone, and the resultant debris avalanche quickly became a mix of pulverized granitic rocks, ice, and mud (Plafker et al. 1971, Cluff 1971). The debris avalanche destroyed all the property in its path. For example, Figs. 3.43 and 3.44 show the condition of the city of Yungay before and after the debris avalanche. These two photographs show the following:

The photograph in Fig. 3.43 was taken before the earthquake-induced debris avalanche with the photographer standing in the Plaza de Armas in the central part of Yungay. Note the palm tree on the left and the large white wall of the cathedral just behind the palm tree. The earthquake-induced debris avalanche originated from the mountains, which are visible in the background.

In Fig. 3.44 this view is almost the same as Fig. 3.43. The debris avalanche triggered by the Peru earthquake on May 31, 1970, caused the devastation. The same palm tree is visible in both figures. The cross marks the location of the former cathedral. The massive cathedral partially diverted the debris avalanche and protected the palm trees. More than 15,000 people lost their lives in the city of Yungay.

FIGURE 3.43 This photograph was taken before the earthquake-induced debris avalanche with the photographer standing in the Plaza de Armas in the central part of Yungay. Note the palm tree on the left and the large white wall of the cathedral just behind the palm tree. Compare this figure with Fig. 3.44. (*Photograph from the Steinbrugge Collection, EERC, University of California, Berkeley.*)

FIGURE 3.44   View almost the same as in Fig. 3.43. A debris avalanche triggered by the Peru earthquake on May 31, 1970, caused the devastation. The same palm tree is visible in both figures. The cross marks the location of the former cathedral. The massive cathedral partially diverted the avalanche and protected the palm trees. (*Photograph from the Steinbrugge Collection, EERC, University of California, Berkeley.*)

March 27, 1964, Prince William Sound Earthquake in Alaska.   As indicated in Sec. 2.4.5, this earthquake was the largest earthquake in North America and the second-largest in this past century (the largest occurred in Chile in 1960). Some details concerning this earthquake are as follows (Pflaker 1972, Christensen 2000, Sokolowski 2000):

- The epicenter was in the northern Prince William Sound about 75 mi (120 km) east of Anchorage and about 55 mi (90 km) west of Valdez. The local magnitude M_L for this earthquake is estimated to be from 8.4 to 8.6. The moment magnitude M_w is reported as 9.2.

- The depth of the main shock was approximately 15 mi (25 km).

- The duration of shaking as reported in the Anchorage area lasted about 4 to 5 min.

- In terms of plate tectonics, the northwestward motion of the Pacific plate at about 2 to 3 in (5 to 7 cm) per year causes the crust of southern Alaska to be compressed and warped, with some areas along the coast being depressed and other areas inland being uplifted. After periods of tens to hundreds of years, the sudden southeastward motion of portions of coastal Alaska relieves this compression as they move back over the subducting Pacific plate.

- There was both uplifting and regional subsidence. For example, some areas east of Kodiak were raised about 30 ft (9 m), and areas near Portage experienced regional subsidence of about 8 ft (2.4 m).

- The maximum intensity per the modified Mercalli intensity scale was XI.

- There were 115 deaths in Alaska and about $300 to $400 million in damages (1964 dollars). The death toll was extremely small for a quake of this size, due to low population density, time of day (holiday), and type of material used to construct many buildings (wood).

During the strong ground shaking from this earthquake, seams of loose saturated sands and sensitive clays suffered a loss of shear strength. This caused entire slopes to move laterally along these weakened seams of soil. These types of landslides devastated the Turnagain Heights residential development and many downtown areas in Anchorage. It has been estimated that 56 percent of the total cost of damage was caused by earthquake-induced landslides (Shannon and Wilson, Inc. 1964, Hansen 1965, Youd 1978, Wilson and Keefer 1985).

Three examples of earthquake-induced landslides and slope movement during this earthquake are as follows:

1. *Turnagain Heights landslide:* An aerial view of this earthquake-induced landslide is shown in Fig. 3.45. The cross sections shown in Fig. 3.46 illustrate the sequence of movement of this landslide during the earthquake. The landslide movement has been described as follows (Nelson 2000):

> During the Good Friday earthquake on March 27, 1964, a suburb of Anchorage, Alaska, known as Turnagain Heights broke into a series of slump blocks that slid toward the ocean. This area was built on sands and gravels overlying marine clay. The upper clay layers were relatively stiff, but the lower layers consisted of sensitive clay. The slide moved about 610 m (2000 ft) toward the ocean, breaking up into a series of blocks. It began at the sea cliffs on the

FIGURE 3.45 Aerial view of the Turnagain Heights landslide caused by the Prince William Sound earthquake in Alaska on March 27, 1964. (*Photograph from the Steinbrugge Collection, EERC, University of California, Berkeley.*)

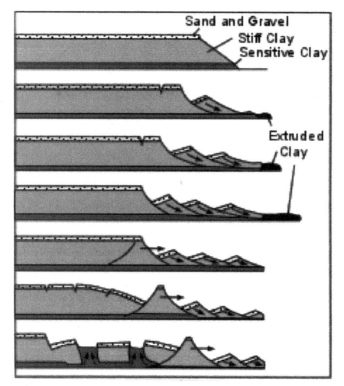

FIGURE 3.46   The above cross sections illustrate the sequence of movement of the Turnagain Heights landslide during the Prince William Sound earthquake in Alaska on March 27, 1964. (*Reproduced from Nelson 2000, based on work by Abbott 1996, with original version by USGS.*)

ocean after about 1.5 minutes of shaking caused by the earthquake, when the lower clay layer became liquefied. As the slide moved into the ocean, clays were extruded from the toe of the slide. The blocks rotating near the front of the slide eventually sealed off the sensitive clay layer preventing further extrusion. This led to pull-apart basins being formed near the rear of the slide and the oozing upward of the sensitive clays into the space created by the extension [see Fig. 3.46]. The movement of the mass of material toward the ocean destroyed 75 homes on the top of the slide.

As mentioned above, the large lateral movement of this earthquake-induced landslide generated numerous slump blocks and pull-apart basins that destroyed about 75 homes located on top of the slide. Examples are shown in Figs. 3.47 to 3.51.

2. *Government Hill landslide:*   The Government Hill School, located in Anchorage, Alaska, was severely damaged by earthquake-induced landslide movement. The school straddled the head of the landslide. When the landslide moved, it caused both lateral and vertical displacement of the school, as shown in Figs. 3.52 and 3.53.

3. *Embankment failure:*   In addition to the movement of massive landslides, such as the Turnagain Heights landslide and the Government Hill landslide, there were smaller

FIGURE 3.47 Damage caused by movement of the Turnagain Heights landslide during the Prince William Sound earthquake in Alaska on March 27, 1964. (*Photograph from the Godden Collection, EERC, University of California, Berkeley.*)

FIGURE 3.48 Damage caused by movement of the Turnagain Heights landslide during the Prince William Sound earthquake in Alaska on March 27, 1964. (*Photograph from the Steinbrugge Collection, EERC, University of California, Berkeley.*)

FIGURE 3.49   Damage caused by movement of the Turnagain Heights landslide during the Prince William Sound earthquake in Alaska on March 27, 1964. (*Photograph from the Steinbrugge Collection, EERC, University of California, Berkeley.*)

FIGURE 3.50   Damage caused by movement of the Turnagain Heights landslide during the Prince William Sound earthquake in Alaska on March 27, 1964. (*Photograph from the Steinbrugge Collection, EERC, University of California, Berkeley.*)

FIGURE 3.51 Damage caused by movement of the Turnagain Heights landslide during the Prince William Sound earthquake in Alaska on March 27, 1964. (*Photograph from the Steinbrugge Collection, EERC, University of California, Berkeley.*)

FIGURE 3.52 Overview of damage to the Government Hill School located at the head of a landslide caused by the Prince William Sound earthquake in Alaska on March 27, 1964. (*Photograph from the Steinbrugge Collection, EERC, University of California, Berkeley.*)

FIGURE 3.53   Close-up view of damage to the Government Hill School located at the head of a landslide caused by the Prince William Sound earthquake in Alaska on March 27, 1964. (*Photograph from the Steinbrugge Collection, EERC, University of California, Berkeley.*)

slides that resulted in substantial damage. For example, Fig. 3.54 shows earthquake-induced lateral deformation of the Anchorage–Portage highway. The relatively small highway embankment was reportedly constructed on a silt deposit (Seed 1970). Both sides of the embankment moved laterally, which resulted in the highway being pulled apart at its centerline.

3.5.3   Seismic Evaluation of Slope Stability

For the seismic evaluation of slope stability, the analysis can be grouped into two general categories, as follows:

1. *Inertia slope stability analysis:*   The inertia slope stability analysis is preferred for those materials that retain their shear strength during the earthquake. There are many different types of inertia slope stability analyses, and two of the most commonly used are the pseudostatic approach and the Newmark method (1965). These two methods are described in Secs. 9.2 and 9.3.
2. *Weakening slope stability analysis:*   The weakening slope stability analysis is preferred for those materials that will experience a significant reduction in shear strength during the earthquake. An example of a weakening landslide is the Turnagain Heights landslide as described in the previous section.

    There are two cases of weakening slope stability analyses involving the liquefaction of soil:

FIGURE 3.54 Cracking of the Anchorage-Portage highway. The small highway embankment experienced lateral movement during the Prince William Sound earthquake in Alaska on March 27, 1964. (*Photograph from the Steinbrugge Collection, EERC, University of California, Berkeley.*)

a. *Flow slide:* As discussed in Sec. 3.4.4, flow failures develop when the static driving forces exceed the shear strength of the soil along the slip surface, and thus the factor of safety is less than 1.0. Figures 3.38 to 3.40 show the flow slide of the Lower San Fernando Dam caused by the San Fernando earthquake on February 9, 1971.

b. *Lateral spreading:* As discussed in Sec. 3.4.5, there could be localized or large-scale lateral spreading of retaining walls and slopes. Examples of large-scale lateral spreading are shown in Figs. 3.41 and 3.42. The concept of cyclic mobility is used to describe large-scale lateral spreading of slopes. In this case, the static driving forces do not exceed the shear strength of the soil along the slip surface, and thus the ground is not subjected to a flow slide. Instead, the driving forces only exceed the resisting forces during those portions of the earthquake that impart net inertial forces in the downslope direction. Each cycle of net inertial forces in the downslope direction causes the driving forces to exceed the resisting forces along the slip surface, resulting in progressive and incremental lateral movement. Often the lateral movement and ground surface cracks first develop at the unconfined toe, and then the slope movement and ground cracks progressively move upslope.

The seismic evaluation for weakening slope stability is discussed further in Secs. 9.4 to 9.6.

3.6 TSUNAMI AND SEICHE

The final secondary effects that are discussed in this chapter are tsunamis and seiches.

Tsunami.   *Tsunami* is a Japanese word that, when translated into English, means "harbor wave." A tsunami is an ocean wave that is created by a disturbance that vertically displaces a column of seawater. Many different types of disturbances can generate a tsunami, such as oceanic meteorite impact, submarine landslide, volcanic island eruption, or earthquake. Specifics concerning earthquake-induced tsunamis are as follows (USGS 2000a):

   1. *Generation of a tsunami:*   Tsunamis can be generated during the earthquake if the seafloor abruptly deforms and vertically displaces the overlying water. When large areas of the seafloor are uplifted or subside, a tsunami can be created. Earthquakes generated at seafloor subduction zones are particularly effective in generating tsunamis. Waves are formed as the displaced water mass, which acts under the influence of gravity, attempts to regain its equilibrium.

   2. *Characteristics of a tsunami:*   A tsunami is different from a normal ocean wave in that it has a long period and wavelength. While typical wind-generated waves may have a wavelength of 150 m (500 ft) and a period of about 10 s, a tsunami can have a wavelength in excess of 100 km (60 mi) and a period on the order of 1 h. In the Pacific Ocean, where the typical water depth is about 4000 m (13,000 ft), a tsunami travels at about 200 m/s (650 ft/s). Because the rate at which a wave loses its energy is inversely related to its wavelength, tsunamis not only propagate at high speeds, but also travel long transoceanic distances with limited energy losses.

   3. *Coastal effect on the tsunami:*   The tsunami is transformed as it leaves the deep water of the ocean and travels into the shallower water near the coast. The tsunami's speed diminishes as it travels into the shallower coastal water and its height grows. While the tsunami may be imperceptible at sea, the shoaling effect near the coast causes the tsunami to grow to be several meters or more in height. When it finally reaches the coast, the tsunami may develop into a rapidly rising or falling tide, a series of breaking waves, or a tidal bore.

   4. *Tsunami run-up height:*   Just like any other ocean wave, a tsunami begins to lose energy as it rushes onshore. For example, part of the wave energy is reflected offshore, and part is dissipated through bottom friction and turbulence. Despite these losses, tsunamis still move inland with tremendous amounts of energy. For example, tsunamis can attain a *run-up height,* defined as the maximum vertical height onshore above sea level, of 10 to 30 m (33 to 100 ft). Figure 3.55 shows a tsunami in the process of moving inland.

   5. *Tsunami damage:*   Tsunamis have great erosional ability, and they can strip beaches of sand and coastal vegetation. Likewise, tsunamis are capable of inundating the land well past the typical high-water level. This fast-moving water associated with the inundating tsunami can destroy houses and other coastal structures. A tsunami generated by the Niigata earthquake in Japan on June 16, 1964, caused the damage shown in Figs. 3.56 and 3.57.

FIGURE 3.55 Tsunami in progress. The site is the village of Kiritoppu, near Kushiro Harbor, Hokkaido, Japan. The Tokachi-oki earthquake in Japan generated the tsunami on March 4, 1952. (*Photograph from the Steinbrugge Collection, EERC, University of California, Berkeley.*)

FIGURE 3.56 Tsunami damage caused by the Niigata earthquake in Japan on June 16, 1964. (*Photograph from the Steinbrugge Collection, EERC, University of California, Berkeley.*)

FIGURE 3.57   Tsunami damage caused by the Niigata earthquake in Japan on June 16, 1964. (*Photograph from the Steinbrugge Collection, EERC, University of California, Berkeley.*)

As discussed in Sec. 2.4.5, the Chile earthquake in 1960 was the largest earthquake in this past century (moment magnitude = 9.5). According to Iida et al. (1967), the tsunami generated by this earthquake killed about 300 people in Chile and 61 people in Hawaii. About 22 h after the earthquake, the tsunami reached Japan and killed an additional 199 people. Figure 3.58 shows an example of tsunami damage in Chile.

Seiche.   An earthquake-induced seiche is very similar to a tsunami, except that it develops in inland waters, such as large lakes. An example of damage caused by a seiche is shown in Fig. 3.59. The building in the water was formerly on the shore about one-quarter mile up the lake. The house was apparently jarred from its foundation by the earthquake and then washed into the lake by a seiche generated by the Hebgen Lake earthquake (magnitude of 7.5) in Montana on August 17, 1959. The house shown in Fig. 3.59 later drifted into the earthquake-created harbor.

Mitigation Measures.   It is usually the responsibility of the engineering geologist to evaluate the possibility of a tsunami or seiche impacting the site. Because of the tremendous destructive forces, options to mitigate damage are often limited. Some possibilities include the construction of walls to deflect the surging water or the use of buildings having weak lower-floor partitions which will allow the water to flow through the building, rather than knocking it down, such as shown in Fig. 3.58.

FIGURE 3.58 Tsunami damage caused by the Chile earthquake in 1960. (*Photograph from the Steinbrugge Collection, EERC, University of California, Berkeley.*)

FIGURE 3.59 The building in the water was formerly on the shore about one-quarter mile up the lake. The house was apparently jarred from its foundation by the earthquake and then washed into the lake by the seiche generated during the Hebgen Lake earthquake in Montana on August 17, 1959. The house later drifted into the earthquake-created harbor shown in the above photograph. (*Photograph from the Steinbrugge Collection, EERC, University of California, Berkeley.*)

CHAPTER 4
EARTHQUAKE STRUCTURAL DAMAGE

4.1 INTRODUCTION

As discussed in Chap. 3, the actual rupture of the ground due to fault movement could damage a structure. Secondary effects, such as the liquefaction of loose granular soil, slope movement or failure, and inundation from a tsunami, could also cause structural damage. This chapter discusses some of the other earthquake-induced effects or structural conditions that can result in damage.

Earthquakes throughout the world cause a considerable amount of death and destruction. Earthquake damage can be classified as being either structural or non-structural. For example, the Federal Emergency Management Agency (1994) states:

> Damage to buildings is commonly classified as either structural or non-structural. Structural damage means the building's structural support has been impaired. Structural support includes any vertical and lateral force resisting systems, such as the building frames, walls, and columns. Non-structural damage does not affect the integrity of the structural support system. Examples of non-structural damage include broken windows, collapsed or rotated chimneys, and fallen ceilings. During an earthquake, buildings get thrown from side to side, and up and down. Heavier buildings are subjected to higher forces than lightweight buildings, given the same acceleration. Damage occurs when structural members are overloaded, or differential movements between different parts of the structure strain the structural components. Larger earthquakes and longer shaking durations tend to damage structures more. The level of damage resulting from a major earthquake can be predicted only in general terms, since no two buildings undergo the exact same motions during a seismic event. Past earthquakes have shown us, however, that some buildings are likely to perform more poorly than others.

There are four main factors that cause structural damage during an earthquake:

1. *Strength of shaking:* For small earthquakes (magnitude less than 6), the strength of shaking decreases rapidly with distance from the epicenter of the earthquake. According to the USGS (2000a), the strong shaking along the fault segment that slips during an earthquake becomes about one-half as strong at a distance of 8 mi, one-quarter as strong at a distance of 17 mi, one-eighth as strong at a distance of 30 mi, and one-sixteenth as strong at a distance of 50 mi.

In the case of a small earthquake, the center of energy release and the point where slip begins are not far apart. But in the case of large earthquakes, which have a significant length of fault rupture, these two points may be hundreds of miles apart. Thus for large earthquakes, the strength of shaking decreases in a direction away from the fault rupture.

2. *Length of shaking:*   The length of shaking depends on how the fault breaks during the earthquake. For example, the maximum shaking during the Loma Prieta earthquake lasted only 10 to 15 s. But during other magnitude earthquakes in the San Francisco Bay area, the shaking may last 30 to 40 s. The longer the ground shakes, the greater the potential for structural damage. In general, the higher the magnitude of an earthquake, the longer the duration of the shaking ground (see Table 2.2).

3. *Type of subsurface conditions:*   Ground shaking can be increased if the site has a thick deposit of soil that is soft and submerged. Many other subsurface conditions can cause or contribute to structural damage. For example, as discussed in Sec. 3.4, there could be structural damage due to liquefaction of loose submerged sands.

4. *Type of building:*   Certain types of buildings and other structures are especially susceptible to the side-to-side shaking common during earthquakes. For example, sites located within approximately 10 mi (16 km) of the epicenter or location of fault rupture are generally subjected to rough, jerky, high-frequency seismic waves that are often more capable of causing short buildings to vibrate vigorously. For sites located at greater distance, the seismic waves often develop into longer-period waves that are more capable of causing highrise buildings and buildings with large floor areas to vibrate vigorously (Federal Emergency Management Agency 1994).

Much as diseases will attack the weak and infirm, earthquakes damage those structures that have inherent weaknesses or age-related deterioration. Those buildings that are not reinforced, poorly constructed, weakened from age or rot, or underlain by soft or unstable soil are most susceptible to damage. This chapter discusses some of these susceptible structures.

4.2   EARTHQUAKE-INDUCED SETTLEMENT

Those buildings founded on solid rock are least likely to experience earthquake-induced differential settlement. However, buildings on soil could be subjected to many different types of earthquake-induced settlement. As discussed in Chap. 3, a structure could settle or be subjected to differential movement from the following conditions:

Tectonic Surface Effects

- Surface fault rupture, which can cause a structure that straddles the fault to be displaced vertically and laterally.
- Regional uplifting or subsidence associated with the tectonic movement.

Liquefaction

- Liquefaction-induced settlement.
- Liquefaction-induced ground loss below the structure, such as the loss of soil through the development of ground surface sand boils.
- Liquefaction-induced bearing capacity failure. Localized liquefaction could also cause limited punching-type failure of individual footings.
- Liquefaction-induced flow slides.
- Liquefaction-induced localized or large-scale lateral spreading.

Seismic-Induced Slope Movement

- Seismic-induced slope movement or failure (Tables 3.1 and 3.2).

- Seismic-induced landslide movement or failure.
- Slumping or minor shear deformations of embankments.

Tsunami or Seiche

- Settlement directly related to a tsunami or seiche. For example, the tsunami could cause erosion of the soil underneath the foundation, leading to settlement of the structure. An example of this condition is shown in Figs. 4.1 and 4.2.

Two additional conditions can cause settlement of a structure:

1. *Volumetric compression, also known as cyclic soil densification:* This type of settlement is due to ground shaking that causes the soil to compress, which is often described as *volumetric compression* or *cyclic soil densification.* An example would be the settlement of dry and loose sands that densify during the earthquake, resulting in ground surface settlement.

2. *Settlement due to dynamic loads caused by rocking:* This type of settlement is due to dynamic structural loads that momentarily increase the foundation pressure acting on the soil. The soil will deform in response to the dynamic structural load, resulting in settlement of the building. This settlement due to dynamic loads is often a result of the structure rocking back and forth.

These two conditions can also work in combination and cause settlement of the foundation. Settlement due to volumetric compression and rocking settlement are discussed in Secs. 7.4 and 7.5.

FIGURE 4.1 Overview of damage caused by a tsunami generated during the Prince William Sound earthquake in Alaska on March 27, 1964. Note the tilted tower in the background. (*Photograph from the Steinbrugge Collection, EERC, University of California, Berkeley.*)

FIGURE 4.2 Close-up view of the tilted tower shown in Fig. 4.1. The tilting of the tower was caused by the washing away of soil due to a tsunami generated during the Prince William Sound earthquake in Alaska on March 27, 1964. (*Photograph from the Steinbrugge Collection, EERC, University of California, Berkeley.*)

4.3 TORSION

Torsional problems develop when the center of mass of the structure is not located at the center of its lateral resistance, which is also known as the center of rigidity. A common example is a tall building that has a first-floor area consisting of a space that is open and supports the upper floors by the use of isolated columns, while the remainder of the first-floor area contains solid load-bearing walls that are interconnected. The open area having isolated columns will typically have much less lateral resistance than that part of the floor containing the interconnected load-bearing walls. While the center of mass of the building may be located at the midpoint of the first-floor area, the center of rigidity is offset toward the area containing the interconnected load-bearing walls. During the earthquake, the center of mass will twist about the center of rigidity, causing torsional forces to be induced into the building frame.

An example is shown in Figs. 4.3 and 4.4. The two views are inside the Hotel Terminal and show the collapse of the second story due to torsional shear failure of the second-floor columns during the Gualan earthquake in Guatemala on February 4, 1976. This torsional failure has been described as follows (EERC 2000):

> Figure 4.3 is a view inside Hotel Terminal showing the collapse of the second story due to shear failure of the second-floor columns. Note the significant lateral displacement (interstory drift to the right) due to the torsional rotation of the upper part of the building.
> Figure 4.4 is a close-up of one of the collapsed columns of Hotel Terminal. Note that the upper floor has displaced to the right and dropped, and the top and bottom sections of the column are now side-by-side. Although the columns had lateral reinforcement (ties), these were

FIGURE 4.3   Torsional failure of the second story of the Hotel Terminal. The torsional failure occurred during the Gualan earthquake in Guatemala on February 4, 1976. (*Photograph from the Godden Collection, EERC, University of California, Berkeley.*)

FIGURE 4.4   Close-up view of a collapsed second-story column at the Hotel Terminal. Note that the upper floor has displaced to the right and dropped, and the top and bottom sections of the column are now side by side. The torsional failure occurred during the Gualan earthquake in Guatemala on February 4, 1976. (*Photograph from the Godden Collection, EERC, University of California, Berkeley.*)

not enough and at inadequate spacing to resist the shear force developed due to the torsional moment which originated in the second story. This failure emphasizes the importance of avoiding large torsional forces and the need for providing an adequate amount of transverse reinforcement with proper detailing.

4.4 SOFT STORY

4.4.1 Definition and Examples

A *soft story,* also known as a *weak story,* is defined as a story in a building that has substantially less resistance, or stiffness, than the stories above or below it. In essence, a soft story has inadequate shear resistance or inadequate ductility (energy absorption capacity) to resist the earthquake-induced building stresses. Although not always the case, the usual location of the soft story is at the ground floor of the building. This is because many buildings are designed to have an open first-floor area that is easily accessible to the public. Thus the first floor may contain large open areas between columns, without adequate shear resistance. The earthquake-induced building movement also causes the first floor to be subjected to the greatest stress, which compounds the problem of a soft story on the ground floor.

Concerning soft stories, the National Information Service for Earthquake Engineering (2000) states:

> In shaking a building, an earthquake ground motion will search for every structural weakness. These weaknesses are usually created by sharp changes in stiffness, strength and/or ductility, and the effects of these weaknesses are accentuated by poor distribution of reactive masses. Severe structural damage suffered by several modern buildings during recent earthquakes illustrates the importance of avoiding sudden changes in lateral stiffness and strength. A typical example of the detrimental effects that these discontinuities can induce is seen in the case of buildings with a "soft story." Inspection of earthquake damage as well as the results of analytical studies have shown that structural systems with a soft story can lead to serious problems during severe earthquake ground shaking. [Numerous examples] illustrate such damage and therefore emphasize the need for avoiding the soft story by using an even distribution of flexibility, strength, and mass.

The following are five examples of buildings having a soft story on the ground floor:

1. *Chi-chi earthquake in Taiwan on September 21, 1999:*   In Taiwan, it is common practice to have an open first-floor area by using columns to support the upper floors. In some cases, the spaces between the columns are filled in with plate-glass windows in order to create ground-floor shops. Figure 4.5 shows an example of this type of construction and the resulting damage caused by the Chi-chi earthquake.

2. *Northridge earthquake in California on January 17, 1994:*   Many apartment buildings in southern California contain a parking garage on the ground floor. To provide an open area for the ground-floor parking area, isolated columns are used to support the upper floors. These isolated columns often do not have adequate shear resistance and are susceptible to collapse during an earthquake. For example, Figs. 4.6 and 4.7 show the collapse of an apartment building during the Northridge earthquake caused by the weak shear resistance of the first-floor garage area.

3. *Loma Prieta earthquake in California on October 19, 1989:*   Another example of a soft story due to a first-floor garage area is shown in Fig. 4.8. The four-story apartment building was located on Beach Street, in the Marina District, San Francisco. The first-floor

FIGURE 4.5 Damage due to a soft story at the ground floor. The damage occurred during the Chi-chi earthquake in Taiwan on September 21, 1999. (*Photograph from the USGS Earthquake Hazards Program, NEIC, Denver.*)

garage area, with its large open areas, had inadequate shear resistance and was unable to resist the earthquake-induced building movements.

4. *Izmit earthquake in Turkey on August 17, 1999:* Details concerning this earthquake have been presented in Sec. 3.3. In terms of building conditions, it has been stated (Bruneau 1999):

> A typical reinforced concrete frame building in Turkey consists of a regular, symmetric floor plan, with square or rectangular columns and connecting beams. The exterior enclosure as well as interior partitioning are of non-bearing unreinforced brick masonry infill walls. These walls contributed significantly to the lateral stiffness of buildings during the earthquake and, in many instances, controlled the lateral drift and resisted seismic forces elastically. This was especially true in low-rise buildings, older buildings where the ratio of wall to floor area was very high, and buildings located on firm soil. Once the brick infills failed, the lateral strength and stiffness had to be provided by the frames alone, which then experienced significant inelasticity in the critical regions. At this stage, the ability of reinforced concrete columns,

FIGURE 4.6 Building collapse caused by a soft story due to the parking garage on the first floor. The building collapse occurred during the Northridge earthquake in California on January 17, 1994.

FIGURE 4.7 View inside the collapsed first-floor parking garage (the arrows point to the columns). The building collapse occurred during the Northridge earthquake in California on January 17, 1994.

FIGURE 4.8   Damage caused by a soft story due to a parking garage on the first floor. The damage occurred during the Loma Prieta earthquake in California on October 17, 1989. (*Photograph from the Loma Prieta Collection, EERC, University of California, Berkeley.*)

beams, and beam-column joints to sustain deformation demands depended on how well the seismic design and detailing requirements were followed both in design and in construction.

A large number of residential and commercial buildings were built with soft stories at the first-floor level. First stories are often used as stores and commercial areas, especially in the central part of cities. These areas are enclosed with glass windows, and sometimes with a single masonry infill at the back. Heavy masonry infills start immediately above the commercial floor. During the earthquake, the presence of a soft story increased deformation demands very significantly, and put the burden of energy dissipation on the first-story columns. Many failures and collapses can be attributed to the increased deformation demands caused by soft stories, coupled with lack of deformability of poorly designed columns. This was particularly evident on a commercial street where nearly all buildings collapsed towards the street.

Examples of this soft story condition are shown in Figs. 4.9 and 4.10.

5. *El Asnam earthquake in Algeria on October 10, 1980:*   An interesting example of damage due to a soft story is shown in Fig. 4.11 and described below (National Information Service for Earthquake Engineering 2000):

> Although most of the buildings in this new housing development [see Fig. 4.11] remained standing after the earthquake, some of them were inclined as much as 20 degrees and dropped up to 1 meter, producing significant damage in the structural and non-structural elements of the first story. The reason for this type of failure was the use of the "Vide Sanitaire," a crawl space about 1 meter above the ground level. This provides space for plumbing and ventilation under the first floor slab and serves as a barrier against transmission of humidity from the ground to the first floor. Unfortunately, the way that the vide sanitaires were constructed created a soft story with inadequate shear resistance. Hence the stubby columns in this crawl space were sheared off by the inertia forces induced by the earthquake ground motion.

Although the above five examples show damage due to a soft story located on the first floor or lowest level of the building, collapse at other stories can also occur depending on

FIGURE 4.9   Damage caused by a soft story at the first-floor level. The damage occurred during the Izmit earthquake in Turkey on August 17, 1999. (*Photograph by Mehmet Celebi, USGS.*)

the structural design. For example, after the Kobe earthquake in Japan on January 17, 1995, it was observed that there were a large number of 20-year and older high-rise buildings that collapsed at the fifth floor. The cause was apparently an older version of the building code that allowed a weaker superstructure beginning at the fifth floor.

While damage and collapse due to a soft story are most often observed in buildings, they can also be developed in other types of structures. For example, Figs. 4.12 and 4.13 show an elevated gas tank that was supported by reinforced concrete columns. The lower level containing the concrete columns behaved as a soft story in that the columns were unable to provide adequate shear resistance during the earthquake.

Concerning the retrofitting of a structure that has a soft story, the National Information Service for Earthquake Engineering (2000) states:

> There are many existing buildings in regions of high seismic risk that, because of their structural systems and/or of the interaction with non-structural components, have soft stories with either inadequate shear resistance or inadequate ductility (energy absorption capacity) in the event of being subjected to severe earthquake ground shaking. Hence they need to be retrofitted. Usually the most economical way of retrofitting such a building is by adding proper shear walls or bracing to the soft stories.

4.4.2 Pancaking

Pancaking occurs when the earthquake shaking causes a soft story to collapse, leading to total failure of the overlying floors. These floors crush and compress together such that the

FIGURE 4.10   Building collapse caused by a soft story at the first-floor level. The damage occurred during the Izmit earthquake in Turkey on August 17, 1999. (*Photograph by Mehmet Celebi, USGS.*)

FIGURE 4.11   Building tilting and damage caused by a soft story due to a ground-floor crawl space. The damage occurred during the El Asnam earthquake in Algeria on October 10, 1980. (*Photograph from the Godden Collection, EERC, University of California, Berkeley.*)

FIGURE 4.12   Overview of a collapsed gas storage tank, located at a gas storage facility near Sabanci Industrial Park, Turkey. The elevated gas storage tank collapsed during the Izmit earthquake in Turkey on August 17, 1999. (*Photograph from the Izmit Collection, EERC, University of California, Berkeley.*)

FIGURE 4.13   Close-up view of the columns that had supported the elevated gas storage tank shown in Fig. 4.12. The columns did not have adequate shear resistance and were unable to support the gas storage tank during the Izmit earthquake in Turkey on August 17, 1999. (*Photograph from the Izmit Collection, EERC, University of California, Berkeley.*)

final collapsed condition of the building consists of one floor stacked on top of another, much like a stack of pancakes.

Pancaking of reinforced concrete multistory buildings was common throughout the earthquake-stricken region of Turkey due to the Izmit earthquake on August 17, 1999. Examples of pancaking caused by this earthquake are shown in Figs. 4.14 to 4.16. Concerning the damage caused by the Izmit earthquake, Bruneau (1999) states:

> Pancaking is attributed to the presence of "soft" lower stories and insufficiently reinforced connections at the column-beam joints. Most of these buildings had a "soft" story—a story with most of its space unenclosed—and a shallow foundation and offered little or no lateral resistance to ground shaking. As many as 115,000 of these buildings—some engineered, some not—were unable to withstand the strong ground shaking and were either badly damaged or collapsed outright, entombing sleeping occupants beneath the rubble. Partial collapses involved the first two stories. The sobering fact is that Turkey still has an existing inventory of several hundred thousand of these highly vulnerable buildings. Some will need to undergo major seismic retrofits; others will be demolished.

Another example of pancaking is shown in Fig. 4.17. The site is located in Mexico City, and the damage was caused by the Michoacan earthquake in Mexico on September 19, 1985. Note in Fig. 4.17 that there was pancaking of only the upper several floors of the parking garage. The restaurant building that abutted the parking garage provided additional lateral support, which enabled the lower three floors of the parking garage to resist the earthquake shaking. The upper floors of the parking garage did not have this additional lateral support and thus experienced pancaking during the earthquake.

FIGURE 4.14 Pancaking of a building during the Izmit earthquake in Turkey on August 17, 1999. (*Photograph by Mehmet Celebi, USGS.*)

FIGURE 4.15 Pancaking of a building, which also partially crushed a bus, during the Izmit earthquake in Turkey on August 17, 1999. (*Photograph from the Izmit Collection, EERC, University of California, Berkeley.*)

FIGURE 4.16 Pancaking of a building, during the Izmit earthquake in Turkey on August 17, 1999. Note that the center of the photograph shows a hole that was excavated through the pancaked building in order to rescue the survivors. (*Photograph from the Izmit Collection, EERC, University of California, Berkeley.*)

FIGURE 4.17   Pancaking of the upper floors of a parking garage during the Michoacan earthquake in Mexico on September 19, 1985. Note that the restaurant building provided additional lateral support which enabled the lower three floors of the parking garage to resist the collapse. (*Photograph from the Steinbrugge Collection, EERC, University of California, Berkeley.*)

4.4.3 Shear Walls

Many different types of structural systems can be used to resist the inertia forces in a building that are induced by the earthquake ground motion. For example, the structural engineer could use braced frames, moment-resisting frames, and shear walls to resist the lateral earthquake-induced forces. Shear walls are designed to hold adjacent columns or vertical support members in place and then transfer the lateral forces to the foundation. The forces resisted by shear walls are predominately shear forces, although a slender shear wall could also be subjected to significant bending (Arnold and Reitherman 1982).

Figure 4.18 shows the failure of a shear wall at the West Anchorage High School caused by the Prince William Sound earthquake in Alaska on March 27, 1964. Although the shear wall shown in Fig. 4.18 contains four small windows, often a shear wall is designed and constructed as a solid and continuous wall, without any window or door openings. The X-shaped cracks between the two lower windows in Fig. 4.18 are 45° diagonal tension cracks,

FIGURE 4.18 Damage to a shear wall at the West Anchorage High School caused by the Prince William Sound earthquake in Alaska on March 27, 1964. (*Photograph from the Steinbrugge Collection, EERC, University of California, Berkeley.*)

which are typical and characteristic of earthquake-induced damage. These diagonal tension cracks are formed as the shear wall moves back and forth in response to the earthquake ground motion.

Common problems with shear walls are that they have inadequate strength to resist the lateral forces and that they are inadequately attached to the foundation. For example, having inadequate shear walls on a particular building level can create a soft story. A soft story could also be created if there is a discontinuity in the shear walls from one floor to the other, such as a floor where its shear walls are not aligned with the shear walls on the upper or lower floors.

Even when adequately designed and constructed, shear walls will not guarantee the survival of the building. For example, Fig. 4.19 shows a comparatively new building that was proclaimed as "earthquake-proof" because of the box-type construction consisting of numerous shear walls. Nevertheless, the structure was severely damaged because of earthquake-induced settlement of the building.

4.4.4 Wood-Frame Structures

It is generally recognized that single-family wood-frame structures that include shear walls in their construction are very resistant to collapse from earthquake shaking. This is due to several factors, such as their flexibility, strength, and light dead loads, which produce low earthquake-induced inertia loads. These factors make the wood-frame construction much better at resisting shear forces and hence more resistant to collapse.

There are exceptions to the general rule that wood-frame structures are resistant to collapse. For example, in the 1995 Kobe earthquake, the vast majority of deaths were due to

FIGURE 4.19 A comparatively new building that was proclaimed as "earthquake-proof" because of the box-type construction consisting of numerous shear walls. Nevertheless, the structure was severely damaged because of earthquake-induced settlement of the building during the Bucharest earthquake on March 4, 1977. (*Photograph from the Steinbrugge Collection, EERC, University of California, Berkeley.*)

the collapse of one- and two-story residential and commercial wood-frame structures. More than 200,000 houses, about 10 percent of all houses in the Hyogo prefecture, were damaged, including more than 80,000 collapsed houses, 70,000 severely damaged, and 7000 consumed by fire. The collapse of the houses has been attributed to several factors, such as (EQE Summary Report, 1995):

- Age-related deterioration, such as wood rot, that weakened structural members.
- Post and beam construction that often included open first-floor areas (i.e., a soft first floor), with few interior partitions that were able to resist lateral earthquake loads.
- Weak connections between the walls and the foundation.
- Inadequate foundations that often consisted of stones or concrete blocks.

- Poor soil conditions consisting of thick deposits of soft or liquefiable soil that settled during the earthquake. Because of the inadequate foundations, the wood-frame structures were unable to accommodate the settlement.

- Inertia loads from heavy roofs that exceeded the lateral earthquake load-resisting capacity of the supporting walls. The heavy roofs were created by using thick mud or heavy tile and were used to resist the winds from typhoons. However, when the heavy roofs collapsed during the earthquake, they crushed the underlying structure.

4.5   POUNDING DAMAGE

Pounding damage can occur when two buildings are constructed close to each other and, as they rock back-and-forth during the earthquake, they collide into each other. Even when two buildings having dissimilar construction materials or different heights are constructed adjacent to each other, it does not necessarily mean that they will be subjected to pounding damage. For example, as shown in Fig. 4.17, the restaurant that was constructed adjacent to the parking garage actually provided lateral support to the garage and prevented the three lower levels from collapsing.

In the common situation for pounding damage, a much taller building, which has a higher period and larger amplitude of vibration, is constructed against a squat and short building that has a lower period and smaller amplitude of vibration. Thus during the earthquake, the buildings will vibrate at different frequencies and amplitudes, and they can collide with each other. The effects of pounding can be especially severe if the floors of one building impact the other building at different elevations, so that, for example, the floor of one building hits a supporting column of an adjacent building.

Figure 4.20 shows an example of pounding damage to the Anchorage-Westward Hotel caused by the Prince William Sound earthquake in Alaska on March 27, 1964. Although not evident in the photograph, the structure shown on the right half of the photograph is a 14-story hotel. The structure visible on the left half of Fig. 4.20 is the hotel ballroom. The pounding damage occurred at the junction of the 14-story hotel and the short and squat ballroom. Note in Fig. 4.20 that the main cracking emanates from the upper left corner of the street-level doorway. The doorway is a structural weak point, which has been exploited during the side-to-side shaking during the earthquake.

Another example of pounding damage and eventual collapse is shown in Fig. 4.21. The buildings were damaged during the Izmit earthquake in Turkey on August 17, 1999. As shown in Fig. 4.21, the pounding damage was accompanied by the collapse of the two buildings into each other.

It is very difficult to model the pounding effects of two structures and hence design structures to resist such damage. As a practical matter, the best design approach to prevent pounding damage is to provide sufficient space between the structures to avoid the problem. If two buildings must be constructed adjacent to each other, then one design feature should be to have the floors of both buildings at the same elevations, so that the floor of one building does not hit a supporting column of an adjacent building.

4.51   Impact Damage from Collapse of Adjacent Structures

Similar to pounding damage, the collapse of a building can affect adjacent structures. For example, Fig. 4.22 shows a building that has lost a corner column due to the collapse of an adjacent building during the Izmit Earthquake in Turkey on August 17, 1999. The buildings were under construction at the time of the earthquake. Note that the roof of the col-

FIGURE 4.20 Pounding damage to the Anchorage-Westward Hotel caused by the Prince William Sound earthquake in Alaska on March 27, 1964. The building on the right half of the photograph is the 14-story hotel, while the building visible on the left half of the photograph is the ballroom. (*Photograph from the Steinbrugge Collection, EERC, University of California, Berkeley.*)

lapsed building now rests on the third story corner of the standing building.

Since the geotechnical engineer and engineering geologist are usually required to discuss any "earthquake hazards" that could affect the planned construction, it may be appropriate for them to evaluate possible collapse of adjacent buildings founded on poor soil or susceptible to geologic hazards.

4.5.2 Asymmetry

Similar to pounding damage, buildings that are asymmetric, such as T- or L-shaped buildings, can experience damage as different parts of the building vibrate at different frequencies and amplitudes. This difference in movement of different parts of the building is due to the relative stiffness of each portion of the building. For example, for the T-shaped build-

FIGURE 4.21 Another example of pounding damage and eventual collapse caused by the Izmit earthquake in Turkey on August 17, 1999. (*Photograph from the Izmit Collection, EERC, University of California, Berkeley.*)

ing, the two segments that make up the building are usually much more stiff in their long directions, then across the segments. Thus damage tends to occur where the two segments of the T join together.

4.6 RESONANCE OF THE STRUCTURE

Resonance is defined as a condition in which the period of vibration of the earthquake-induced ground shaking is equal to the natural period of vibration of the building. When resonance occurs, the shaking response of the building is enhanced, and the amplitude of vibration of the building rapidly increases. Tall buildings, bridges, and other large structures respond most to ground shaking that has a high period of vibration, and small structures respond most to low-period shaking. For example, a rule of thumb is that the period of vibration is about equal to 0.1 times the number of stories in a building. Thus a 10-story

FIGURE 4.22   The building shown has lost a corner column due to the collapse of an adjacent building during the Izmit earthquake in Turkey on August 17, 1999. Note that the roof of the collapsed building now rests on the third-story corner of the standing building. (*Photograph from the Izmit Collection, EERC, University of California, Berkeley.*)

building would have a natural period of vibration of about 1 s, and if the earthquake-induced ground motion also has a period of vibration of about 1 s, then resonance is expected to occur for the 10-story building.

A response spectrum can be used to directly assess the nature of the earthquake ground motion on the structure. A response spectrum is basically a plot of the maximum displacement, velocity, or acceleration versus the natural period of a single-degree-of-freedom system. Different values of system damping can be used, and thus a family of such curves could be obtained. This information can then be used by the structural engineer in the design of the building. The response spectrum is discussed further in Sec. 11.5.

4.6.1   Soft Ground Effects

If the site is underlain by soft ground, such as a soft and saturated clay deposit, then there could be an increased peak ground acceleration a_{max} and a longer period of vibration of the ground. The following two examples illustrate the effect of soft clay deposits.

Michoacan Earthquake in Mexico on September 19, 1985.   There was extensive damage to Mexico City caused by the September 19, 1985, Michoacan earthquake. The greatest damage in Mexico City occurred to those buildings underlain by 125 to 164 ft (39 to 50 m) of soft clays, which are within the part of the city known as the Lake Zone (Stone et al. 1987). Because the epicenter of the earthquake was so far from Mexico City, the peak ground acceleration a_{max} recorded in the foothills of Mexico City (rock site) was about

FIGURE 4.23 Building collapse in Mexico City caused by the Michoacan earthquake in Mexico on September 19, 1985. (*Photograph from the Steinbrugge Collection, EERC, University of California, Berkeley.*)

FIGURE 4.24 Building collapse in Mexico City caused by the Michoacan earthquake in Mexico on September 19, 1985. (*Photograph from the Steinbrugge Collection, EERC, University of California, Berkeley.*)

0.04g. However, at the Lake Zone, the peak ground accelerations a_{max} were up to 5 times greater than at the rock site (Kramer 1996). In addition, the characteristic site periods were estimated to be 1.9 to 2.8 s (Stone et al. 1987). This longer period of vibration of the ground tended to coincide with the natural period of vibration of the taller buildings in the 5- to 20-story range. The increased peak ground acceleration and the effect of resonance caused either collapse or severe damage of these taller buildings, such as shown in Figs. 4.23 and 4.25. To explain this condition of an increased peak ground acceleration and a longer period of surface vibration, an analogy is often made between the shaking of these soft clays and the shaking of a bowl of jelly.

Loma Prieta Earthquake in San Francisco Bay Area on October 17, 1989. A second example of soft ground effects is the Loma Prieta earthquake on October 17, 1989. Figure 4.26 presents the ground accelerations (east-west direction) at Yerba Buena Island and at Treasure Island (R. B. Seed et al. 1990). Both sites are about the same distance from the epicenter of the Loma Prieta earthquake. However, the Yerba Buena Island seismograph is located directly on a rock outcrop, while the Treasure Island seismograph is underlain by

FIGURE 4.25 Building damage and tilting in Mexico City caused by the Michoacan earthquake in Mexico on September 19, 1985. (*Photograph from the Steinbrugge Collection, EERC, University of California, Berkeley.*)

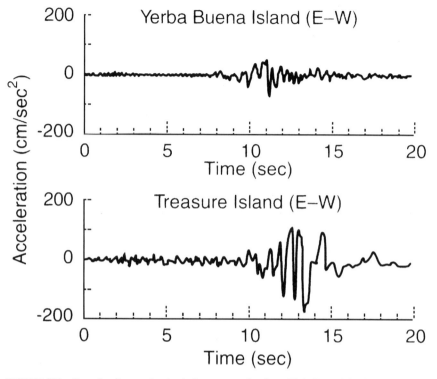

FIGURE 4.26   Ground surface acceleration in the east-west direction at Yerba Buena Island and at Treasure Island for the Loma Prieta earthquake in California on October 17, 1989. (*From Seed et al. 1990.*)

45 ft (13.7 m) of loose sandy soil over 55 ft (16.8 m) of San Francisco Bay mud (a normally consolidated silty clay). Note the significantly different ground acceleration plots for these two sites. The peak ground acceleration in the east-west direction at Yerba Buena Island was only 0.06g, while at Treasure Island the peak ground acceleration in the east-west direction was 0.16g (Kramer 1996). Thus the soft clay site had a peak ground acceleration that was 2.7 times that of the hard rock site.

The amplification of the peak ground acceleration by soft clay also contributed to damage of structures throughout the San Francisco Bay area. For example, the northern portion of the Interstate 880 highway (Cypress Street Viaduct) that collapsed was underlain by the San Francisco Bay mud (see Figs. 4.27 to 4.29). The southern portion of the Interstate 880 highway was not underlain by the bay mud, and it did not collapse.

As these two examples illustrate, local soft ground conditions can significantly increase the peak ground acceleration a_{max} by a factor of 3 to 5 times. The soft ground can also increase the period of ground surface shaking, leading to resonance of taller structures. The geotechnical engineer and engineering geologist will need to evaluate the possibility of increasing the peak ground acceleration a_{max} and increasing the period of ground shaking for sites that contain thick deposits of soft clay. This is discussed further in Sec. 5.6.

FIGURE 4.27   Overview of the collapse of the Cypress Street Viaduct caused by the Loma Prieta earthquake in California on October 17, 1989. (*From USGS.*)

FIGURE 4.28   Close-up view of the collapse of the Cypress Street Viaduct caused by the Loma Prieta earthquake in California on October 17, 1989. (*From USGS.*)

FIGURE 4.29 Close-up view of the collapse of the Cypress Street Viaduct caused by the Loma Prieta earthquake in California on October 17, 1989. (*From USGS.*)

P · A · R · T · 2

GEOTECHNICAL EARTHQUAKE ENGINEERING ANALYSES

CHAPTER 5
SITE INVESTIGATION FOR GEOTECHNICAL EARTHQUAKE ENGINEERING

The following notation is used in this chapter:

SYMBOL DEFINITION

a_{max}	Peak ground acceleration
c	Cohesion based on a total stress analysis
c'	Cohesion based on an effective stress analysis
C_b	Borehole diameter correction
C_N	Correction factor to account for the overburden pressure
C_r	Rod length correction
D	Inside diameter of the SPT sampler
D_r	Relative density
e	Void ratio of soil
e_{max}	Void ratio corresponding to loosest possible state of soil
e_{min}	Void ratio corresponding to densest possible state of soil
E_m	Hammer efficiency
F	Outside diameter of the SPT sampler
FS_L	Factor of safety against liquefaction
g	Acceleration of gravity
h	Depth below ground surface
N	Measured SPT blow count (that is, N value in blows per foot)
N_{60}	N value corrected for field testing procedures
$(N_1)_{60}$	N value corrected for field testing procedures and overburden pressure
q_c	Cone resistance
q_{c1}	Cone resistance corrected for overburden pressure
r_u	Pore water pressure ratio
s_u	Undrained shear strength of soil
S_t	Sensitivity of cohesive soil
u	Pore water pressure
Z	Seismic zone factor
ϕ	Friction angle of sand (Sec. 5.4)
ϕ	Friction angle based on a total stress analysis (Sec. 5.5)
ϕ'	Friction angle based on an effective stress analysis
ϕ_r'	Drained residual friction angle
γ_t	Total unit weight of soil

σ Total stress
σ' Effective stress ($\sigma' = \sigma - u$)
σ_{v0}' Vertical effective stress

5.1 INTRODUCTION

Part 2 of the book describes the different types of geotechnical earthquake engineering analyses. Specific items that are included in Part 2 are as follows:

• Site investigation for geotechnical earthquake engineering (Chap. 5)
• Liquefaction (Chap. 6)
• Settlement of structures (Chap. 7)
• Bearing capacity (Chap. 8)
• Slope stability (Chap. 9)
• Retaining walls (Chap. 10)
• Other earthquake effects (Chap. 11)

It is important to recognize that without adequate and meaningful data from the site investigation, the engineering analyses presented in the following chapters will be of doubtful value and may even lead to erroneous conclusions. In addition, when performing the site investigation, the geotechnical engineer may need to rely on the expertise of other specialists. For example, as discussed in this chapter, geologic analyses are often essential for determining the location of active faults and evaluating site-specific impacts of the design earthquake.

The purpose of this chapter is to discuss the site investigation that may be needed for geotechnical earthquake engineering analyses. The focus of this chapter is on the information that is needed for earthquake design, and not on the basic principles of subsurface exploration and laboratory testing. For information on standard subsurface exploration and laboratory testing, see Day (1999, 2000).

In terms of the investigation for assessing seismic hazards, *Guidelines for Evaluating and Mitigating Seismic Hazards in California* (Division of Mines and Geology 1997) states: "the working premise for the planning and execution of a site investigation within seismic hazard zones is that the suitability of the site should be demonstrated. This premise will persist until either: (a) the site investigation satisfactorily demonstrates the absence of liquefaction or landslide hazard, or (b) the site investigation satisfactorily defines the liquefaction or landslide hazard and provides a suitable recommendation for its mitigation." Thus the purpose of the site investigation should be to demonstrate the absence of seismic hazards or to adequately define the seismic hazards so that suitable recommendations for mitigation can be developed. The scope of the site investigation is discussed next.

5.1.1 Scope of the Site Investigation

The scope of the site investigation depends on many different factors such as the type of facility to be constructed, the nature and complexity of the geologic hazards that could impact the site during the earthquake, economic considerations, level of risk, and specific requirements such as local building codes or other regulatory specifications. The most rigorous geotechnical earthquake investigations would be required for critical facilities. For example, the Federal Emergency Management Agency (1994) states:

> Critical facilities are considered parts of a community's infrastructure that must remain operational after an earthquake, or facilities that pose unacceptable risks to public safety if severely damaged. Essential facilities are needed during an emergency, such as hospitals, fire and police stations, emergency operation centers and communication centers. High-risk facilities, if severely damaged, may result in a disaster far beyond the facilities themselves. Examples include nuclear power plants, dams and flood control structures, freeway interchanges and bridges, industrial plants that use or store explosives, toxic materials or petroleum products. High-occupancy facilities have the potential of resulting in a large number of casualties or crowd control problems. This category includes high-rise buildings, large assembly facilities, and large multifamily residential complexes. Dependent care facilities house populations with special evacuation considerations, such as preschools and schools, rehabilitation centers, prisons, group care homes, and nursing and convalescent homes. Economic facilities are those facilities that should remain operational to avoid severe economic impacts, such as banks, archiving and vital record keeping facilities, airports and ports, and large industrial and commercial centers.
>
> It is essential that critical facilities designed for human occupancy have no structural weaknesses that can lead to collapse. The Federal Emergency Management Agency has suggested the following seismic performance goals for health care facilities:
>
> 1. The damage to the facilities should be limited to what might be reasonably expected after a destructive earthquake and should be repairable and not life-threatening.
> 2. Patients, visitors, and medical, nursing, technical and support staff within and immediately outside the facility should be protected during an earthquake.
> 3. Emergency utility systems in the facility should remain operational after an earthquake.
> 4. Occupants should be able to evacuate the facility safely after an earthquake.
> 5. Rescue and emergency workers should be able to enter the facility immediately after an earthquake and should encounter only minimum interference and danger.
> 6. The facility should be available for its planned disaster response role after an earthquake.

As previously mentioned, in addition to the type of facility, the scope of the investigation may be dependent on the requirements of the local building codes or other regulatory specifications. Prior to initiating a site investigation for seismic hazards, the geotechnical engineer and engineering geologist should obtain the engineering and geologic requirements of the governing review agency. For example, *Guidelines for Evaluating and Mitigating Seismic Hazards in California* (Division of Mines and Geology 1997) states that geotechnical engineers and engineering geologists:

> May save a great deal of time (and the client's money), and possibly misunderstandings, if they contact the reviewing geologist or engineer at the initiation of the investigation. Reviewers typically are familiar with the local geology and sources of information and may be able to provide additional guidance regarding their agency's expectations and review practices. Guidelines for geologic or geotechnical reports have been prepared by a number of agencies and are available to assist reviewers in their evaluation of reports. Distribution of copies of written policies and guidelines adopted by the agency usually alerts the applicants and consultants about procedures, report formats, and levels of investigative detail that will expedite review and approval of the project.

The scope of the investigation for geotechnical earthquake engineering is usually divided into two parts: (1) the screening investigation and (2) the quantitative evaluation of the seismic hazards (Division of Mines and Geology 1997). These two items are individually discussed in the next two sections.

5.2 SCREENING INVESTIGATION

The first step in geotechnical earthquake engineering is to perform a screening investigation. The purpose of the screening investigation is to assess the severity of the seismic hazards at the site, or in other words to screen out those sites that do not have seismic hazards. If it can be clearly demonstrated that a site is free of seismic hazards, then the quantitative evaluation could be omitted. On the other hand, if a site is likely to have seismic hazards, then the screening investigation can be used to define those hazards before proceeding with the quantitative evaluation.

An important consideration for the screening investigation is the effect that the new construction will have on potential seismic hazards. For example, as a result of grading or construction at the site, the groundwater table may be raised or adverse bedding planes may be exposed that result in a landslide hazard. Thus when a screening investigation is performed, both the existing condition and the final constructed condition must be evaluated for seismic hazards. Another important consideration is off-site seismic hazards. For example, the city of Yungay was devastated by an earthquake-induced debris avalanche that originated at a source located many miles away, as discussed in Sec. 3.5.2 (see Fig. 3.44).

The screening investigation should be performed on both a regional and a site-specific basis. The first step in the screening investigation is to review available documents, such as the following:

1. *Preliminary design information:* The documents dealing with preliminary design and proposed construction of the project should be reviewed. For example, the structural engineer or architect may have design information, such as the building location, size, height, loads, and details on proposed construction materials and methods. Preliminary plans may even have been developed that show the proposed construction.

2. *History of prior site development:* If the site had prior development, it is also important to obtain information on the history of the site. The site could contain old deposits of fill, abandoned septic systems and leach fields, buried storage tanks, seepage pits, cisterns, mining shafts, tunnels, and other artificial and subsurface works that could impact the proposed development. There may also be old reports that document seismic hazards at the site.

3. *Seismic history of the area:* There may be many different types of documents and maps that provide data on the seismic history of the area. For example, there may be seismic history information on the nature of past earthquake-induced ground shaking. This information could include the period of vibration, ground acceleration, magnitude, and intensity (isoseismal maps) of past earthquakes. This data can often be obtained from seismology maps and reports that illustrate the differences in ground shaking intensity based on geologic type; 50-, 100-, and 250-year acceleration data; and type of facilities or landmarks.

Geographical maps and reports are important because they can identify such items as the pattern, type, and movement of nearby potentially active faults or fault systems, and the distance of the faults to the area under investigation. Historical earthquake records should also be reviewed to determine the spatial and temporal distribution of historic earthquake epicenters.

4. *Aerial photographs and geologic maps:* During the screening investigation, the engineering geologist and geotechnical engineer should check aerial photographs and geologic maps. Aerial photographs and geologic maps can be useful in identifying existing and potential slope instability, fault ground rupture, liquefaction, and other geologic hazards. The type of observed features includes headwall scarps, debris chutes, fissures, grabens, and sand boils. By comparing older aerial photographs with newer ones, the engineering geologist can also observe any artificial or natural changes that have occurred at the site.

Geologic reports and maps can be especially useful to the geotechnical engineer and engineering geologist because they often indicate seismic hazards such as faults and landslides. Geologic reports and maps may indicate the geometry of the fault systems, the subsoil profile, and the amplification of seismic waves due to local conditions, which are important factors in the evaluation of seismic risk. For example, Fig. 5.1 presents a portion of a geologic map, and Fig. 5.2 shows cross sections through the area shown in Fig. 5.1 (from Kennedy 1975). Note that the geologic map and cross sections indicate the location of several faults and the width of the faults, and often state whether the faults are active or inactive. For example, Fig. 5.2 shows the Rose Canyon fault zone, an active fault having a ground shear zone about 300 m (1000 ft) wide. The cross sections in Fig. 5.2 also show fault-related displacement of various rock layers.

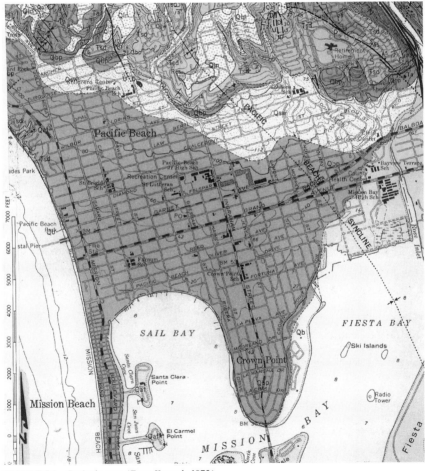

FIGURE 5.1 Geologic map. (*From Kennedy 1975.*)

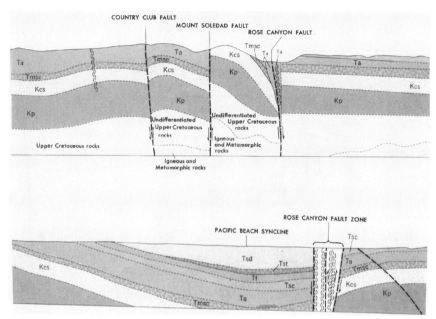

FIGURE 5.2   Geologic cross sections. (*From Kennedy 1975.*)

A major source for geologic maps in the United States is the U.S. Geological Survey (USGS). The USGS prepares many different geologic maps, books, and charts; a list of USGS publications is provided in *Index of Publications of the Geological Survey* (USGS 1997). The USGS also provides an *Index to Geologic Mapping in the United States,* which shows a map of each state and indicates the areas where a geologic map has been published.

5. *Special study maps:*   For some areas, special study maps may have been developed that indicate local seismic hazards. For example, Fig. 5.3 presents a portion of the Seismic Safety Study (1995) that shows the location of the Rose Canyon fault zone. Special study maps may also indicate other geologic and seismic hazards, such as potentially liquefiable soil, landslides, and abandoned mines.

6. *Topographic maps:*   Both old and recent topographic maps can provide valuable site information. Figure 5.4 presents a portion of the topographic map for the Encinitas Quadrangle, California (USGS 1975). As shown in Fig. 5.4, the topographic map is drawn to scale and shows the locations of buildings, roads, freeways, train tracks, and other civil engineering works as well as natural features such as canyons, rivers, lagoons, sea cliffs, and beaches. The topographic map in Fig. 5.4 even shows the locations of sewage disposal ponds, and water tanks; and by using different colors and shading, it indicates older versus newer development. But the main purpose of the topographic map is to indicate ground surface elevations or elevations of the seafloor, such as shown in Fig. 5.4. This information can be used to determine the major topographic features at the site and to evaluate potential seismic hazards.

7. *Building codes or other regulatory specifications:*   A copy of the most recently adopted local building code should be reviewed. Investigation and design requirements for ordinary structures, critical facilities, and lifelines may be delineated in building codes or

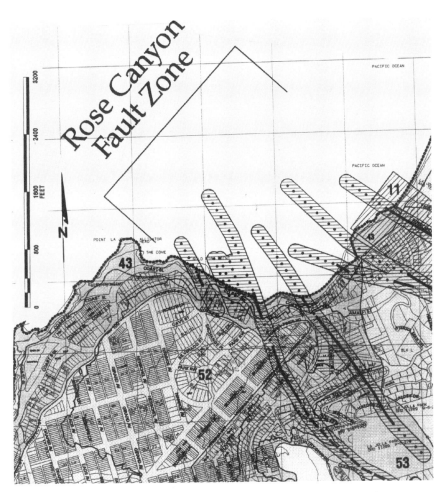

FIGURE 5.3   Portion of Seismic Safety Study, 1995. (*Developed by the City of San Diego.*)

other regulatory documents. For example, the *Uniform Building Code* (1997) provides seismic requirements that have been adopted by many building departments in the United States. These seismic code specifications have also been incorporated into the building codes in other countries.

8. *Other available documents:*   There are many other types of documents and maps that may prove useful during the screening investigation. Examples include geologic and soils engineering maps and reports used for the development of adjacent properties (often available at the local building department), water well logs, and agricultural soil survey reports.

After the site research has been completed, the next step in the screening investigation is a field reconnaissance. The purpose is to observe the site conditions and document any recent changes to the site that may not be reflected in the available documents. The field reconnaissance should also be used to observe surface features and other details that may not be readily evident from the available documents. Once the site research and field recon-

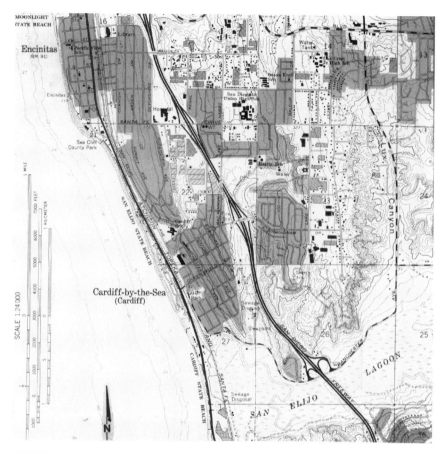

FIGURE 5.4   Topographic map. (*From USGS 1975.*)

naissance are completed, the engineering geologist and geotechnical engineer can then complete the screening investigation. The results should either clearly demonstrate the lack of seismic hazards or indicate the possibility of seismic hazards, in which case a quantitative evaluation is required.

It should be mentioned that even if the result of the screening investigation indicates no seismic hazards, the governing agency might not accept this result for critical facilities. It may still require that subsurface exploration demonstrate the absence of seismic hazards for critical facilities.

5.3   QUANTITATIVE EVALUATION

The purpose of the quantitative evaluation is to obtain sufficient information on the nature and severity of the seismic hazards so that mitigation recommendations can be developed. The quantitative evaluation consists of the following:

- *Geologic mapping:*   The first step is to supplement the results of the field reconnaissance (see Sec. 5.2) with geologic mapping, which can be used to further identify such features as existing landslides and surficial deposits of unstable soil.

- *Subsurface exploration:*   The results of the screening investigation and geologic mapping are used to plan the subsurface exploration, which could consist of the excavation of borings, test pits, or trenches. During the subsurface exploration, soil samples are often retrieved from the excavations. Field testing could also be performed in the excavations. Subsurface exploration is discussed in Sec. 5.4.

- *Laboratory testing:*   The purpose of the laboratory testing is to determine the engineering properties of the soil to be used in the seismic hazard analyses. Laboratory testing is discussed in Sec. 5.5.

- *Engineering and geologic analyses:*   An important parameter for the engineering and geologic analysis of seismic hazards is the peak ground acceleration. This is discussed in Sec. 5.6.

- *Report preparation:*   The results of the screening investigation and quantitative evaluation are often presented in report form that describes the seismic hazards and presents the geologic and geotechnical recommendations. Section 5.7 presents guidelines on the report content for seismic hazards.

5.4   SUBSURFACE EXPLORATION

There are many different aspects of subsurface exploration. The most important part of the subsurface exploration typically consists of the excavation of borings, test pits, and trenches. Soil samples are usually retrieved from these excavations and then tested in the laboratory to determine their engineering properties. In addition, field tests, such as the standard penetration test (SPT) or cone penetration test (CPT) could also be performed. These aspects of the subsurface exploration are individually discussed in the following sections. In addition, App. A (Glossary 1) presents a list of field testing terms and definitions.

5.4.1   Borings, Test Pits, and Trenches

Objectives of the Excavations.   The main objectives of the borings, test pits, and trenches are to determine the nature and extent of the seismic hazards. In this regard, the Division of Mines and Geology (1997) states:

> The subsurface exploration should extend to depths sufficient to expose geologic and subsurface water conditions that could affect slope stability or liquefaction potential. A sufficient quantity of subsurface information is needed to permit the engineering geologist and/or civil engineer to extrapolate with confidence the subsurface conditions that might affect the project, so that the seismic hazard can be properly evaluated, and an appropriate mitigation measure can be designed by the civil engineer. The preparation of engineering geologic maps and geologic cross sections is often an important step into developing an understanding of the significance and extent of potential seismic hazards. These maps and/or cross sections should extend far enough beyond the site to identify off-site hazards and features that might affect the site.

Excavation Layout.   The required number and spacing of borings, test pits, and trenches for a particular project must be based on judgment and experience. Obviously the more test excavations that are performed, the more knowledge will be obtained about the subsurface conditions and the seismic hazards. This can result in a more economical foundation design and less risk of the project being impacted by geologic and seismic hazards.

In general, boring layouts should not be random. Instead, if an approximate idea of the location of the proposed structure is known, then the borings should be concentrated in that area. For example, borings could be drilled at the four corners of a proposed building, with an additional (and deepest) boring located at the center of the proposed building. If the building location is unknown, then the borings should be located in lines, such as across the valley floor, in order to develop soil and geologic cross sections.

If geologic or seismic hazards may exist outside the building footprint, then they should also be investigated with borings. For example, if there is an adjacent landslide or fault zone that could impact the site, then it will also need to be investigated with subsurface exploration.

Some of the factors that influence the decisions on the number and spacing of borings include the following:

- *Relative costs of the investigation:*   The cost of additional borings must be weighed against the value of additional subsurface information.
- *Type of project:*   A more detailed and extensive subsurface investigation is required for an essential facility as compared to a single-family dwelling.
- *Topography (flatland versus hillside):*   A hillside project usually requires more subsurface investigation than a flatland project because of the slope stability requirements.
- *Nature of soil deposits (uniform versus erratic):*   Fewer borings may be needed when the soil deposits are uniform as compared to erratic deposits.
- *Geologic and seismic hazards:*   The more known or potential geologic and seismic hazards at the site, the greater the need for subsurface exploration.
- *Access:*   In many cases, the site may be inaccessible, and access roads will have to be constructed. Creating access roads throughout the site can be expensive and disruptive and may influence decisions on the number and spacing of borings.
- *Government or local building department requirements:*   For some projects, there may be specifications on the required number and spacing of borings.

Often a preliminary subsurface plan is developed to perform a limited number of exploratory borings. The purpose is just to obtain a rough idea of the soil, rock, and groundwater conditions and the potential geologic and seismic hazards at the site. Then once the preliminary subsurface data are analyzed, additional borings as part of a detailed seismic exploration are performed. The detailed subsurface exploration can be used to better define the soil profile, explore geologic and seismic hazards, and obtain further data on the critical subsurface conditions and seismic hazards that will likely have the greatest impact on the design and construction of the project.

Depth of Excavations.   In terms of the depth of the subsurface exploration, R. B. Seed (1991) states:

> Investigations should extend to depths below which liquefiable soils cannot reasonably be expected to occur (e.g., to bedrock, or to hard competent soils of sufficient geologic age that possible underlying units could not reasonably be expected to pose a liquefaction hazard). At most sites where soil is present, such investigation will require either borings or trench/test pit excavation. Simple surface inspection will suffice only when bedrock is exposed over essentially the full site, or in very unusual cases when the local geology is sufficiently well-documented as to fully ensure the complete lack of possibility of occurrence of liquefiable soils (at depth) beneath the exposed surface soil unit(s).

Down-Hole Logging. For geologic hazards such as landslides, a common form of sub-surface exploration is large-diameter bucket-auger borings that are down-hole logged by the geotechnical engineer or engineering geologist. Figure 5.5 shows a photograph of the top of the boring with the geologist descending into the hole in a steel cage. Note in Fig. 5.5 that a collar is placed around the top of the hole to prevent loose soil or rocks from being accidentally knocked down the hole. The process of down-hole logging is a valuable tech-nique because it allows the geotechnical engineer or engineering geologist to observe the subsurface materials as they exist in place. Usually the process of excavation of the boring smears the side of the hole, and the surface must be chipped away to observe intact soil or rock. Going down-hole is dangerous because of the possibility of a cave-in of the hole as well as "bad air" (presence of poisonous gases or lack of oxygen) and should only be attempted by an experienced geotechnical engineer or engineering geologist.

The down-hole observation of soil and rock can lead to the discovery of important geo-logic and seismic hazards. For example, Figs. 5.6 and 5.7 provide an example of the type of conditions observed down-hole. Figure 5.6 shows a knife that has been placed in an open fracture in bedrock. The open fracture in the rock was caused by massive landslide move-ment. Figure 5.7 is a side view of the same condition.

Trench Excavations. Backhoe trenches are an economical means of performing subsur-face exploration. The backhoe can quickly excavate the trench, which can then be used to observe and test the in situ soil. In many subsurface explorations, backhoe trenches are used to evaluate near-surface and geologic conditions (i.e., up to 15 ft deep), with borings being used to investigate deeper subsurface conditions. Backhoe trenches are especially useful for performing fault studies. For example, Figs. 5.8 and 5.9 show two views of the excava-

FIGURE 5.5   Down-hole logging. Note that the arrow points to the top of the steel cage used for the down-hole logging.

FIGURE 5.6   Knife placed in an open fracture in bedrock caused by landslide movement. The photograph was taken down-hole in a large-diameter auger boring.

FIGURE 5.7   Side view of the condition shown in Fig. 5.6.

FIGURE 5.8 Backhoe trench for a fault study.

tion of a trench that is being used to investigate the possibility of an on-site active fault. Figure 5.9 is a close-up view of the conditions in the trench and shows the fractured and disrupted nature of the rock. Note in Fig. 5.9 that metal shoring has been installed to prevent the trench from caving in. Often the fault investigations are performed by the engineering geologist with the objective of determining if there are active faults that cross the site. In addition, the width of the shear zone of the fault can often be determined from the trench excavation studies. If there is uncertainty as to whether a fault is active, then often datable material must be present in the trench excavation in order to determine the date of the most recent fault movement. Krinitzsky et al. (1993) present examples of datable materials, as follows:

- Displacements of organic matter or other datable horizons across faults
- Sudden burials of marsh soils
- Killed trees
- Disruption of archaeological sites
- Liquefaction intrusions cutting older liquefaction

5.4.2 Soil Sampling

To study the potential seismic hazards of a soil deposit, the ideal situation would be to obtain an undisturbed soil specimen, apply the same stress conditions that exist in the field, and then subject the soil specimen to the anticipated earthquake-induced cyclic shear stress. The resulting soil behavior could then be used to evaluate the seismic hazards. The disadvantages of this approach are that undisturbed soil specimens and sophisticated laboratory equipment would be required. Usually in engineering practice, this approach is not practical or is too expensive, and other options are used as described below.

FIGURE 5.9 Close-up view of trench excavation.

Cohesive Soils. Although undisturbed cohesive soil samples can often be obtained during the subsurface exploration, the usual approach in practice is to obtain the soil engineering properties from standard laboratory tests. In terms of the undrained shear strength of the soil, the unconfined compression test (ASTM D 2166-98, 2000) or the consolidated undrained triaxial compression test (ASTM D 4767-95, 2000) is usually performed. Typically standard soil sampling practices, such as the use of thin-walled Shelby tubes, are used to obtain undisturbed cohesive soil specimens (see Day 1999). Section 5.5.1 describes the interpretation of this data for use in geotechnical earthquake engineering analyses.

Granular Soils. There are three different methods that can be used to obtain undisturbed soil specimens of granular soil (Poulos et al. 1985, Ishihara 1985, Hofmann et al. 2000):

1. *Tube sampling:* Highly sophisticated techniques can be employed to obtain undisturbed soil specimens from tube samplers. For example, a fixed-piston sampler consists of a piston that is fixed at the bottom of the borehole by a rod that extends to the ground surface. A thin-walled tube is then pushed into the ground past the piston, while the piston rod is held fixed.

Another approach is to temporarily lower the groundwater table in the borehole and allow the water to drain from the soil. The partially saturated soil will then be held together by capillarity, which will enable the soil strata to be sampled. When brought to the ground surface, the partially saturated soil specimen is frozen. Because the soil is only partially saturated, the volume increase of water as it freezes should not significantly disturb the soil structure. The frozen soil specimen is then transported to the laboratory for testing.

Although the soil specimen may be considered to be an undisturbed specimen, there could still be disruption of the soil structure during all phases of the sampling operation. The greatest disturbance will probably occur during the physical pushing of the sampler into the soil.

2. *Block sampling:*   Another approach for near-surface soil is to temporarily lower the groundwater table. Then a test pit or trench is excavated into the soil. Because the groundwater table has been lowered, the partially saturated soil will be held together by capillarity. A block sample is then cut from the sides of the test pit or trench, and the block sample is transported to the laboratory for testing.

If the soil does not have enough capillarity to hold itself together, then this method will not work. In addition, the soil could be disturbed due to stress relief when making the excavation or when extracting the soil specimen.

3. *Freezing technique:*   The essential steps in the freezing technique are to first freeze the soil and then cut or core the frozen soil from the ground. The freezing is accomplished by installing pipes in the ground and then circulating ethanol and crushed dry ice or liquid nitrogen through the pipes. Because water increases in volume upon freezing, it is important to establish a slow freezing front so that the freezing water can slowly expand and migrate out of the soil pores. This process can minimize the sample disturbance associated with the increase in volume of freezing water.

From a practical standpoint, the three methods described above are usually not economical for most projects. Thus laboratory testing is not practical, and the analyses of earthquake hazards (such as liquefaction) are normally based on field testing that is performed during the subsurface exploration. The two most commonly used field tests are the standard penetration test (SPT) and the cone penetration test (CPT), as discussed in the next two sections.

5.4.3 Standard Penetration Test

Test Procedure.   The standard penetration test can be used for all types of soil, but in general the SPT should only be used for granular soils (Coduto 1994). The SPT can be especially valuable for clean sand deposits where the sand falls or flows out from the sampler when retrieved from the ground. Without a soil sample, other types of tests, such as the standard penetration test, must be used to assess the engineering properties of the sand. Often when a borehole is drilled, if subsurface conditions indicate a sand stratum and sampling tubes come up empty, the sampling gear can be quickly changed to perform standard penetration tests.

The standard penetration test consists of driving a thick-walled sampler into the granular soil deposit. The test parameters are as follows:

- *Sampler:*   Per ASTM D 1586-99 (2000), the SPT sampler must have an inside barrel diameter $D = 3.81$ cm (1.5 in) and an outside diameter $F = 5.08$ cm (2 in), as shown in Fig. 5.10.

- *Driving hammer:*   The SPT sampler is driven into the sand by using a 63.5-kg (140-lb) hammer falling a distance of 0.76 m (30 in).

- *Driving distance:*   The SPT sampler is driven a total of 45 cm (18 in), with the number of blows recorded for each 15-cm (6-in) interval.

- *N value:*   The *measured SPT N value* (blows per foot) is defined as the penetration resistance of the soil, which equals the sum of the number of blows required to drive the SPT

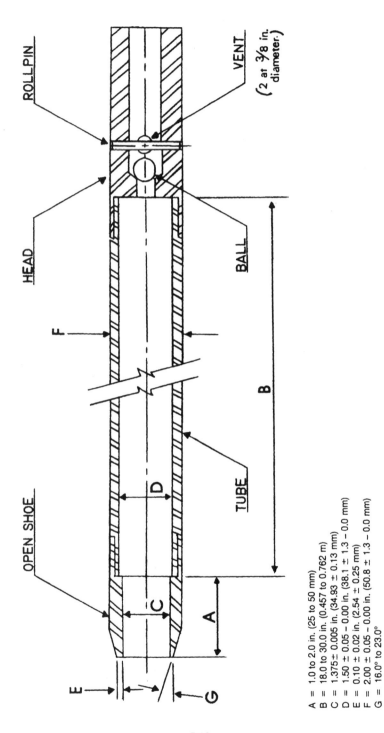

A = 1.0 to 2.0 in. (25 to 50 mm)
B = 18.0 to 30.0 in. (0.457 to 0.762 m)
C = 1.375± 0.005 in. (34.93 ± 0.13 mm)
D = 1.50 ± 0.05 − 0.00 in. (38.1 ± 1.3 − 0.0 mm)
E = 0.10 ± 0.02 in. (2.54 ± 0.25 mm)
F = 2.00 ± 0.05 − 0.00 in. (50.8 ± 1.3 − 0.0 mm)
G = 16.0° to 23.0°

FIGURE 5.10 Standard penetration test sampler. *(Reprinted with permission from the American Society for Testing and Materials 2000.)*

sampler over the depth interval of 15 to 45 cm (6 to 18 in). The reason the number of blows required to drive the SPT sampler for the first 15 cm (6 in) is not included in the N value is that the drilling process often disturbs the soil at the bottom of the borehole, and the readings at 15 to 45 cm (6 to 18 in) are believed to be more representative of the in situ penetration resistance of the granular soil.

Factors That Could Affect the Test Results.   The measured SPT N value can be influenced by the type of soil, such as the amount of fines and gravel-size particles in the soil. Saturated sands that contain appreciable fine soil particles, such as silty or clayey sands, could give abnormally high N values if they have a tendency to dilate or abnormally low N values if they have a tendency to contract during the undrained shear conditions associated with driving the SPT sampler. Gravel-size particles increase the driving resistance (hence increased N value) by becoming stuck in the SPT sampler tip or barrel.

A factor that could influence the measured SPT N value is groundwater. It is important to maintain a level of water in the borehole at or above the in situ groundwater level. This is to prevent groundwater from rushing into the bottom of the borehole, which could loosen the granular soil and result in low measured N values.

Besides the soil and groundwater conditions described above, many different testing factors can influence the accuracy of the SPT readings. For example, the measured SPT N value could be influenced by the hammer efficiency, the rate at which the blows are applied, the borehole diameter, and the rod lengths. The different factors that can affect the standard penetration test results are presented in Table 5.1.

Corrections for Testing and Overburden Pressure.   Corrections can be applied to the test results to compensate for the testing procedures (Skempton 1986):

$$N_{60} = 1.67 E_m C_b C_r N \tag{5.1}$$

where N_{60} = standard penetration test N value corrected for field testing procedures
   E_m = hammer efficiency (for U.S. equipment, E_m is 0.6 for a safety hammer and 0.45 for a doughnut hammer)
   C_b = borehole diameter correction (C_b = 1.0 for boreholes of 65- to 115-mm diameter, 1.05 for 150-mm diameter, and 1.15 for 200-mm diameter hole)
   C_r = rod length correction (C_r = 0.75 for up to 4 m of drill rods, 0.85 for 4 to 6 m of drill rods, 0.95 for 6 to 10 m of drill rods, and 1.00 for drill rods in excess of 10 m)
   N = measured standard penetration test N value

For many geotechnical earthquake engineering evaluations, such as liquefaction analysis, the standard penetration test N_{60} value [Eq. (5.1)] is corrected for the vertical effective stress σ'_{v0}. When a correction is applied to the N_{60} value to account for the vertical effective pressure, these values are referred to as $(N_1)_{60}$ values. The procedure consists of multiplying the N_{60} value by a correction C_N in order to calculate the $(N_1)_{60}$ value. Figure 5.11 presents a chart that is commonly used to obtain the correction factor C_N. Another option is to use the following equation:

$$(N_1)_{60} = C_N N_{60} = (100/\sigma'_{v0})^{0.5} N_{60} \tag{5.2}$$

where $(N_1)_{60}$ = standard penetration test N value corrected for both field testing procedures and overburden pressure
   C_N = correction factor to account for overburden pressure. As indicated in Eq. (5.2), C_N is approximately equal to $(100/\sigma'_{v0})^{0.5}$, where σ'_{v0} is the vertical

effective stress, also known as the effective overburden pressure, in kilopascals. Suggested maximum values of C_N range from 1.7 to 2.0 (Youd and
Idriss 1997, 2001).

N_{60} = standard penetration test N value corrected for field testing procedures.
Note that N_{60} is calculated by using Eq. (5.1).

Correlations between SPT Results and Soil Properties. Commonly used correlations
between the SPT results and various soil properties are as follows:

- *Table 5.2:* This table presents a correlation between the measured SPT N value (blows
  per foot) and the density condition of a clean sand deposit. Note that this correlation is

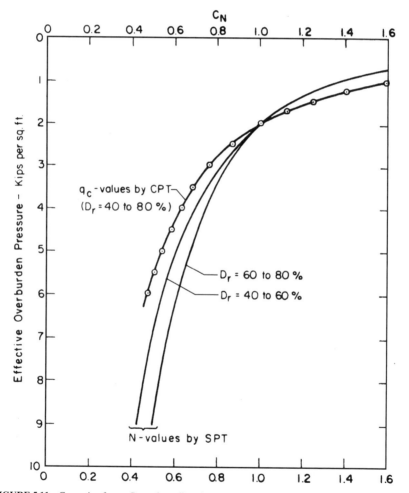

FIGURE 5.11 Correction factor C_N used to adjust the standard penetration test N value and cone penetration test q_c value for the effective overburden pressure. The symbol D_r refers to the relative density of the
sand. (*Reproduced from Seed et al. 1983, with permission from the American Society of Civil Engineers.*)

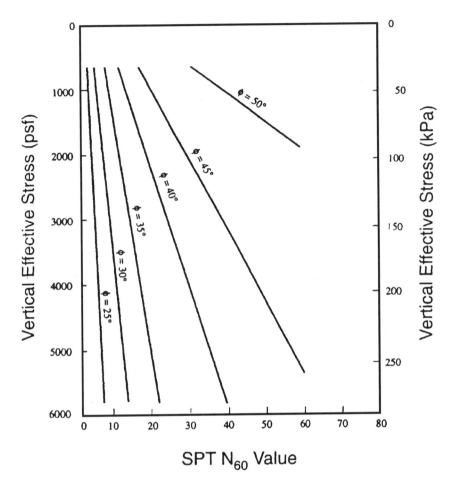

FIGURE 5.12 Empirical correlation between the standard penetration test N_{60} value, vertical effective stress, and friction angle for clean quartz sand deposits. (*Adapted from de Mello 1971, reproduced from Coduto 1994.*)

very approximate and the boundaries between different density conditions are not as distinct as implied by the table. As indicated in Table 5.2, if it only takes 4 blows or less to drive the SPT sampler, then the sand should be considered to be very loose and could be subjected to significant settlement due to the weight of a structure or due to earthquake shaking. On the other hand, if it takes more than 50 blows to drive the SPT sampler, then the sand is considered to be in a very dense condition and would be able to support high bearing loads and would be resistant to settlement from earthquake shaking.

- *Table 5.3:* This table is based on the work by Tokimatsu and Seed (1987) and is similar to Table 5.2, except that it provides a correlation between $(N_1)_{60}$ and the relative density.

• *Figure 5.12:*   This figure is based on the work by de Mello (1971) and presents an empirical correlation between the standard penetration test N_{60} value [Eq. (5.1)], the vertical effective stress σ'_{v0}, and the friction angle ϕ of clean, quartz sand.

Popularity of SPT Test.   Even with the limitations and all the corrections that must be applied to the measured N value, the standard penetration test is probably the most widely used field test in the United States. This is because it is relatively easy to use, the test is economical compared to other types of field testing, and the SPT equipment can be quickly adapted and included as part of almost any type of drilling rig.

5.4.4 Cone Penetration Test

The idea for the cone penetration test is similar to the standard penetration test except that instead of driving a thick-walled sampler into the soil, a steel cone is pushed into the soil. There are many different types of cone penetration devices, such as the mechanical cone, mechanical-friction cone, electric cone, and piezocone (see App. A, Glossary 1, for descriptions). The simplest type of cone is shown in Fig. 5.13 (from ASTM D 3441-98, 2000). First the cone is pushed into the soil to the desired depth (initial position), and then a force is applied to the inner rods which moves the cone downward into the extended position. The force required to move the cone into the extended position (Fig. 5.13) divided by the horizontally projected area of the cone is defined as the *cone resistance* q_c. By continually repeating the two-step process shown in Fig. 5.13, the cone resistance data are obtained at increments of depth. A continuous record of the cone resistance versus depth can be obtained by using the electric cone, where the cone is pushed into the soil at a rate of 10 to 20 mm/s (2 to 4 ft/min).

Figure 5.14 (adapted from Robertson and Campanella 1983) presents an empirical correlation between the cone resistance q_c, vertical effective stress, and friction angle ϕ of clean, quartz sand. Note that Fig. 5.14 is similar in appearance to Fig. 5.12, which should be the case because both the SPT and the CPT involve basically the same process of forcing an object into the soil and then measuring the resistance of the soil to penetration by the object.

For many geotechnical earthquake engineering evaluations, such as liquefaction analysis, the cone penetration test q_c value is corrected for the vertical effective stress σ'_{v0}. When a correction is applied to the q_c value to account for the vertical effective pressure, these values are referred to as q_{c1} values. The procedure consists of multiplying the q_c value by a correction C_N in order to calculate the q_{c1} value. Figure 5.11 presents a chart that is commonly used to obtain the correction factor C_N. Another option is to use the following equation:

$$q_{c1} = C_N q_c = \frac{1.8 q_c}{0.8 + \sigma'_{v0}/100} \tag{5.3}$$

where q_{c1} = corrected CPT tip resistance (corrected for overburden pressure)
  C_N = correction factor to account for overburden pressure. As indicated in Eq. (5.3), C_N is approximately equal to $1.8/(0.8 + \sigma'_{v0}/100)$, where σ'_{v0} is the vertical effective stress in kilopascals.
  q_c = cone penetration tip resistance

A major advantage of the cone penetration test is that by using the electric cone, a continuous subsurface record of the cone resistance q_c can be obtained. This is in contrast to

TABLE 5.1 Factors That Can Affect the Standard Penetration Test Results

Factors that can affect the standard penetration test results	Comments
Inadequate cleaning of the borehole	SPT is only partially made in original soil. Sludge may be trapped in the sampler and compressed as the sampler is driven, increasing the blow count. This may also prevent sample recovery.
Not seating the sampler spoon on undisturbed material	Incorrect N value is obtained.
Driving of the sample spoon above the bottom of the casing	The N value is increased in sands and reduced in cohesive soil.
Failure to maintain sufficient hydrostatic head in boring	The water table in the borehole must be at least equal to the piezometric level in the sand; otherwise the sand at the bottom of the borehole may be transformed to a loose state.
Attitude of operators	Blow counts for the same soil using the same rig can vary, depending on who is operating the rig and perhaps the mood of operator and time of drilling.
Overdriven sample	Higher blow counts usually result from overdriven sampler.
Sampler plugged by gravel	Higher blow counts result when gravel plugs the sampler. The resistance of loose sand could be highly overestimated.
Plugged casing	High N values may be recorded for loose sand when sampling below the groundwater table. Hydrostatic pressure causes sand to rise and plug the casing.
Overwashing ahead of casing	Low blow count may result for dense sand since sand is loosened by overwashing.
Drilling method	Drilling technique (e.g., cased holes versus mud-stabilized holes) may result in different N values for the same soil.
Not using the standard hammer drop	Energy delivered per blow is not uniform. European countries have adopted an automatic trip hammer not currently in use in North America.
Free fall of the drive weight is not attained	Using more than 1.5 turns of rope around the drum and/or using wire cable will restrict the fall of the drive weight.
Not using the correct weight	Driller frequently supplies drive hammers with weights varying from the standard by as much as 10 lb.
Weight does not strike the drive cap concentrically	Impact energy is reduced, increasing the N value.
Not using a guide rod	Incorrect N value is obtained.
Not using a good tip on the sampling spoon	If the tip is damaged and reduces the opening or increases the end area, the N value can be increased.

TABLE 5.1 Factors That Can Affect the Standard Penetration Test Results (*Continued*)

Factors that can affect the standard penetration test results	Comments
Use of drill rods heavier than standard	With heavier rods, more energy is absorbed by the rods, causing an increase in the blow count.
Not recording blow counts and penetration accurately	Incorrect N values are obtained.
Incorrect drilling	The standard penetration test was originally developed from wash boring techniques. Drilling procedures which seriously disturb the soil will affect the N value, for example, drilling with cable tool equipment.
Using large drill holes	A borehole correction is required for large-diameter boreholes. This is because larger diameters often result in a decrease in the blow count.
Inadequate supervision	Frequently a sampler will be impeded by gravel or cobbles, causing a sudden increase in blow count. This is often not recognized by an inexperienced observer. Accurate recording of drilling sampling and depth is always required.
Improper logging of soils	The sample is not described correctly.
Using too large a pump	Too high a pump capacity will loosen the soil at the base of the hole, causing a decrease in blow count.

Source: NAVFAC DM-7.1 (1982).

TABLE 5.2 Correlation between Uncorrected SPT N Value and Density of Clean Sand

Uncorrected N value (blows per foot)	Sand density	Relative density D_r, percent
0–4	Very loose condition	0–15
4–10	Loose condition	15–35
10–30	Medium condition	35–65
30–50	Dense condition	65–85
Over 50	Very dense condition	85–100

Note: Relative density $D_r = 100(e_{max} - e)/(e_{max} - e_{min})$, where e_{max} = void ratio corresponding to the loosest possible state of the soil, usually obtained by pouring the soil into a mold of known volume (ASTM D 4254-96, 2000), e_{min} = void ratio corresponding to the densest possible state of the soil, usually obtained by vibrating the soil particles into a dense state (ASTM D 4253-96, 2000), and e = the natural void ratio of the soil.
Sources: Terzaghi and Peck (1967) and Lambe and Whitman (1969).

TABLE 5.3 Correlation between $(N_1)_{60}$ and Density of Sand

$(N_1)_{60}$ (blows per foot)	Sand density	Relative density D_r, percent
0–2	Very loose condition	0–15
2–5	Loose condition	15–35
5–20	Medium condition	35–65
20–35	Dense condition	65–85
Over 35	Very dense condition	85–100

Source: Tokimatsu and Seed (1987).

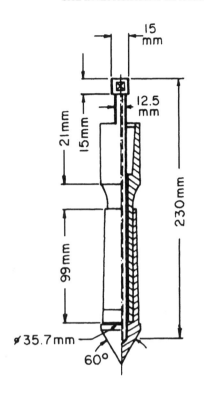
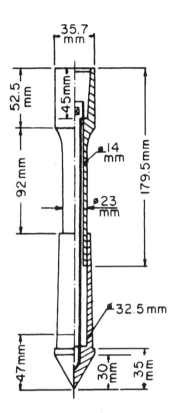

INITIAL POSITION   **EXTENDED POSITION**

FIGURE 5.13 Example of mechanical cone penetrometer tip (Dutch mantle cone). (*Reprinted with permission from the American Society for Testing and Materials 2000.*)

the standard penetration test, which obtains data at intervals in the soil deposit. Disadvantages of the cone penetration test are that soil samples cannot be recovered and special equipment is required to produce a steady and slow penetration of the cone. Unlike the SPT, the ability to obtain a steady and slow penetration of the cone is not included as part of conventional drilling rigs. Because of these factors, in the United States, the CPT is used less frequently than the SPT.

5.5 LABORATORY TESTING

As discussed in Sec. 5.4.2, soil engineering properties that are used in earthquake analyses are usually obtained from field tests (SPT and CPT) or from standard laboratory tests (see Day

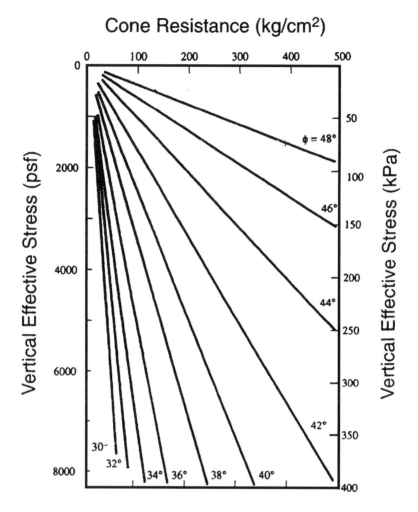

FIGURE 5.14 Empirical correlation between cone resistance, vertical effective stress, and friction angle for clean quartz sand deposits. *Note:* 1 kg/cm² approximately equals 1 ton/ft² (*Adapted from Robertson and Campanella 1983; reproduced from Coduto 1994.*)

1999, 2001a). Special laboratory tests used to model the engineering behavior of the soil subjected to earthquake loading are typically not performed in practice. For example, in terms of assessing liquefaction potential, Seed (1987) states: "In developing solutions to practical problems involving the possibility of soil liquefaction, it is the writer's judgment that field case studies and in situ tests provide the most useful and practical tools at the present time."

Section 5.5.1 discusses the shear strength of the soil, which is an important parameter needed for earthquake analyses of foundations, slopes, and retaining walls. Section 5.5.2 briefly discusses the cyclic triaxial test, which is a valuable laboratory test used for the research of the dynamic properties of soil. Appendix A (Glossary 2) presents a list of laboratory testing terms and definitions.

5.5.1 Shear Strength

The shear strength is an essential soil engineering property that is needed for many types of earthquake evaluations. There are two basic types of analyses that utilize the shear strength of the soil: (1) the total stress analysis and (2) the effective stress analysis.

Under no circumstances can a total stress analysis and an effective stress analysis be combined. For example, suppose a slope stability analysis is needed for a slope consisting of alternating sand and clay layers. The factor of safety of the slope must be determined by using either a total stress analysis or an effective stress analysis, as follows:

1. Total stress analysis
   - Use total stress shear strength parameters (s_u or c and ϕ).
   - Use total unit weight of the soil γ_t.
   - Ignore the groundwater table.
2. Effective stress analysis
   - Use effective stress shear strength parameters (c' and ϕ').
   - Determine the earthquake-induced pore water pressures u_e.

Further discussions of the total stress analysis and effective stress analysis are provided next.

Total Stress Analysis.   The total stress analysis uses the undrained shear strength of the soil. The total stress analysis is often performed for cohesive soil, such as silts and clays. Total stress analyses are used for the design of foundations, slopes, and retaining walls that are subjected to earthquake shaking. The actual analysis is performed for rapid loading or unloading conditions that usually develop during the earthquake. This analysis is ideally suited for earthquakes, because there is a change in shear stress which occurs quickly enough that soft cohesive soil does not have time to consolidate; or in the case of heavily overconsolidated cohesive soils, the negative pore water pressures do not have time to dissipate. The total stress analysis uses the total unit weight γ_t of the soil, and the location of the groundwater table is not considered in the analysis.

To perform a total stress analysis, the undrained shear strength of the soil must be determined. The undrained shear strength s_u of the cohesive soil is often obtained from unconfined compression tests (ASTM D 2166-98, 2000) or from vane shear tests. An alternative approach is to use the total stress parameters (c and ϕ) from triaxial tests, such as the unconsolidated undrained triaxial compression test (ASTM D 2850-95, 2000) or the consolidated undrained triaxial compression test (ASTM D 4767-95, 2000).

An advantage of the total stress analysis is that the undrained shear strength could be obtained from tests (such as the unconfined compression test or vane shear test) that are easy to perform. A major disadvantage of this approach is that the accuracy of the undrained shear strength is always in doubt because it depends on the shear-induced pore water pressures (which are not measured), which in turn depend on the many details (i.e., sample disturbance, strain rate effects, and anisotropy) of the test procedures (Lambe and Whitman 1969).

Effective Stress Analysis.   The effective stress analysis uses the drained shear strength parameters (c' and ϕ'). Most earthquake analyses of granular soils, such as sands and gravels, are made using the effective stress analysis (with the possible except of liquefaction-induced flow slides). For cohesionless soil, $c' = 0$, and the effective friction angle ϕ' is often obtained from drained direct shear tests or from empirical correlations, such as the standard penetration test (Fig. 5.12) or the cone penetration test (Fig. 5.14).

The effective stress analysis could be used for earthquake-induced loading, provided the earthquake-induced pore water pressures can be estimated. In other words, the effective stress generated during the earthquake must be determined. An advantage of the effective stress analysis is that it more fundamentally models the shear strength of the soil, because shear strength is directly related to effective stress. A major disadvantage of the effective stress analysis is that the pore water pressures must be included in the earthquake analysis. The accuracy of the pore water pressure is often in doubt because of the many factors which affect the magnitude of pore water pressure changes, such as the determination of changes in pore water pressure resulting from changes in earthquake loads. For effective stress analysis, assumptions are frequently required concerning the pore water pressures that will be generated by the earthquake.

Cohesionless Soil.   These types of soil are nonplastic, and they include such soils as gravels, sands, and nonplastic silt, such as rock flour. A cohesionless soil develops its shear strength as a result of the frictional and interlocking resistance between the individual soil particles. A cohesionless soil can be held together only by a confining pressure, and it will fall apart when the confining pressure is released. For the earthquake analysis of cohesionless soil, it is often easier to perform an effective stress analysis, as discussed below:

1. *Cohesionless soil above the groundwater table:*   Often the cohesionless soil above the groundwater table will have negative pore water pressures due to capillary tension of pore water fluid. The capillary tension tends to hold together the soil particles and to provide additional shear strength to the soil. For geotechnical engineering analyses, it is common to assume that the pore water pressures are equal to zero, which ignores the capillary tension. This conservative assumption is also utilized for earthquake analyses. Thus the shear strength of soil above the groundwater table is assumed to be equal to the effective friction angle ϕ' from empirical correlations (such as Figs. 5.12 and 5.14), or it is equal to the effective friction angle ϕ' from drained direct shear tests performed on saturated soil (ASTM D 3080-98, 2000).

2. *Dense cohesionless soil below the groundwater table:*   As discussed in Chap. 6, dense cohesionless soil tends to dilate during the earthquake shaking. This causes the excess pore water pressures to become negative, and the shear strength of the soil is actually momentarily increased. Thus for dense cohesionless soil below the groundwater table, the shear strength is assumed to be equal to the effective friction angle ϕ' from empirical correlations (such as Figs. 5.12 and 5.14); or it is equal to the effective friction angle ϕ' from drained direct shear tests performed on saturated soil (ASTM D 3080-98, 2000). In the effective stress analysis, the negative excess pore water pressures are ignored, and the pore water pressure is assumed to be hydrostatic. Once again, this is a conservative approach.

3. *Loose cohesionless soil below the groundwater table:*   As discussed in Chap. 6, loose cohesionless soil tends to contract during the earthquake shaking. This causes the development of pore water pressures, and the shear strength of the soil is decreased. If liquefaction occurs, the shear strength of the soil can be decreased to essentially zero. For any cohesionless soil that is likely to liquefy during the earthquake, one approach is to assume that ϕ' is equal to zero (i.e., no shear strength).

For those loose cohesionless soils that have a factor of safety against liquefaction greater than 1.0, the analysis will usually need to take into account the reduction in shear strength due to the increase in pore water pressure as the soil contracts. One approach is to use the effective friction angle ϕ' from empirical correlations (such as Figs. 5.12 and 5.14) or the effective friction angle ϕ' from drained direct shear tests performed on saturated soil (ASTM D 3080-98, 2000). In addition, the earthquake-induced pore water pressures must be used in the effective stress analysis.

The disadvantage of this approach is that it is very difficult to estimate the pore water pressures generated by the earthquake-induced contraction of the soil. One option is to use Fig. 5.15, which presents a plot of the factor of safety against liquefaction FS_L versus pore water pressure ratio r_u, defined as $r_u = u/(\gamma_t h)$, where u = pore water pressure, γ_t = total unit weight of the soil, and h = depth below the ground surface.

As indicated in Fig. 5.15, at a factor of safety against liquefaction FS_L equal to 1.0 (i.e., liquefied soil), $r_u = 1.0$. Using a value of $r_u = 1.0$, then $r_u = 1.0 = u/(\gamma_t h)$. This means that the pore water pressure u must be equal to the total stress ($\sigma = \gamma_t h$), and hence the effective stress σ' is equal to zero ($\sigma' = \sigma - u$). For a granular soil, an effective stress equal to zero means that the soil will not possess any shear strength (i.e., it has liquefied). Chapter 6 presents the analyses that are used to determine the factor of safety against liquefaction.

4. *Flow failures in cohesionless soil:*   As indicated above, the earthquake analyses for cohesionless soil will often be performed using an effective stress analysis, using ϕ' and assumptions concerning the earthquake-induced pore water pressure. Flow failures are also often analyzed using an effective stress analysis with a value of the pore water pressure ratio = 1.0, or by using a shear strength of the liquefied soil equal to zero (that is, $\phi' = 0$ and $c' = 0$). This is discussed further in Sec. 9.4.

Cohesive Soil.   These types of soil are plastic, they include such soils as silts and clays, and have the ability to be rolled and molded (hence they have a plasticity index). For the earthquake analysis of cohesive soil, it is often easier to perform a total stress analysis, as discussed below:

1. *Cohesive soil above the groundwater table:*   Often the cohesive soil above the groundwater table will have negative pore water pressures due to capillary tension of the

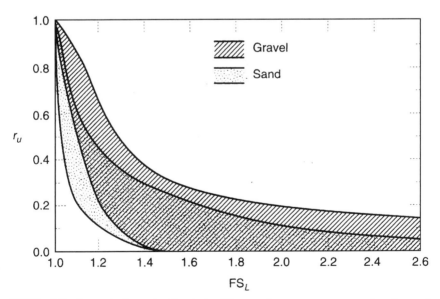

FIGURE 5.15   Factor of safety against liquefaction FS_L versus the pore water pressure ratio r_u for gravel and sand. (*Developed by Marcuson and Hynes 1990, reproduced from Kramer 1996.*)

pore water fluid. In some cases, the cohesive soil may even be dry and desiccated. The capillary tension tends to hold together the soil particles and provide additional shear strength to the soil. For a total stress analysis, the undrained shear strength s_u of the cohesive soil could be determined from unconfined compression tests or vane shear tests. As an alternative, total stress parameters (c and ϕ) could be determined from triaxial tests (e.g., ASTM D 2850-95 and ASTM D 4767-95, 2000).

Because of the negative pore water pressures, a future increase in water content would tend to decrease the undrained shear strength s_u of partially saturated cohesive soil above the groundwater table. Thus a possible change in water content in the future should be considered. In addition, a triaxial test performed on a partially saturated cohesive soil often has a stress-strain curve that exhibits a peak shear strength which then reduces to an ultimate value. If there is a significant drop-off in shear strength with strain, it may be prudent to use the ultimate value in earthquake analyses.

2. *Cohesive soil below the groundwater table having low sensitivity:* The *sensitivity* S_t of a cohesive soil is defined as the undrained shear strength of an undisturbed soil specimen divided by the undrained shear strength of a completely remolded soil specimen. The sensitivity thus represents the loss of undrained shear strength as a cohesive soil specimen is remolded. An earthquake also tends to shear a cohesive soil back and forth, much as the remolding process does. For cohesive soil having low sensitivity ($S_t \leq 4$), the reduction in the undrained shear strength during the earthquake should be small.

3. *Cohesive soil below the groundwater table having a high sensitivity:* For highly sensitive and quick clays ($S_t > 8$), there could be a significant shear strength loss during the earthquake shaking. An example was the Turnagain Heights landslide discussed in Sec. 3.5.2.

The stress-strain curve from a triaxial test performed on a highly sensitive or quick clay often exhibits a peak shear strength that develops at a low vertical strain, followed by a dramatic drop-off in strength with continued straining of the soil specimen. The analysis needs to include the estimated reduction in undrained shear strength due to the earthquake shaking. In general, the most critical conditions exist when the highly sensitive or quick clay is subjected to a high static shear stress (such as the Turnagain Heights landslide). If, during the earthquake, the sum of the static shear stress and the seismic-induced shear stress exceeds the undrained shear strength of the soil, then a significant reduction in shear strength is expected to occur.

Cohesive soils having a medium sensitivity ($4 < S_t \leq 8$) would tend to be an intermediate case.

4. *Drained residual shear strength ϕ_r' for cohesive soil:* As indicated above, the earthquake analyses for cohesive soil will often be performed using a total stress analysis (that is, s_u from unconfined compression tests and vane shear tests, or c and ϕ from triaxial tests).

An exception is cohesive slopes that have been subjected to a significant amount of shear deformation. For example, the stability analysis of ancient landslides, slopes in overconsolidated fissured clays, and slopes in fissured shales will often be based on the drained residual shear strength of the failure surface (Bjerrum 1967, Skempton and Hutchinson 1969, Skempton 1985, Hawkins and Privett 1985, Ehlig 1992). When the stability of such a slope is to be evaluated for earthquake shaking, then the drained residual shear strength ϕ_r' should be used in the analysis. The drained residual shear strength can be determined from laboratory tests by using the torsional ring shear or direct shear apparatus (Day 2001a).

In order to perform the effective stress analysis, the pore water pressures are usually assumed to be unchanged during the earthquake shaking. The slope or landslide mass will also be subjected to additional destabilizing forces due to the earthquake shaking. These destabilizing forces can be included in the effective stress slope stability analysis, and this approach is termed the *pseudostatic method* (see Sec. 9.2.5).

Analysis for Subsoil Profiles Consisting of Cohesionless and Cohesive Soil.   For earthquake analysis where both cohesionless soil and cohesive soil must be considered, either a total stress analysis or an effective stress analysis could be performed. As indicated above, usually the effective shear strength parameters are known for the cohesionless soil. Thus subsoil profiles having layers of sand and clay are often analyzed using an effective stress analysis (c' and ϕ') with an estimation of the earthquake-induced pore water pressures.

If the sand layers will liquefy during the anticipated earthquake, then a total stress analysis could be performed using the undrained shear strength s_u for the clay and assuming the undrained shear strength of the liquefied sand layer is equal to zero ($s_u = 0$). Bearing capacity or slope stability analyses using total stress parameters can then be performed so that the circular or planar slip surface passes through or along the liquefied sand layer.

Summary of Shear Strength for Geotechnical Earthquake Engineering.   Table 5.4 presents a summary of the soil type versus type of analysis and shear strength that should be used for earthquake analyses.

5.5.2   Cyclic Triaxial Test

The cyclic triaxial test has been used extensively in the study of soil subjected to simulated earthquake loading. For example, the cyclic triaxial test has been used for research studies on the liquefaction behavior of soil. The laboratory test procedures are as follows (ASTM D 5311-96, 2000):

1. A cylindrical soil specimen is placed in the triaxial apparatus and sealed in a watertight rubber membrane (see Fig. 5.16).

2. A backpressure is used to saturate the soil specimen.

3. An isotropic effective confining pressure is applied to the soil specimen, and the soil specimen is allowed to equilibrate under this effective stress. Tubing, such as shown in Fig. 5.16, allows for the flow of water during saturation and equilibration as well as the measurement of pore water pressure during the test.

4. Following saturation and equilibration at the effective confining pressure, the valve to the drainage measurement system is shut, and the soil specimen is subjected to an undrained loading. To simulate the earthquake loading, a constant-amplitude sinusoidally varying axial load (i.e., cyclic axial load) is applied to the top of the specimen. The cyclic axial load simulates the change in shear stress induced by the earthquake.

5. During testing, the cyclic axial load, specimen axial deformation, and pore water pressure in the soil specimen are recorded. For the testing of loose sand specimens, the cyclic axial loading often causes an increase in the pore water pressure in the soil specimen, which results in a decrease in the effective stress and an increase in the axial deformation.

The cyclic triaxial test is a very complicated test, it requires special laboratory equipment, and there are many factors the affect the results (Townsend 1978, Mulilis et al. 1978). Actual laboratory test data from the cyclic triaxial test are presented in Sec. 6.2.

TABLE 5.4 Soil Type versus Type of Analysis and Shear Strength for Earthquake Engineering

Soil type	Type of analysis	Field condition	Shear strength
Cohesionless soil	Use an effective stress analysis	Cohesionless soil above the groundwater table	Assume pore water pressures are equal to zero, which ignores the capillary tension. Use ϕ' from empirical correlations or from laboratory tests such as drained direct shear tests.
		Dense cohesionless soil below the groundwater table	Dense cohesionless soil dilates during the earthquake shaking (hence negative excess pore water pressure). Assume earthquake-induced negative excess pore water pressures are zero, and use ϕ' from empirical correlations or from laboratory tests such as drained direct shear tests.
		Loose cohesionless soil below the groundwater table	Excess pore water pressures u_e generated during the contraction of soil structure. For $FS_L \leq 1.0$, use $\phi' = 0$ or $r_u = 1.0$. For $FS_L > 1$, use r_u from Fig. 5.15 and ϕ' from empirical correlations or from laboratory tests such as drained direct shear tests.
		Flow failures	Flow failures are also often analyzed using an effective stress analysis with a value of the pore water pressure ratio = 1.0, or by using a shear strength of the liquefied soil equal to zero ($\phi' = 0$ and $c' = 0$).
Cohesive soil	Use a total stress analysis	Cohesive soil above the groundwater table	Determine s_u from unconfined compression tests or vane shear tests. As an alternative, use total stress parameters (c and ϕ) from triaxial tests. Consider shear strength decrease due to increase in water content. For a significant drop-off in strength with strain, consider using ultimate shear strength for earthquake analysis.
		Cohesive soil below the groundwater table with $S_t \leq 4$	Determine s_u from unconfined compression tests or vane shear tests. As an alternative, use total stress parameters (c and ϕ) from triaxial tests.
		Cohesive soil below the groundwater table with $S_t > 8$	Include an estimated reduction in undrained shear strength due to earthquake shaking. Most significant strength loss occurs when the sum of the static shear stress and the seismic-induced shear stress exceeds the undrained shear strength of the soil. Cohesive soils having a medium sensitivity ($4 < S_t \leq 8$) are an intermediate case.
	Possible exception	Existing landslides	Use an effective stress analysis and the drained residual shear strength (ϕ_r') for the slide plane. Assume pore water pressures are unchanged during earthquake shaking. Include destabilizing earthquake forces in slope stability analyses (pseudostatic method).

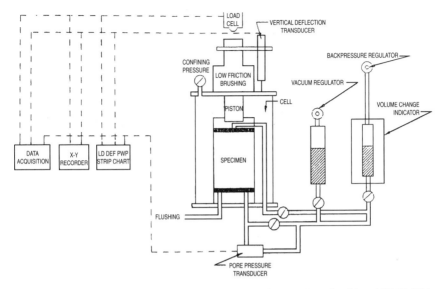

FIGURE 5.16 Schematic diagram of the cyclic triaxial test equipment. (*Reproduced from ASTM D 5311-96, 2000. Reproduced with permission from the American Society for Testing and Materials.*)

5.6 PEAK GROUND ACCELERATION

5.6.1 Introduction

As indicated in Fig. 2.14, the ground motion caused by earthquakes is generally characterized in terms of ground surface displacement, velocity, and acceleration. Geotechnical engineers traditionally use acceleration, rather than velocity or displacement, because acceleration is directly related to the dynamic forces that earthquakes induce on the soil mass. For geotechnical analyses, the measure of the cyclic ground motion is represented by the maximum horizontal acceleration at the ground surface a_{max}. The maximum horizontal acceleration at ground surface is also known as the *peak horizontal ground acceleration*. For most earthquakes, the horizontal acceleration is greater than the vertical acceleration, and thus the peak horizontal ground acceleration also turns out to be the peak ground acceleration (PGA).

For earthquake engineering analyses, the peak ground acceleration a_{max} is one of the most difficult parameters to determine. It represents an acceleration that will be induced sometime in the future by an earthquake. Since it is not possible to predict earthquakes, the value of the peak ground acceleration must be based on prior earthquakes and fault studies.

Often attenuation relationships are used in the determination of the peak ground acceleration. An *attenuation relationship* is defined as a mathematical relationship that is used to estimate the peak ground acceleration at a specified distance from the earthquake. Numerous attenuation relationships have been developed. Many attenuation equations relate the peak ground acceleration to (1) the earthquake magnitude and (2) the distance between the site and the seismic source (the causative fault). The increasingly larger pool of seismic data recorded in the world, and particularly in the western United States, has allowed researchers to develop reliable empirical attenuation equations that are used to model the ground motions generated during an earthquake (Federal Emergency Management Agency 1994).

5.6.2   Methods Used to Determine the Peak Ground Acceleration

The engineering geologist is often the best individual to determine the peak ground acceleration a_{max} at the site based on fault, seismicity, and attenuation relationships. Some of the more commonly used methods to determine the peak ground acceleration at a site are as follows:

- *Historical earthquake:*   One approach is to consider the past earthquake history of the site. For the more recent earthquakes, data from seismographs can be used to determine the peak ground acceleration. For older earthquakes, the location of the earthquake and its magnitude are based on historical accounts of damage.

  Computer programs, such as the EQSEARCH computer program (Blake 2000b), have been developed that incorporate past earthquake data. By inputting the location of the site, the peak ground acceleration a_{max} could be determined. For example, Figs. B.1 to B.11 (App. B) present an example of the determination of a_{max} based on the history of seismic activity in the southwestern United States and northern Mexico.

  The peak horizontal ground acceleration a_{max} should never be based solely on the history of seismic activity in an area. The reason is because the historical time frame of recorded earthquakes is usually too small. Thus the value of a_{max} determined from historical studies should be compared with the value of a_{max} determined from the other methods described below.

- *Code or other regulatory requirements:*   There may be local building code or other regulatory requirements that specify design values of peak ground acceleration. For example, by using Fig. 5.17 to determine the seismic zone for a given site, the peak ground acceleration coefficient a_{max}/g can be obtained from Table 5.5. Depending on the distance to active faults and the underlying subsoil profile, the values in Table 5.5 could underestimate or overestimate the peak ground acceleration.

- *Maximum credible earthquake:*   The maximum credible earthquake (MCE) is often considered to be the largest earthquake that can reasonably be expected to occur based on known geologic and seismologic data. In essence, the maximum credible earthquake is the maximum earthquake that an active fault can produce, considering the geologic evidence of past movement and recorded seismic history of the area. According to Kramer (1996), other terms that have been used to describe similar worst-case levels of shaking include *safe shutdown earthquake* (used in the design of nuclear power plants), *maximum capable earthquake, maximum design earthquake, contingency level earthquake, safe level earthquake, credible design earthquake,* and *contingency design earthquake.* In general, these terms are used to describe the uppermost level of earthquake forces in the design of essential facilities.

  The maximum credible earthquake is determined for particular earthquakes or levels of ground shaking. As such, the analysis used to determine the maximum credible earthquake is typically referred to as a *deterministic method.*

- *Maximum probable earthquake:*   There are many different definitions of the maximum probable earthquake. The maximum probable earthquake is based on a study of nearby active faults. By using attenuation relationships, the maximum probable earthquake magnitude and maximum probable peak ground acceleration can be determined.

  A commonly used definition of maximum probable earthquake is the largest predicted earthquake that a fault is capable of generating within a specified time period, such as 50 or 100 years. Maximum probable earthquakes are most likely to occur within the design life of the project, and therefore, they have been commonly used in assessing seismic risk (Federal Emergency Management Agency 1994).

  Another commonly used definition of a maximum probable earthquake is an earthquake that will produce a peak ground acceleration a_{max} with a 50 percent probability of exceedance in 50 years (USCOLD 1985).

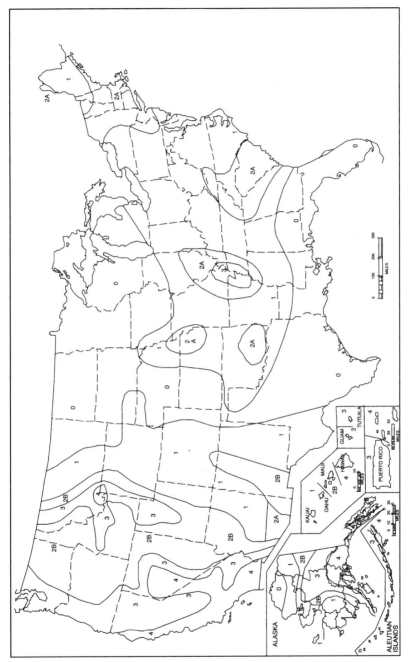

FIGURE 5.17 Seismic zone map of the United States. *(Reproduced with permission from the Uniform Building Code 1997.)*

5.35

According to Kramer (1996), other terms that have been used to describe earthquakes of similar size are *operating basis earthquake, operating level earthquake, probable design earthquake,* and *strength level earthquake.*

- *USGS earthquake maps:*   Another method for determining the peak ground acceleration is to determine the value of a_{max} that has a certain probability of exceedance in a specific number of years. The design basis ground motion can often be determined by a site-specific hazard analysis, or it may be determined from a hazard map.

  An example of a hazard map for California and Nevada is shown Fig. 5.18. This map was developed by the USGS (1996) and shows the peak ground acceleration for California and Nevada. There are similar maps for the entire continental United States, Alaska, and Hawaii. Note that the locations of the highest peak ground acceleration in Fig. 5.19 are similar to the locations of the highest seismic zones shown in Fig. 5.17, and vice versa. The USGS (1996) has also prepared maps that show peak ground acceleration with a 5 percent and 2 percent probability of exceedance in 50 years. These maps are easily accessible on the Internet (see U.S. Geological Survey, National Seismic Hazard Mapping Project).

  The various USGS maps showing peak ground acceleration with a 10, 5, and 2 percent probability of exceedance in 50 years provide the user with the choice of the appropriate level of hazard or risk. Such an approach is termed a *probabilistic method,* with the choice of the peak ground acceleration based on the concept of acceptable risk.

A typical ranking of the value of peak ground acceleration a_{max} obtained from the different methods described above, from the least to greatest value, is as follows:

1. Maximum probable earthquake (deterministic method)
2. USGS earthquake map: 10 percent probability of exceedance in 50 years (probabilistic method)
3. USGS earthquake map: 5 percent probability of exceedance in 50 years (probabilistic method)
4. USGS earthquake map: 2 percent probability of exceedance in 50 years (probabilistic method)
5. Maximum credible earthquake (deterministic method)

5.6.3 Example of the Determination of Peak Ground Acceleration

This example deals with the proposed W. C. H. Medical Library in La Mesa, California. The different methods used to determine the peak ground acceleration for this project were as follows:

- *Historical earthquake:*   The purpose of the EQSEARCH (Blake 2000b) computer program is to perform a historical search of earthquakes. For this computer program, the input data are shown in Fig. B.1 (App. B) and include the job number, job name, site coordinates in terms of latitude and longitude, search parameters, attenuation relationship, and other earthquake parameters. The output data are shown in Figs. B.2 to B.11. As indicated in Fig. B.4, the largest earthquake site acceleration from 1800 to 1999 is $a_{max} = 0.189g$.

  The EQSEARCH computer program also indicates the number of earthquakes of a certain magnitude that have affected the site. For example, from 1800 to 1999, there were two earthquakes of magnitude 6.5 or larger that impacted the site (see Fig. B.5).

- *Largest maximum earthquake:*   The EQFAULT computer program (Blake 2000a) was developed to determine the largest maximum earthquake site acceleration. For this com

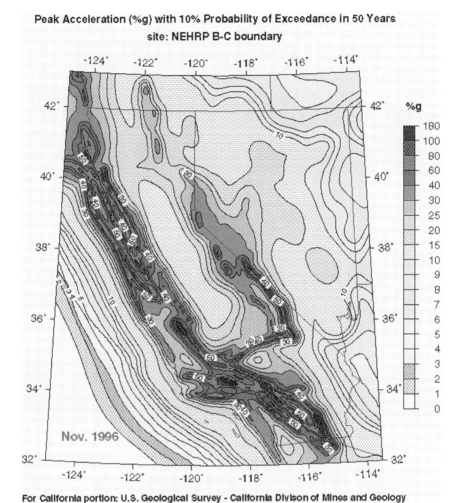

Peak Acceleration (%g) with 10% Probability of Exceedance in 50 Years

site: NEHRP B-C boundary

For California portion: U.S. Geological Survey - California Divison of Mines and Geology

For Nevada and surrounding states: USGS

FIGURE 5.18   Peak ground acceleration ($\%g$) with a 10 percent probability of exceedance in 50 years for California and Nevada. (*USGS 1996.*)

puter program, the input data are shown in Fig. B.12 and include the job number, job name, site coordinates in terms of latitude and longitude, search radius, attenuation relationship, and other earthquake parameters. The output data are shown in Figs. B.13 to B.19. As indicated in Fig. B.13, the largest maximum earthquake site acceleration a_{max} is $0.4203g$.

- *Probability analysis:*   Figures B.20 to B.25 present a probabilistic analysis for the determination of the peak ground acceleration at the site using the FRISKSP computer program (Blake 2000c). Two probabilistic analyses were performed using different attenuation relationships. As shown in Figs. B.21 and B.23, the data are plotted in terms of the peak ground acceleration versus probability of exceedance for a specific design life of the structure.

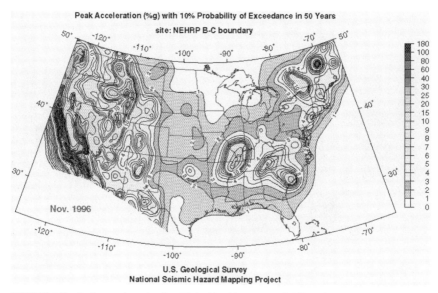

FIGURE 5.19   Peak ground acceleration (%g) with a 10 percent probability of exceedance in 50 years for the continental United States. (*USGS 1996.*)

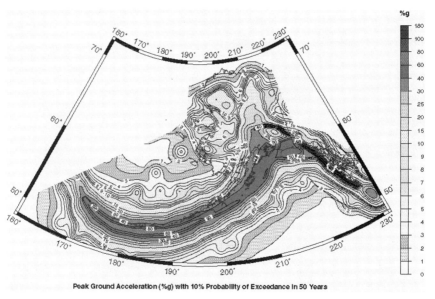

FIGURE 5.20   Peak ground acceleration (%g) with a 10 percent probability of exceedance in 50 years for Alaska. (*USGS 1996.*)

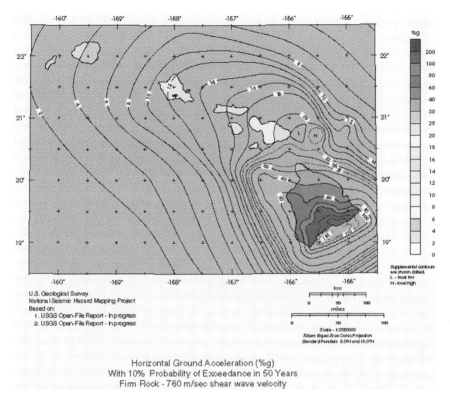

Horizontal Ground Acceleration (%g)
With 10% Probability of Exceedance in 50 Years
Firm Rock - 760 m/sec shear wave velocity

FIGURE 5.21   Peak ground acceleration (%g) with a 10 percent probability of exceedance in 50 years for Hawaii. (*USGS 1996.*)

- *USGS earthquake maps:*   Instead of using seismic maps such as shown in Figs. 5.18 to 5.21, the USGS enables the Internet user to obtain the peak ground acceleration (PGA) for a specific Zip code location (see Fig. 5.22). In Fig. 5.22, PGA is the peak ground acceleration, PE is the probability of exceedance, and SA is the spectral acceleration.

For this project (i.e., the W. C. H. Medical Library), a summary of the different values of peak ground acceleration a_{max} is provided below:

$a_{max} = 0.189g$ (historical earthquakes, see Fig. B.4)

$a_{max} = 0.212g$ (10% probability of exceedance in 50 years, see Fig. 5.22)

$a_{max} = 0.280g$ (5% probability of exceedance in 50 years, see Fig. 5.22)

$a_{max} = 0.389g$ (2% probability of exceedance in 50 years, see Fig. 5.22)

$a_{max} = 0.40g$ (seismic zone 4, see Table 5.5)

$a_{max} = 0.420g$ (largest maximum earthquake, see Fig. B.13)

TABLE 5.5   Seismic Zone Factor Z

Seismic zone	Seismic zone factor Z
0	0
1	0.075
2A	0.15
2B	0.20
3	0.30
4	0.40

Notes:
1. Data obtained from Table 16-I of the *Uniform Building Code* (1997).
2. See Fig. 5.17 (seismic zone map) for specific locations of the seismic zones 0 through 4.
3. Section 1804.5 of the *Uniform Building Code* (1997) states: "Peak ground acceleration may be determined based on a site-specific study taking into account soil amplification effects. In the absence of such a study, peak ground acceleration may be assumed equal to the seismic zone factor in Table 16-I" (that is, $Z = a_{max}/g$). In structural analysis, Z is also used in combination with other factors to determine the design seismic load acting on the structure.

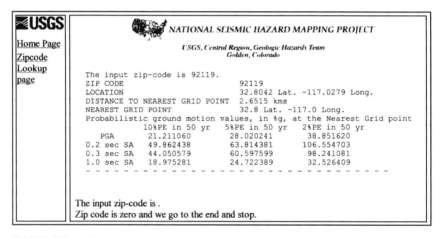

FIGURE 5.22   Peak ground acceleration for a specific Zip code location. (*USGS 1996.*)

There is a considerable variation in values for a_{max} as indicated above, from a low of $0.189g$ to a high of $0.420g$. The geotechnical engineer should work with the engineering geologist in selecting the most appropriate value of a_{max}. For the above data, based on a design life of 50 years and recognizing that the library is not an essential facility, an appropriate range of a_{max} to be used for the earthquake analyses is $0.189g$ to $0.212g$. Using a probabilistic approach, a value of $0.21g$ would seem appropriate.

If the project was an essential facility or had a design life in excess of 50 years, then a higher peak ground acceleration should be selected. For example, if the project had a 75-year design life and used a 10 percent probability of exceedance, then a peak ground acceleration a_{max} of about $0.25g$ should be used in the earthquake analyses (see Fig. B.21). On the other hand, if the project was an essential facility that must be able to resist the largest maximum earthquake, then an appropriate value of peak ground acceleration a_{max} would be $0.42g$. As these examples illustrate, it takes considerable experience and judgment in selecting the value of a_{max} to be used for the earthquake analyses.

5.6.4 Local Soil and Geologic Conditions

For the determination of the peak ground acceleration a_{max} as discussed in the previous sections, local soil and geologic conditions were not included in the analysis. USGS recommends that the final step in the determination of a_{max} for a particular site be to adjust the value (if needed) for such factors as these:

1. Directivity of ground motion, which can cause stronger shaking in certain directions.
2. Soft soils, which can increase the peak ground acceleration (see Sec. 4.6.1). Often a site that may be susceptible to liquefaction will also contain thick deposits of soft soil. The local soil condition of a thick deposit of soft clay is the most common reason for increasing the peak ground acceleration a_{max}.
3. Basin effects, such as the conversion to surface waves and reverberation experienced by sites in an alluvial basin.

5.7   REPORT PREPARATION

The results of the screening investigation and the quantitative evaluation will often need to be summarized in report form for review by the client and the governing agency. The items that should be included in the report, per the *Guidelines for Evaluating and Mitigating Seismic Hazards in California* (Division of Mines and Geology 1997), are as follows:

- Description of the proposed project's location, topographic relief, drainage, geologic and soil materials, and any proposed grading
- Site plan map of the project showing the locations of all explorations, including test pits, borings, penetration test locations, and soil or rock samples ,
- Description of the seismic setting, historic seismicity, nearest pertinent strong-motion records, and methods used to estimate (or source of) earthquake ground motion parameters used in liquefaction and landslide analysis
- A geologic map, at a scale of 1 : 24,000 or larger, that shows bedrock, alluvium, colluvium, soil material, faults, shears, joint systems, lithologic contacts, seeps or springs, soil or bedrock slumps, and other pertinent geologic and soil features existing on and adjacent to the project site
- Logs of borings, test pits, or other data obtained during the subsurface exploration
- Geologic cross sections depicting the most critical (least stable) slopes, geologic structure, stratigraphy, and subsurface water conditions, supported by boring and/or trench logs at appropriate locations
- Laboratory test results, soil classification, shear strength, and other pertinent geotechnical data
- Specific recommendations for mitigation alternatives necessary to reduce known and/or anticipated geologic/seismic hazards to an acceptable level of risk.

Not all the above information in the list may be relevant or required. On the other hand, some investigations may require additional types of data or analyses, which should also be included in the report. For example, usually both the on-site and off-site geologic and seismic hazards that could affect the site will need to be addressed. An example of a geotechnical engineering report that includes the results of the screening investigation and quantitative evaluation for seismic hazards is provided in App. D.

5.8 PROBLEMS

The problems have been divided into basic categories as indicated below:

Standard Penetration and Cone Penetration Tests

5.1 A standard penetration test (SPT) was performed on a near-surface deposit of clean sand where the number of blows to drive the sampler 18 in was 5 for the first 6 in, 8 for the second 6 in, and 9 for the third 6 in. Calculate the measured SPT N value (blows per foot) and indicate the in situ density condition of the sand per Table 5.2. *Answer:* Measured SPT N value = 17, and per Table 5.2, the sand has a medium density.

5.2 A clean sand deposit has a level ground surface, a total unit weight γ_t above the groundwater table of 18.9 kN/m³ (120 lb/ft³), and a submerged unit weight γ_b of 9.84 kN/m³ (62.6 lb/ft³). The groundwater table is located 1.5 m (5 ft) below ground surface. Standard penetration tests were performed in a 10-cm-diameter (4-in) borehole. At a depth of 3 m (10 ft) below ground surface, a standard penetration test was performed using a doughnut hammer with a blow count of 3 blows for the first 15 cm (6 in), 4 blows for the second 15 cm (6 in), and 5 blows for the third 15 cm (6 in) of diving penetration. Assuming hydrostatic pore water pressures, determine the vertical effective stress (σ'_{v0}) at a depth of 3 m (10 ft) and the corrected N value [that is, N_{60}, Eq. (5.1)]. *Answers:* σ'_{v0} = 43 kPa (910 lb/ft²) and N_{60} = 5.1.

5.3 Using the data from Prob. 5.2, determine the N value corrected for both field testing and overburden pressure, and indicate the in situ condition of the sand per Table 5.3. *Answers:* $(N_1)_{60}$ = 7.8 and per Table 5.3, the sand has a medium density.

5.4 Use the data from Prob. 5.2 and assume a cone penetration test was performed at a depth of 3 m (10 ft) and the cone resistance q_c = 40 kg/cm² (3900 kPa). Determine the CPT tip resistance corrected for overburden pressure. *Answer:* q_{c1} = 59 kg/cm² (5800 kPa).

Shear Strength Correlations

5.5 Using the data from Prob. 5.2, determine the friction angle φ of the sand using Fig. 5.12. *Answer:* φ = 30°

5.6 Using the data from Prob. 5.4, determine the friction angle φ of the sand using Fig. 5.14. *Answer:* φ = 40°.

CHAPTER 6
LIQUEFACTION

The following notation is used in this chapter:

SYMBOL	DEFINITION
a	Acceleration
a_{max}	Maximum horizontal acceleration at ground surface (also known as peak ground acceleration)
C_b	Borehole diameter correction
C_N, C_v	Correction factor to account for overburden pressure
C_r	Rod length correction
CRR	Cyclic resistance ratio
CSR, SSR	Cyclic stress ratio, also known as the seismic stress ratio
D_r	Relative density
e_i	Initial void ratio
E_m	Hammer efficiency
F	Horizontal earthquake force
FS	Factor of safety against liquefaction
g	Acceleration of gravity
k_0	Coefficient of lateral earth pressure at rest
LL	Liquid limit
m	Mass of the soil column
M_L	Local magnitude of earthquake
M_s	Surface wave magnitude of earthquake
M_w	Moment magnitude of earthquake
N_{60}	N value corrected for field testing procedures
$(N_1)_{60}$	N value corrected for field testing procedures and overburden pressure
q_{c1}	Cone resistance corrected for overburden pressure
r_d	Depth reduction factor
u_e	Excess pore water pressure
V_s	Shear wave velocity measured in field
V_{s1}	Shear wave velocity corrected for overburden pressure
w	Water content
W	Weight of soil column
z	Depth below ground surface
γ_t	Total unit weight of soil
σ_{dc}	Cyclic deviator stress (cyclic triaxial test)
σ_{v0}	Total vertical stress
σ_0'	Effective confining pressure
σ_{v0}'	Vertical effective stress
τ_{cyc}	Uniform cyclic shear stress amplitude of earthquake
τ_d	Cyclic shear stress
τ_{max}	Maximum shear stress

6.1 INTRODUCTION

This chapter deals with the liquefaction of soil. An introduction to liquefaction was presented in Sec. 3.4. The concept of liquefaction was first introduced by Casagrande in the late 1930s (also see Casagrande 1975).

As mentioned in Sec. 3.4, the typical subsurface soil condition that is susceptible to liquefaction is a loose sand, which has been newly deposited or placed, with a groundwater table near ground surface. During an earthquake, the application of cyclic shear stresses induced by the propagation of shear waves causes the loose sand to contract, resulting in an increase in pore water pressure. Because the seismic shaking occurs so quickly, the cohesionless soil is subjected to an undrained loading (total stress analysis). The increase in pore water pressure causes an upward flow of water to the ground surface, where it emerges in the form of mud spouts or sand boils. The development of high pore water pressures due to the ground shaking and the upward flow of water may turn the sand into a liquefied condition, which has been termed *liquefaction*. For this state of liquefaction, the effective stress is zero, and the individual soil particles are released from any confinement, as if the soil particles were floating in water (Ishihara 1985).

Structures on top of the loose sand deposit that has liquefied during an earthquake will sink or fall over, and buried tanks will float to the surface when the loose sand liquefies (Seed 1970). Section 3.4 has shown examples of damage caused by liquefaction. Sand boils, such as shown in Fig. 3.19, often develop when there has been liquefaction at a site.

After the soil has liquefied, the excess pore water pressure will start to dissipate. The length of time that the soil will remain in a liquefied state depends on two main factors: (1) the duration of the seismic shaking from the earthquake and (2) the drainage conditions of the liquefied soil. The longer and the stronger the cyclic shear stress application from the earthquake, the longer the state of liquefaction persists. Likewise, if the liquefied soil is confined by an upper and a lower clay layer, then it will take longer for the excess pore water pressures to dissipate by the flow of water from the liquefied soil. After the liquefaction process is complete, the soil will be in a somewhat denser state.

This chapter is devoted solely to *level-ground liquefaction*. Liquefaction can result in ground surface settlement (Sec. 7.2) or even a bearing capacity failure of the foundation (Sec. 8.2). Liquefaction can also cause or contribute to lateral movement of slopes, which is discussed in Secs. 9.4 and 9.5.

6.2 LABORATORY LIQUEFACTION STUDIES

The liquefaction of soils has been extensively studied in the laboratory. There is a considerable amount of published data concerning laboratory liquefaction testing. This section presents examples of laboratory liquefaction data from Ishihara (1985) and Seed and Lee (1965).

6.2.1 Laboratory Data from Ishihara

Figures 6.1 and 6.2 (from Ishihara 1985) present the results of laboratory tests performed on hollow cylindrical specimens of saturated Fuji River sand tested in a torsional shear test apparatus. Figure 6.1 shows the results of laboratory tests on a saturated sand having a medium density (D_r = 47 percent), and Fig. 6.2 shows the results of laboratory tests on a saturated sand in a dense state (D_r = 75 percent). Prior to the cyclic shear testing, both soil specimens were subjected to an effective confining pressure σ_0' of 98 kN/m^2 (2000 lb/ft^2). The saturated sand specimens were then subjected to undrained conditions during the application of the cyclic shear stress. Several different plots are shown in Figs. 6.1 and 6.2, as follows:

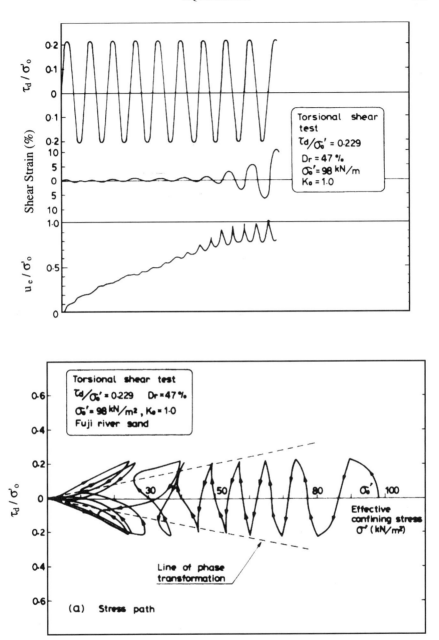

FIGURE 6.1   Laboratory test data from cyclic torsional shear tests performed on Fuji River sand having a medium density (D_r = 47 percent). (*Reproduced from Ishihara 1985.*)

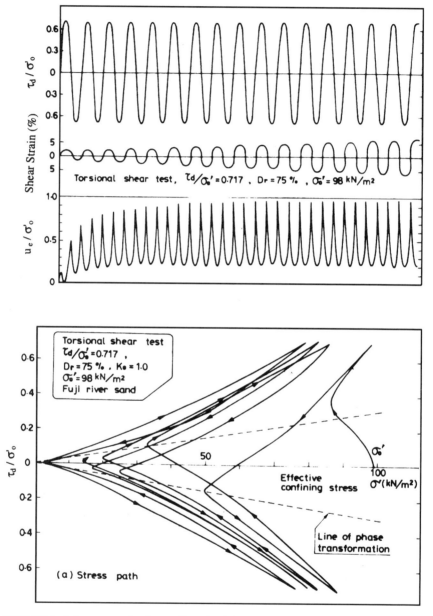

FIGURE 6.2   Laboratory test data from cyclic torsional shear tests performed on Fuji River sand having a dense state (D_r = 75 percent). (*Reproduced from Ishihara 1985.*)

1. *Plot of normalized cyclic shear stress* τ_d/σ_0': The uppermost plot shows the constant-amplitude cyclic shear stress that is applied to the saturated sand specimens. The applied cyclic shear stress has a constant amplitude and a sinusoidal pattern. The constant-amplitude cyclic shear stress τ_d has been normalized by dividing it by the initial effective confining pressure σ_0'. Note in Figs. 6.1 and 6.2 that the sand having a medium density $(D_r = 47$ percent) was subjected to a much lower constant-amplitude cyclic stress than the dense sand $(D_r = 75$ percent); that is, $\tau_d/\sigma_0' = 0.229$ for the sand having a medium density and $\tau_d/\sigma_0' = 0.717$ for the sand in a dense state.

2. *Plot of percent shear strain:* This plot shows the percent shear strain as the constant-amplitude cyclic shear stress is applied to the soil specimen. Note that for the sand having a medium density $(D_r = 47$ percent) there is a sudden and rapid increase in shear strain as high as 20 percent. For the dense sand $(D_r = 75$ percent), there is not a sudden and dramatic increase in shear strain, but rather the shear strain slowly increases with applications of the cyclic shear stress.

3. *Plot of normalized excess pore water pressure* u_e/σ_0': The normalized excess pore water pressure is also known as the *cyclic pore pressure ratio*. Because the soil specimens were subjected to undrained conditions during the application of the cyclic shear stress, excess pore water pressures u_e will develop as the constant-amplitude cyclic shear stress is applied to the soil. The excess pore water pressure u_e has been normalized by dividing it by the initial effective confining pressure σ_0'. When the excess pore water pressure u_e becomes equal to the initial effective confining pressure σ_0', the effective stress will become zero. Thus the condition of zero effective stress occurs when the ratio u_e/σ_0' is equal to 1.0. Note in Fig. 6.1 that the shear strain dramatically increases when the effective stress is equal to zero. As previously mentioned, liquefaction occurs when the effective stress becomes zero during the application of cyclic shear stress. Thus, once the sand having a medium density $(D_r = 47$ percent) liquefies, there is a significant increase in shear strain.

For the dense sand $(D_r = 75$ percent), u_e/σ_0' also becomes equal to 1.0 during the application of the cyclic shear stress. But the dense sand does not produce large shear displacements. This is because on reversal of the cyclic shear stress, the dense sand tends to dilate, resulting in an increased undrained shear resistance. Although the dense sand does reach a liquefaction state (that is, $u_e/\sigma_0' = 1.0$), it is only a momentary condition. Thus, this state has been termed *peak cyclic pore water pressure ratio of 100 percent with limited strain potential* (Seed 1979a). This state is also commonly referred to as *cyclic mobility* (Casagrande 1975, Castro 1975). The term *cyclic mobility* can be used to describe a state where the soil may only momentarily liquefy, with a limited potential for undrained deformation.

4. *Stress paths:* The lower plot in Figs. 6.1 and 6.2 shows the stress paths during application of the constant-amplitude cyclic shear stress. For the sand having a medium density $(D_r = 47$ percent), there is a permanent loss in shear strength as the stress path moves to the left with each additional cycle of constant-amplitude shear stress.

For the dense sand (Fig. 6.2), there is not a permanent loss in shear strength during the application of additional cycles of constant-amplitude shear stress. Instead, the stress paths tend to move up and down the shear strength envelope as the cycles of shear stress are applied to the soil.

It should be recognized that earthquakes will not subject the soil to uniform constant-amplitude cyclic shear stresses such as shown in the upper plot of Figs. 6.1 and 6.2. Nevertheless, this type of testing provides valuable insight into soil behavior.

In summary, the test results shown in Figs. 6.1 and 6.2 indicate that the sand having a medium density $(D_r = 47$ percent) has a sudden and dramatic increase in shear strain when the soil liquefies (i.e., when u_e/σ_0' becomes equal to 1.0). If the sand had been tested in a loose or very loose state, the loss of shear strength upon liquefaction would be even more sudden and dramatic. For loose sand, this initial liquefaction when u_e/σ_0' becomes equal to 1.0 coincides with the contraction of the soil structure, subsequent liquefaction, and large deformations. As such, for loose sands, the terms *initial liquefaction* and *liquefaction* have been used interchangeably.

For dense sands, the state of initial liquefaction ($u_e/\sigma_0' = 1.0$) does not produce large deformations because of the dilation tendency of the sand upon reversal of the cyclic stress. However, there could be some deformation at the onset of initial liquefaction, which is commonly referred to as cyclic mobility.

6.2.2　Laboratory Data from Seed and Lee

Figure 6.3 (from Seed and Lee 1965) shows a summary of laboratory data from cyclic triaxial tests performed on saturated specimens of Sacramento River sand. Cylindrical sand specimens were first saturated and subjected in the triaxial apparatus to an isotropic effective confining pressure of 100 kPa (2000 lb/ft^2). The saturated sand specimens were then subjected to undrained conditions during the application of the cyclic deviator stress in the triaxial apparatus (see Sec. 5.5.2 for discussion of cyclic triaxial test).

Numerous sand specimens were prepared at different void ratios (e_i = initial void ratio). The sand specimens were subjected to different values of cyclic deviator stress σ_{dc}, and the number of cycles of deviator stress required to produce initial liquefaction and 20 percent axial strain was recorded. The laboratory data shown in Fig. 6.3 indicate the following:

1. For sand having the same initial void ratio e_i and same effective confining pressure, the higher the cyclic deviator stress σ_{dc}, the lower the number of cycles of deviator stress required to cause initial liquefaction.

2. Similar to item 1, for a sand having the same initial void ratio e_i and same effective confining pressure, the cyclic deviator stress σ_{dc} required to cause initial liquefaction will decrease as the number of cycles of deviator stress is increased.

3. For sand having the same effective confining pressure, the denser the soil (i.e., the lower the value of the initial void ratio), the greater the resistance to liquefaction. Thus a dense soil will require a higher cyclic deviator stress σ_{dc} or more cycles of the deviator stress in order to cause initial liquefaction, as compared to the same soil in a loose state.

4. Similar to item 3, the looser the soil (i.e., the higher the value of the initial void ratio), the lower the resistance to liquefaction. Thus a loose soil will require a lower cyclic deviator stress σ_{dc} or fewer cycles of the deviator stress in order to cause initial liquefaction, as compared to the same soil in a dense state.

6.3　MAIN FACTORS THAT GOVERN LIQUEFACTION IN THE FIELD

There are many factors that govern the liquefaction process for in situ soil. Based on the results of laboratory tests (Sec. 6.2) as well as field observations and studies, the most important factors that govern liquefaction are as follows:

1. *Earthquake intensity and duration:* In order to have liquefaction of soil, there must be ground shaking. The character of the ground motion, such as acceleration and duration of shaking, determines the shear strains that cause the contraction of the soil particles and the development of excess pore water pressures leading to liquefaction. The

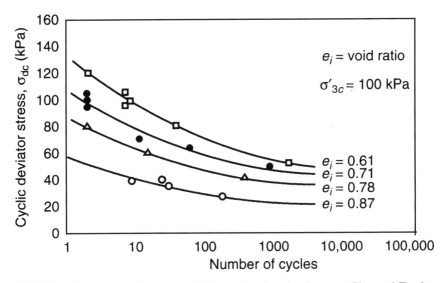

FIGURE 6.3   Laboratory test data from cyclic triaxial tests performed on Sacramento River sand. The plotted data represent the cyclic deviator stress versus number of cycles of deviator stress required to cause initial liquefaction and 20 percent axial strain. (*Initially developed by Seed and Lee 1965, reproduced from Kramer 1996.*)

most common cause of liquefaction is due to the seismic energy released during an earthquake. The potential for liquefaction increases as the earthquake intensity and duration of shaking increase. Those earthquakes that have the highest magnitude will produce both the largest ground acceleration and the longest duration of ground shaking (see Table 2.2).

Although data are sparse, there would appear to be a shaking threshold that is needed to produce liquefaction. These threshold values are a peak ground acceleration a_{max} of about $0.10g$ and local magnitude M_L of about 5 (National Research Council 1985, Ishihara 1985). Thus, a liquefaction analysis would typically not be needed for those sites having a peak ground acceleration a_{max} less than $0.10g$ or a local magnitude M_L less than 5.

Besides earthquakes, other conditions can cause liquefaction, such as subsurface blasting, pile driving, and vibrations from train traffic.

2. Groundwater table: The condition most conducive to liquefaction is a near-surface groundwater table. Unsaturated soil located above the groundwater table will not liquefy. If it can be demonstrated that the soils are currently above the groundwater table and are highly unlikely to become saturated for given foreseeable changes in the hydrologic regime, then such soils generally do not need to be evaluated for liquefaction potential.

At sites where the groundwater table significantly fluctuates, the liquefaction potential will also fluctuate. Generally, the historic high groundwater level should be used in the liquefaction analysis unless other information indicates a higher or lower level is appropriate (Division of Mines and Geology 1997).

Poulos et al. (1985) state that liquefaction can also occur in very large masses of sands or silts that are dry and loose and loaded so rapidly that the escape of air from the voids is restricted. Such movement of dry and loose sands is often referred to as *running soil* or *running ground.* Although such soil may flow as liquefied soil does, in this text, such soil deformation will not be termed liquefaction. It is best to consider that liquefaction only occurs for soils that are located below the groundwater table.

3. *Soil type:*   In terms of the soil types most susceptible to liquefaction, Ishihara (1985) states: "The hazard associated with soil liquefaction during earthquakes has been known to be encountered in deposits consisting of fine to medium sand and sands containing low-plasticity fines. Occasionally, however, cases are reported where liquefaction apparently occurred in gravelly soils."

Thus, the soil types susceptible to liquefaction are nonplastic (cohesionless) soils. An approximate listing of cohesionless soils from least to most resistant to liquefaction is clean sands, nonplastic silty sands, nonplastic silt, and gravels. There could be numerous exceptions to this sequence. For example, Ishihara (1985, 1993) describes the case of tailings derived from the mining industry that were essentially composed of ground-up rocks and were classified as rock flour. Ishihara (1985, 1993) states that the rock flour in a water-saturated state did not possess significant cohesion and behaved as if it were a clean sand. These tailings were shown to exhibit as low a resistance to liquefaction as clean sand.

Seed et al. (1983) stated that based on both laboratory testing and field performance, the great majority of cohesive soils will not liquefy during earthquakes. Using criteria originally stated by Seed and Idriss (1982) and subsequently confirmed by Youd and Gilstrap (1999), in order for a cohesive soil to liquefy, it must meet *all* the following three criteria:

- The soil must have less than 15 percent of the particles, based on dry weight, that are finer than 0.005 mm (i.e., percent finer at 0.005 mm < 15 percent).
- The soil must have a liquid limit (LL) that is less than 35 (that is, LL < 35).
- The water content w of the soil must be greater than 0.9 of the liquid limit [that is, $w > 0.9$ (LL)].

If the cohesive soil does not meet all three criteria, then it is generally considered to be not susceptible to liquefaction. Although the cohesive soil may not liquefy, there could still be a significant undrained shear strength loss due to the seismic shaking.

4. *Soil relative density D_r:*   Based on field studies, cohesionless soils in a loose relative density state are susceptible to liquefaction. Loose nonplastic soils will contract during the seismic shaking which will cause the development of excess pore water pressures. As indicated in Sec. 6.2, upon reaching initial liquefaction, there will be a sudden and dramatic increase in shear displacement for loose sands.

For dense sands, the state of initial liquefaction does not produce large deformations because of the dilation tendency of the sand upon reversal of the cyclic shear stress. Poulos et al. (1985) state that if the in situ soil can be shown to be dilative, then it need not be evaluated because it will not be susceptible to liquefaction. In essence, dilative soils are not susceptible to liquefaction because their undrained shear strength is greater than their drained shear strength.

5. *Particle size gradation:*   Uniformly graded nonplastic soils tend to form more unstable particle arrangements and are more susceptible to liquefaction than well-graded soils. Well-graded soils will also have small particles that fill in the void spaces between the large particles. This tends to reduce the potential contraction of the soil, resulting in less excess pore water pressures being generated during the earthquake. Kramer (1996) states that field evidence indicates that most liquefaction failures have involved uniformly graded granular soils.

6. *Placement conditions or depositional environment:*   Hydraulic fills (fill placed under water) tend to be more susceptible to liquefaction because of the loose and segregated soil structure created by the soil particles falling through water. Natural soil deposits formed in lakes, rivers, or the ocean also tend to form a loose and segregated soil structure and are more susceptible to liquefaction. Soils that are especially susceptible to liquefaction are formed in lacustrine, alluvial, and marine depositional environments.

7. *Drainage conditions:*   If the excess pore water pressure can quickly dissipate, the soil may not liquefy. Thus highly permeable gravel drains or gravel layers can reduce the liquefaction potential of adjacent soil.

8. *Confining pressures:*   The greater the confining pressure, the less susceptible the soil is to liquefaction. Conditions that can create a higher confining pressure are a deeper groundwater table, soil that is located at a deeper depth below ground surface, and a surcharge pressure applied at ground surface. Case studies have shown that the possible zone of liquefaction usually extends from the ground surface to a maximum depth of about 50 ft (15 m). Deeper soils generally do not liquefy because of the higher confining pressures.

This does not mean that a liquefaction analysis should not be performed for soil that is below a depth of 50 ft (15 m). In many cases, it may be appropriate to perform a liquefaction analysis for soil that is deeper than 50 ft (15 m). An example would be sloping ground, such as a sloping berm in front of a waterfront structure or the sloping shell of an earth dam (see Fig. 3.38). In addition, a liquefaction analysis should be performed for any soil deposit that has been loosely dumped in water (i.e., the liquefaction analysis should be performed for the entire thickness of loosely dumped fill in water, even if it exceeds 50 ft in thickness). Likewise, a site where alluvium is being rapidly deposited may also need a liquefaction investigation below a depth of 50 ft (15 m). Considerable experience and judgment are required in the determination of the proper depth to terminate a liquefaction analysis.

9. *Particle shape:*   The soil particle shape can also influence liquefaction potential. For example, soils having rounded particles tend to densify more easily than angular-shape soil particles. Hence a soil containing rounded soil particles is more susceptible to liquefaction than a soil containing angular soil particles.

10. *Aging and cementation:*   Newly deposited soils tend to be more susceptible to liquefaction than older deposits of soil. It has been shown that the longer a soil is subjected to a confining pressure, the greater the liquefaction resistance (Ohsaki 1969, Seed 1979a, Yoshimi et al. 1989). Table 6.1 presents the estimated susceptibility of sedimentary deposits to liquefaction versus the geologic age of the deposit.

The increase in liquefaction resistance with time could be due to the deformation or compression of soil particles into more stable arrangements. With time, there may also be the development of bonds due to cementation at particle contacts.

11. *Historical environment:*   It has also been determined that the historical environment of the soil can affect its liquefaction potential. For example, older soil deposits that have already been subjected to seismic shaking have an increased liquefaction resistance compared to a newly formed specimen of the same soil having an identical density (Finn et al. 1970, Seed et al. 1975).

Liquefaction resistance also increases with an increase in the overconsolidation ratio (OCR) and the coefficient of lateral earth pressure at rest k_0 (Seed and Peacock 1971, Ishihara et al. 1978). An example would be the removal of an upper layer of soil due to erosion. Because the underlying soil has been preloaded, it will have a higher overconsolidation ratio and it will have a higher coefficient of lateral earth pressure at rest k_0. Such a soil that has been preloaded will be more resistant to liquefaction than the same soil that has not been preloaded.

12. *Building load:*   The construction of a heavy building on top of a sand deposit can decrease the liquefaction resistance of the soil. For example, suppose a mat slab at ground surface supports a heavy building. The soil underlying the mat slab will be subjected to shear stresses caused by the building load. These shear stresses induced into the soil by the building load can make the soil more susceptible to liquefaction. The reason is that a smaller additional shear stress will be required from the earthquake in order to cause contraction and hence liquefaction of the soil. For level-ground liquefaction discussed in this chapter, the effect of the building load is ignored. Although building loads are not considered in the liquefaction analysis in this chapter, the building loads must be included in all liquefaction-induced settlement, bearing capacity, and stability analyses, as discussed in Chaps. 7 through 9.

In summary, the site conditions and soil type most susceptible to liquefaction are as follows:

Site Conditions

- Site that is close to the epicenter or location of fault rupture of a major earthquake
- Site that has a groundwater table close to ground surface

Soil Type Most Susceptible to Liquefaction for Given Site Conditions

- Sand that has uniform gradation and rounded soil particles, very loose or loose density state, recently deposited with no cementation between soil grains, and no prior preloading or seismic shaking

6.4 LIQUEFACTION ANALYSIS

6.4.1 Introduction

The first step in the liquefaction analysis is to determine if the soil has the ability to liquefy during an earthquake. As discussed in Sec. 6.3 (item number 3), the vast majority of soils that are susceptible to liquefaction are cohesionless soils. Cohesive soils should not be considered susceptible to liquefaction unless they meet all three criteria listed in Sec. 6.3 (see item 3, soil type).

The most common type of analysis to determine the liquefaction potential is to use the standard penetration test (SPT) (Seed et al. 1985, Stark and Olson 1995). The analysis is based on the simplified method proposed by Seed and Idriss (1971). This method of liquefaction analysis proposed by Seed and Idriss (1971) is often termed the *simplified procedure*. This is the most commonly used method to evaluate the liquefaction potential of a site. The steps are as follows:

1. *Appropriate soil type:*   As discussed above, the first step is to determine if the soil has the ability to liquefy during an earthquake. The soil must meet the requirements listed in Sec. 6.3 (item 3).

2. *Groundwater table:*   The soil must be below the groundwater table. The liquefaction analysis could also be performed if it is anticipated that the groundwater table will rise in the future, and thus the soil will eventually be below the groundwater table.

3. *CSR induced by earthquake:*   If the soil meets the above two requirements, then the simplified procedure can be performed. The first step in the simplified procedure is to determine the cyclic stress ratio (CSR) that will be induced by the earthquake (Sec. 6.4.2).

A major unknown in the calculation of the CSR induced by the earthquake is the peak horizontal ground acceleration a_{max} that should be used in the analysis. The peak horizontal ground acceleration is discussed in Sec. 5.6. Threshold values needed to produce liquefaction are discussed in Sec. 6.3 (item 1). As previously mentioned, a liquefaction analysis would typically not be needed for those sites having a peak ground acceleration a_{max} less than $0.10g$ or a local magnitude M_L less than 5.

4. *CRR from standard penetration test:*   By using the standard penetration test, the cyclic resistance ratio (CRR) of the in situ soil is then determined (Sec. 6.4.3). If the CSR induced by the earthquake is greater than the CRR determined from the standard penetration test, then it is likely that liquefaction will occur during the earthquake, and vice versa.

5. *Factor of safety* (FS):   The final step is to determine the factor of safety against liquefaction (Sec. 6.4.4), which is defined as FS = CRR/CSR.

TABLE 6.1   Estimated Susceptibility of Sedimentary Deposits to Liquefaction during Strong Seismic Shaking Based on Geologic Age and Depositional Environment

Type of deposit	General distribution of cohesionless sediments in deposits	Likelihood that cohesionless sediments, when saturated, would be susceptible to liquefaction (by age of deposit)			
		<500 years	Holocene	Pleistocene	Pre-Pleistocene
(a) Continental deposits					
Alluvial fan and plain	Widespread	Moderate	Low	Low	Very low
Delta and fan-delta	Widespread	High	Moderate	Low	Very low
Dunes	Widespread	High	Moderate	Low	Very low
Marine terrace/plain	Widespread	Unknown	Low	Very low	Very low
Talus	Widespread	Low	Low	Very low	Very low
Tephra	Widespread	High	High	Unknown	Unknown
Colluvium	Variable	High	Moderate	Low	Very low
Glacial till	Variable	Low	Low	Very low	Very low
Lacustrine and playa	Variable	High	Moderate	Low	Very low
Loess	Variable	High	High	High	Unknown
Floodplain	Locally variable	High	Moderate	Low	Very low
River channel	Locally variable	Very high	High	Low	Very low
Sebka	Locally variable	High	Moderate	Low	Very low
Residual soils	Rare	Low	Low	Very low	Very low
Tuff	Rare	Low	Low	Very low	Very low
(b) Coastal zone					
Beach—large waves	Widespread	Moderate	Low	Very low	Very low
Beach—small waves	Widespread	High	Moderate	Low	Very low
Delta	Widespread	Very high	High	Low	Very low
Estuarine	Locally variable	High	Moderate	Low	Very low
Foreshore	Locally variable	High	Moderate	Low	Very low
Lagoonal	Locally variable	High	Moderate	Low	Very low
(c) Artificial					
Compacted fill	Variable	Low	Unknown	Unknown	Unknown
Uncompacted fill	Variable	Very high	Unknown	Unknown	Unknown

Source: Data from Youd and Hoose (1978), reproduced from R. B. Seed (1991).

6.4.2   Cyclic Stress Ratio Caused by the Earthquake

If it is determined that the soil has the ability to liquefy during an earthquake and the soil is below or will be below the groundwater table, then the liquefaction analysis is performed. The first step in the simplified procedure is to calculate the cyclic stress ratio, also commonly referred to as the *seismic stress ratio* (SSR), that is caused by the earthquake.

To develop the CSR earthquake equation, it is assumed that there is a level ground surface and a soil column of unit width and length, and that the soil column will move horizontally as a rigid body in response to the maximum horizontal acceleration a_{max} exerted by the earthquake at ground surface. Figure 6.4 shows a diagram of these assumed conditions. Given these assumptions, the weight W of the soil column is equal to $\gamma_t z$, where γ_t = total unit weight of the soil and z = depth below ground surface. The horizontal earthquake force F acting on the soil column (which has a unit width and length) is:

$$F = ma = \left(\frac{W}{g}\right)a = \left(\frac{\gamma_t z}{g}\right)a_{max} = \sigma_{vo}\left(\frac{a_{max}}{g}\right) \tag{6.1}$$

where F = horizontal earthquake force acting on soil column that has a unit width and length, lb or kN

m = total mass of soil column, lb or kg, which is equal to W/g

W = total weight of soil column, lb or kN. For the assumed unit width and length of soil column, the total weight of the soil column is $\gamma_t z$.

γ_t = total unit weight of soil, lb/ft^3 or kN/m^3.

z = depth below ground surface of soil column, as shown in Fig. 6.4.

a = acceleration, which in this case is the maximum horizontal acceleration at ground surface caused by the earthquake ($a = a_{max}$), ft/s^2 or m/s^2

a_{max} = maximum horizontal acceleration at ground surface that is induced by the earthquake, ft/s^2 or m/s^2. The maximum horizontal acceleration is also commonly referred to as the peak ground acceleration (see Sec. 5.6).

σ_{v0} = total vertical stress at bottom of soil column, lb/ft^2 or kPa. The total vertical stress = $\gamma_t z$.

As shown in Fig. 6.4, by summing forces in the horizontal direction, the force F acting on the rigid soil element is equal to the maximum shear force at the base on the soil element. Since the soil element is assumed to have a unit base width and length, the maximum shear force F is equal to the maximum shear stress τ_{max}, or from Eq. (6.1):

$$\tau_{max} = F = \sigma_{v0}\left(\frac{a_{max}}{g}\right) \tag{6.2}$$

Dividing both sides of the equation by the vertical effective stress σ'_{vo} gives

$$\frac{\tau_{max}}{\sigma'_{v0}} = \left(\frac{\sigma_{v0}}{\sigma'_{v0}}\right)\left(\frac{a_{max}}{g}\right) \tag{6.3}$$

Since the soil column does not act as a rigid body during the earthquake, but rather the soil is deformable, Seed and Idriss (1971) incorporated a depth reduction factor r_d into the right side of Eq. (6.3), or

$$\frac{\tau_{max}}{\sigma'_{v0}} = r_d\left(\frac{\sigma_{v0}}{\sigma'_{v0}}\right)\left(\frac{a_{max}}{g}\right) \tag{6.4}$$

For the simplified method, Seed et al. (1975) converted the typical irregular earthquake record to an equivalent series of uniform stress cycles by assuming the following:

$$\tau_{cyc} = 0.65\tau_{max} \tag{6.5}$$

where τ_{cyc} = uniform cyclic shear stress amplitude of the earthquake (lb/ft^2 or kPa).

In essence, the erratic earthquake motion was converted to an equivalent series of uniform cycles of shear stress, referred to as τ_{cyc}. By substituting Eq. (6.5) into Eq. (6.4), the earthquake-induced cyclic stress ratio is obtained.

$$CSR = \frac{\tau_{cyc}}{\sigma'_{v0}} = 0.65r_d\left(\frac{\sigma_{v0}}{\sigma'_{v0}}\right)\left(\frac{a_{max}}{g}\right) \tag{6.6}$$

where CSR = cyclic stress ratio (dimensionless), also commonly referred to as seismic stress ratio

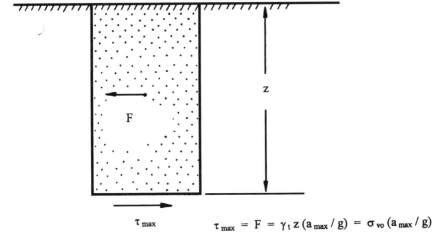

$$\tau_{max} = F = \gamma_t\, z\, (a_{max}\,/\,g) = \sigma_{vo}\,(a_{max}\,/\,g)$$

FIGURE 6.4 Conditions assumed for the derivation of the CSR earthquake equation.

a_{max} = maximum horizontal acceleration at ground surface that is induced by the earthquake, ft/s^2 or m/s^2, also commonly referred to as the peak ground acceleration (see Sec. 5.6)

g = acceleration of gravity (32.2 ft/s^2 or 9.81 m/s^2)

σ_{v0} = total vertical stress at a particular depth where the liquefaction analysis is being performed, lb/ft^2 or kPa. To calculate total vertical stress, total unit weight γ_t of soil layer (s) must be known

σ'_{v0} = vertical effective stress at that same depth in soil deposit where σ_{v0} was calculated, lb/ft^2 or kPa. To calculate vertical effective stress, location of groundwater table must be known

r_d = depth reduction factor, also known as stress reduction coefficient (dimensionless)

As previously mentioned, the depth reduction factor was introduced to account for the fact that the soil column shown in Fig. 6.4 does not behave as a rigid body during the earthquake. Figure 6.5 presents the range in values for the depth reduction factor r_d versus depth below ground surface. Note that with depth, the depth reduction factor decreases to account for the fact that the soil is not a rigid body, but is rather deformable. As indicated in Fig. 6.5, Idriss (1999) indicates that the values of r_d depend on the magnitude of the earthquake. As a practical matter, the r_d values are usually obtained from the curve labeled "Average values by Seed & Idriss (1971)" in Fig. 6.5.

Another option is to assume a linear relationship of r_d versus depth and use the following equation (Kayen et al. 1992):

$$r_d = 1 - 0.012z \qquad (6.7)$$

where z = depth in meters below the ground surface where the liquefaction analysis is being performed (i.e., the same depth used to calculate σ_{v0} and σ'_{v0}).

For Eq. (6.6), the vertical total stress σ_{v0} and vertical effective stress σ'_{v0} can be readily calculated using basic geotechnical principles. Equation (6.7) or Fig. 6.5 could be used to determine the depth reduction factor r_d. Thus all parameters in Eq. (6.6) can be readily calculated, except for the peak ground acceleration a_{max}, which is discussed in Sec. 5.6.

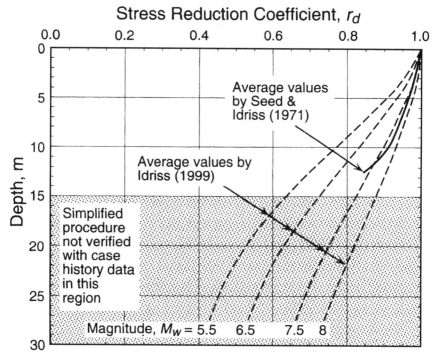

FIGURE 6.5   Reduction factor r_d versus depth below level or gently sloping ground surfaces. (*From Andrus and Stokoe 2000, reproduced with permission from the American Society of Civil Engineers.*)

6.4.3 Cyclic Resistance Ratio from the Standard Penetration Test

The second step in the simplified procedure is to determine the cyclic resistance ratio of the in situ soil. The cyclic resistance ratio represents the liquefaction resistance of the in situ soil. The most commonly used method for determining the liquefaction resistance is to use the data obtained from the standard penetration test. The standard penetration test is discussed in Sec. 5.4.3. The advantages of using the standard penetration test to evaluate the liquefaction potential are as follows:

1. *Groundwater table:*   A boring must be excavated in order to perform the standard penetration test. The location of the groundwater table can be measured in the borehole. Another option is to install a piezometer in the borehole, which can then be used to monitor the groundwater level over time.

2. *Soil type:*   In clean sand, the SPT sampler may not be able to retain a soil sample. But for most other types of soil, the SPT sampler will be able to retrieve a soil sample. The soil sample retrieved in the SPT sampler can be used to visually classify the soil and to estimate the percent fines in the soil. In addition, the soil specimen can be returned to the laboratory, and classification tests can be performed to further assess the liquefaction susceptibility of the soil (see item 3, Sec. 6.3).

3. *Relationship between N value and liquefaction potential:*   In general, the factors that increase the liquefaction resistance of a soil will also increase the $(N_1)_{60}$ from the standard penetration test [see Sec. 5.4.3 for the procedure to calculate $(N_1)_{60}$]. For example, a well-graded dense soil that has been preloaded or aged will be resistant to liquefaction and will have high values of $(N_1)_{60}$. Likewise, a uniformly graded soil with a loose and segregated soil structure will be more susceptible to liquefaction and will have much lower values of $(N_1)_{60}$.

Based on the standard penetration test and field performance data, Seed et al. (1985) concluded that there are three approximate potential damage ranges that can be identified:

$(N_1)_{60}$	Potential damage
0–20	High
20–30	Intermediate
>30	No significant damage

As indicated in Table 5.3, an $(N_1)_{60}$ value of 20 is the approximate boundary between the medium and dense states of the sand. Above an $(N_1)_{60}$ of 30, the sand is in either a dense or a very dense state. For this condition, initial liquefaction does not produce large deformations because of the dilation tendency of the sand upon reversal of the cyclic shear stress. This is the reason that such soils produce no significant damage, as indicated by the above table.

Figure 6.6 presents a chart that can be used to determine the cyclic resistance ratio of the in situ soil. This figure was developed from investigations of numerous sites that had liquefied or did not liquefy during earthquakes. For most of the data used in Fig. 6.6, the earthquake magnitude was close to 7.5 (Seed et al. 1985). The three lines shown in Fig. 6.6 are for soil that contains 35, 15, or ≤5 percent fines. The lines shown in Fig. 6.6 represent approximate dividing lines, where data to the left of each individual line indicate field liquefaction, while data to the right of the line indicate sites that generally did not liquefy during the earthquake.

Use Fig. 6.6 to determine the cyclic resistance ratio of the in situ soil, as follows:

1. *Standard penetration test $(N_1)_{60}$ value:* Note in Fig. 6.6 that the horizontal axis represents data from the standard penetration test, which must be expressed in terms of the $(N_1)_{60}$ value. In the liquefaction analysis, the standard penetration test N_{60} value [Eq. (5.1)] is corrected for the overburden pressure [see Eq. (5.2)]. As discussed in Sec. 5.4.3, when a correction is applied to the N_{60} value to account for the effect of overburden pressure, this value is referred to as $(N_1)_{60}$.

2. *Percent fines:* Once the $(N_1)_{60}$ value has been calculated, the next step is to determine or estimate the percent fines in the soil. For a given $(N_1)_{60}$ value, soils with more fines have a higher liquefaction resistance. Figure 6.6 is applicable for nonplastic silty sands or for plastic silty sands that meet the criteria for cohesive soils listed in Sec. 6.3 (see item 3, soil type).

3. *Cyclic resistance ratio for an anticipated magnitude 7.5 earthquake:* Once the $(N_1)_{60}$ value and the percent fines in the soil have been determined, then Fig. 6.6 can be used to obtain the cyclic resistance ratio of the soil. To use Fig. 6.6, the figure is entered with the corrected standard penetration test $(N_1)_{60}$ value from Eq. (5.2), and then by intersecting the appropriate fines content curve, the cyclic resistance ratio is obtained.

As shown in Fig. 6.6, for a magnitude 7.5 earthquake, clean sand will not liquefy if the $(N_1)_{60}$ value exceeds 30. For an $(N_1)_{60}$ value of 30, the sand is in either a dense or a very dense state (see Table 5.3). As previously mentioned, dense sands will not liquefy because they tend to dilate during shearing.

4. *Correction for other magnitude earthquakes:* Figure 6.6 is for a projected earthquake that has a magnitude of 7.5. The final factor that must be included in the analysis is

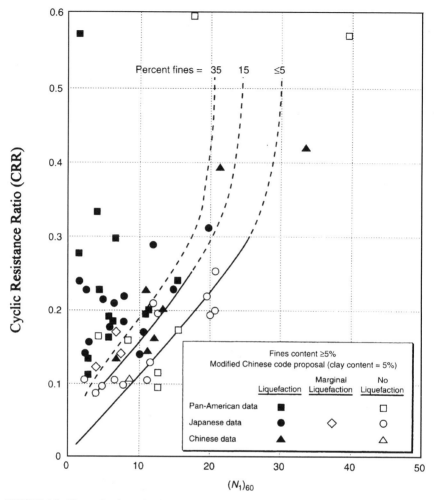

FIGURE 6.6 Plot used to determine the cyclic resistance ratio for clean and silty sands for $M = 7.5$ earthquakes. (*After Seed et al. 1985, reprinted with permission of the American Society of Civil Engineers.*)

the magnitude of the earthquake. As indicated in Table 2.2, the higher the magnitude of the earthquake, the longer the duration of ground shaking. A higher magnitude will thus result in a higher number of applications of cyclic shear strain, which will decrease the liquefaction resistance of the soil. Figure 6.6 was developed for an earthquake magnitude of 7.5; and for other different magnitudes, the CRR values from Fig. 6.6 would be multiplied by the magnitude scaling factor indicated in Table 6.2. Figure 6.7 presents other suggested magnitude scaling factors.

As discussed in Sec. 2.4.4, it could be concluded that the local magnitude M_L, the surface wave magnitude M_s, and moment magnitude M_w scales are reasonably close to one another below a value of about 7. Thus for a magnitude of 7 or below, any one of these magnitude scales can be used to determine the magnitude scaling factor. At high magnitude val-

TABLE 6.2   Magnitude Scaling Factors

Anticipated earthquake magnitude	Magnitude scaling factor (MSF)
8½	0.89
7½	1.00
6¾	1.13
6	1.32
5¼	1.50

Note: To determine the cyclic resistance ratio of the in situ soil, multiply the magnitude scaling factor indicated above by the cyclic resistance ratio determined from Fig. 6.6.

Source: Seed et al. (1985).

ues, the moment magnitude M_w tends to significantly deviate from the other magnitude scales, and the moment magnitude M_w should be used to determine the magnitude scaling factor from Table 6.2 or Fig. 6.7.

Two additional correction factors may need to be included in the analysis. The first correction factor is for the liquefaction of deep soil layers (i.e., depths where $\sigma'_{v0} > 100$ kPa, in which liquefaction has not been verified by the Seed and Idriss simplified procedure, see Youd and Idriss 2001). The second correction factor is for sloping ground conditions, which is discussed in Sec. 9.4.2.

As indicated in Secs. 4.6.1 and 5.6.4, both the peak ground acceleration a_{max} and the length of ground shaking increase for sites having soft, thick, and submerged soils. In a sense, the earthquake magnitude accounts for the increased shaking at a site; that is, the higher the magnitude, the longer the ground is subjected to shaking. Thus for sites having soft, thick, and submerged soils, it may be prudent to increase both the peak ground acceleration a_{max} and the earthquake magnitude to account for local site effects.

6.4.4   Factor of Safety against Liquefaction

The final step in the liquefaction analysis is to calculate the factor of safety against liquefaction. If the cyclic stress ratio caused by the anticipated earthquake [Eq. (6.6)] is greater than the cyclic resistance ratio of the in situ soil (Fig. 6.6), then liquefaction could occur during the earthquake, and vice versa. The factor of safety against liquefaction (FS) is defined as follows:

$$\text{FS} = \frac{\text{CRR}}{\text{CSR}} \tag{6.8}$$

The higher the factor of safety, the more resistant the soil is to liquefaction. However, soil that has a factor of safety slightly greater than 1.0 may still liquefy during an earthquake. For example, if a lower layer liquefies, then the upward flow of water could induce liquefaction of the layer that has a factor of safety slightly greater than 1.0.

In the above liquefaction analysis, there are many different equations and corrections that are applied to both the cyclic stress ratio induced by the anticipated earthquake and the cyclic resistance ratio of the in situ soil. For example, there are four different corrections (that is, E_m, C_b, C_r, and σ'_{v0}) that are applied to the standard penetration test N value in order to calculate the $(N_1)_{60}$ value. All these different equations and various corrections may provide the engineer with a sense of high accuracy, when in fact the entire analysis is only a gross approximation. The analysis should be treated as such, and engineering experience

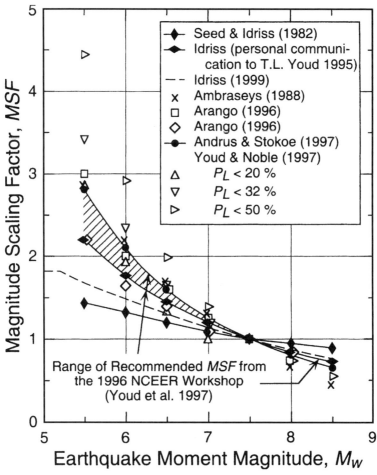

FIGURE 6.7   Magnitude scaling factors derived by various investigators. (*From Andrus and Stokoe 2000, reprinted with permission of the American Society of Civil Engineers.*)

and judgment are essential in the final determination of whether a site has liquefaction potential.

6.4.5   Example Problem

The following example problem illustrates the procedure that is used to determine the factor of safety against liquefaction: It is planned to construct a building on a cohesionless sand deposit (fines < 5 percent). There is a nearby major active fault, and the engineering geologist has determined that for the anticipated earthquake, the peak ground acceleration a_{max} will be equal to 0.40g. Assume the site conditions are the same as stated in Problems 5.2 and 5.3, that is, a level ground surface with the groundwater table located 1.5 m below ground surface and the standard penetration test performed at a depth of 3 m. Assuming an anticipated earthquake magnitude of 7.5, calculate the factor of safety against liquefaction for the saturated clean sand located at a depth of 3 m below ground surface.

Solution. Per Probs. 5.2 and 5.3, σ'_{v0} = 43 kPa and $(N_1)_{60}$ = 7.7. Using the soil unit weights from Prob. 5.2, we have

$$\sigma_{v0} = (1.5 \text{ m}) (18.9 \text{ kN/m}^3) + (1.5 \text{ m}) (9.84 + 9.81 \text{ kN/m}^3) = 58 \text{ kPa}$$

Using Eq. (6.7) with z = 3 m gives r_d = 0.96. Use the following values:

$$r_d = 0.96$$

$$\frac{\sigma_{v0}}{\sigma'_{v0}} = \frac{58}{43} = 1.35$$

$$\frac{a_{max}}{g} = 0.40$$

And inserting the above values into Eq. (6.6), we see that the cyclic stress ratio due to the anticipated earthquake is 0.34.

The next step is to determine the cyclic resistance ratio of the in situ soil. Entering Fig. 6.6 with $(N_1)_{60}$ = 7.7 and intersecting the curve labeled less than 5 percent fines, we find that the cyclic resistance ratio of the in situ soil at a depth of 3 m is 0.09.

The final step is to calculate the factor of safety against liquefaction by using Eq. (6.8):

$$FS = \frac{CRR}{CSR} = \frac{0.09}{0.34} = 0.26$$

Based on the factor of safety against liquefaction, it is probable that during the anticipated earthquake the in situ sand located at a depth of 3 m below ground surface will liquefy.

6.4.6 Cyclic Resistance Ratio from the Cone Penetration Test

As an alternative to using the standard penetration test, the cone penetration test can be used to determine the cyclic resistance ratio of the in situ soil. The first step is to determine the corrected CPT tip resistance q_{c1} by using Eq. (5.3). Then Fig. 6.8 can be used to determine the cyclic resistance ratio of the in situ soil. The final step is to determine the factor of safety against liquefaction by using Eq. (6.8).

Note that Fig. 6.8 was developed for an anticipated earthquake that has a magnitude of 7.5. The magnitude scaling factors in Table 6.2 or Fig. 6.7 can be used if the anticipated earthquake magnitude is different from 7.5. Figure 6.8 also has different curves that are to be used depending on the percent fines in the soil (F.C. = percent fines in the soil). For a given q_{c1} value, soils with more fines have a higher cyclic resistance ratio. Figure 6.9 presents a chart that can be used to assess the liquefaction of clean gravels (5 percent or less fines) and silty gravels.

6.4.7 Cyclic Resistance Ratio from the Shear Wave Velocity

The shear wave velocity of the soil can also be used to determine the factor of safety against liquefaction. The shear wave velocity can be measured in situ by using several different geophysical techniques, such as the uphole, down-hole, or cross-hole methods. Other methods that can be used to determine the in situ shear wave velocity include the seismic cone penetrometer and suspension logger (see Woods 1994).

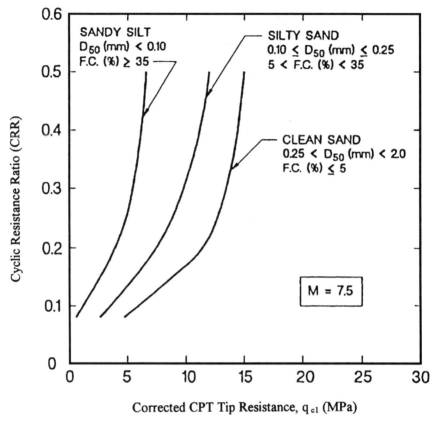

FIGURE 6.8 Relationship between cyclic resistance ratio (CRR) and corrected CPT tip resistance values for clean sand, silty sand, and sandy silt for $M = 7.5$ earthquakes. (*From Stark and Olson 1995, reprinted with permission of the American Society of Civil Engineers.*)

Much like the SPT and CPT, the shear wave velocity is corrected for the overburden pressure by using the following equation (Sykora 1987, Robertson et al. 1992):

$$V_{s1} = V_s C_v = V_s \left(\frac{100}{\sigma'_{v0}} \right)^{0.25} \tag{6.9}$$

where V_{s1} = corrected shear wave velocity (corrected for overburden pressure)
C_v = correction factor to account for overburden pressure. As indicated in the above equation, C_v is approximately equal to $(100/\sigma'_{v0})^{0.25}$, where σ'_{v0} is the vertical effective stress, kPa
V_s = shear wave velocity measured in field

When the shear wave velocity is used to determine the cyclic resistance ratio, Fig. 6.10 is used instead of Figs. 6.6 and 6.8. The curves in Fig. 6.10 are based on field performance data (i.e., sites with liquefaction versus no liquefaction). Figure 6.10 is entered with the corrected shear wave velocity V_{s1} from Eq. (6.9), and then by intersecting the appropriate fines content

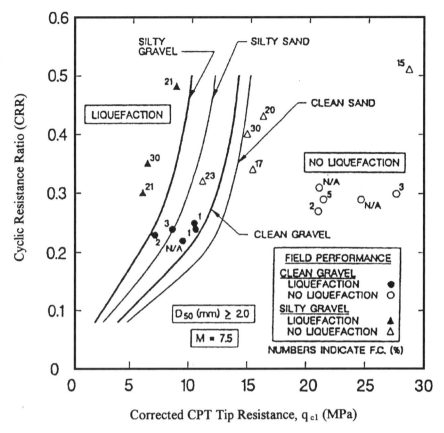

FIGURE 6.9   Relationship between cyclic resistance ratio and corrected CPT tip resistance values for clean gravel and silty gravel for $M = 7.5$ earthquakes. (*From Stark and Olson 1995, reprinted with permission of the American Society of Civil Engineers.*)

curve, the cyclic resistance ratio is obtained. The factor of safety against liquefaction is then calculated by using Eq. (6.8) (or FS = CRR/CSR). Note that Fig. 6.10 was developed for a moment magnitude M_w earthquake of 7.5. The magnitude scaling factors in Table 6.2 or Fig. 6.7 can be used if the anticipated earthquake magnitude is different from 7.5.

An advantage of using the shear wave velocity to determine the factor of safety against liquefaction is that it can be used for very large sites where an initial evaluation of the lique-faction potential is required. Disadvantages of this method are that soil samples are often not obtained as part of the testing procedure, thin strata of potentially liquefiable soil may not be identified, and the method is based on small strains of the soil, whereas the liquefaction process actually involves high strains.

In addition, as indicated in Fig. 6.10, there are few data to accurately define the curves above a CRR of about 0.3. Furthermore, the curves are very steep above a shear wave veloc-ity of 200 m/s, and a small error in measuring the shear wave velocity could result in a sig-nificant error in the factor of safety. For example, an increase in shear wave velocity from 190 to 210 m/s will essentially double the CRR. Because of the limitations of this method, it is best to use the shear wave velocity as a supplement for the SPT and CPT methods.

6.5 *REPORT PREPARATION*

The results of the liquefaction analysis will often need to be summarized in report form for review by the client and governing agency. A listing of the information that should be included in the report, per the *Guidelines for Evaluating and Mitigating Seismic Hazards in California* (Division of Mines and Geology 1997), is as follows:

- If methods other than the standard penetration test (ASTM D 1586-99) and cone penetration test (ASTM D 3441-98) are used, include a description of pertinent equipment and procedural details of field measurements of penetration resistance (i.e., borehole type, hammer type and drop mechanism, sampler type and dimensions, etc.).
- Include boring logs that show raw (unmodified) N values if SPTs are performed or CPT probe logs showing raw q_c values and plots of raw sleeve friction if CPTs are performed.
- Provide an explanation of the basis and methods used to convert raw SPT, CPT, and/or other nonstandard data to "corrected" and "standardized" values [e.g., Eqs. (5.2) and (5.3)].
- Tabulate and/or plot the corrected SPT or corrected CPT values that were used in the liquefaction analysis.
- Provide an explanation of the method used to develop estimates of the design earthquake-induced cyclic stress ratio [e.g., CSR from Eq. (6.6)].
- Similarly, provide an explanation of the method used to develop estimates of the cyclic resistance ratio of the in situ soil (e.g., CRR from Figs. 6.6 to 6.9).
- Determine factors of safety against liquefaction for the design earthquake [e.g. Eq. (6.8)].
- Show the factors of safety against liquefaction at various depths and/or within various potentially liquefiable soil units.
- State conclusions regarding the potential for liquefaction and its likely impact on the proposed project.
- If needed, provide a discussion of mitigation measures necessary to reduce potential damage caused by liquefaction to an acceptable level of risk.
- For projects where remediation has been performed, show criteria for SPT-based or CPT-based acceptable testing that will be used to demonstrate that the site has had satisfactory remediation (see example in Fig. 6.11).

An example of a geotechnical engineering report that includes the results of the liquefaction analysis is provided in App. D.

6.6 *PROBLEMS*

The problems have been divided into basic categories as indicated below:

Soil Type versus Liquefaction Potential

6.1 Figure 6.12 shows laboratory classification data for eight different soils. Note in Fig. 6.12 that W_l is the liquid limit, W_p is the plastic limit, and PI is the plasticity index of the soil. Based on the soil properties summarized in Fig. 6.12, determine if each soil could be susceptible to liquefaction. *Answer:* Soil types 1 through 4 and 7 could be susceptible to liquefaction: soil types 5, 6, and 8 are not susceptible to liquefaction.

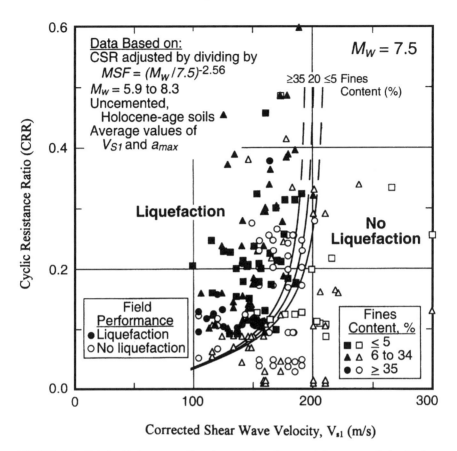

FIGURE 6.10 Relationship between cyclic resistance ratio and corrected shear wave velocity for clean sand, silty sand, and sandy silt for $M = 7.5$ earthquakes. (*From Andrus and Stokoe 2000, reprinted with permission of the American Society of Civil Engineers.*)

Factor of Safety against Liquefaction

6.2 Use the data from the example problem in Sec. 6.4.5, but assume that there is a vertical surcharge pressure applied at ground surface that equals 20 kPa. Determine the cyclic stress ratio induced by the design earthquake. *Answer:* CSR = 0.31.

6.3 Use the data from the example problem in Sec. 6.4.5, but assume that $a_{max}/g = 0.1$ and the sand contains 15 percent nonplastic fines. Calculate the factor of safety against liquefaction. *Answer:* See Table 6.3.

6.4 Use the data from the example problem in Sec. 6.4.5, but assume that $a_{max}/g = 0.2$ and the earthquake magnitude $M = 5\frac{1}{4}$. Calculate the factor of safety against liquefaction. *Answer:* See Table 6.3.

6.5 Use the data from the example problem in Sec. 6.4.5, but assume at a depth of 3 m that $q_c = 3.9$ MPa. Calculate the factor of safety against liquefaction. *Answer:* See Table 6.3.

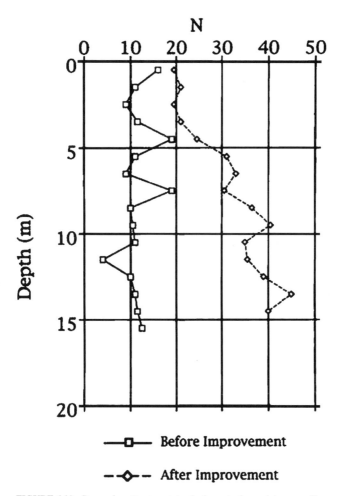

N

—□— **Before Improvement**

—◇— **After Improvement**

FIGURE 6.11   Pre- and posttreatment standard penetration resistance profiles at a warehouse site. (*Reproduced from the Kobe Geotechnical Collection, EERC, University of California, Berkeley.*)

6.6   Use the data from the example problem in Sec. 6.4.5, but assume that the shear wave velocity $V_s = 150$ m/s. Calculate the factor of safety against liquefaction. *Answer:* See Table 6.3.

6.7   Use the data from the example problem in Sec. 6.4.5, but assume that the soil type is crushed limestone (i.e., soil type 1, see Fig. 6.12) and at a depth of 3 m, $q_{c1} = 5.0$ MPa. Calculate the factor of safety against liquefaction. *Answer:* See Table 6.3.

6.8   Use the data from the example problem in Sec. 6.4.5, but assume that the soil type is silty gravel (i.e., soil type 2, see Fig. 6.12) and at a depth of 3 m, $q_{c1} = 7.5$ MPa. Calculate the factor of safety against liquefaction. *Answer:* See Table 6.3.

6.9   Use the data from the example problem in Sec. 6.4.5, but assume that the soil type is gravelly sand (i.e., soil type 3, see Fig. 6.12) and at a depth of 3 m, $q_{c1} = 14$ MPa. Calculate the factor of safety against liquefaction. *Answer:* See Table 6.3.

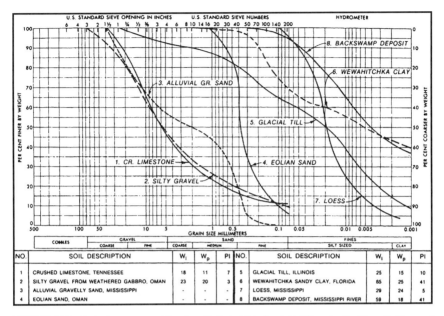

FIGURE 6.12   Grain size curves and Atterberg limits test data for eight different soils. (*From Rollings and Rollings 1996, reproduced with permission of McGraw-Hill, Inc.*)

6.10   Use the data from the example problem in Sec. 6.4.5, but assume that the soil type is eolian sand (i.e., soil type 4, see Fig. 6.12). Calculate the factor of safety against liquefaction. *Answer:* See Table 6.3.

6.11   Use the data from the example problem in Sec. 6.4.5, but assume that the soil type is noncemented loess (i.e., soil type 7, see Fig. 6.12). Calculate the factor of safety against liquefaction. *Answer:* See Table 6.3.

Subsoil Profiles

6.12   Figure 6.13 shows the subsoil profile at Kawagishi-cho in Niigata. Assume a level-ground site with the groundwater table at a depth of 1.5 m below ground surface; the medium sand and medium-fine sand have less than 5 percent fines; the total unit weight γ_t of the soil above the groundwater table is 18.3 kN/m^3; and the buoyant unit weight γ_b of the soil below the groundwater table is 9.7 kN/m^3.

The standard penetration data shown in Fig. 6.13 are uncorrected N values. Assume a hammer efficiency E_m of 0.6 and a boring diameter of 100 mm, and the length of drill rods is equal to the depth of the SPT test below ground surface. The earthquake conditions are a peak ground acceleration a_{max} of 0.16g and a magnitude of 7.5. Using the standard penetration test data, determine the factor of safety against liquefaction versus depth. *Answer:* See App. E for the solution and Fig. 6.14 for a plot of the factor of safety against liquefaction versus depth.

6.13   In Fig. 6.13, assume the cyclic resistance ratio (labeled *cyclic strength* in Fig. 6.13) for the soil was determined by modeling the earthquake conditions in the laboratory (i.e., the amplitude and number of cycles of the sinusoidal load are equivalent to $a_{max} = 0.16g$ and magnitude = 7.5). Using the laboratory cyclic strength tests performed on large-diameter samples, determine the factor of safety against liquefaction versus depth. *Answer:* See

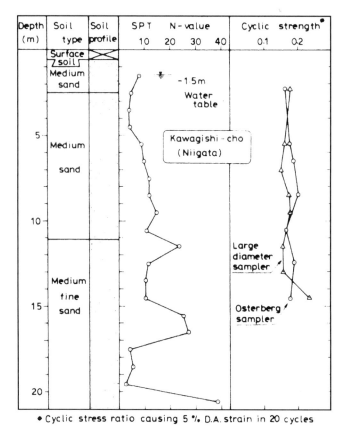

Depth (m)	Soil type	Soil profile	SPT N-value 10 20 30 40	Cyclic strength* 0.1 0.2

* Cyclic stress ratio causing 5 % D.A. strain in 20 cycles

FIGURE 6.13   Subsoil profile, Kawagishi-cho, Niigata. (*Reproduced from Ishihara 1985.*)

App. E for the solution and Fig. 6.14 for a plot of the factor of safety against liquefaction versus depth.

6.14   Based on the results from Probs. 6.12 and 6.13, what zones of soil will liquefy during the earthquake? *Answer:* Per Fig. 6.14, the standard penetration test data indicate that there are three zones of liquefaction from about 2 to 11 m, 12 to 15 m, and 17 to 20 m below ground surface. Per Fig. 6.14, the laboratory cyclic strength tests indicate that there are two zones of liquefaction from about 6 to 8 m and 10 to 14 m below ground surface.

6.15   Figure 6.15 shows the subsoil profile at a sewage disposal site in Niigata. Assume a level-ground site with the groundwater table at a depth of 0.4 m below ground surface, the medium to coarse sand has less than 5 percent fines, the total unit weight γ_t of the soil above the groundwater table is 18.3 kN/m³, and the buoyant unit weight γ_b of the soil below the groundwater table is 9.7 kN/m³.

The standard penetration data shown in Fig. 6.15 are uncorrected N values. Assume a hammer efficiency E_m of 0.6 and a boring diameter of 100 mm, and the length of drill rods is equal to the depth of the SPT test below ground surface. The earthquake conditions are a peak ground acceleration a_{max} of 0.16g and a magnitude of 7.5. Using the standard penetration test data, determine the factor of safety against liquefaction versus depth. *Answer:*

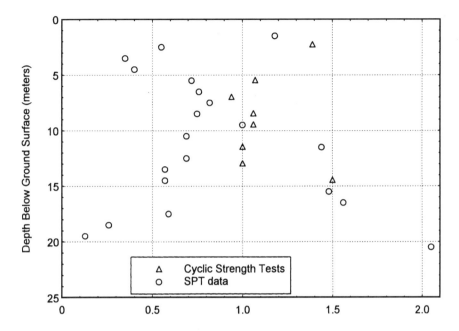

FIGURE 6.14   Solution plot for Probs. 6.12 and 6.13.

See App. E for the solution and Fig. 6.16 for a plot of the factor of safety against liquefaction versus depth.

6.16   In Fig. 6.15, assume the cyclic resistance ratio (labeled *cyclic strength* in Fig. 6.15) for the soil was determined by modeling the earthquake conditions in the laboratory (i.e., the amplitude and number of cycles of the sinusoidal load are equivalent to $a_{max} = 0.16g$ and magnitude = 7.5). Using the laboratory cyclic strength tests performed on block samples, determine the factor of safety against liquefaction versus depth. *Answer:* See App. E for the solution and Fig. 6.16 for a plot of the factor of safety against liquefaction versus depth.

6.17   Based on the results from Probs. 6.15 and 6.16, what zones of soil would be most likely to liquefy? *Answer:* Per Fig. 6.16, the standard penetration test data indicate that there are two zones of liquefaction from about 1.2 to 6.7 m and from 12.7 to 13.7 m below ground surface. Per Fig. 6.16, the laboratory cyclic strength tests indicate that the soil has a factor of safety against liquefaction in excess of 1.0.

Remediation Analysis

6.18   Figure 6.11 presents "before improvement" and "after improvement" standard penetration resistance profiles at a warehouse site. Assume a level-ground site with the groundwater table at a depth of 0.5 m below ground surface, the soil type is a silty sand with an average of 15 percent fines, the total unit weight γ_t of the soil above the groundwater table is 18.9 kN/m³, and the buoyant unit weight γ_b of the soil below the groundwater table is 9.8 kN/m³. Neglect any increase in unit weight of the soil due to the improvement process.

The standard penetration data shown in Fig. 6.11 are uncorrected N values. Assume a hammer efficiency E_m of 0.6 and a boring diameter of 100 mm, and the length of drill rods is equal to the depth of the SPT test below ground surface. The design earthquake condi-

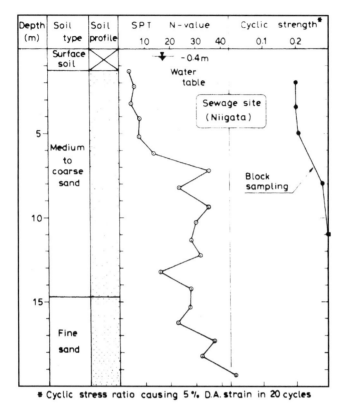

FIGURE 6.15   Subsoil profile, sewage site, Niigata. (*Reproduced from Ishihara
1985.*)

tions are a peak ground acceleration a_{max} of 0.40g and moment magnitude M_w of 8.5.
Determine the factor of safety against liquefaction for the before-improvement and after-
improvement conditions. Was the improvement process effective in reducing the potential
for liquefaction at the warehouse site? *Answer:* See App. E for the solution and Fig. 6.17
for a plot of the factor of safety against liquefaction versus depth. Since the after improvement factor of safety against liquefaction exceeds 1.0 for the design earthquake, the
improvement process was effective in eliminating liquefaction potential at the site.

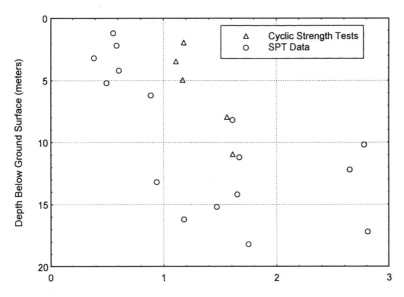

FIGURE 6.16 Solution plot for Probs. 6.15 and 6.16.

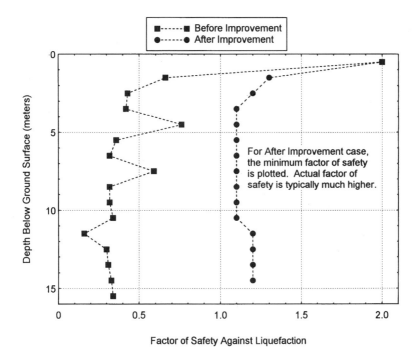

FIGURE 6.17 Solution plot for Prob. 6.18.

TABLE 6.3 Summary of Answers for Probs. 6.3 to 6.11

Problem no.	Soil type	a_{max}/g	Earthquake magnitude	$(N_1)_{60}$ blows/ft; q_{c1}, MPa; V_{s1}, m/s	Cyclic stress ratio	Cyclic resistance ratio	FS = CRR/CSR
Section 6.4.5	Clean sand	0.40	7½	7.7 blows/ft	0.34	0.09	0.26
Problem 6.3	Sand—15% fines	0.10	7½	7.7 blows/ft	0.084	0.14	1.67
Problem 6.4	Clean sand	0.20	5¼	7.7 blows/ft	0.17	0.14	0.82
Problem 6.5	Clean sand	0.40	7½	5.8 MPa	0.34	0.09	0.26
Problem 6.6	Clean sand	0.40	7½	185 m/s	0.34	0.16	0.47
Problem 6.7	Crushed limestone	0.40	7½	5.0 MPa	0.34	0.18	0.53
Problem 6.8	Silty gravel	0.40	7½	7.5 MPa	0.34	0.27	0.79
Problem 6.9	Clean gravelly sand	0.40	7½	14 MPa	0.34	0.44	1.29
Problem 6.10	Eolian sand	0.40	7½	7.7 blows/ft	0.34	0.09	0.26
Problem 6.11	Loess	0.40	7½	7.7 blows/ft	0.34	0.18	0.53

Note: See App. E for solutions.

CHAPTER 7
EARTHQUAKE-INDUCED SETTLEMENT

The following notation is used in this chapter:

SYMBOL   *DEFINITION*

a_{max}, a_p Maximum horizontal acceleration at ground surface (also known as peak ground acceleration)

CSR   Cyclic stress ratio

D_r Relative density

E_m Hammer efficiency

F Lateral force reacting to earthquake-induced base shear

FS, FS_L Factor of safety against liquefaction

g Acceleration of gravity

G_{eff} Effective shear modulus at induced strain level

G_{max} Shear modulus at a low strain level

H Initial thickness of soil layer

H_1 Thickness of surface layer that does not liquefy

H_2 Thickness of soil layer that will liquefy during earthquake

ΔH Change in height of soil layer

k_0 Coefficient of earth pressure at rest

N_{corr} Value added to $(N_1)_{60}$ to account for fines in soil

N Uncorrected SPT blow count (blows per foot)

N_1 Japanese standard penetration test value for Fig. 7.1

$(N_1)_{60}$ N value corrected for field testing procedures and overburden pressure

OCR   Overconsolidation ratio $= \sigma'_{vm}/\sigma'_{v0}$

q_{c1} Cone resistance corrected for overburden pressure

r_d Depth reduction factor

u_e Excess pore water pressure

V Base shear induced by earthquake

Δ Earthquake-induced maximum differential settlement of foundation

ε_v Volumetric strain

γ_{eff} Effective shear strain

γ_{max} Maximum shear strain

γ_t Total unit weight of soil

ρ_{max} Earthquake-induced total settlement of foundation

σ_{v0} Total vertical stress

σ'_m Mean principal effective stress

σ'_{vm} Maximum past pressure, also known as preconsolidation pressure

σ'_{v0} Vertical effective stress

σ'_1 Major principal effective stress

σ'_2 Intermediate principal effective stress

σ'_3 Minor principal effective stress

$\Delta\sigma_v$ Increase in foundation pressure due to earthquake

τ_{cyc} Uniform cyclic shear stress amplitude of earthquake

τ_{eff} Effective shear stress induced by earthquake

τmax Maximum shear stress induced by earthquake

7.1 INTRODUCTION

As discussed in Sec. 4.2, those buildings founded on solid rock are least likely to experience earthquake-induced differential settlement. However, buildings on soil could be subjected to many different types of earthquake-induced settlement. This chapter deals with only settlement of soil for a level-ground surface condition. The types of earthquake-induced settlement discussed in this chapter are as follows:

• *Settlement versus the factor of safety against liquefaction (Sec. 7.2):* This section discusses two methods that can be used to estimate the ground surface settlement for various values of the factor of safety against liquefaction (FS). If FS is less than or equal to 1.0, then liquefaction will occur, and the settlement occurs as water flows from the soil in response to the earthquake-induced excess pore water pressures. Even for FS greater than 1.0, there could still be the generation of excess pore water pressures and hence settlement of the soil. However, the amount of settlement will be much greater for the liquefaction condition compared to the nonliquefied state.

• *Liquefaction-induced ground damage (Sec. 7.3):* There could also be liquefaction-induced ground damage that causes settlement of structures. For example, there could be liquefaction-induced ground loss below the structure, such as the loss of soil through the development of ground surface sand boils. The liquefied soil could also cause the development of ground surface fissures that cause settlement of structures.

• *Volumetric compression (Sec. 7.4):* Volumetric compression is also known as *soil densification.* This type of settlement is due to ground shaking that causes the soil to compress together, such as dry and loose sands that densify during the earthquake.

• *Settlement due to dynamic loads caused by rocking (Sec. 7.5):* This type of settlement is due to dynamic structural loads that momentarily increase the foundation pressure acting on the soil. The soil will deform in response to the dynamic structural load, resulting in settlement of the building. This settlement due to dynamic loads is often a result of the structure rocking back and forth.

The usual approach for settlement analyses is to first estimate the amount of earthquake-induced total settlement ρ_{max} of the structure. Because of variable soil conditions and structural loads, the earthquake-induced settlement is rarely uniform. A common assumption is that the maximum differential settlement Δ of the foundation will be equal to 50 to 75 percent of ρ_{max} (that is, $0.5\rho_{max} \leq \Delta \leq 0.75\rho_{max}$). If the anticipated total settlement ρ_{max} and/or the maximum differential settlement Δ is deemed to be unacceptable, then soil improvement or the construction of a deep foundation may be needed. Chapters 12 and 13 deal with mitigation measures such as soil improvement or the construction of deep foundations.

7.2 SETTLEMENT VERSUS FACTOR OF SAFETY AGAINST LIQUEFACTION

7.2.1 Introduction

This section discusses two methods that can be used to estimate the ground surface settlement for various values of the factor of safety against liquefaction. A liquefaction analysis (Chap. 6) is first performed to determine the factor of safety against liquefaction. If FS is less than or equal to 1.0, then liquefaction will occur, and the settlement occurs as water flows from the soil in response to the earthquake-induced excess pore water pressures. Even for FS greater than 1.0, there could still be the generation of excess pore water pressures and hence settlement of the soil. However, the amount of settlement will be much greater for the liquefaction condition compared to the nonliquefied state.

This section is solely devoted to an estimation of ground surface settlement for various values of the factor of safety. Other types of liquefaction-induced movement, such as bearing capacity failures, flow slides, and lateral spreading, are discussed in Chaps. 8 and 9.

7.2.2 Methods of Analysis

Method by Ishihara and Yoshimine (1992). Figure 7.1 shows a chart developed by Ishihara and Yoshimine (1992) that can be used to estimate the ground surface settlement of saturated clean sands for a given factor of safety against liquefaction. The procedure for using Fig. 7.1 is as follows:

1. *Calculate the factor of safety against liquefaction* FS_L: The first step is to calculate the factor of safety against liquefaction, using the procedure outlined in Chap. 6 [i.e., Eq. (6.8)].

2. *Soil properties:* The second step is to determine one of the following properties: relative density D_r of the in situ soil, maximum shear strain to be induced by the design earthquake γ_{max}, corrected cone penetration resistance q_{c1} kg/cm^2, or Japanese standard penetration test N_1 value.

Kramer (1996) indicates that the Japanese standard penetration test typically transmits about 20 percent more energy to the SPT sampler, and the equation $N_1 = 0.83(N_1)_{60}$ can be used to convert the $(N_1)_{60}$ value to the Japanese N_1 value. However, R. B. Seed (1991) states that Japanese SPT results require corrections for blow frequency effects and hammer release, and that these corrections are equivalent to an overall effective energy ratio E_m of 0.55 (versus $E_m = 0.60$ for U.S. safety hammer). Thus R. B. Seed (1991) states that the $(N_1)_{60}$ values should be increased by about 10 percent (that is, 0.6/0.55) when using Fig. 7.1 to estimate volumetric compression, or $N_1 = 1.10(N_1)_{60}$. As a practical matter, it can be

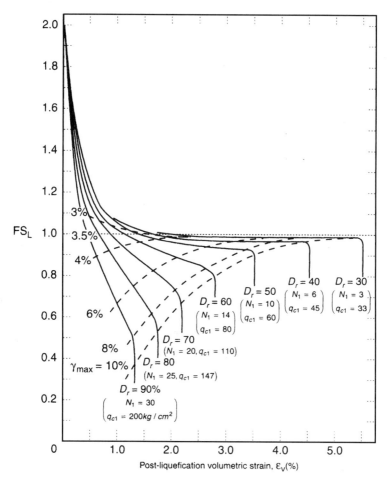

FIGURE 7.1 Chart for estimating the ground surface settlement of clean sand as a function of the factor of safety against liquefaction FS_L. To use this figure, one of the following properties must be determined: relative density D_r of the in situ soil, maximum shear strain to be induced by the design earthquake γ_{max}, corrected cone penetration resistance q_{c1} (kg/cm²), or Japanese standard penetration test N_1 value. For practical purposes, assume the Japanese standard penetration test N_1 value is equal to the $(N_1)_{60}$ value from Eq. (5.2). (*Reproduced from Kramer 1996, originally developed by Ishihara and Yoshimine 1992.*)

assumed that the Japanese N_1 value is approximately equivalent to the $(N_1)_{60}$ value calculated from Eq. (5.2) (Sec. 5.4.3).

3. *Volumetric strain:* In Fig. 7.1, enter the vertical axis with the factor of safety against liquefaction, intersect the appropriate curve corresponding to the Japanese N_1 value [assume Japanese $N_1 = (N_1)_{60}$ from Eq. (5.2)], and then determine the volumetric strain ε_v from the horizontal axis. Note in Fig. 7.1 that each N_1 curve can be extended straight downward to obtain the volumetric strain for very low values of the factor of safety against liquefaction.

4. *Settlement:* The settlement of the soil is calculated as the volumetric strain, expressed as a decimal, times the thickness of the liquefied soil layer.

Note in Fig. 7.1 that the volumetric strain can also be calculated for clean sand that has a factor of safety against liquefaction in excess of 1.0. For FS_L greater than 1.0 but less than 2.0, the contraction of the soil structure during the earthquake shaking results in excess pore water pressures that will dissipate and cause a smaller amount of settlement. At FS_L equal to or greater than 2.0, Fig. 7.1 indicates that the volumetric strain will be essentially equal to zero. This is because for FS_L higher than 2.0, only small values of excess pore water pressures u_e will be generated during the earthquake shaking (i.e., see Fig. 5.15).

Method by Tokimatsu and Seed (1984, 1987).   Figure 7.2 shows a chart developed by Tokimatsu and Seed (1984, 1987) that can be used to estimate the ground surface settlement of saturated clean sands. The solid lines in Fig. 7.2 represent the volumetric strain for liquefied soil (i.e., factor of safety against liquefaction less than or equal to 1.0). Note that the solid line labeled 1 percent volumetric strain in Fig. 7.2 is similar to the dividing line in Fig. 6.6 between liquefiable and nonliquefiable clean sand.

The dashed lines in Fig. 7.2 represent the volumetric strain for a condition where excess pore water pressures are generated during the earthquake, but the ground shaking is not sufficient to cause liquefaction (that is, FS > 1.0). This is similar to the data in Fig. 7.1, in that

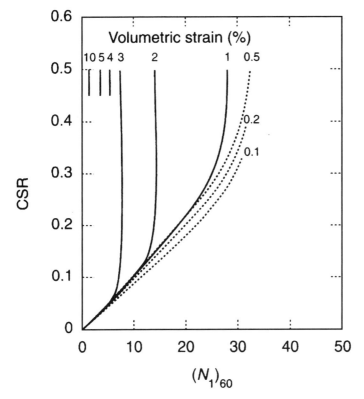

FIGURE 7.2   Chart for estimating the ground surface settlement of clean sand for factor of safety against liquefaction less than or equal to 1.0 (solid lines) and greater than 1.0 (dashed lines). To use this figure, the cyclic stress ratio from Eq. (6.6) and the $(N_1)_{60}$ value from Eq. (5.2) must be determined. (*Reproduced from Kramer 1996, originally developed by Tokimatsu and Seed 1984.*)

the contraction of the soil structure during the earthquake shaking could cause excess pore water pressures that will dissipate and result in smaller amounts of settlement. Thus by using the dashed lines in Fig. 7.2, the settlement of clean sands having a factor of safety against liquefaction in excess of 1.0 can also be calculated. The procedure for using Fig. 7.2 is as follows:

1. *Calculate the cyclic stress ratio:* The first step is to calculate the cyclic stress ratio (CSR) by using Eq. (6.6). Usually a liquefaction analysis (Chap. 6) is first performed, and thus the value of CSR should have already been calculated.

2. *Adjusted CSR value:* Figure 7.2 was developed for a magnitude 7.5 earthquake. Tokimatsu and Seed (1987) suggest that the cyclic stress ratio calculated from Eq. (6.6) be adjusted if the magnitude of the anticipated earthquake is different from 7.5. The corrected CSR value is obtained by dividing the CSR value from Eq. (6.6) by the magnitude scaling factor from Table 6.2. The chart in Fig. 7.2 is entered on the vertical axis by using this corrected CSR value.

As will be illustrated by the following example problem, applying an earthquake magnitude correction factor to the cyclic stress ratio is usually unnecessary. The reason is that once liquefication has occurred, a higher magnitude earthquake will not result in any additional settlement of the liquefied soil. Thus as a practical matter, the chart in Fig. 7.2 can be entered on the vertical axis with the CSR value from Eq. (6.6).

3. $(N_1)_{60}$ *value:* Now calculate the $(N_1)_{60}$ value [Eq. (5.2), see Sec. 5.4.3]. Usually a liquefaction analysis (Chap. 6) is first performed, and thus the value of $(N_1)_{60}$ should have already been calculated.

4. *Volumetric strain:* In Fig. 7.2, the volumetric strain is determined by entering the vertical axis with the CSR from Eq. (6.6) and entering the horizontal axis with the $(N_1)_{60}$ value from Eq. (5.2).

5. *Settlement:* The settlement of the soil is calculated as the volumetric strain, expressed as a decimal, times the thickness of the liquefied soil layer.

Example Problem. This example problem illustrates the procedure used to determine the ground surface settlement of soil using Figs. 7.1 and 7.2.

Use the data from the example problem in Sec. 6.4.5. Assume that the liquefied soil layer is 1.0 m thick. As indicated in Sec. 6.4.5, the factor of safety against liquefaction is 0.26, and the calculated value of $(N_1)_{60}$ determined at a depth of 3 m below ground surface is equal to 7.7.

- *Solution using Fig. 7.1:* For Fig. 7.1, assume that the Japanese N_1 value is approximately equal to the $(N_1)_{60}$ value from Eq. (5.2), or use Japanese $N_1 = 7.7$. The Japanese N_1 curves labeled 6 and 10 are extended straight downward to FS = 0.26, and then by extrapolating between the curves for an N_1 value of 7.7, the volumetric strain is equal to 4.1 percent. Since the in situ liquefied soil layer is 1.0 m thick, the ground surface settlement of the liquefied soil is equal to 1.0 m times 0.041, or a settlement of 4.1 cm.

- *Solution using Fig. 7.2:* Per the example problem in Sec. 6.4.5, the cyclic stress ratio from Eq. (6.6) is equal to 0.34, and the calculated value of $(N_1)_{60}$ determined at a depth of 3 m below ground surface is equal to 7.7. Entering Fig. 7.2 with CSR = 0.34 and $(N_1)_{60}$ = 7.7, the volumetric strain is equal to 3.0 percent. Since the in situ liquefied soil layer is 1.0 m thick, the ground surface settlement of the liquefied soil is equal to 1.0 m times 0.030, or a settlement of 3.0 cm.

Suppose instead of assuming the earthquake will have a magnitude of 7.5, the example problem is repeated for a magnitude $5^1/4$ earthquake. As indicated in Table 6.2, the magnitude scaling factor $= 1.5$, and thus the corrected CSR is equal to 0.34 divided by 1.5, or 0.23. Entering Fig. 7.2 with the modified CSR $= 0.23$ and $(N_1)_{60} = 7.7$, the volumetric strain is still equal to 3.0 percent. Thus, provided the sand liquefies for both the magnitude $5\frac{1}{4}$ and magnitude 7.5 earthquakes, the settlement of the liquefied soil is the same.

• *Summary of values:*   Based on the two methods, the ground surface settlement of the 1.0-m-thick liquefied sand layer is expected to be on the order of 3 to 4 cm.

Silty Soils.   Figures 7.1 and 7.2 were developed for clean sand deposits (fines ≤ 5 percent). For silty soils, R. B. Seed (1991) suggests that the most appropriate adjustment is to increase the $(N_1)_{60}$ values by adding the values of N_{corr} indicated below:

Percent fines	N_{corr}
≤ 5	0
10	1
25	2
50	4
75	5

7.2.3 Limitations

The methods presented in Figs. 7.1 and 7.2 can only be used for the following cases:

• *Lightweight structures:*   Settlement of lightweight structures, such as wood-frame buildings bearing on shallow foundations

• *Low net bearing stress:*   Settlement of any other type of structure that imparts a low net bearing pressure onto the soil

• *Floating foundation:*   Settlement of floating foundations, provided the zone of liquefaction is below the bottom of the foundation and the floating foundation does not impart a significant net stress upon the soil

• *Heavy structures with deep liquefaction:*   Settlement of heavy structures, such as massive buildings founded on shallow foundations, provided the zone of liquefaction is deep enough that the stress increase caused by the structural load is relatively low

• *Differential settlement:*   Differential movement between a structure and adjacent appurtenances, where the structure contains a deep foundation that is supported by strata below the zone of liquefaction

The methods presented in Figs. 7.1 and 7.2 cannot be used for the following cases:

• *Foundations bearing on liquefiable soil:*   Do not use Figs. 7.1 and 7.2 when the foundation is bearing on soil that will liquefy during the design earthquake. Even lightly loaded foundations will sink into the liquefied soil.

• *Heavy buildings with underlying liquefiable soil:*   Do not use Figs. 7.1 and 7.2 when the liquefied soil is close to the bottom of the foundation and the foundation applies a large net load onto the soil. In this case, once the soil has liquefied, the foundation load will cause it to punch or sink into the liquefied soil. There could even be a bearing capacity type of failure. Obviously these cases will lead to settlement well in excess of the values obtained from Figs. 7.1 and 7.2. It is usually very difficult to determine the settlement for these conditions, and the best engineering solution is to provide a sufficiently high static factor of safety so that there is ample resistance against a bearing capacity failure. This is discussed further in Chap. 8.

- *Buoyancy effects:* Consider possible buoyancy effects. Examples include buried storage tanks or large pipelines that are within the zone of liquefied soil. Instead of settling, the buried storage tanks and pipelines may actually float to the surface when the ground liquefies.

- *Sloping ground condition:* Do not use Figs. 7.1 and 7.2 when there is a sloping ground condition. If the site is susceptible to liquefaction-induced flow slide or lateral spreading, the settlement of the building could be well in excess of the values obtained from Figs. 7.1 and 7.2. This is discussed further in Chap. 9.

- *Liquefaction-induced ground damage:* The calculations using Figs. 7.1 and 7.2 do not include settlement that is related to the loss of soil through the development of ground surface sand boils or the settlement of shallow foundations caused by the development of ground surface fissures. These types of settlement are discussed in the next section.

7.3   LIQUEFACTION-INDUCED GROUND DAMAGE

7.3.1   Types of Damage

As previously mentioned, there could also be liquefaction-induced ground damage that causes settlement of structures. This liquefaction-induced ground damage is illustrated in Fig. 7.3. As shown, there are two main aspects to the ground surface damage:

1. *Sand boils:* There could be liquefaction-induced ground loss below the structure, such as the loss of soil through the development of ground surface sand boils. Often a line of sand boils, such as shown in Fig. 7.4, is observed at ground surface. A row of sand boils often develops at the location of cracks or fissures in the ground.

2. *Surface fissures:* The liquefied soil could also cause the development of ground surface fissures which break the overlying soil into blocks that open and close during the earthquake. Figure 7.5 shows the development of one such fissure. Note in Fig. 7.5 that liquefied soil actually flowed out of the fissure.

The liquefaction-induced ground conditions illustrated in Fig. 7.3 can damage all types of structures, such as buildings supported on shallow foundations, pavements, flatwork, and utilities. In terms of the main factor influencing the liquefaction-induced ground dam-

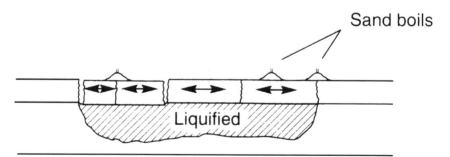

FIGURE 7.3   Ground damage caused by the liquefaction of an underlying soil layer. (*Reproduced from Kramer 1996, originally developed by Youd 1984.*)

FIGURE 7.4 Line of sand boils caused by liquefaction during the Niigata (Japan) earthquake of June 16, 1964. (*Photograph from the Steinbrugge Collection, EERC, University of California, Berkeley.*)

age, Ishihara (1985) states:

> One of the factors influencing the surface manifestation of liquefaction would be the thickness of a mantle of unliquefied soils overlying the deposit of sand which is prone to liquefaction. Should the mantle near the ground surface be thin, the pore water pressure from the underlying liquefied sand deposit will be able to easily break through the surface soil layer, thereby bringing about the ground rupture such as sand boiling and fissuring. On the other hand, if the mantle of the subsurface soil is sufficiently thick, the uplift force due to the excess water pressure will not be strong enough to cause a breach in the surface layer, and hence, there will be no surface manifestation of liquefaction even if it occurs deep in the deposit.

7.3.2 Method of Analysis

Based on numerous case studies, Ishihara (1985) developed a chart (Fig. 7.6*a*) that can be used to determine the thickness of the unliquefiable soil surface layer H_1 in order to prevent damage due to sand boils and surface fissuring. Three different situations were used by

FIGURE 7.5 Surface fissure caused by the Izmit earthquake in Turkey on August 17, 1999. Note that liquefied soil flowed out of the fissure. (*Photograph from the Izmit Collection, EERC, University of California, Berkeley.*)

Ishihara (1985) in the development of the chart, and they are shown in Fig. 7.6*b*.

Since it is very difficult to determine the amount of settlement due to liquefaction-induced ground damage (Fig. 7.3), one approach is to ensure that the site has an adequate surface layer of unliquefiable soil by using Fig. 7.6. If the site has an inadequate surface layer of unliquefiable soil, then mitigation measures such as the placement of fill at ground surface, soil improvement, or the construction of deep foundations may be needed (Chaps. 12 and 13).

To use Fig. 7.6, the thickness of layers H_1 and H_2 must be determined. Guidelines are as follows:

1. *Thickness of the unliquefiable soil layer H_1:* For two of the three situations in Fig. 7.6*b,* the unliquefiable soil layer is defined as that thickness of soil located above the groundwater table. As previously mentioned in Sec. 6.3, soil located above the groundwater table will not liquefy.

One situation in Fig. 7.6*b* is for a portion of the unliquefiable soil below the groundwater table. Based on the case studies, this soil was identified as unliquefiable cohesive soil (Ishihara 1985). As a practical matter, it would seem the "unliquefiable soil" below the

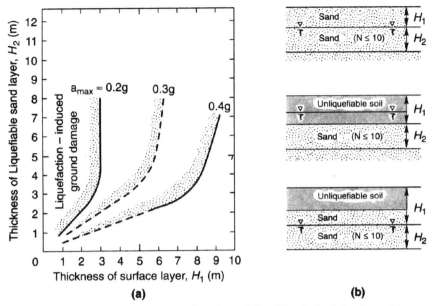

FIGURE 7.6 (*a*) Chart that can be used to evaluate the possibility of liquefaction-induced ground damage based on H_1, H_2, and the peak ground acceleration a_{max}. (*b*) Three situations used for the development of the chart, where H_1 = thickness of the surface layer that will not liquefy during the earthquake and H_2 = thickness of the liquefiable soil layer. (*Reproduced from Kramer 1996, originally developed by Ishihara 1985.*)

groundwater table that is used to define the layer thickness H_1 would be applicable for any soil that has a factor of safety against liquefaction in excess of 1.0. However, if the factor of safety against liquefaction is only slightly in excess of 1.0, it could still liquefy due to the upward flow of water from layer H_2. Considerable experience and judgment are required in determining the thickness H_1 of the unliquefiable soil when a portion of this layer is below the groundwater table.

2. *Thickness of the liquefied soil layer H_2:*   Note in Fig. 7.6*b* that for all three situations, the liquefied sand layer H_2 has an uncorrected N value that is less than or equal to 10. These N value data were applicable for the case studies evaluated by Ishihara (1985). It would seem that irrespective of the N value, H_2 could be the thickness of the soil layer which has a factor of safety against liquefaction that is less than or equal to 1.0.

7.3.3 Example Problem

This example problem illustrates the use of Fig. 7.6. Use the data from Prob. 6.15, which deals with the subsurface conditions shown in Fig. 6.15 for the sewage disposal site. Based on the standard penetration test data, the zone of liquefaction extends from a depth of 1.2 to 6.7 m below ground surface. Assume the surface soil (upper 1.2 m) shown in Fig. 6.15 consists of an unliquefiable soil. Using a peak ground acceleration a_{max} of 0.20g, will there be liquefaction-induced ground damage at this site?

Solution.　Since the zone of liquefaction extends from a depth of 1.2 to 6.7 m, the thickness of the liquefiable sand layer H_2 is equal to 5.5 m. By entering Fig. 7.6 with $H_2 = 5.5$ m and intersecting the $a_{max} = 0.2g$ curve, the minimum thickness of the surface layer H_1 needed to prevent surface damage is 3 m. Since the surface layer of unliquefiable soil is only 1.2 m thick, there will be liquefaction-induced ground damage.

Some appropriate solutions would be as follows: (1) At ground surface, add a fill layer that is at least 1.8 m thick, (2) densify the sand and hence improve the liquefaction resistance of the upper portion of the liquefiable layer, or (3) use a deep foundation supported by soil below the zone of liquefaction.

7.4 VOLUMETRIC COMPRESSION

7.4.1 Main Factors Causing Volumetric Compression

Volumetric compression is also known as soil densification. This type of settlement is due to earthquake-induced ground shaking that causes the soil particles to compress together. Noncemented cohesionless soils, such as dry and loose sands or gravels, are susceptible to this type of settlement. Volumetric compression can result in a large amount of ground surface settlement. For example, Grantz et al. (1964) describe an interesting case of ground vibrations from the 1964 Alaskan earthquake that caused 0.8 m (2.6 ft) of alluvium settlement.

Silver and Seed (1971) state that the earthquake-induced settlement of dry cohesionless soil depends on three main factors:

1. *Relative density D_r of the soil:*　The looser the soil, the more susceptible it is to volumetric compression. Those cohesionless soils that have the lowest relative densities will be most susceptible to soil densification. Often the standard penetration test is used to assess the density condition of the soil.

2. *Maximum shear strain γ_{max} induced by the design earthquake:*　The larger the shear strain induced by the earthquake, the greater the tendency for a loose cohesionless soil to compress. The amount of shear strain will depend on the peak ground acceleration a_{max}. A higher value of a_{max} will lead to a greater shear strain of the soil.

3. *Number of shear strain cycles:*　The more cycles of shear strain, the greater the tendency for the loose soil structure to compress. For example, it is often observed that the longer a loose sand is vibrated, the greater the settlement. The number of shear strain cycles can be related to the earthquake magnitude. As indicated in Table 2.2, the higher the earthquake magnitude, the longer the duration of ground shaking.

In summary, the three main factors that govern the settlement of loose and dry cohesionless soil are the relative density, amount of shear strain, and number of shear strain cycles. These three factors can be accounted for by using the standard penetration test, peak ground acceleration, and earthquake magnitude.

7.4.2 Simple Settlement Chart

Figure 7.7 presents a simple chart that can be used to estimate the settlement of dry sand (Krinitzsky et al. 1993). The figure uses the standard penetration test N value and the peak ground acceleration a_p to calculate the earthquake-induced volumetric strain (that is, $\Delta H/H$, expressed as a percentage). Figure 7.7 accounts for two of the three main factors causing volumetric compression: the looseness of the soil based on the standard penetration test and the amount of shear strain based on the peak ground acceleration a_p.

Note in Fig. 7.7 that the curves are labeled in terms of the uncorrected N values. As a practical matter, the curves should be in terms of the standard penetration test $(N_1)_{60}$ values [i.e., Eq. (5.2), Sec. 5.4.3]. This is because the $(N_1)_{60}$ value more accurately represents the density condition of the sand. For example, given two sand layers having the same uncorrected N value, the near-surface sand layer will be in a much denser state than the sand layer located at a great depth.

To use Fig. 7.7, both the $(N_1)_{60}$ value of the sand and the peak ground acceleration a_p must be known. Then by entering the chart with the a_p/g value and intersecting the desired $(N_1)_{60}$ curve, the volumetric strain ($\Delta H/H$, expressed as a percentage) can be determined. The volumetric compression (i.e., settlement) is then calculated by multiplying the volumetric strain, expressed as a decimal, by the thickness of the soil layer H.

7.4.3 Method by Tokimatsu and Seed

A much more complicated method for estimating the settlement of dry sand has been proposed by Tokimatsu and Seed (1987), based on the prior work by Seed and Silver (1972) and Pyke et al. (1975). The steps in using this method are as follows:

1. *Determine the earthquake-induced effective shear strain* γ_{eff}. The first step is to determine the shear stress induced by the earthquake and then to convert this shear stress to an effective shear strain γ_{eff}. Using Eq. (6.6) and deleting the vertical effective stress σ'_{v0} from both sides of the equation gives

$$\tau_{cyc} = 0.65 r_d \sigma_{v0} (a_{max}/g) \qquad (7.1)$$

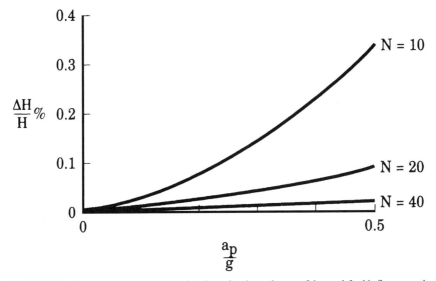

FIGURE 7.7 Simple chart that can be used to determine the settlement of dry sand. In this figure, use the peak ground acceleration a_p and assume that N refers to $(N_1)_{60}$ values from Eq. (5.2). (*Reproduced from Krinitzsky et al. 1993, with permission from John Wiley & Sons.*)

where τ_{cyc} = uniform cyclic shear stress amplitude of the earthquake

r_d = depth reduction factor, also known as *stress reduction coefficient* (dimensionless). Equation (6.7) or Fig. 6.5 can be used to obtain the value of r_d.

σ_{v0} = total vertical stress at a particular depth where the settlement analysis is being performed, lb/ft^2 or kPa. To calculate total vertical stress, total unit weight γ_t of soil layer (s) must be known.

a_{max} = maximum horizontal acceleration at ground surface that is induced by the earthquake, ft/s^2 or m/s^2, which is also commonly referred to as the peak ground acceleration (see Sec. 5.6)

g = acceleration of gravity (32.2 ft/s^2 or 9.81 m/s^2)

As discussed in Chap. 6, Eq. (7.1) was developed by converting the typical irregular earthquake record to an equivalent series of uniform stress cycles by assuming that τ_{cyc} = $0.65\tau_{max}$, where τ_{max} is equal to the maximum earthquake-induced shear stress. Thus τ_{cyc} is the amplitude of the uniform stress cycles and is considered to be the effective shear stress induced by the earthquake (that is, $\tau_{eff} = \tau_{cyc}$). To determine the earthquake-induced effective shear strain, the relationship between shear stress and shear strain can be utilized:

$$\tau_{cyc} = \tau_{eff} = \gamma_{eff}G_{eff} \tag{7.2}$$

where τ_{eff} = effective shear stress induced by the earthquake, which is considered to be equal to the amplitude of uniform stress cycles used to model earthquake motion ($\tau_{cyc} = \tau_{eff}$), lb/ft^2 or kPa

γ_{eff} = effective shear strain that occurs in response to the effective shear stress (dimensionless)

G_{eff} = effective shear modulus at induced strain level, lb/ft^2 or kPa

Substituting Eq. (7.2) into (7.1) gives

$$\gamma_{eff}G_{eff} = 0.65r_d\sigma_{v0}\,(a_{max}/g) \tag{7.3}$$

And finally, dividing both sides of the equation by G_{max}, which is defined as the shear modulus at a low strain level, we get as the final result

$$\gamma_{eff}\left(\frac{G_{eff}}{G_{max}}\right) = 0.65r_d\left(\frac{\sigma_{v0}}{G_{max}}\right)\left(\frac{a_{max}}{g}\right) \tag{7.4}$$

Similar to the liquefaction analysis in Chap. 6, all the parameters on the right side of the equation can be determined except for G_{max}. Based on the work by Ohta and Goto (1976) and Seed et al. (1984, 1986), Tokimatsu and Seed (1987) recommend that the following equation be used to determine G_{max}:

$$G_{max} = 20{,}000\,[(N_1)_{60}]^{0.333}\,(\sigma'_m)^{0.50} \tag{7.5}$$

where G_{max} = shear modulus at a low strain level, lb/ft^2

$(N_1)_{60}$ = standard penetration test N value corrected for field testing procedures and overburden pressure [i.e., Eq. (5.2)]

σ'_m = mean principal effective stress, defined as the average of the sum of the three principal effective stresses, or $(\sigma_1' + \sigma_2' + \sigma_3')/3$. For a geostatic condition and a sand deposit that has not been preloaded (i.e., OCR = 1.0), the coefficient of earth pressure at rest $k_0 \cong 0.5$. Thus the value of $\sigma'_m \cong 0.67\sigma\sigma'_{v0}$. Note in Eq. (7.5) that the value of σ'_m must be in terms of pounds per square foot.

After the value of G_{max} has been determined from Eq. (7.5), the value of γ_{eff} (G_{eff}/G_{max}) can be calculated by using Eq. (7.4). To determine the effective shear strain γ_{eff} of the soil, Fig. 7.8 is entered with the value of $\gamma_{eff}(G_{eff}/G_{max})$ and upon intersecting the appropriate value of mean principal effective stress (σ'_m in ton/ft²), the effective shear strain γ_{eff} is obtained from the vertical axis.

2. *Determine the volumetric strain ε_v.* Figure 7.9 can be used to determine the volumetric strain ε_v of the soil. This figure was developed for cases involving 15 equivalent uniform strain cycles, which is representative of a magnitude 7.5 earthquake. In Fig. 7.9, the cyclic shear strain γ_{cyc} is equivalent to the effective shear strain γ_{eff} calculated from step 1, except that the cyclic shear strain γ_{cyc} is expressed as a percentage (%γ_{cyc} = 100 γ_{eff}). To determine the volumetric strain ε_v in percent, either the relative density D_r of the in situ soil or data from the standard penetration test must be known. For Fig. 7.9, assume the N_1 in the figure refers to $(N_1)_{60}$ values from Eq. (5.2).

To use Fig. 7.9, first convert γ_{eff} from step 1 to percent cyclic shear strain (%γ_{cyc} = 100γ_{eff}). Then enter the horizontal axis with percent γ_{cyc}, and upon intersecting the relative

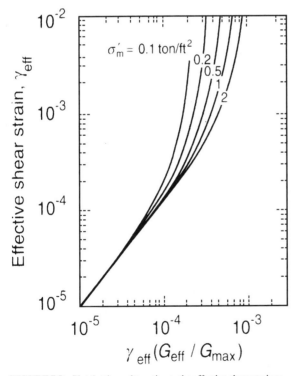

FIGURE 7.8 Plot that is used to estimate the effective shear strain γ_{eff} for values of $\gamma_{eff}(G_{eff}/G_{max})$ from Eq. (7.4) and the mean principal effective stress σ'_m. *(Reproduced from Tokimatsu and Seed 1987, with permission from the American Society of Civil Engineers.)*

density D_r curve or the $(N_1)_{60}$ curve, the value of the volumetric strain ε_v is obtained from the vertical axis.

3. *Multidirectional shear:*   The development of Fig. 7.9 was based on unidirectional simple shear conditions, or in other words, shear strain in only one direction. However, actual earthquake shaking conditions are multidirectional, where the soil is strained back and forth. Based on unidirectional and multidirectional tests, Pyke et al. (1975) conclude that "the settlements caused by combined horizontal motions are about equal to the sum of the settlements caused by the components acting alone." Therefore, the unidirectional volumetric strains determined from Fig. 7.9 must be doubled to account for the multidirectional shaking effects of the earthquake.

4. *Magnitude of the earthquake:*   Figure 7.9 was developed for a magnitude 7.5 earthquake (that is, 15 cycles at $0.65\tau_{max}$). Table 7.1 presents the volumetric strain ratio that can be used to determine the volumetric strain ε_v for different-magnitude earthquakes. The procedure is to multiply the volumetric strain ε_v from step 3 by the volumetric strain ratio VSR from Table 7.1.

Note that the volumetric strain ratio is similar in concept to the magnitude scaling factor (MSF) in Table 6.2. It would seem that the volumetric strain ratio in Table 7.1 should be equal to the inverse of the magnitude scaling factors in Table 6.2 (that is, VSR = 1.0/MSF). However, they do not equate because the correction in Table 7.1 is made for volumetric strain, while the correction in Table 6.2 is made for shear stress.

5. *Settlement:*   Because of the variations in soil properties with depth, the soil profile should be divided into several different layers. The volumetric strain from step 4 is then calculated for each layer. The settlement for each layer is the volumetric strain, expressed as a decimal, times the thickness of the layer. The total settlement is calculated as the sum of the settlement calculated for each soil layer.

Section 7.4.4 presents an example problem illustrating the various steps outlined above. This method proposed by Tokimatsu and Seed (1987) is most applicable for dry sands that

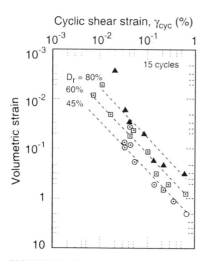
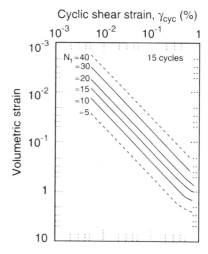

FIGURE 7.9 Plots that can be used to estimate the volumetric strain ε_v based on the cyclic shear strain γ_{cyc} and relative density D_r or N_1 value. Assume that N_1 in this figure refers to the $(N_1)_{60}$ values from Eq. (5.2). (*Reproduced from Tokimatsu and Seed 1987, with permission from the American Society of Civil Engineers.*)

TABLE 7.1 Earthquake Magnitude versus Volumetric Strain Ratio for Dry Sands

Earthquake magnitude	Number of representative cycles at $0.65\tau_{max}$	Volumetric strain ratio
$8\frac{1}{2}$	26	1.25
$7\frac{1}{2}$	15	1.00
$6\frac{3}{4}$	10	0.85
6	5	0.60
$5\frac{1}{4}$	2–3	0.40

Notes: To account for the earthquake magnitude, multiply the volumetric strain ε_v from Fig. 7.9 by the VSR. Data were obtained from Tokimatsu and Seed (1987).

have 5 percent or less fines. For dry sands (i.e., water content = 0 percent), capillary action does not exist between the soil particles. As the water content of the sand increases, capillary action produces a surface tension that holds together the soil particles and increases their resistance to earthquake-induced volumetric settlement. As a practical matter, clean sands typically have low capillarity and thus the method by Tokimatsu and Seed (1987) could also be performed for damp and moist sands.

For silty soils, R. B. Seed (1991) suggests that the most appropriate adjustment is to increase the $(N_1)_{60}$ values by adding the values of N_{corr} indicated in Sec. 7.2.2.

7.4.4 Example Problem

Silver and Seed (1972) investigated a 50-ft- (15-m-) thick deposit of dry sand that experienced about $2\frac{1}{2}$ in (6 cm) of volumetric compression caused by the San Fernando earthquake of 1971. They indicated that the magnitude 6.6 San Fernando earthquake subjected the site to a peak ground acceleration a_{max} of $0.45g$. The sand deposit has a total unit weight $\gamma_t = 95$ lb/ft^3 (15 kN/m^3) and an average $(N_1)_{60} = 9$. Estimate the settlement of this 50-ft- (15-m-) thick sand deposit using the methods outlined in Secs. 7.4.2 and 7.4.3.

Solution Using Fig. 7.7. As shown in Fig. 7.7, the volumetric compression rapidly increases as the $(N_1)_{60}$ value decreases. Since the peak ground acceleration $a_p = 0.45g$, the horizontal axis is entered at 0.45. For an $(N_1)_{60}$ value of 9, the volumetric strain $\Delta H/H$ is about equal to 0.35 percent. The ground surface settlement is obtained by multiplying the volumetric strain, expressed as a decimal, by the thickness of the sand layer, or $0.0035 \times$ 50 ft = 0.18 ft or 2.1 in (5.3 cm).

Solution Using the Tokimatsu and Seed (1987) Method. Table 7.2 presents the solution using the Tokimatsu and Seed (1987) method as outlined in Sec. 7.4.3. The steps are as follows:

1. *Layers:* The soil was divided into six layers.

2. *Thickness of the layers:* The upper two layers are 5.0 ft (1.5 m) thick, and the lower four layers are 10 ft (3.0 m) thick.

3. *Vertical effective stress:* For dry sand, the pore water pressures are zero and the vertical effective stress σ'_{v0} is equal to the vertical total stress σ_v. This stress was calculated by multiplying the total unit weight ($\gamma_t = 95$ lb/ft^3) by the depth to the center of each layer.

TABLE 7.2 Settlement Calculations Using the Tokimatsu and Seed (1987) Method

Layer number (1)	Layer thickness, ft (2)	$\sigma'_{v0} = \sigma_v$, lb/ft^2 (3)	$(N_1)_{60}$ (4)	G_{max} [Eq. (7.5)] kip/ft^2 (5)	$\gamma_{eff}(G_{eff}/G_{max})$ [Eq. (7.4)] (6)	γ_{eff} (Fig. 7.8) (7)	$\%\gamma_{cyc} = 100\gamma_{eff}$ (8)	ε_v (Fig. 7.9) (9)	Multi-directional shear = $2\varepsilon_v$ (10)	Multiply by VSR (11)	Settlement, in (12)
1	5	238	9	517	1.3×10^{-4}	5×10^{-4}	5×10^{-2}	0.14	0.28	0.22	0.13
2	5	713	9	896	2.3×10^{-4}	1.0×10^{-3}	1.0×10^{-1}	0.29	0.58	0.46	0.28
3	10	1425	9	1270	3.1×10^{-4}	1.3×10^{-3}	1.3×10^{-1}	0.40	0.80	0.64	0.77
4	10	2375	9	1630	3.9×10^{-4}	1.4×10^{-3}	1.4×10^{-1}	0.43	0.86	0.69	0.83
5	10	3325	9	1930	4.4×10^{-4}	1.3×10^{-3}	1.3×10^{-1}	0.40	0.80	0.64	0.77
6	10	4275	9	2190	4.8×10^{-4}	1.3×10^{-3}	1.3×10^{-1}	0.40	0.80	0.64	0.77

Total = 3.5 in

4. $(N_1)_{60}$ *values:*   As previously mentioned, the average $(N_1)_{60}$ value for the sand deposit was determined to be 9.

5. G_{max}:   Equation (7.5) was used to calculate the value of G_{max}. It was assumed that the mean principal effective stress σ'_m was equal to $0.65\sigma'_{v0}$. Note that G_{max} is expressed in terms of kips per square foot (ksf) in Table 7.2.

6. *Equation (7.4):*   The value of $\gamma_{eff}(G_{eff}/G_{max})$ was calculated by using Eq. (7.4). A peak ground acceleration a_{max} of $0.45g$ and a value of r_d from Eq. (6.7) were used in the analysis.

7. *Effective shear strain* γ_{eff}:   Based on the values of $\gamma_{eff}(G_{eff}/G_{max})$ and the mean principal effective stress (σ'_m in ton/ft^2), Fig. 7.8 was used to obtain the effective shear strain.

8. *Percent cyclic shear strain* $\%\gamma_{cyc}$:   The percent cyclic shear strain was calculated as γ_{eff} times 100.

9. *Volumetric strain* ε_v: Entering Fig. 7.9 with the percent cyclic shear strain and using $(N_1)_{60} = 9$, the percent volumetric strain ε_v was obtained from the vertical axis.

10. *Multidirectional shear:*   The values of percent volumetric strain ε_v from step 9 were doubled to account for the multidirectional shear.

11. *Earthquake magnitude:*   The earthquake magnitude is equal to 6.6. Using Table 7.1, the volumetric strain ratio is approximately equal to 0.8. To account for the earthquake magnitude, the percent volumetric strain ε_v from step 10 was multiplied by the VSR.

12. *Settlement:*   The final step was to multiply the volumetric strain ε_v from step 11, expressed as a decimal, by the layer thickness. The total settlement was calculated as the sum of the settlement from all six layers (i.e., total settlement = 3.5 in).

Summary of Values.   Based on the two methods, the ground surface settlement of the 50-ft- (15-m-) thick sand layer is expected to be on the order of 2 to 3½ in (5 to 9 cm). As previously mentioned, the actual settlement as reported by Seed and Silver (1972) was about 2½ in (6 cm).

7.4.5 Limitations

The methods for the calculation of volumetric compression as presented in Sec. 7.4 can only be used for the following cases:

- *Lightweight structures:*   Settlement of lightweight structures, such as wood-frame buildings bearing on shallow foundations

- *Low net bearing stress:*   Settlement of any other type of structure that imparts a low net bearing pressure onto the soil

- *Floating foundation:*   Settlement of floating foundations, provided the floating foundation does not impart a significant net stress upon the soil

- *Heavy structures with deep settlement:*   Settlement of heavy structures, such as massive buildings founded on shallow foundations, provided the zone of settlement is deep enough that the stress increase caused by the structural load is relatively low

- *Differential settlement:*   Differential movement between a structure and adjacent appurtenances, where the structure contains a deep foundation that is supported by strata below the zone of volumetric compression

The methods for the calculation of volumetric compression as presented in Sec. 7.4 cannot be used for the following cases:

- *Heavy buildings bearing on loose soil:*   Do not use the methods when the foundation applies a large net load onto the loose soil. In this case, the heavy foundation will punch downward into the loose soil during the earthquake. It is usually very difficult to determine the settlement for these conditions, and the best engineering solution is to provide a sufficiently high static factor of safety so that there is ample resistance against a bearing capacity failure. This is further discussed in Chap. 8.

- *Sloping ground condition:*   These methods will underestimate the settlement for a sloping ground condition. The loose sand may deform laterally during the earthquake, and the settlement of the building could be well in excess of the calculated values.

7.5 SETTLEMENT DUE TO DYNAMIC LOADS CAUSED BY ROCKING

Details on this type of settlement are as follows:

- *Settlement mechanism:*   This type of settlement is caused by dynamic structural loads that momentarily increase the foundation pressure acting on the soil, such as illustrated in Fig. 7.10. The soil will deform in response to the dynamic structural load, resulting in settlement of the building. This settlement due to dynamic loads is often a result of the structure rocking back and forth.

- *Vulnerable soil types:*   Both cohesionless soil and cohesive soil are susceptible to rocking settlement. For cohesionless soils, loose sands and gravels are prone to rocking settlement. In addition, rocking settlement and volumetric compression (Sec. 7.4) often work in combination to cause settlement of the structure.

  Cohesive soils can also be susceptible to rocking settlement. The types of cohesive soils most vulnerable are normally consolidated soils (OCR = 1.0), such as soft clays and organic soils. There can be significant settlement of foundations on soft saturated clays and organic soils because of undrained plastic flow when the foundations are overloaded during the seismic shaking. Large settlement can also occur if the existing vertical effective stress σ'_{v0} plus the dynamic load $\Delta\sigma_v$ exceeds the maximum past pressure σ'_{vm} of the cohesive soil, or $\sigma'_{v0} + \Delta\sigma_v > \sigma'_{vm}$.

  Another type of cohesive soil that can be especially vulnerable to rocking settlement is sensitive clays. These soils can lose a portion of their shear strength during the cyclic loading. The higher the sensitivity, the greater the loss of shear strength for a given shear strain.

- *Susceptible structures:*   Lightly loaded structures would be least susceptible to rocking settlement. On the other hand, tall and heavy buildings that have shallow foundations bearing on vulnerable soils would be most susceptible to this type of settlement.

- *Example:*   Figure 7.11 presents an example of damage caused by rocking settlement. The rocking settlement occurred to a tall building located in Mexico City. The rocking settlement was caused by the September 19, 1985, Michoacan earthquake, which is described in Sec. 4.6.1.

In terms of the analysis for rocking settlement, R. B. Seed (1991) states:

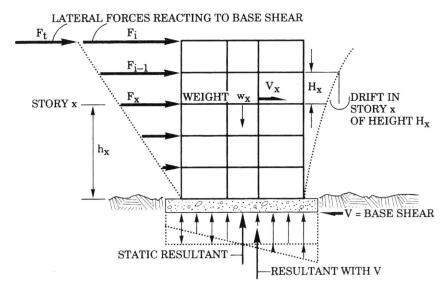

FIGURE 7.10   Diagram illustrating lateral forces F in response to the base shear V caused by the earthquake. Note that the uniform static bearing pressure is altered by the earthquake such that the pressure is increased along one side of the foundation. (*Reproduced from Krinitzsky et al. 1993, with permission from John Wiley & Sons.*)

Vertical accelerations during earthquake seldom produce sufficient vertical thrust to cause significant foundation settlements. Horizontal accelerations, on the other hand, can cause "rocking" of a structure, and the resulting structural overturning moments can produce significant cyclic vertical thrusts on the foundation elements. These can, in turn, result in cumulative settlements, with or without soil liquefaction or other strength loss. This is generally a potentially serious concern only for massive, relatively tall structures. Structures on deep foundations are not necessarily immune to this hazard; structures founded on "friction piles" (as opposed to more solidly-based end-bearing piles) may undergo settlements of up to several inches or more in some cases. It should be noted that the best engineering solution is generally simply to provide a sufficiently high static factor of safety in bearing in order to allow for ample resistance to potential transient seismic loading.

As indicated above, the best engineering solution is to provide a sufficiently high factor of safety against a bearing capacity failure, which is discussed in Chap. 8.

7.6 PROBLEMS

The problems have been divided into basic categories as indicated below:

Liquefaction-Induced Settlement

7.1   Use the data from the example problem in Sec. 7.2.2, but assume that $a_{max}/g = 0.1$ and the sand contains 15 percent nonplastic fines. Calculate the settlement, using Figs. 7.1 and 7.2. *Answer:* See Table 7.3.

FIGURE 7.11 Settlement caused by the building rocking back and forth during the Michoacan earthquake in Mexico on September 19, 1985. (*Photograph from the Steinbrugge Collection, EERC, University of California, Berkeley.*)

7.2   Use the data from the example problem in Sec. 7.2.2, but assume that $a_{max}/g = 0.2$ and the earthquake magnitude $M = 5\frac{1}{4}$. Calculate the liquefaction-induced settlement, using Figs. 7.1 and 7.2. *Answer:* See Table 7.3.

7.3   Use the data from the example problem in Sec. 7.2.2, but assume at a depth of 3 m that $q_c = 3.9$ MPa. Calculate the liquefaction-induced settlement, using Figs. 7.1 and 7.2. *Answer:* See Table 7.3.

7.4   Use the data from the example problem in Sec. 7.2.2, but assume that the shear wave velocity $V_s = 150$ m/s. Calculate the liquefaction-induced settlement, using Figs. 7.1 and 7.2. *Answer:* See Table 7.3.

7.5   Use the data from the example problem in Sec. 7.2.2, but assume that the soil type is crushed limestone (i.e., soil type 1, see Fig. 6.12) and at a depth of 3 m, $q_{c1} = 5.0$ MPa. Calculate the liquefaction-induced settlement, using Figs. 7.1 and 7.2. *Answer:* See Table 7.3.

7.6   Use the data from the example problem in Sec. 7.2.2, but assume that the soil type is silty gravel (i.e., soil type 2, see Fig. 6.12) and at a depth of 3 m, $q_{c1} = 7.5$ MPa. Calculate the liquefaction-induced settlement, using Figs. 7.1 and 7.2. *Answer:* See Table 7.3.

TABLE 7.3 Summary of Answers for Probs. 7.1 to 7.9

Problem no.	Soil type	a_{max}/g	Earthquake magnitude	$(N_1)_{60}$ bl./ft q_{c1}, MPa V_{s1}, m/s	Cyclic stress ratio (CSR)	Cyclic resistance ratio (CRR)	FS = CRR / CSR	Settlement, cm (Fig. 7.1)	Settlement, cm (Fig. 7.2)
Section 7.2.2	Clean sand	0.40	$7\frac{1}{2}$	7.7 blows/ft	0.34	0.09	0.26	4.1	3.0
Problem 7.1	Sand—15% fines	0.10	$7\frac{1}{2}$	7.7 blows/ft	0.084	0.14	1.67	0.15	0.15
Problem 7.2	Clean sand	0.20	$5\frac{1}{4}$	7.7 blows/ft	0.17	0.14	0.82	4.1	2.9
Problem 7.3	Clean sand	0.40	$7\frac{1}{2}$	5.8 MPa	0.34	0.09	0.26	3.6	3.0
Problem 7.4	Clean sand	0.40	$7\frac{1}{2}$	185 m/s	0.34	0.16	0.47	2.8	2.1
Problem 7.5	Crushed limestone	0.40	$7\frac{1}{2}$	5.0 MPa	0.34	0.18	0.53	4.2	3.1
Problem 7.6	Silty gravel	0.40	$7\frac{1}{2}$	7.5 MPa	0.34	0.27	0.79	3.0	2.2
Problem 7.7	Gravelly sand	0.40	$7\frac{1}{2}$	14 MPa	0.34	0.44	1.29	0.3	1.2
Problem 7.8	Eolian sand	0.40	$7\frac{1}{2}$	7.7 blows/ft	0.34	0.09	0.26	4.1	3.0
Problem 7.9	Loess	0.40	$7\frac{1}{2}$	7.7 blows/ft	0.34	0.18	0.53	3.0	2.3

Note: See App. E for solutions.

7.23

7.7 Use the data from the example problem in Sec. 7.2.2, but assume that the soil type is gravelly sand (i.e., soil type 3, see Fig. 6.12) and at a depth of 3 m, $q_{c1} = 14$ MPa. Calculate the settlement, using Figs. 7.1 and 7.2. *Answer:* See Table 7.3.

7.8 Use the data from the example problem in Sec. 7.2.2, but assume that the soil type is eolian sand (i.e., soil type 4, see Fig. 6.12). Calculate the liquefaction-induced settlement, using Figs. 7.1 and 7.2. *Answer:* See Table 7.3.

7.9 Use the data from the example problem in Sec. 7.2.2, but assume that the soil type is noncemented loess (i.e., soil type 7, see Fig. 6.12). Calculate the liquefaction-induced settlement, using Figs. 7.1 and 7.2. *Answer:* See Table 7.3.

7.10 Assume a site has clean sand and a groundwater table near ground surface. The following data are determined for the site:

Layer depth, m	Cyclic stress ratio	$(N_1)_{60}$
2–3	0.18	10
3–5	0.20	5
5–7	0.22	7

Using Figs. 7.1 and 7.2, calculate the total liquefaction-induced settlement of these layers caused by a magnitude 7.5 earthquake. *Answer:* Per Fig. 7.1, 22 cm; per Fig. 7.2, 17 cm.

Liquefaction-Induced Settlement, Subsoil Profiles

7.11 Use the data from Prob. 6.12 and the subsoil profile shown in Fig. 6.13. Ignore any possible settlement of the soil above the groundwater table (i.e., ignore settlement from ground surface to a depth of 1.5 m). Also ignore any possible settlement of the soil located below a depth of 21 m. Using Figs. 7.1 and 7.2, calculate the earthquake-induced settlement of the sand located below the groundwater table. *Answer:* Per Fig. 7.1, 61 cm; per Fig. 7.2, 53 cm.

7.12 Use the data from Prob. 6.15 and the subsoil profile shown in Fig. 6.15. Ignore any possible settlement of the surface soil (i.e., ignore settlement from ground surface to a depth of 1.2 m). Also ignore any possible settlement of soil located below a depth of 20 m. Using Figs. 7.1 and 7.2, calculate the earthquake-induced settlement of the sand located below the groundwater table. *Answer:* Per Fig. 7.1, 22 cm; per Fig. 7.2, 17 cm.

7.13 Figure 7.12 shows the subsoil profile at the Agano River site in Niigata. Assume a level-ground site with the groundwater table at a depth of 0.85 m below ground surface. The medium sand, medium to coarse sand, and coarse sand layers have less than 5 percent fines. The fine to medium sand layers have an average of 15 percent fines. The total unit weight γ_t of the soil above the groundwater table is 18.5 kN/m³, and the buoyant unit weight γ_b of the soil below the groundwater table is 9.8 kN/m³.

The standard penetration data shown in Fig. 7.12 are uncorrected N values. Assume a hammer efficiency E_m of 0.6 and a boring diameter of 100 mm; and the length of drill rods is equal to the depth of the SPT below ground surface. The design earthquake conditions are a peak ground acceleration a_{max} of 0.20g and magnitude of 7.5. Based on the standard penetration test data and using Figs. 7.1 and 7.2, calculate the earthquake-induced settlement of the soil located at a depth of 0.85 to 15.5 m below ground surface. *Answer:* Per Fig. 7.1, 30 cm; per Fig. 7.2, 24 cm.

7.14 Figure 7.13 shows the subsoil profile at a road site in Niigata. Assume a level-ground site with the groundwater table at a depth of 2.5 m below ground surface. Also assume that all the soil types located below the groundwater table meet the criteria for potentially liquefiable soil. The medium sand layers have less than 5 percent fines, the

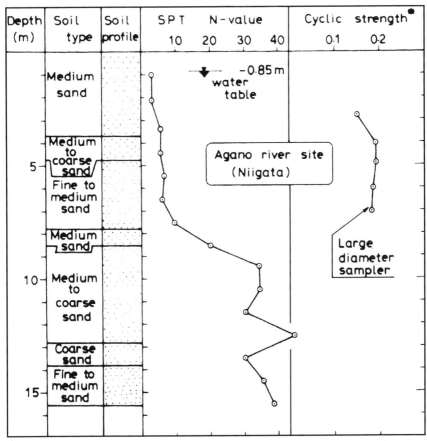

Depth (m)	Soil type	Soil profile	SPT N-value 10 20 30 40	Cyclic strength[*] 0.1 0.2

FIGURE 7.12 Subsoil profile, Agano River site, Niigata. (*Reproduced from Ishihara, 1985.*)

♦ Cyclic stress ratio causing 5 % D.A. strain in 20 cycles

sandy silt layer has 50 percent fines, and the silt layers have 75 percent fines. The total unit weight γ_t of the soil above the groundwater table is 18.5 kN/m³, and the buoyant unit weight γ_b of the soil below the groundwater table is 9.8 kN/m³.

The standard penetration data shown in Fig. 7.13 are uncorrected N values. Assume a hammer efficiency E_m of 0.6 and a boring diameter of 100 mm; and the length of drill rods is equal to the depth of the SPT below ground surface. The design earthquake conditions are a peak ground acceleration a_{max} of 0.20g and magnitude of 7.5. Based on the standard penetration test data and using Figs. 7.1 and 7.2, calculate the earthquake-induced settlement of the soil located at a depth of 2.5 to 15 m below ground surface. *Answer:* Per Fig. 7.1, 34 cm; per Fig. 7.2, 27 cm.

7.15 Use the data from Prob. 6.18 and Fig. 6.11. Based on Figs. 7.1 and 7.2, calculate the earthquake-induced settlement of the soil located at a depth of 0.5 to 16 m below ground surface for the before-improvement and after-improvement conditions. *Answers:* Before

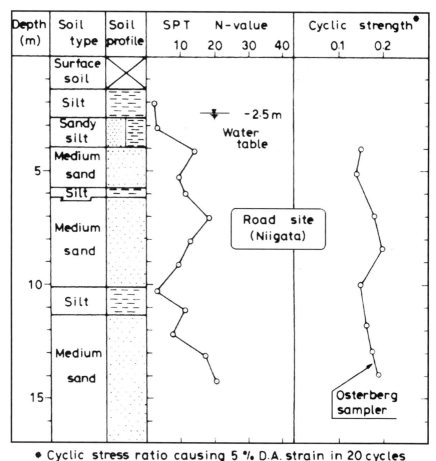

• Cyclic stress ratio causing 5 % D.A. strain in 20 cycles

FIGURE 7.13   Subsoil profile, road site, Niigata. (*Reproduced from Ishihara, 1985.*)

improvement: per Fig. 7.1, 45 cm; per Fig. 7.2, 35 cm. After improvement: per Fig. 7.1, 0.3 cm; and per Fig. 7.2, 2.7 cm.

7.16   Use the data from Prob. 6.12 and Fig. 6.13. Assume that there has been soil improvement from ground surface to a depth of 15 m, and for the zone of soil having soil improvement (0 to 15-m depth), the factor of safety against liquefaction is greater than 2.0. A mat foundation for a heavy building will be constructed such that the bottom of the mat is at a depth of 1.0 m. The mat foundation is 20 m long and 10 m wide, and according to the structural engineer, the foundation will impose a net stress of 50 kPa onto the soil (the 50 kPa includes earthquake-related seismic load). Calculate the earthquake-induced settlement of the heavy building, using Figs. 7.1 and 7.2. *Answer:* Per Fig. 7.1, 17 cm; per Fig. 7.2, 19 cm.

7.17   Use the data from Prob. 6.15 and Fig. 6.15. A sewage disposal tank will be installed at a depth of 2 to 4 m below ground surface. Assuming the tank is empty at the time of the design earthquake, calculate the liquefaction-induced settlement of the tank. *Answer:* Since the tank is in the middle of a liquefied soil layer, it is expected that the empty tank will not settle, but rather will float to the ground surface.

Liquefaction-Induced Ground Damage

7.18   A soil deposit has a 6-m-thick surface layer of unliquefiable soil underlain by a 4-m-thick layer that is expected to liquefy during the design earthquake. The design earthquake has a peak ground acceleration a_{max} equal to 0.40g. Will there be liquefaction-induced ground damage for this site? *Answer:* Based on Fig. 7.6, liquefaction-induced ground damage is expected for this site.

7.19   Use the data from Prob. 6.12 and Fig. 6.13. Assume that the groundwater table is unlikely to rise above its present level. Using a peak ground acceleration a_{max} equal to 0.20g and the standard penetration test data, will there be liquefaction-induced ground damage for this site? *Answer:* Based on Fig. 7.6, liquefaction-induced ground damage is expected for this site.

7.20   Use the data from Prob. 7.13 and Fig. 7.12. Assume that the groundwater table is unlikely to rise above its present level. Using a peak ground acceleration a_{max} equal to 0.20g and the standard penetration test data, determine the minimum thickness of a fill layer that must be placed at the site in order to prevent liquefaction-induced ground damage for this site. *Answer:* Based on Fig. 7.6, minimum thickness of fill layer = 2.2 m.

7.21   Use the data from Prob. 7.14 and Fig. 7.13. Assume that the groundwater table is unlikely to rise above its present level. Using a peak ground acceleration a_{max} equal to 0.20g and the standard penetration test data, will there be liquefaction-induced ground damage for this site? *Answer:* The solution depends on the zone of assumed liquefaction (see App. E).

Volumetric Compression

7.22   Solve the example problem in Sec. 7.4.4, using the Tokimatsu and Seed (1987) method and assuming that the 50-ft-thick deposit of sand has $(N_1)_{60}$ = 5. *Answer:* 11 in (28 cm).

7.23   Solve the example problem in Sec. 7.4.4, using the Tokimatsu and Seed (1987) method and the chart shown in Fig. 7.7, assuming that the 50-ft-thick deposit of sand has $(N_1)_{60}$ = 15. *Answer:* Using the Tokimatsu and Seed (1987) method, settlement = 1.3 in (3.3 cm). Using the chart shown in Fig. 7.7, settlement = 0.9 in (2 cm).

7.24   Solve the example problem in Sec. 7.4.4, using the Tokimatsu and Seed (1987) method and the chart shown in Fig. 7.7, assuming that the 50-ft-thick deposit of sand will be subjected to a peak ground acceleration of 0.20g and the earthquake magnitude = 7.5. *Answer:* Using the Tokimatsu and Seed (1987) method, settlement = 0.9 in (2.3 cm). Using the chart shown in Fig. 7.7, settlement = 0.6 in (1.5 cm).

7.25   Solve the example problem in Sec. 7.4.4, using the Tokimatsu and Seed (1987) method and assuming that the 50-ft-thick deposit of sand has $(N_1)_{60}$ = 5, a peak ground acceleration of 0.20g, and the earthquake magnitude = 7.5. *Answer:* Settlement = 2 in (5 cm).

CHAPTER 8
BEARING CAPACITY ANALYSES FOR EARTHQUAKES

The following notation is used in this chapter:

SYMBOL	DEFINITION
B	Width of footing
B'	Reduced footing width to account for eccentricity of load
c	Cohesion based on total stress analysis
c'	Cohesion based on effective stress analysis
D_f	Depth below ground surface to bottom of footing
D_r	Relative density
e	Eccentricity of vertical load Q
e_1, e_2	Eccentricities along and across footing (Fig. 8.9)
FS	Factor of safety
H_1	Thickness of surface layer that does not liquefy
k_0	Coefficient of earth pressure at rest
L	Length of footing
L'	Reduced footing length to account for eccentricity of load
N	Measured SPT blow count (N value in blows per foot)
N_c, N_γ, N_q	Dimensionless bearing capacity factors
$(N_1)_{60}$	N value corrected for field testing procedures and overburden pressure
P, Q	Footing load
q_{all}	Allowable bearing pressure
q_{ult}	Ultimate bearing capacity
q'	Largest bearing pressure exerted by eccentrically loaded footing
q''	Lowest bearing pressure exerted by eccentrically loaded footing
Q_{ult}	Load causing a bearing capacity failure
r_u	Pore water pressure ratio
R	Shear resistance of soil
s_u	Undrained shear strength of soil
S_t	Sensitivity of soil
T	Vertical distance from bottom of footing to top of liquefied soil layer
u_e	Excess pore water pressure generated during earthquake
w_l	Liquid limit
w_p	Plastic limit
ϕ	Friction angle based on total stress analysis
ϕ'	Friction angle based on effective stress analysis
γ_b	Buoyant unit weight of saturated soil below groundwater table
γ_t	Total unit weight of soil

σ'	Initial effective stress acting on shear surface
σ_h	Horizontal total stress
σ_h'	Horizontal effective stress
σ_v	Vertical total stress
σ_{vm}'	Maximum past pressure, also known as preconsolidation pressure
σ_{v0}'	Vertical effective stress
τ_f	Shear strength of soil

8.1 INTRODUCTION

8.1.1 General, Punching, and Local Shear

A *bearing capacity failure* is defined as a foundation failure that occurs when the shear stresses in the soil exceed the shear strength of the soil. For both the static and seismic cases, bearing capacity failures of foundations can be grouped into three categories, (Vesic 1963, 1967, 1975):

  1. *General shear (Fig. 8.1):*   As shown in Fig. 8.1, a general shear failure involves total rupture of the underlying soil. There is a continuous shear failure of the soil (solid lines) from below the footing to the ground surface. When the load is plotted versus settlement of the footing, there is a distinct load at which the foundation fails (solid circle), and this is designated Q_{ult}. The value of Q_{ult} divided by the width B and length L of the footing is considered to be the *ultimate bearing capacity* q_{ult} of the footing. The ultimate bearing capacity has been defined as the bearing stress that causes a sudden catastrophic failure of the foundation (Lambe and Whitman 1969).
  Note in Fig. 8.1 that a general shear failure ruptures and pushes up the soil on both sides of the footing. For actual failures in the field, the soil is often pushed up on only one side of the footing with subsequent tilting of the structure. A general shear failure occurs for soils that are in a dense or hard state.

  2. *Punching shear (Fig. 8.2):*   As shown in Fig. 8.2, a punching shear failure does not develop the distinct shear surfaces associated with a general shear failure. For punching shear, the soil outside the loaded area remains relatively uninvolved, and there is minimal movement of soil on both sides of the footing.
  The process of deformation of the footing involves compression of soil directly below the footing as well as the vertical shearing of soil around the footing perimeter. As shown in Fig. 8.2, the load-settlement curve does not have a dramatic break, and for punching shear, the bearing capacity is often defined as the first major nonlinearity in the load-settlement curve (open circle). A punching shear failure occurs for soils that are in a loose or soft state.

  3. *Local shear failure (Fig. 8.3):*   As shown in Fig. 8.3, local shear failure involves rupture of the soil only immediately below the footing. There is soil bulging on both sides of the footing, but the bulging is not as significant as in general shear. Local shear failure can be considered as a transitional phase between general shear and punching shear. Because of the transitional nature of local shear failure, the bearing capacity could be defined as the first major nonlinearity in the load-settlement curve (open circle) or at the point where the settlement rapidly increases (solid circle). A local shear failure occurs for soils that are in a medium or firm state.

  Table 8.1 presents a summary of the type of bearing capacity failure that would most likely develop based on soil type and soil properties.

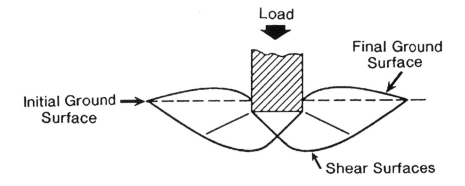

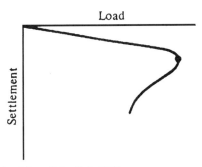

FIGURE 8.1   General shear foundation failure. (*After Vesic 1963.*)

8.1.2 Bearing Capacity Failures

Compared to the number of structures damaged by earthquake-induced settlement, there are far fewer structures that have earthquake-induced bearing capacity failures. This is because of the following factors:

1. *Settlement governs:*   The foundation design is based on several requirements. Two of the main considerations are that (1) settlement due to the building loads must not exceed tolerable values and (2) there must be an adequate factor of safety against a bearing capacity failure. In most cases, settlement governs and the foundation bearing pressures recommended by the geotechnical engineer are based on limiting the amount of expected settlement due to the static or seismic cases. In other cases where the settlement is too high, the building is often constructed with a deep foundation, which also reduces the possibility of a bearing capacity failure.

2. *Extensive studies:*   There have been extensive studies of both static and seismic bearing capacity failures, which have led to the development of bearing capacity equations that are routinely used in practice to determine the ultimate bearing capacity of the foundation.

3. *Factor of safety:*   To determine the allowable bearing pressure q_{all}, the ultimate bearing capacity q_{ult} is divided by a factor of safety. The normal factor of safety used for static bearing capacity analyses is 3. For the evaluation of the bearing capacity for seismic

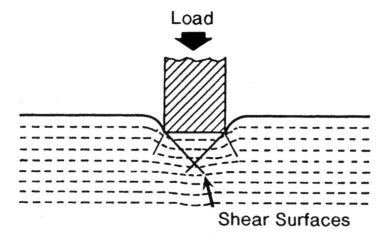

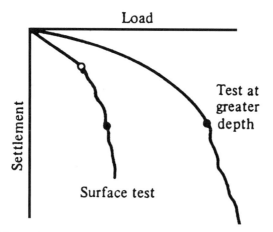

FIGURE 8.2 Punching shear foundation failure. (*After Vesic 1963.*)

analysis, the factor of safety is often in the range of 5 to 10 (Krinitzsky et al. 1993). These are high factors of safety compared to other factors of safety, such as only 1.5 for slope stability analyses (Chap. 9).

4. *Minimum footing sizes:* Building codes often require minimum footing sizes and embedment depths. Larger footing sizes will lower the bearing pressure on the soil and reduce the potential for static or seismic bearing capacity failures.

5. *Allowable bearing pressures:* In addition, building codes often have maximum allowable bearing pressures for different soil and rock conditions. Table 8.2 presents maximum allowable bearing pressures based on the *Uniform Building Code* (Table 18-I-A, 1997). Especially in the case of dense or stiff soils, these allowable bearing pressures often have adequate factors of safety for both static and seismic cases.

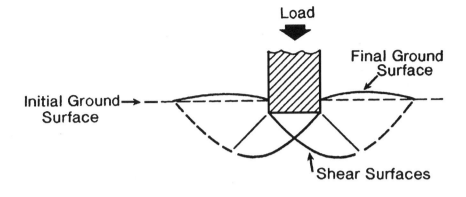

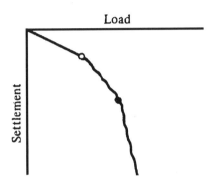

FIGURE 8.3 Local shear foundation failure. (*After Vesic 1963.*)

6. *Footing dimensions:* Usually the structural engineer will determine the size of the footing by dividing the maximum footing load by the allowable bearing pressure. Typically the structural engineer uses values of dead, live, and seismic loads that also contain factors of safety. For example, the live load may be from the local building code, which specifies minimum live load requirements for specific building uses (e.g., see Table 16-A, *Uniform Building Code,* 1997). Thus the load that is used to proportion the footing also contains a factor of safety, which is in addition to the factor of safety that was used to determine the allowable bearing pressure.

The documented cases of bearing capacity failures during earthquakes indicate that usually the following three factors (separately or in combination) are the cause of the failure:

1. *Soil shear strength:* Common problems include an overestimation of the shear strength of the underlying soil. Another common situation leading to a bearing capacity failure is the loss of shear strength during the earthquake, because of the liquefaction of the soil or the loss of shear strength for sensitive clays.

2. *Structural load:* Another common problem is that the structural load at the time of the bearing capacity failure was greater than that assumed during the design phase. This can often occur when the earthquake causes rocking of the structure, and the resulting structural overturning moments produce significant cyclic vertical thrusts on the foundation elements and underlying soil.

TABLE 8.1 Summary of Type of Bearing Capacity Failure versus Soil Properties

Type of bearing capacity failure	Cohesionless soil (e.g., sands)			Cohesive soil (e.g., clays)	
	Density condition	Relative density D_r, percent	$(N_1)_{60}$	Consistency	Undrained shear strength s_u
General shear failure (Fig. 8.1)	Dense to very dense	65–100	>20	Very stiff to hard	>2000 lb/ft^2 (>100 kPa)
Local shear failure (Fig. 8.3)	Medium	35–65	5–20	Medium to stiff	500–2000 lb/ft^2 (25–100 kPa)
Punching shear failure (Fig. 8.2)	Loose to very loose	0–35	<5	Soft to very soft	<500 lb/ft^2 (<25 kPa)

3. *Change in site conditions:* An altered site can produce a bearing capacity failure. For example, if the groundwater table rises, then the potential for liquefaction is increased. Another example is the construction of an adjacent excavation, which could result in a reduction in support and a bearing capacity failure.

The most common cause of a seismic bearing capacity failure is liquefaction of the underlying soil. Section 3.4.2 presents an introduction to bearing capacity failures caused by liquefaction during the earthquake. Figures 3.20 to 3.22 show examples of bearing capacity failures caused by the Niigata earthquake on June 16, 1964. Figure 8.4 shows a bearing capacity failure due to liquefaction during the Izmit earthquake in Turkey on August 17, 1999. Another example is shown in Fig. 8.5, where rather than falling over, the building has literally punched downward into the liquefied soil.

Although bearing capacity failures related to liquefaction of underlying soils are most common, there could also be localized failures due to punching shear when the footing is overloaded, such as by the building's rocking back and forth. Figure 8.6 presents an example of a punching-type failure. The building foundation shown in Fig. 8.6 was constructed of individual spread footings that were interconnected with concrete tie beams. The building collapsed during the Caracas earthquake in Venezuela on July 29, 1967, and when the foundation was exposed, it was discovered that the spread footings had punched downward into the soil. Note in Fig. 8.6 that the tie beam at the center of the photograph was bent and pulled downward when the footing punched into the underlying soil.

8.1.3 Shear Strength

Because the bearing capacity failure involves a shear failure of the underlying soil (Figs. 8.1 to 8.3), the analysis will naturally include the shear strength of the soil (Sec. 5.5.1). As shown in Figs. 8.1 to 8.3, the depth of the bearing capacity failure tends to be rather shallow. For static bearing capacity analyses, it is often assumed that the soil involved in the bearing capacity failure can extend to a depth equal to B (footing width) below the bottom of the footing. However, for cases involving earthquake-induced liquefaction failures or punching shear failures, the depth of soil involvement could exceed the footing width. For buildings with numerous spread footings that occupy a large portion of the building area, the individual pressure bulbs from each footing may combine, and thus the entire width of the building could be involved in a bearing capacity failure.

Either a total stress analysis or an effective stress analysis must be used to determine the bearing capacity of a foundation. These two types of analyses are discussed in Sec. 5.5.1. Table 5.4 presents a summary of the type of analyses and the shear strength parameters that should be used for the bearing capacity analyses.

FIGURE 8.4   The building suffered a liquefaction-induced bearing capacity failure during the Izmit earthquake in Turkey on August 17, 1999. (*Photograph from the Izmit Collection, EERC, University of California, Berkeley.*)

8.1.4   One-Third Increase in Bearing Pressure for Seismic Conditions

When the recommendations are presented for the allowable bearing pressures at a site, it is common practice for the geotechnical engineer to recommend that the allowable bearing pressure be increased by a factor of one-third when performing seismic analyses. For example, in soil reports, it is commonly stated: "For the analysis of earthquake loading, the allowable bearing pressure and passive resistance may be increased by a factor of one-third." The rational behind this recommendation is that the allowable bearing pressure has an ample factor of safety, and thus for seismic analyses, a lower factor of safety would be acceptable.

Usually the above recommendation is appropriate for the following materials:

1. Massive crystalline bedrock and sedimentary rock that remains intact during the earthquake
2. Dense to very dense granular soil
3. Heavily overconsolidated cohesive soil, such as very stiff to hard clays

These materials do not lose shear strength during the seismic shaking, and therefore an increase in bearing pressure is appropriate.

A one-third increase in allowable bearing pressure should not be recommended for the following materials:

1. Foliated or friable rock that fractures apart during the earthquake
2. Loose soil subjected to liquefaction or a substantial increase in excess pore water pressure
3. Sensitive clays that lose shear strength during the earthquake
4. Soft clays and organic soils that are overloaded and subjected to plastic flow

These materials have a reduction in shear strength during the earthquake. Since the materials are weakened by the seismic shaking, the static values of allowable bearing pres-

FIGURE 8.5 The building suffered a liquefaction-induced punching shear failure during the Izmit earthquake in Turkey on August 17, 1999. (*Photograph from the Izmit Collection, EERC, University of California, Berkeley.*)

sure should not be increased for the earthquake analyses. In fact, the allowable bearing pressure may actually have to be reduced to account for the weakening of the soil during the earthquake. The remainder of this chapter deals with the determination of the bearing capacity of soils that are weakened by seismic shaking.

8.2 BEARING CAPACITY ANALYSES FOR LIQUEFIED SOIL

8.2.1 Introduction

Section 8.2 deals with the bearing capacity of foundations underlain by liquefied soil. The liquefaction analysis presented in Chap. 6 can be used to determine those soil layers that will liquefy during the design earthquake.

Table 8.3 summarizes the requirements and analyses for soil susceptible to liquefaction. The steps are as follows:

FIGURE 8.6　The building foundation shown above was constructed of individual footings that were interconnected with concrete tie beams. The building collapsed during the Caracas earthquake in Venezuela on July 29, 1967. When the foundation was exposed, it was discovered that the spread footings had punched downward into the soil. Note that the tie beam at the center of the photograph was bent and pulled downward when the footing punched into the underlying soil. (*Photograph from the Steinbrugge Collection, EERC, University of California, Berkeley.*)

1. *Requirements:*　The first step is to determine whether the two requirements listed in Table 8.3 are met. If these two requirements are not met, then the foundation is susceptible to failure during the design earthquake, and special design considerations, such as the use of deep foundations or soil improvement, are required.

2. *Settlement analysis:*　Provided that the two design requirements are met, the next step is to perform a settlement analysis using Figs. 7.1 and 7.2. Note that in some cases, the settlement analysis is unreliable (e.g., heavy buildings with an underlying liquefied soil layer close to the bottom of the foundation).

3. *Bearing capacity analysis:*　There are two different types of bearing capacity analysis that can be performed. The first deals with a shear failure where the footing punches into the liquefied soil layer (Sec. 8.2.2). The second case uses the traditional Terzaghi bearing

TABLE 8.2 Allowable Bearing Pressures

Material type	Allowable bearing pressure*	Maximum allowable bearing pressure[†]
Massive crystalline bedrock	4,000 lb/ft² (200 kPa)	12,000 lb/ft² (600 kPa)
Sedimentary and foliated rock	2,000 lb/ft² (100 kPa)	6,000 lb/ft² (300 kPa)
Gravel and sandy gravel (GW, GP)[‡]	2,000 lb/ft² (100 kPa)	6,000 lb/ft² (300 kPa)
Nonplastic soil: sands, silty gravel, and nonplastic silt (GM, SW, SP, SM)[‡]	1,500 lb/ft² (75 kPa)	4,500 lb/ft² (220 kPa)
Plastic soil: silts and clays (ML, MH, SC, CL, CH)[‡]	1,000 lb/ft² (50 kPa)	3,000 lb/ft² (150 kPa)[§]

*Minimum footing width and embedment depth equal 1 ft (0.3 m).
[†]An increase of 20 percent of the allowable bearing pressure is allowed for each additional 1 ft (0.3 m) of width or depth up to the maximum allowable bearing pressures listed in the rightmost column. An exception is plastic soil; see last note.
[‡]Group symbols from the Unified Soil Classification System.
[§]No increase in the allowable bearing pressure is allowed for an increase in width of the footing.

For dense or stiff soils, allowable bearing values are generally conservative. For very loose or very soft soils, allowable bearing values may be too high.
Source: Data from *Uniform Building Code* (1997).

capacity equation, with a reduction in the bearing capacity factors to account for the loss of shear strength of the underlying liquefied soil layer (Sec. 8.2.3).

4. *Special considerations:* Special considerations may be required if the structure is subjected to buoyancy or if there is a sloping ground condition.

8.2.2 Punching Shear Analysis

Illustration of Punching Shear. Figure 8.7 illustrates the earthquake-induced punching shear analysis. The soil layer portrayed by dashed lines represents unliquefiable soil which is underlain by a liquefied soil layer. For the punching shear analysis, it is assumed that the load will cause the foundation to punch straight downward through the upper unliquefiable soil layer and into the liquefied soil layer. As shown in Fig. 8.7, this assumption means that there will be vertical shear surfaces in the soil that start at the sides of the footing and extend straight downward to the liquefied soil layer. It is also assumed that the liquefied soil has no shear strength.

Factor of Safety. Using the assumptions outlined above, the factor of safety (FS) can be calculated as follows:
For strip footings:

$$\text{FS} = \frac{R}{P} = \frac{2T\tau_f}{P} \tag{8.1a}$$

For spread footings:

$$\text{FS} = \frac{R}{P} = \frac{2\,(B + L)\,(T\tau_f)}{P} \tag{8.1b}$$

TABLE 8.3   Requirements and Analyses for Soil Susceptible to Liquefaction

Requirements and analyses	Design conditions
Requirements	1. *Bearing location of foundation:* The foundation must not bear on soil that will liquefy during the design earthquake. Even lightly loaded foundations will sink into the liquefied soil. 2. *Surface layer H_1:* As discussed in Sec. 7.3, there must be an adequate thickness of an unliquefiable soil surface layer H_1 to prevent damage due to sand boils and surface fissuring (see Fig. 7.3). Without this layer, there could be damage to shallow foundations, pavements, flatwork, and utilities.
Settlement analysis	Use Figs 7.1 and 7.2 for the following conditions: 1. *Lightweight structures:* Settlement of lightweight structures, such as wood-frame buildings bearing on shallow foundations. 2. *Low net bearing stress:* Settlement of any other type of structure that imparts a low net bearing pressure onto the soil. 3. *Floating foundation:* Settlement of floating foundations, provided the zone of liquefaction is below the bottom of the foundation and the floating foundation does not impart a significant net stress upon the soil. 4. *Heavy structures with deep liquefaction:* Settlement of heavy structures, such as massive buildings founded on shallow foundations, provided the zone of liquefaction is deep enough that the stress increase caused by the structural load is relatively low. 5. *Differential settlement:* Differential movement between a structure and adjacent appurtenances, where the structure contains a deep foundation that is supported by strata below the zone of liquefaction.
Bearing capacity analysis	Use the analyses presented in Secs. 8.2 and 8.3 for the following conditions: 1. *Heavy buildings with underlying liquefied soil:* Use a bearing capacity analysis when there is a soil layer below the bottom of the foundation that will be susceptible to liquefaction during the design earthquake. In this case, once the soil has liquefied, the foundation load could cause it to punch or sink into the liquefied soil, resulting in a bearing capacity failure (see Sec. 8.2). 2. *Check bearing capacity:* Perform a bearing capacity analysis whenever the footing imposes a net pressure onto the soil and there is an underlying soil layer that will be susceptible to liquefaction during the design earthquake (see Sec. 8.2). 3. *Positive induced pore water pressures:* For cases where the soil will not liquefy during the design earthquake, but there will be the development of excess pore water pressures, perform a bearing capacity analysis (see Sec. 8.3).
Special considerations	1. *Buoyancy effects:* Consider possible buoyancy effects. Examples include buried storage tanks or large pipelines that are within the zone of liquefied soil. Instead of settling, the buried storage tanks and pipelines may actually float to the surface when the ground liquefies. 2. *Sloping ground condition:* Determine if the site is susceptible to liquefaction-induced flow slide or lateral spreading (see Chap. 9).

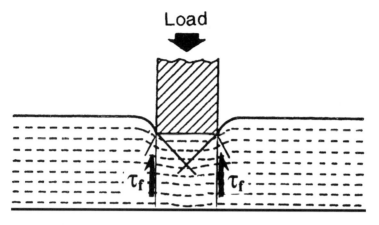

Liquefied Soil Layer

FIGURE 8.7   Illustration of a punching shear analysis. The dashed lines represent unliquefiable soil that is underlain by a liquefied soil layer. In the analysis, the footing will punch vertically downward and into the liquefied soil.

where R = shear resistance of soil. For strip footings, R is the shear resistance per unit length of footing, lb/ft or kN/m. For spread footings, R is the shear resistance beneath entire footing perimeter, lb or kN.

P = footing load. For strip footings, P is the load per unit length of footing, lb/ft or kN/m. For spread footings, P is total load of footing, lb or kN. The footing load includes dead, live, and seismic loads acting on footing as well as weight of footing itself. Typically the value of P would be provided by the structural engineer.

T = vertical distance from the bottom of footing to top of liquefied soil layer, ft or m
τ_f = shear strength of unliquefiable soil layer, lb/ft^2 or kPa
B = width of footing, ft or m
L = length of footing, ft or m

Note in Eq. (8.1b) that the term $2(B + L)$ represents the entire perimeter of the spread footing. When this term is multiplied by T, it represents the total perimeter area that the footing must push through in order to reach the liquefied soil layer. For an assumed footing size and given loading condition, the only unknowns in Eqs. (8.1a) and (8.1b) are the vertical distance from the bottom of the footing to the top of the liquefied soil layer T and the shear strength of the unliquefiable soil layer τ_f. The value of T would be based on the liquefaction analysis (Chap. 6) and the proposed depth of the footing. The shear strength of the unliquefiable soil layer τ_f can be calculated as follows:

1. For an unliquefiable soil layer consisting of cohesive soil (e.g., clays), use a total stress analysis:

$$\tau_f = s_u \qquad (8.2a)$$

or

$$\tau_f = c + \sigma_h \tan \phi \qquad (8.2b)$$

where s_u = undrained shear strength of cohesive soil (total stress analysis), lb/ft^2 or kPa. As discussed in Sec. 5.5.1, often undrained shear strength is obtained from unconfined compression tests or vane shear tests.

c, ϕ = undrained shear strength parameters (total stress analysis). As discussed in Sec. 5.5.1, these undrained shear strength parameters are often obtained from triaxial tests, such as unconsolidated undrained triaxial compression test (ASTM D 2850-95, 2000) or consolidated undrained triaxial compression tests (ASTM D 4767-95, 2000).

σ_h = horizontal total stress, lb/ft^2 or kPa. Since vertical shear surfaces are assumed (see Fig. 8.7), normal stress acting on shear surfaces will be the horizontal total stress. For cohesive soil, σ_h is often assumed to be equal to $\frac{1}{2}\sigma_v$.

2. For an unliquefiable soil layer consisting of cohesionless soil (e.g., sands), use an effective stress analysis:

$$\tau_f = \sigma_h' \tan \phi' = k_0 \sigma_{v0}' \tan \phi' \qquad (8.2c)$$

where σ_h' = horizontal effective stress, lb/ft^2 or kPa. Since vertical shear surfaces are assumed (see Fig. 8.7), the normal stress acting on the shear surface will be the horizontal effective stress. The horizontal effective stress σ_h' is equal to the coefficient of earth pressure at rest k_0 times the vertical effective stress σ_{v0}', or $\sigma_h' = k_0 \sigma_{v0}'$.

ϕ' = effective friction angle of cohesionless soil (effective stress analysis). Effective friction angle could be determined from drained direct shear tests or from empirical correlations such as shown in Figs. 5.12 and 5.14.

Example Problems. The following example problems illustrate the use of Eqs. (8.1) and (8.2).

Example Problem for Cohesive Surface Layer (Total Stress Analysis). Use the data from Prob. 6.15, which deals with the subsurface conditions shown in Fig. 6.15 (i.e., the sewage disposal site). Based on the standard penetration test data, the zone of liquefaction extends from a depth of 1.2 to 6.7 m below ground surface. Assume the surface soil (upper 1.2 m) shown in Fig. 6.15 consists of an unliquefiable cohesive soil and during construction, an additional 1.8-m-thick layer of cohesive soil will be placed at ground surface. Use a peak ground acceleration a_{max} of 0.20g.

Assume that after the 1.8-m-thick layer is placed at ground surface, it is proposed to construct a sewage disposal plant. The structural engineer would like to use shallow strip footings to support exterior walls and interior spread footings to support isolated columns. It is proposed that the bottom of the footings be at a depth of 0.5 m below ground surface. The structural engineer has also indicated that the maximum total loads (including the weight of the footing and the dynamic loads) are 50 kN/m for the strip footings and 500 kN for the spread footings. It is desirable to use 1-m-wide strip footings and square spread footings that are 2 m wide.

For both the existing 1.2-m-thick unliquefiable cohesive soil layer and the proposed additional 1.8-m-thick fill layer, assume that the undrained shear strength s_u of the soil is equal to 50 kPa. Calculate the factor of safety of the footings, using Eq. (8.1).

Solution. The first step is to check the two requirements in Table 8.3. Since the footings will be located within the upper unliquefiable cohesive soil, the first requirement is met. As indicated in the example problem in Sec. 7.3.3, the surface unliquefiable soil layer must be at least 3 m thick to prevent liquefaction-induced ground damage. Since a fill layer equal to 1.8 m is proposed for the site, the final thickness of the unliquefiable soil will be equal to 3 m. Thus the second requirement is met.

To calculate the factor of safety in terms of a bearing capacity failure for the strip and spread footings, the following values are used:

$P = 50$ kN/m for strip footing and 500 kN for spread footing

$T = 2.5$ m   i.e., total thickness of unliquefiable soil layer minus footing embedment depth $= 3$ m $- 0.5$ m $= 2.5$ m

$\tau_f = s_u = 50$ kPa $= 50$ kN/m^2

$B = L = 2$ m

Substituting the above values into Eqs. (8.1a) and (8.1b) yields

$$FS = \frac{2\,T\tau_f}{P} = \frac{2\,(2.5\text{ m})\,(50\text{ kN/m}^2)}{50\text{ kN/m}} = 5.0 \qquad \text{strip footing}$$

$$FS = \frac{2\,(B + L)\,T\tau_f}{P} = \frac{2\,(2 + 2)\,(2.5\text{ m})\,(50\text{ kN/m}^2)}{500\text{ kN}} = 2.0 \qquad \text{spread footing}$$

For a seismic analysis, a factor of safety of 5.0 would be acceptable, but the factor of safety of 2.0 would probably be too low.

Example Problem for Cohesionless Surface Layer (Effective Stress Analysis). Use the same data, but assume the surface soil and the proposed 1.8-m-thick fill layer are sands with an effective friction angle ϕ' equal to 32° and a coefficient of earth pressure at rest k_0 equal to 0.5. Also assume that instead of the groundwater table being at a depth of 0.4 m (see Fig. 6.15), it is at a depth of 1.2 m below the existing ground surface. Calculate the factor of safety of the footings, using Eq. (8.1).

Solution. To calculate the factor of safety in terms of a bearing capacity failure for the strip and spread footings, the following values are used:

$P = 50$ kN/m for strip footing and 500 kN for spread footing

$T = 2.5$ m   i.e., total thickness of unliquefiable soil layer minus footing embedment depth $= 3$ m $- 0.5$ m $= 2.5$ m

$\sigma'_{v0} = \sigma_v - u$ Since soil is above groundwater table, assume $u = 0$. Use a total unit weight of 18.3 kN/m^3 (Prob. 6.15) and an average depth of 1.75 m [$(0.5 + 3.0)/2 = 1.75$ m] or $\sigma'_{v0} = 18.3 \times 1.75 = 32$ kPa.

$\tau_f = k_0\,\sigma'_{v0}\,\tan\phi' = (0.5)\,(32\text{ kPa})\,(\tan 32°) = 10$ kPa $= 10$ kN/m^2 [Eq. (8.2c)]

$B = L = 2$ m

Substituting the above values into Eqs. (8.1a) and (8.1b) gives

$$FS = \frac{2T\tau_f}{P} = \frac{2\,(2.5\text{ m})\,(10\text{ kN/m}^2)}{50\text{ kN/m}} = 1.0 \qquad \text{strip footing}$$

$$FS = \frac{2\,(B + L)\,T\tau_f}{P} = \frac{2\,(2 + 2)\,(2.5\text{ m})\,(10\text{ kN/m}^2)}{500\text{ kN}} = 0.4 \qquad \text{spread footing}$$

For the seismic bearing capacity analyses, these factors of safety would indicate that both the strip and spread footings would punch down through the upper sand layer and into the liquefied soil layer.

As a final check, the FS calculated from the earthquake-induced punching shear analysis must be compared with the FS calculated from the static bearing capacity analysis (i.e., nonearthquake condition). The reason for this comparison is that FS for the earthquake punching shear case [Eq. (8.1)] could exceed the FS calculated from the static condition. This often occurs when the liquefied soil layer is at a significant depth below the bottom of the footing, or in other words at high values of T/B. In any event, the lower value of FS from either the earthquake punching shear analysis or the static bearing capacity analysis would be considered the critical condition.

8.2.3 Terzaghi Bearing Capacity Equation

Introduction.   The most commonly used bearing capacity equation is that equation developed by Terzaghi (1943). For a uniform vertical loading of a strip footing, Terzaghi (1943) assumed a shallow footing and general shear failure (Fig. 8.1) in order to develop the following bearing capacity equation:

$$q_{ult} = \frac{Q_{ult}}{BL} = cN_c + \tfrac{1}{2}\gamma_t BN_\gamma + \gamma_t D_f N_q \qquad (8.3)$$

where
$\quad q_{ult}$ = ultimate bearing capacity for a strip footing, kPa or lb/ft^2
$\quad Q_{ult}$ = vertical load causing a general shear failure of underlying soil (Fig. 8.1)
$\qquad B$ = width of strip footing, m or ft
$\qquad L$ = length of strip footing, m or ft
$\qquad \gamma_t$ = total unit weight of soil, kN/m^3 or lb/ft^3
$\qquad D_f$ = vertical distance from ground surface to bottom of strip footing, m or ft
$\qquad c$ = cohesion of soil underlying strip footing, kPa or lb/ft^2
$N_c, N_\gamma,$ and N_q = dimensionless bearing capacity factors

As indicated in Eq. (8.3), three terms are added to obtain the ultimate bearing capacity of the strip footing. These terms represent the following:

cN_c The first term accounts for the cohesive shear strength of the soil located below the strip footing. If the soil below the footing is cohesionless (that is, $c = 0$), then this term is zero.

$\tfrac{1}{2}\gamma_t BN_\gamma$ The second term accounts for the frictional shear strength of the soil located below the strip footing. The friction angle ϕ is not included in this term, but is accounted for by the bearing capacity factor N_γ. Note that γ_t represents the total unit weight of the soil located below the footing.

$\gamma_t D_f N_q$ This third term accounts for the soil located above the bottom of the footing. The value of γ_t times D_f represents a surcharge pressure that helps to increase the bearing capacity of the footing. If the footing were constructed at ground surface (that is, $D_f = 0$), then this term would equal zero. This third term indicates that the deeper the footing, the greater the ultimate bearing capacity of the footing. In this term, γ_t represents the total unit weight of the soil located above the bottom of the footing. The total unit weights above and below the footing bottom may be different, in which case different values are used in the second and third terms of Eq. (8.3).

As previously mentioned, Eq. (8.3) was developed by Terzaghi (1943) for strip footings. For other types of footings and loading conditions, corrections need to be applied to the bearing capacity equation. Many different types of corrections have been proposed (e.g., Meyerhof 1951, 1953, 1965). One commonly used form of the bearing capacity equation for spread (square footings) and combined footings (rectangular footings) subjected to uniform vertical loading is as follows (NAVFAC DM-7.2, 1982):

$$q_{ult} = \frac{Q_{ult}}{BL} = cN_c \left(1 + 0.3 \frac{B}{L} \right) + 0.4\gamma_t BN_\gamma + \gamma_t D_f N_q \qquad (8.4)$$

Equation (8.4) is similar to Eq. (8.3), and the terms have the same definitions. An important consideration is that for the strip footing, the shear strength is actually based on a plane strain condition (soil is confined along the long axis of the footing). It has been stated that the friction angle ϕ is about 10 percent higher in the plane strain condition than the friction angle ϕ measured in the triaxial apparatus (Meyerhof 1961, Perloff and Baron 1976). Ladd et al. (1977) indicated that the friction angle ϕ in plane strain is larger than ϕ in triaxial shear by 4° to 9° for dense sands. A difference in friction angle of 4° to 9° has a significant impact on the bearing capacity factors. In practice, plane strain shear strength tests are not performed, and thus there is an added factor of safety for the strip footing compared to the analysis for spread or combined footings.

Bearing Capacity Equation for a Cohesive Soil Layer Underlain by Liquefied Soil. For the situation of a cohesive soil layer overlying a sand that will be susceptible to liquefaction, a total stress analysis can be performed. This type of analysis uses the undrained shear strength of the cohesive soil (Sec. 5.5.1). The undrained shear strength s_u could be determined from field tests, such as the vane shear test (VST), or in the laboratory from unconfined compression tests. Using a total stress analysis, $s_u = c$ and $\phi = 0$ for Eqs. (8.3) and (8.4). For $\phi = 0$, the Terzaghi bearing capacity factors are $N_\gamma = 0$ and $N_q = 1$ (Terzaghi 1943). The bearing capacity equations, (8.3) and (8.4), thus reduce to the following:
For strip footings:

$$q_{ult} = cN_c + \gamma_t D_f = s_u N_c + \gamma_t D_f \qquad (8.5a)$$

For spread footings:

$$q_{ult} = cN_c \left(1 + 0.3 \frac{B}{L} \right) + \gamma_t D_f = s_u N_c \left(1 + 0.3 \frac{B}{L} \right) + \gamma_t D_f \qquad (8.5b)$$

In dealing with shallow footings, the second term $(\gamma_t D_f)$ in Eq. 8.5 tends to be rather small. Thus by neglecting the second term in Eq. (8.5), the final result is as follows:
For strip footings:

$$q_{ult} = cN_c = s_u N_c \qquad (8.6a)$$

For spread footings:

$$q_{ult} = cN_c \left(1 + 0.3 \frac{B}{L} \right) = s_u N_c \left(1 + 0.3 \frac{B}{L} \right) \qquad (8.6b)$$

In order to use Eq. (8.6) to evaluate the ability of a footing to shear through a cohesive soil layer and into a liquefied soil layer, the undrained shear strength of the cohesive soil must be known (that is, $c = s_u$). In addition, the bearing capacity factor N_c must be determined. The presence of an underlying liquefied soil layer will tend to decrease the values for N_c. Figure 8.8 can be used to determine the values of N_c for the condition of a unliquefiable cohesive soil layer overlying a soil layer that is expected to liquefy during the design earthquake. In Fig. 8.8, the terms are defined as follows:

Layer 1 = upper cohesive soil layer that has a uniform undrained shear strength, lb/ft² or kPa, or $s_u = c = c_1$

STRENGTH PROFILE

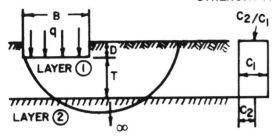

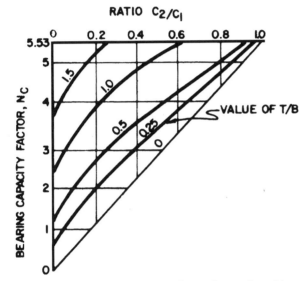

FIGURE 8.8   Bearing capacity factor N_c for two layer soil conditions. (*Reproduced from NAVFAC DM-7.2, 1982.*)

Layer 2 = lower soil layer that will liquefy during the design earthquake. The usual assumption is that the liquefied soil does not possess any shear strength, or $c_2 = 0$.

T = vertical distance from the bottom of the footing to top of the liquefied soil layer, ft or m

B = width of footing, ft or m

Since the liquefied soil layer (i.e., layer 2) has zero shear strength (that is, $c_2 = 0$), the ratio of c_2/c_1 will also be equal to zero. By entering Fig. 8.8 with $c_2/c_1 = 0$ and intersecting the desired T/B curve, the value of N_c can be determined. Using Fig. 8.8, values of N_c for different T/B ratios are as follows:

T/B	N_c	Percent reduction in N_c
0	0	100
0.25	0.7	87
0.50	1.3	76
1.00	2.5	55
1.50	3.8	31
∞	5.5	0

Example Problem for Cohesive Surface Layer. This example problem illustrates the use of Eq. (8.6). Use the data from the example problem in Sec. 8.2.2.

Solution. To calculate the factor of safety in terms of a bearing capacity failure for the strip and spread footings, the following values are used:

$P = 50$ kN/m for strip footing and 500 kN for spread footing

$T = 2.5$ m   i.e., total thickness of unliquefiable soil layer minus footing embedment depth $= 3$ m $- 0.5$ m $= 2.5$ m

$c_1 = s_u = 50$ kPa $= 50$ kN/m^2 upper cohesive soil layer

$c_2 = 0$ kPa $= 0$ kN/m^2 liquefied soil layer

$B = 1$ m   strip footing

$B = L = 2$ m   spread footing

$N_c = 5.5$ for strip footing, using Fig. 8.8 with $T/B = 2.5/1 = 2.5$ and $c_2/c_1 = 0$

$N_c = 3.2$ for spread footing, using Fig. 8.8 with $T/B = 2.5/2 = 1.25$ and $c_2/c_1 = 0$

Substituting the above values into Eqs. (8.6a) and (8.6b), gives

$q_{ult} = cN_c = s_u N_c = (50$ kN/m$^2)$ $(5.5) = 275$ kN/m^2 for strip footing

$q_{ult} = s_u N_c \left(1 + 0.3 \dfrac{B}{L}\right) = 1.3 s_u N_c = (1.3)$ $(50$ kN/m$^2)$ $(3.2) = 208$ kN/m^2 for spread footing

The ultimate load is calculated as follows:

$Q_{ult} = q_{ult} B = (275$ kN/m$^2)$ $(1$ m$) = 275$ kN/m   for strip footing

$Q_{ult} = q_{ult} B^2 = (208$ kN/m$^2)$ $(2$ m$)^2 = 832$ kN   for spread footing

And finally the factor of safety is calculated as follows:

$$\text{FS} = \frac{Q_{ult}}{P} = \frac{275 \text{ kN/m}}{50 \text{ kN/m}} = 5.5 \quad \text{for strip footing}$$

$$\text{FS} = \frac{Q_{ult}}{P} = \frac{832 \text{ kN}}{500 \text{ kN}} = 1.7 \quad \text{for spread footing}$$

These values are similar to the values calculated in Sec. 8.2.2 (that is, FS $= 5.0$ for the strip footing and FS $= 2.0$ for the spread footing).

8.2.4 Deep Foundations

Deep foundations are used when the upper soil stratum is too soft, weak, or compressible to support the static and earthquake-induced foundation loads. Deep foundations are also used when there is a possibility of undermining of the foundation. For example, bridge piers are often founded on deep foundations to prevent a loss of support due to flood conditions which could cause river bottom scour. In addition, deep foundations are used when the expected settlement is excessive (Chap. 7), to prevent ground surface damage of the structure (Sec. 7.3), or to prevent a bearing capacity failure caused by the liquefaction of an underlying soil deposit.

Types of Deep Foundations.   The most common types of deep foundations are piles and piers that support individual footings or mat foundations. Piles are defined as relatively long, slender, columnlike members often made of steel, concrete, or wood that are either driven into place or cast in place in predrilled holes. Common types of piles are as follows:

- *Batter pile:*   This pile is driven in at an angle inclined to the vertical to provide high resistance to lateral loads. If the soil should liquefy during an earthquake, then the lateral resistance of the batter pile may be significantly reduced.

- *End-bearing pile:*   The support capacity of this pile is derived principally from the resistance of the foundation material on which the pile tip rests. End-bearing piles are often used when a soft upper layer is underlain by a dense or hard stratum. If the upper soft layer should settle or liquefy during an earthquake, the pile could be subjected to downdrag forces, and the pile must be designed to resist these soil-induced forces.

- *Friction pile:*   The support capacity of this pile is derived principally from the resistance of the soil friction and/or adhesion mobilized along the side of the pile. Friction piles are often used in soft clays where the end-bearing resistance is small because of punching shear at the pile tip. If the soil is susceptible to liquefaction during an earthquake, then both the frictional resistance and the lateral resistance of the pile may be lost during the earthquake.

- *Combined end-bearing and friction pile:*   This pile derives its support capacity from combined end-bearing resistance developed at the pile tip and frictional and/or adhesion resistance on the pile perimeter.

A *pier* is defined as a deep foundation system, similar to a cast-in-place pile, that consists of a columnlike reinforced concrete member. Piers are often of large enough diameter to enable down-hole inspection. Piers are also commonly referred to as drilled shafts, bored piles, or drilled caissons.

There are many other methods available for forming deep foundation elements. Examples include earth stabilization columns, such as (NAVFAC DM-7.2, 1982):

- *Mixed-in-place piles:*   A mixed-in-place soil-cement or soil-lime pile.

- *Vibroflotation-replacement stone columns:*   Vibroflotation or another method is used to make a cylindrical, vertical hole which is filled with compacted open-graded gravel or crushed rock. The stone columns also have the additional capability of reducing the potential for soil liquefaction by allowing the earthquake-induced pore water pressures to rapidly dissipate as water flows into the highly permeable open-graded gravel or crushed rock.

- *Grouted stone columns:*   These are similar to the above but include filling voids with bentonite-cement or water-sand-bentonite cement mixtures.

- *Concrete Vibroflotation columns:*   These are similar to stone columns, but concrete is used instead of gravel.

Design Criteria.   Several different items are used in the design and construction of piles:

1. *Engineering analysis:*   Based on the results of engineering analysis, a deep foundation could be designed and constructed such that it penetrates all the soil layers that are expected to liquefy during the design earthquake. In this case, the deep foundation will derive support from the unliquefiable soil located below the potentially troublesome soil strata. However, the presence of down-drag loads as well as the loss of lateral resistance due to soil liquefaction must be considered in the engineering analysis.

If a liquefiable soil layer is located below the bottom of the deep foundation, then Sec. 8.2.2 could be used to analyze the possibility of the deep foundation's punching into the underlying liquefied soil layer. For end-bearing piles, the load applied to the pile cap can be assumed to be transferred to the pile tips. Then based on the shear strength of the unliquefiable soil below the bottom of the piles as well as the vertical distance from the pile tip to the liquefiable soil layer, the factor of safety can be calculated using Eq. (8.1*b*). Note that B and L in Eq. (8.1*b*) represent the width and length, respectively, of the pile group.

2. *Field load tests:*   Prior to the construction of the foundation, a pile or pier could be load-tested in the field to determine its carrying capacity. Because of the uncertainties in the design of piles based on engineering analyses, pile load tests are common. The pile load test can often result in a more economical foundation then one based solely on engineering analyses. Pile load tests can even be performed to evaluate dynamic loading conditions. For example, ASTM provides guidelines on the dynamic testing of piles (for example, D 4945-96, "Standard Test Method for High-Strain Dynamic Testing of Piles" 2000). In this test method, ASTM states:

> This test method is used to provide data on strain or force and acceleration, velocity or displacement of a pile under impact force. The data are used to estimate the bearing capacity and the integrity of the pile, as well as hammer performance, pile stresses, and soil dynamics characteristics, such as soil damping coefficients and quake values.

A limitation of field load tests is that they cannot simulate the response of the pile for those situations where the soil is expected to liquefy during the design earthquake. Thus the results of the pile load tests would have to be modified for the expected liquefaction conditions.

3. *Application of pile driving resistance:*   In the past, the pile capacity was estimated based on the driving resistance during the installation of the pile. Pile driving equations, such as the *Engineering News formula* (Wellington 1888), were developed that related the pile capacity to the energy of the pile driving hammer and the average net penetration of the pile per blow of the pile hammer. But studies have shown that there is no satisfactory relationship between the pile capacity from pile driving equations and the pile capacity measured from load tests. Based on these studies, it has been concluded that use of pile driving equations is no longer justified (Terzaghi and Peck 1967).

Especially for high displacement piles that are closely spaced, the vibrations and soil displacement associated with driving the piles will densify granular soil. Thus the liquefaction resistance of the soil is often increased due the pile driving (see compaction piles in Sec. 12.3.3).

4. *Specifications and experience:*   Other factors that should be considered in the deep foundation design include the governing building code or agency requirements and local experience. Local experience, such as the performance of deep foundations during prior earthquakes, can be a very important factor in the design and construction of pile foundations.

The use of pile foundations is discussed further in Chap. 13.

8.2.5 Other Design Considerations

There are many other possible considerations in the determination of the bearing capacity of soil that will liquefy during the design earthquake. Some important items are as follows:

Determination of T. An essential part of the bearing capacity analysis is the determination of T, which is the distance from the bottom of the footing to the top of the liquefied soil layer. This distance may be easy to determine if the upper unliquefiable soil layer is a cohesive soil, such as a fat clay.

It is much more difficult to determine T for soil that is below the groundwater table and has a factor of safety against liquefaction that is slightly greater than 1.0. This is because if a lower layer liquefies, an upward flow of water could induce liquefaction of the layer that has a factor of safety slightly greater than 1.0. In addition, the shear stress induced on the soil by the foundation can actually reduce the liquefaction resistance of loose soil (see Sec. 9.4.2). Because of these effects, considerable experience and judgment are required in the determination of T.

Lateral Loads. In addition to the vertical load acting on the footing, it may also be subjected to both static and dynamic lateral loads. A common procedure is to treat lateral loads separately and resist the lateral loads by using the soil pressure acting on the sides of the footing (passive pressure) and by using the frictional resistance along the bottom of the footing.

Moments and Eccentric Loads. It is always desirable to design and construct shallow footings so that the vertical load is applied at the center of gravity of the footing. For combined footings that carry more than one vertical load, the combined footing should be designed and constructed so that the vertical loads are symmetric. For earthquake loading, the footing is often subjected to a moment. This moment can be represented by a load P that is offset a certain distance (known as the *eccentricity*) from the center of gravity of the footing.

There are many different methods to evaluate eccentrically loaded footings. Because an eccentrically loaded footing will create a higher bearing pressure under one side than under the opposite side, one approach is to evaluate the actual pressure distribution beneath the footing. The usual procedure is to assume a rigid footing (hence linear pressure distribution) and use the section modulus ($\frac{1}{6} B^2$) in order to calculate the largest and smallest bearing pressures. For a footing having a width B, the largest q' and smallest q'' bearing pressures are as follows:

$$q' = \frac{Q (B + 6e)}{B^2} \qquad\qquad (8.7a)$$

$$q'' = \frac{Q (B - 6e)}{B^2} \qquad\qquad (8.7b)$$

where q' = largest bearing pressure underneath footing, which is located along the same side of footing as the eccentricity, kPa or lb/ft^2

q'' = smallest bearing pressure underneath footing, which is located at the opposite side of footing, kPa or lb/ft^2

$Q = P$ = footing load, lb/ft or kN/m. For both strip footings and spread footings, Q is the load per unit length of footing. Footing load includes dead, live, and seismic loads acting on the footing as well as the weight of the footing itself. Typically the value of Q would be provided by the structural engineer.

e = eccentricity of the load Q, that is, the lateral distance from Q to the center of gravity of footing, m or ft

B = width of footing, m or ft

A usual requirement is that the load Q be located within the middle one-third of the footing, and the above equations are valid only for this condition. The value of q' must not exceed the allowable bearing pressure q_{all}.

Figure 8.9 presents another approach for footings subjected to moments. As indicated in Fig. 8.9a, the moment M is converted to a load Q that is offset from the center of gravity of the footing by an eccentricity e. This approach is identical to the procedure outlined for Eq. (8.7).

The next step is to calculate a reduced area of the footing. As indicated in Fig. 8.9b, the new footing dimensions are calculated as $L' = L - 2e_1$ and $B' = B - 2e_2$. A reduction in footing dimensions in both directions would be applicable only for the case where the footing is subjected to two moments, one moment in the long direction of the footing (hence e_1) and the other moment across the footing (hence e_2). If the footing is subjected to only one moment in either the long or short direction of the footing, then the footing is reduced in only one direction. Similar to Eq. (8.7), this method should be utilized only if the load Q is located within the middle one-third of the footing.

Once the new dimensions L' and B' of the footing have been calculated, the procedure outlined in Sec. 8.2.3 is used by substituting L' for L and B' for B.

Sloping Ground Conditions. Although methods have been developed to determine the allowable bearing capacity of foundations at the top of slopes (e.g., NAVFAC DM-7.2, 1982, page 7.2-135), these methods should be used with caution when dealing with earthquake analyses of soil that will liquefy during the design earthquake. This is because, as shown in Sec. 3.4, the site could be impacted by liquefaction-induced lateral spreading and flow slides. Even if the general vicinity of the site is relatively level, the effect of liquefaction on adjacent slopes or retaining walls must be included in the analysis. For example, Fig. 8.10 shows an example of a warehouse that experienced 2 m of settlement due to lateral movement of a quay wall caused by the liquefaction of a sand layer. If the site consists of sloping ground or if there is a retaining wall adjacent to the site, then in addition to a bearing capacity analysis, a slope stability analysis (Chap. 9) or a retaining wall analysis (Chap. 10) should also be performed.

Inclined Base of Footing. Charts have been developed to determine the bearing capacity factors for footings having inclined bottoms. However, it has been stated that inclined bases should never be constructed for footings (AASHTO 1996). During the earthquake, the inclined footing could translate laterally along the sloping soil or rock contact. If a sloping contact of underlying hard material will be encountered during the excavation of the footing, then the hard material should be excavated in order to construct a level footing that is entirely founded within the hard material.

8.2.6 Example Problem

This example problem for cohesive surface layer illustrates the use of Eq. (8.7) and Fig. 8.9. Use the data from the example problem in Sec. 8.2.2. Assume that in addition to the vertical loads, the strip footing and spread footing will experience an earthquake-induced moment equal to 5 kN·m/m and 150 kN·m, respectively. Furthermore, assume that these moments act in a single direction (i.e., in the B direction).

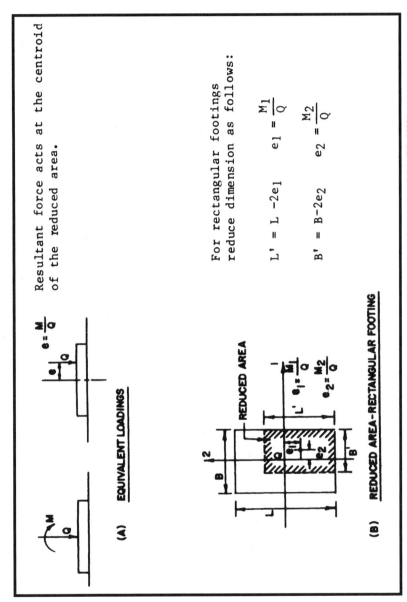

Resultant force acts at the centroid of the reduced area.

For rectangular footings reduce dimension as follows:

$$L' = L - 2e_1 \qquad e_1 = \frac{M_1}{Q}$$

$$B' = B - 2e_2 \qquad e_2 = \frac{M_2}{Q}$$

(A) EQUIVALENT LOADINGS

$$e = \frac{M}{Q}$$

(B) REDUCED AREA-RECTANGULAR FOOTING

REDUCED AREA

$$e_1 = \frac{M_1}{Q}$$

$$e_2 = \frac{M_2}{Q}$$

FIGURE 8.9 Reduced-area method for a footing subjected to a moment. (*Reproduced from NAVFAC DM-7.2, 1982.*)

8.23

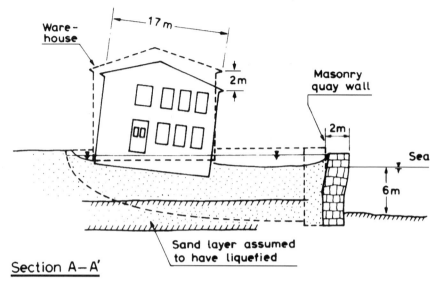

Section A—A'

FIGURE 8.10 Damage to a warehouse due to lateral movement of a quay wall in Zelenica. The liquefaction of the sand layer was caused by the Monte Negro earthquake on April 15, 1979. (*Reproduced from Ishihara 1985.*)

Solution for Strip Footing Using Eq. (8.7). To calculate the factor of safety in terms of a bearing capacity failure for the strip footing, the following values are used:

$Q = P = 50$ kN/m   for strip footing

$e = \dfrac{M}{Q} = \dfrac{5 \text{ kN} \cdot \text{m/m}}{50 \text{ kN/m}} = 0.10$ m   for middle one-third of footing, e cannot exceed

0.17 m, and therefore e is within middle one-third of footing

$q' = \dfrac{Q \, (B + 6e)}{B^2} = \dfrac{50 \, [1 + (6) \, (0.1)]}{1^2} = 80$ kN/m^2 [Eq. (8.7)]

$T = 2.5$ m   i.e., total thickness of unliquefiable soil layer minus footing embedment depth $= 3$ m $- 0.5$ m $= 2.5$ m

$c_1 = s_u = 50$ kPa $= 50$ kN/m^2 upper cohesive soil layer

$c_2 = 0$ kPa $= 0$ kN/m^2 liquefied soil layer

$B = 1$ m

$N_c = 5.5$ using Fig. 8.8 with T/B $= 2.5/1.0 = 2.5$ and $c_2/c_1 = 0$

Using the Terzaghi bearing capacity equation to calculate q_{ult} yields

$$q_{ult} = cN_c = s_u N_c = (50 \text{ kN/m}^2)(5.5) = 275 \text{ kN/m}^2 \qquad [\text{Eq. (8.6}a)\ ]$$

And finally the factor of safety is calculated as follows:

$$FS = \frac{q_{ult}}{q'} = \frac{275 \text{ kN/m}^2}{80 \text{ kN/m}^2} = 3.4$$

Solution for Strip Footing Using Fig. 8.9.   To calculate the factor of safety in terms of a bearing capacity failure for the strip footing, the following values are used:

$$Q = P = 50 \text{ kN/m} \qquad \text{for strip footing}$$

$$e = \frac{M}{Q} = \frac{5 \text{ kN} \cdot \text{m/m}}{50 \text{ kN/m}} = 0.10 \text{ m} \qquad \text{for middle one-third of footing, } e \text{ cannot}$$

exceed 0.17 m, and therefore e is within middle one-third of footing

$$B' = B - 2e = 1 - 2(0.10) = 0.8 \text{ m} \qquad \text{Fig. 8.9}$$

$T = 2.5 \text{ m}$ i.e., total thickness of unliquefiable soil layer minus footing embedment depth = 3 m − 0.5 m = 2.5 m

$c_1 = s_u = 50 \text{ kPa} = 50 \text{ kN/m}^2$ upper cohesive soil layer

$c_2 = 0 \text{ kPa} = 0 \text{ kN/m}^2$ liquefied soil layer

$N_c = 5.5$ using Fig. 8.8 with $T/B = 2.5/1.0 = 2.5$ and $c_2/c_1 = 0$

Using the Terzaghi bearing capacity equation to calculate q_{ult} gives

$$q_{ult} = cN_c = s_u N_c = (50 \text{ kN/m}^2)(5.5) = 275 \text{ kN/m}^2 \qquad [\text{Eq. (8.6}a)\ ]$$

$$Q_{ult} = q_{ult} B' = (275 \text{ kN/m}^2)(0.8 \text{ m}) = 220 \text{ kN/m}$$

And finally the factor of safety is calculated as follows:

$$FS = \frac{Q_{ult}}{Q} = \frac{220 \text{ kN/m}}{50 \text{ kN/m}} = 4.4$$

Solution for Spread Footing Using Eq. (8.7).   To calculate the factor of safety in terms of a bearing capacity failure for the spread footing, the following values are used:

$$Q = P = 500 \text{ kN} \qquad \text{for spread footing}$$

$$e = \frac{M}{Q} = \frac{150 \text{ kN} \cdot \text{m}}{500 \text{ kN}} = 0.30 \text{ m} \qquad \text{for middle one-third of footing, } e \text{ cannot}$$

exceed 0.33 m, and therefore e is within middle one-third of footing

Converting Q to a load per unit length of the footing yields

$$Q = \frac{500 \text{ kN}}{2 \text{ m}} = 250 \text{ kN/m}$$

$$q' = \frac{Q(B + 6e)}{B^2} = \frac{250[2 + (6)(0.3)]}{2^2} = 238 \text{ kN/m}^2 \quad \text{[Eq. (8.7)]}$$

$T = 2.5$ m   i.e., total thickness of unliquefiable soil layer minus footing embedment depth = 3 m − 0.5 m = 2.5 m

$c_1 = s_u = 50$ kPa = 50 kN/m²   upper cohesive soil layer

$c_2 = 0$ kPa = 0 kN/m²   liquefied soil layer

$B = 2$ m

$N_c = 3.2$ for spread footing, using Fig. 8.8 with $T/B = 2.5/2 = 1.25$ and $c_2/c_1 = 0$

Using the Terzaghi bearing capacity equation to calculate q_{ult} results in

$$q_{ult} = s_u N_c \left(1 + 0.3 \frac{B}{L}\right) = 1.3 \, s_u N_c = (1.3)(50 \text{ kN/m}^2)(3.2) = 208 \text{ kN/m}^2$$

And finally the factor of safety is calculated as follows:

$$\text{FS} = \frac{q_{ult}}{q'} = \frac{208 \text{ kN/m}^2}{238 \text{ kN/m}^2} = 0.87$$

Solution for Spread Footing Using Fig. 8.9.   To calculate the factor of safety in terms of a bearing capacity failure for the spread footing, the following values are used:

$Q = P = 500$ kN   for spread footing

$e = \dfrac{M}{Q} = \dfrac{150 \text{ kN} \cdot \text{m}}{500 \text{ kN}} = 0.30$ m   for middle one-third of footing, e cannot

    exceed 0.33 m, and therefore e is within middle one-third of footing

$B' = B - 2e = 2 - 2(0.30) = 1.4$ m   Fig. 8.9

$L' = L = 2$ m   moment only in B direction of footing

$T = 2.5$ m   i.e., total thickness of unliquefiable soil layer minus footing embedment depth = 3 m − 0.5 m = 2.5 m

$c_1 = s_u = 50$ kPa = 50 kN/m²   upper cohesive soil layer

$c_2 = 0$ kPa = 0 kN/m²   liquefied soil layer

$N_c = 3.2$ for spread footing, using Fig. 8.8 with $T/B = 2.5/2 = 1.25$ and $c_2/c_1 = 0$

Using the Terzaghi bearing capacity equation to calculate q_{ult} gives

$$q_{ult} = s_u N_c \left(1 + 0.3 \frac{B'}{L'}\right) = 1.2 \, s_u N_c = (1.2)(50 \text{ kN/m}^2)(3.2) = 190 \text{ kN/m}^2$$

$$Q_{ult} = q_{ult} B'L' = (190 \text{ kN/m}^2)(1.4 \text{ m})(2 \text{ m}) = 530 \text{ kN}$$

And finally the factor of safety is calculated as follows:

$$FS = \frac{Q_{ult}}{Q} = \frac{530 \text{ kN}}{500 \text{ kN}} = 1.06$$

In summary, the factors of the safety factor in terms of a bearing capacity failure for the strip and spread footings are as follows:

Method	Factor of safety	
	Strip footing	Spread footing
Using Eq. (8.7a)	3.4	0.87
Using Fig. 8.9	4.4	1.06
No moment (i.e., values from Sec. 8.2.3)	5.5	1.7

8.3 GRANULAR SOIL WITH EARTHQUAKE-INDUCED PORE WATER PRESSURES

8.3.1 Introduction

Section 8.2 deals with soil that is weakened during the earthquake due to liquefaction. This section deals with granular soil that does not liquefy; rather, there is a reduction in shear strength due to an increase in pore water pressure. Examples include sands and gravels that are below the groundwater table and have a factor of safety against liquefaction that is greater than 1.0 but less than 2.0. If the factor of safety against liquefaction is greater than 2.0, the earthquake-induced excess pore water pressures will typically be small enough that their effect can be neglected.

8.3.2 Bearing Capacity Equation

Using the Terzaghi bearing capacity equation and an effective stress analysis, and recognizing that sands and gravels are cohesionless (that is, $c' = 0$), we see that Eq. (8.3) reduces to the following:

$$q_{ult} = \tfrac{1}{2}\gamma_t B N_\gamma + \gamma_t D_f N_q \tag{8.8}$$

For shallow foundations, it is best to neglect the second term ($\gamma_t D_f N_q$) in Eq. 8.8. This is because this term represents the resistance of the soil located above the bottom of the footing, which may not be mobilized for a punching shear failure into the underlying weakened granular soil layer. Thus by neglecting the second term in Eq. (8.8):

$$q_{ult} = \tfrac{1}{2}\gamma_t B N_\gamma \tag{8.9}$$

Assuming that the location of groundwater table is close to the bottom of the footing, the buoyant unit weight γ_b is used in place of the total unit weight γ_t in Eq. (8.9). In addition, since this is an effective stress analysis, the increase in excess pore water pressures that are generated during the design earthquake must be accounted for in Eq. (8.9). Using Fig. 5.15 can accomplish this, which is a plot of the pore water pressure ratio $r_u = u_e/\sigma'$ versus the factor of safety against liquefaction (Chap. 6). Using the buoyant unit weight γ_b in

place of the total unit weight γ_t and inserting the term $1 - r_u$ to account for the effect of the excess pore water pressures generated by the design earthquake, we get the final result for the ultimate bearing capacity q_{ult} as follows:
For strip footings,

$$q_{ult} = \frac{1}{2} (1 - r_u) \gamma_b B N_\gamma \qquad (8.10a)$$

For spread footings based on Eq. (8.4),

$$q_{ult} = 0.4 (1 - r_u) \gamma_b B N_\gamma \qquad (8.10b)$$

where r_u = pore water pressure ratio from Fig. 5.15 (dimensionless). To determine r_u, the factor of safety against liquefaction of soil located below the bottom of the footing must be determined (see Chap. 6). As previously mentioned, Eq. (8.10) is valid only if the factor of safety against liquefaction is greater than 1.0. When factor of safety against liquefaction is greater than 2.0, Terzaghi bearing capacity equation can be utilized, taking into account the location of groundwater table (see section 8.2.1 of Day 1999).

γ_b = buoyant unit weight of soil below footing, lb/ft^3 or kN/m^3. As previously mentioned, Eq. (8.10) was developed based on an assumption that the groundwater table is located near the bottom of footing or it is anticipated that the groundwater table could rise so that it is near the bottom of the footing.

B = width of footing, ft or m

N_γ = bearing capacity factor (dimensionless). Figure 8.11 presents a chart that can be used to determine the value of N_γ based on the effective friction angle ϕ' of the granular soil.

8.3.3 Example Problem

This example problem illustrates the use of Eq. (8.10). A site consists of a sand deposit with a fluctuating groundwater table. The proposed development will consist of buildings having shallow strip footings to support bearing walls and interior spread footings to support isolated columns. The expected depth of the footings will be 0.5 to 1.0 m. Assume that the groundwater table could periodically rise to a level that is close to the bottom of the footings. Also assume the following parameters: buoyant unit weight of the sand is 9.7 kN/m^3, the sand below the groundwater table has a factor of safety against liquefaction of 1.3, the effective friction angle of the sand $\phi' = 32°$, and the footings will have a minimum width of 1.5 and 2.5 m for the strip and spread footings, respectively. Using a factor of safety of 5, determine the allowable bearing capacity of the footings.

Solution. We use the following values:

$\gamma_b = 9.7$ kN/m^3

$N_\gamma = 21$ entering Fig. 8.11 with $\phi' = 32°$ and intersecting N_γ curve, the value of N_γ from the vertical axis is 21

$B = 1.5$ m for strip footings and 2.5 m for spread footings

$r_u = 0.20$ entering Fig. 5.15 with a factor of safety against liquefaction of 1.3,

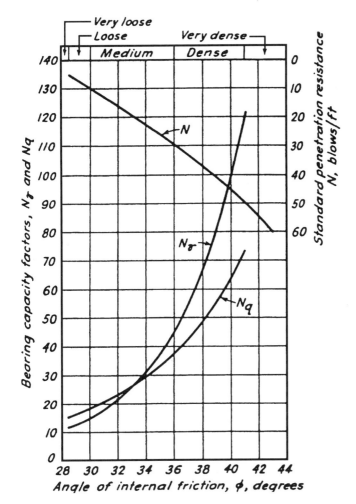

FIGURE 8.11 Bearing capacity factors N_γ and N_q, which automatically incorporate allowance for punching and local shear failure. The standard penetration resistance N value indicated in this chart refers to the uncorrected N value. (*From Peck et al. 1974; reproduced with permission of John Wiley & Sons.*)

value of r_u for sand varies from 0.05 to 0.35. Using an average value, $r_u = 0.20$.

Inserting the above values into Eq. (8.10) yields
For the strip footings:

$$q_{ult} = \tfrac{1}{2}(1 - r_u)\,\gamma_b B N_\gamma = \tfrac{1}{2}(1 - 0.20)(9.7\ \text{kN/m}^3)(1.5\ \text{m})(21) = 120\ \text{kPa}$$

And using a factor of safety of 5.0 gives

$$q_{all} = \frac{q_{ult}}{FS} = \frac{120 \text{ kPa}}{5.0} = 24 \text{ kPa}$$

For the spread footings:

$$q_{ult} = 0.4 (1 - r_u) \gamma_b B N_\gamma = 0.4 (1 - 0.20) (9.7 \text{ kN/m}^3) (2.5 \text{ m}) (21) = 160 \text{ kPa}$$

And using a factor of safety of 5.0 gives

$$q_{all} = \frac{q_{ult}}{FS} = \frac{160 \text{ kPa}}{5.0} = 32 \text{ kPa}$$

Thus provided the strip and spread footings are at least 1.5 and 2.5 m wide, respectively, the allowable bearing capacity is equal to 24 kPa for the strip footings and 32 kPa for the spread footings. These allowable bearing pressures would be used to determine the size of the footings based on the anticipated dead, live, and seismic loads.

8.4 BEARING CAPACITY ANALYSIS FOR COHESIVE SOIL WEAKENED BY THE EARTHQUAKE

8.4.1 Introduction

As discussed in Sec. 5.5.1, cohesive soils and organic soils can also be susceptible to a loss of shear strength during the earthquake. Examples include sensitive clays, which lose shear strength when they are strained back and forth. In dealing with such soils, it is often desirable to limit the stress exerted by the footing during the earthquake so that it is less than the maximum past pressure σ'_{vm} of the cohesive or organic soils. This is to prevent the soil from squeezing out or deforming laterally from underneath the footing.

8.4.2 Bearing Capacity Equation

As mentioned in Sec. 7.5, it is often very difficult to predict the amount of earthquake-induced settlement for foundations bearing on cohesive and organic soils. One approach is to ensure that the foundation has an adequate factor of safety in terms of a bearing capacity failure. To perform a bearing capacity analysis, a total stress analysis can be performed by assuming that $c = s_u$. Using Eq. (8.6), and for a relatively constant undrained shear strength versus depth below the footing, the ultimate bearing capacity is as follows:
For strip footings:

$$q_{ult} = cN_c = 5.5s_u \qquad (8.11a)$$

For spread footings:

$$q_{ult} = cN_c \left(1 + 0.3 \frac{B}{L}\right) = 5.5s_u \left(1 + 0.3 \frac{B}{L}\right) \qquad (8.11b)$$

For a given footing size, the only unknown in Eq. (8.11) is the undrained shear strength s_u. Table 5.4 presents guidelines in terms of the undrained shear strength that should be utilized for earthquake engineering analyses. These guidelines for the selection of the undrained shear strength s_u as applied to bearing capacity analyses are as follows:

1. *Cohesive soil above the groundwater table:* Often the cohesive soil above the groundwater table will have negative pore water pressures due to capillary tension of the pore water fluid. In some cases, the cohesive soil may even be dry and desiccated. The capillary tension tends to hold together the soil particles and to provide additional shear strength to the soil. For the total stress analysis, the undrained shear strength s_u of the cohesive soil could be determined from unconfined compression tests or vane shear tests.

Because of the negative pore water pressures, a future increase in water content would tend to decrease the undrained shear strength s_u of partially saturated cohesive soil above the groundwater table. Thus a possible change in water content in the future should be considered. In addition, an unconfined compression test performed on a partially saturated cohesive soil often has a stress-strain curve that exhibits a peak shear strength which then reduces to an ultimate value. If there is a significant drop-off in shear strength with strain, it may be prudent to use the ultimate value in the bearing capacity analysis.

2. *Cohesive soil below the groundwater table having low sensitivity:* The *sensitivity* S_t of a cohesive soil is defined as the undrained shear strength of an undisturbed soil specimen divided by the undrained shear strength of a completely remolded soil specimen. The sensitivity thus represents the loss of undrained shear strength as a cohesive soil specimen is remolded. An earthquake also tends to shear a cohesive soil back and forth, much as the remolding process does. For cohesive soil having low sensitivity ($S_t \leq 4$), the reduction in the undrained shear strength during the earthquake should be small. Thus the undrained shear strength from the unconfined compression test or vane shear tests could be used in the bearing capacity analysis (for field vane tests, consider a possible reduction in shear strength due to strain rate and anisotropy effects, see Table 7.13 in Day 2000).

3. *Cohesive soil below the groundwater table having a high sensitivity:* For highly sensitive and quick clays ($S_t > 8$), the earthquake-induced ground shaking will tend to shear the soil back and forth, much as the remolding process does. For these types of soils, there could be a significant shear strength loss during the earthquake shaking.

The stress-strain curve from an unconfined compression test performed on a highly sensitive or quick clay often exhibits a peak shear strength that develops at a low vertical strain, followed by a dramatic drop-off in strength with continued straining of the soil specimen. An example of this type of stress-strain curve is shown in Fig. 8.12. The analysis will need to include the estimated reduction in undrained shear strength due to the earthquake shaking. In general, the most critical conditions exist when the highly sensitive or quick clay is subjected to a high static shear stress (such as the high bearing pressure acting on the soil). If, during the earthquake, the sum of the static shear stress and the seismic induced shear stress exceeds the undrained shear strength of the soil, then a significant reduction in shear strength is expected to occur.

Cohesive soils having a medium sensitivity ($4 < S_t \leq 8$) tend to be an intermediate case.

Some of the other factors that may need to be considered in the bearing capacity analysis are as follows:

1. *Earthquake parameters:* The nature of the design earthquake, such as the peak ground acceleration a_{max} and earthquake magnitude, is a factor. The higher the peak ground acceleration and the higher the magnitude of the earthquake, the greater the tendency for the cohesive soil to be strained and remolded by the earthquake shaking.

2. *Soil behavior:* As mentioned above, the important soil properties for the bearing capacity analysis are the undrained shear strength s_u, sensitivity S_t, maximum past pressure σ'_{vm}, and the stress-strain behavior of the soil (e.g., Fig. 8.12).

3. *Rocking:* The increase in shear stress caused by the dynamic loads acting on the foundation must be considered in the analysis. Lightly loaded foundations tend to produce the smallest dynamic loads, while heavy and tall buildings subject the foundation to high dynamic loads due to rocking.

Given the many variables as outlined above, it takes considerable experience and judgment in the selection of the undrained shear strength s_u to be used in Eq. (8.11).

8.4.3 Example Problem

This example problem illustrates the use of Eq. (8.13). Assume that a site has a subsoil profile shown in Fig. 8.13. Suppose that a tall building will be constructed at the site. In addition, during the life of the structure, it is anticipated that the building will be subjected to significant earthquake-induced ground shaking.

Because of the desirability of underground parking, a mat foundation will be constructed such that the bottom of the mat is located at a depth of 20 ft (6 m) below ground surface. Assuming that the mat foundation will be 100 ft long and 100 ft wide (30 m by 30 m), determine the allowable bearing pressure that the mat foundation can exert on the underlying clay layer. Further assume that the clay below the bottom of the mat will not be disturbed (i.e., lose shear strength) during construction of the foundation.

Solution. Based on the sensitivity values S_t listed in Fig. 8.11, this clay would be classified as a quick clay. The analysis has been divided into two parts.

Part A. To prevent the soil from being squeezed out or deforming laterally from underneath the foundation due to rocking of the structure during the earthquake, the allowable bearing pressure should not exceed the maximum past pressure (also known as the preconsolidation stress). Recognizing that the building pressure will decrease with depth, the critical condition is just below the bottom of the foundation (i.e., depth = 20 ft). At a depth of 20 ft (6 m), the preconsolidation stress is about 1.2 kg/cm² (2500 lb/ft²), and it increases with depth. Thus the allowable bearing pressure should not exceed 120 kPa (2500 lb/ft²).

Part B. The next step is to consider a bearing capacity failure. As indicated in Fig. 8.13, the average undrained shear strength s_u from field vane shear tests below a depth of 20 ft (6 m) is about 0.6 kg/cm² (1200 lb/ft²). Field vane shear tests tend to overestimate the undrained shear strength because of the fast strain rate and anisotropy effects, and thus a correction should be applied. Using Bjerrum's (1972) recommended correction (see Fig. 7.19 of Day 2000), the correction factor = 0.85 for a plasticity index = 40 (the plasticity index is from Fig. 8.13, where the liquid limit w_l is about 65 and the plastic limit w_p is about 25). Thus the corrected undrained shear strength is equal to 0.6 kg/cm² times 0.85, or s_u = 0.5 kg/cm² (50 kPa).

Using Eq. (8.11b) gives

$$q_{ult} = 5.5 s_u \left(1 + 0.3\, \frac{B}{L} \right) = 7.1 s_u = (7.1)\,(50\text{ kPa}) = 350\text{ kPa}$$

Using a factor of safety of 5.0 to account for the possibility of a loss of shear strength during the earthquake yields

$$q_{all} = \frac{q_{ult}}{FS} = \frac{350\text{ kPa}}{5.0} = 70\text{ kPa or }1400\text{ lb/ft}^2$$

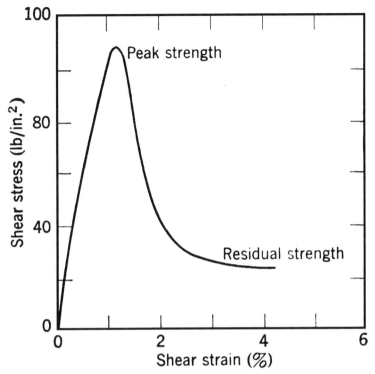

FIGURE 8.12   Stress-strain curve from a shear strength test performed on Cucaracha clay. (*From Lambe and Whitman 1969, reprinted with permission of John Wiley & Sons*)

This allowable bearing pressure does not include any factor to account for the depth of the footing below the ground surface. Usually with shallow foundations, the depth effect is small and could be neglected. However, in this case the bottom of the foundation will be at a depth of 20 ft (6 m). As indicated in Fig. 8.13, the existing vertical effective stress at this depth is equal to 0.5 kg/cm² (50 kPa). Thus the allowable pressure that the foundation can exert on the soil is equal to 70 kPa plus 50 kPa, or 120 kPa (2500 lb/ft²), which is equal to the maximum value calculated from part A.

In summary, the allowable bearing pressure is 120 kPa (2500 lb/ft²). This allowable bearing pressure is the maximum pressure that the foundation can exert on the soil for the condition of a mat foundation located at a depth below ground surface of 20 ft (6 m). Note that the foundation pressure calculated from the structural dead, live, and seismic loads, as well as any eccentricity of loads caused by rocking of the structure during the earthquake, should not exceed this allowable bearing value.

8.5   REPORT PREPARATION

Based on the results of the settlement analysis (Chap. 7) and the bearing capacity analysis (Chap. 8) for both the static and dynamic conditions, the geotechnical engineer would typ-

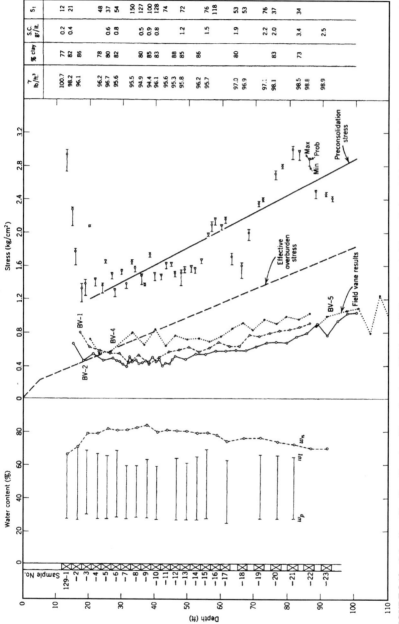

FIGURE 8.13 Subsoil profile, Canadian clay. (*From Lambe and Whitman 1969, reprinted with permission of John Wiley & Sons*)

ically provide design recommendations such as the minimum footing dimensions, embedment requirements, and allowable bearing capacity values. These recommendations would normally be included in a soils report. Appendix D presents an example of a geotechnical engineering report.

An example of typical wording for a bearing material that is not expected to be weakened by the earthquake is as follows:

> The subject site consists of intact Mission Valley formation (siltstone and sandstone) bedrock. For the static design condition, the allowable bearing pressure for spread footings is 8000 lb/ft^2 (400 kPa) provided that the footing is at least 5 ft (1.5 m) wide with a minimum of 2-ft (0.6-m) embedment in firm, intact bedrock. For continuous wall footings, the allowable bearing pressure is 4000 lb/ft^2 (200 kPa) provided that the footing is at least 2 ft (0.6 m) wide with a minimum of 2-ft (0.6-m) embedment in firm, intact bedrock. It is recommended that the structures be entirely supported by bedrock.
>
> Because of cut-fill transition conditions, it is anticipated that piers will be needed for the administrative building. Belled piers can be designed for an allowable end-bearing pressure of 12,000 lb/ft^2 (600 kPa) provided that the piers have a diameter of at least 2 ft (0.6 m), length of at least 10 ft (3 m), with a minimum embedment of 3 ft (0.9 m) in firm, intact bedrock. It is recommended that the geotechnical engineer observe pier installation to confirm embedment requirements.
>
> In designing to resist lateral loads, passive resistance of 1200 lb/ft^2 per foot of depth (200 kPa per meter of depth) to a maximum value of 6000 lb/ft^2 per foot of depth (900 kPa per meter of depth) and a coefficient of friction equal to 0.35 may be utilized for embedment within firm bedrock.
>
> For the analysis of earthquake loading, the above values of allowable bearing pressure and passive resistance may be increased by a factor of one-third. This material is not expected to be weakened by the earthquake-induced ground motion.

An example of typical wording for a bearing material that is expected to be weakened by the earthquake is as follows:

> The subject site consists of a 10-ft-thick upper layer of cohesive soil that is underlain by a 15-ft-thick layer of submerged loose sand. Based on our analysis, it is anticipated that the 15-ft-thick sand layer will liquefy during the design earthquake. Since the site is essentially level, lateral movement due to a liquefaction-induced flow failure or lateral spreading is not anticipated to occur. In addition, the upper 10-ft-thick clay layer should be adequate to prevent liquefaction-induced ground damage (i.e., sand boils, surface fissuring, etc.).
>
> It is our recommendation that the lightly loaded structures be supported by the 10-ft-thick upper cohesive soil layer. For the design condition of lightly loaded shallow foundations, the allowable bearing pressure is 1000 lb/ft^2 (50 kPa). It is recommended that the shallow footings be embedded at a depth of 1 ft (0.3 m) below ground surface and be at least 1 ft (0.3 m) wide.
>
> It is anticipated that piles or piers will be needed for the heavily loaded industrial building. The piles or piers should be founded in the unliquefiable soil stratum which is located at a depth of 25 ft. The piles or piers can be designed for an allowable end-bearing pressure of 4000 lb/ft^2 (200 kPa), provided that the piles or piers have a diameter of at least 1 ft (0.3 m) and are embedded at least 5 ft (1.5 m) into the unliquefiable soil strata. It is recommended that the geotechnical engineer observe pile and pier installation to confirm embedment requirements. The piles or piers should also be designed for down-drag loads during the anticipated earthquake-induced liquefaction of the loose sand layer.
>
> In designing to resist lateral loads, the upper 10-ft-thick clay layer can provide passive resistance of 100 lb/ft^2 per foot of depth (equivalent fluid pressure). For seismic analysis, the underlying 15-ft-thick sand layer should be assumed to have zero passive resistance.
>
> The above values of allowable bearing pressure and passive resistance should not be increased for the earthquake conditions. As previously mentioned, the loose sand layer from a depth of 10 to 25 ft below ground surface is expected to liquefy during the design earthquake (i.e., weakened soil conditions).

8.6 PROBLEMS

The problems have been divided into basic categories as indicated below.

Bearing Capacity for Shallow Foundations Underlain by Liquefied Soil

8.1 Use the data from the example problem for the cohesive surface layer in Sec. 8.2.2. Calculate the spread footing size so that the factor of safety is equal to 5. *Answer:* 5-m by 5-m spread footing.

8.2 Use the data from the example problem for the cohesive surface layer in Sec. 8.2.2. Calculate the maximum concentric load that can be exerted to the 2-m by 2-m spread footing such that the factor of safety is equal to 5. *Answer: P* = 200 kN.

8.3 Use the data from the example problem for the cohesionless surface layer in Sec. 8.2.2. Calculate the maximum concentric load that can be exerted to the strip footing such that the factor of safety is equal to 5. *Answer: P* = 10 kN/m.

8.4 Use the data from the example problem for the cohesionless surface layer in Sec. 8.2.2. Calculate the maximum concentric load that can be exerted to the 2-m by 2-m spread footing such that the factor of safety is equal to 5. *Answer: P* = 40 kN.

8.5 Use the data from Prob. 6.12 and the subsoil profile shown in Fig. 6.13. Assume the following: Peak ground acceleration is equal to $0.20g$, the groundwater table is unlikely to rise above its present level, a 1.5-m-thick fill layer will be constructed at the site, the soil above the groundwater table (including the proposed fill layer) is sand with an effective friction angle ϕ' equal to 33° and $k_0 = 0.5$, and the foundation will consist of shallow strip and spread footings that are 0.3 m deep. Using a factor of safety equal to 5 and footing widths of 1 m, determine the allowable bearing pressure for the strip and spread footings. *Answer:* q_{all} = 10 kPa for 1-m-wide strip footings and q_{all} = 20 kPa for 1-m by 1-m spread footings.

8.6 Solve Prob. 8.5, except assume that the soil above the groundwater table (including the proposed fill layer) is cohesive soil that has undrained shear strength of 20 kPa. *Answer:* q_{all} = 20 kPa for the 1-m-wide strip footings, and q_{all} = 40 kPa for the 1-m by 1-m spread footings.

8.7 Solve Prob. 8.5, using the Terzaghi bearing capacity equation. What values should be used in the design of the footings? *Answer:* q_{all} = 48 kPa for the 1-m-wide strip footings, and q_{all} = 38 kPa for the 1-m by 1-m spread footings. For the design of the footings, use the lower values calculated in Prob. 8.5.

8.8 Solve Prob. 8.6, using the Terzaghi bearing capacity equation. What values should be used in the design of the footings? *Answer:* q_{all} = 22 kPa for the 1-m-wide strip footings, and q_{all} = 30 kPa for the 1-m by 1-m spread footings. For the design of the strip footings, use the value from Prob. 8.6 (q_{all} = 20 kPa). For the design of the spread footings, use the lower value calculated in this problem (q_{all} = 30 kPa).

8.9 Solve Prob. 8.6, assuming that the spread footing is 3 m by 3 m. Use the methods outlined in Secs. 8.2.2 and 8.2.3. What bearing capacity values should be used in the design of the footings? *Answer:* From Eq. (8.1), q_{all} = 14 kPa. From the method in Sec. 8.2.3, q_{all} = 12 kPa. Use the lower value of 12 kPa for the design of the 3-m by 3-m spread footing.

Bearing Capacity for Deep Foundations Underlain by Liquefied Soil

8.10 Use the data from Prob. 6.12 and Fig. 6.13. Assume that a 20-m by 20-m mat foundation is supported by piles, with the tip of the piles located at a depth of 15 m. The piles are evenly spaced along the perimeter and interior portion of the mat. The structural engineer has determined that the critical design load (sum of live, dead, and seismic loads) is equal to 50 MN, which can be assumed to act at the center of the mat and will be transferred to the pile tips. The effective friction angle ϕ' of the sand from a depth of 15 to 17 m is equal to 34° and $k_0 = 0.60$. Calculate the factor of safety [using Eq. (8.1)] for an earthquake-induced punching shear failure into the liquefied soil located at a depth of 17 to 20 m below ground surface. *Answer:* FS = 0.20 and therefore the pile foundation will punch down into the liquefied soil layer located at a depth of 17 to 20 m below ground surface.

8.11   Use the data from Prob. 8.10, but assume that high-displacement friction piles are used to support the mat. The friction piles will densify the upper 15 m of soil and prevent liquefaction of this soil. In addition, the piles will primarily resist the 50-MN load by soil friction along the pile perimeters. Using the 2 : 1 approximation and assuming it starts at a depth of $2/3L$ (where L = pile length), determine the factor of safety [using Eq. (8.1)] for an earthquake-induced punching shear failure into the liquefied soil located at a depth of 17 to 20 m below ground surface. *Answer:* FS = 0.27 and therefore the pile foundation will punch down into the liquefied soil layer located at a depth of 17 to 20 m below ground surface.

Eccentrically Loaded Foundations

8.12   Use the data from the example problem in Sec. 8.2.6. Assume that the eccentricity e is 0.10 m for the strip footing and 0.3 m for the spread footing. Determine the values of Q and M for a factor of safety of 5. *Answer:* For the strip footing, $Q = 34$ kN/m and $M = 3.4$ kN·m/m. For the spread footing, $Q = 88$ kN and $M = 26$ kN·m.

8.13   Use the data for the spread footing from the example problem in Sec. 8.2.6. Assume that there are 150 kN·m moments acting in both the B and L directions. Calculate the factor of safety in terms of a bearing capacity failure. *Answer:* FS = 0.82.

8.14   Use the data from the example problem in Sec. 8.4.3. Assume that the structural engineer has determined that the design load (dead, live, plus rocking seismic load) is 15,000 kips with an eccentricity of 5 ft (eccentricity only in the B direction). Is this an acceptable design based on the allowable bearing values provided in Sec. 8.4.3? *Answer:* Yes, the bearing pressure exerted by the mat is less than the allowable bearing pressure.

Granular Soil with Earthquake-Induced Excess Pore Water Pressures

8.15   Using the data from the example problem in Sec. 8.3.3, determine the allowable load (dead, live, plus seismic) that the footings can support. Assume concentric loading conditions (i.e., no eccentricity). *Answer:* $Q = 36$ kN/m for the strip footing and $Q = 200$ kN for the spread footing.

8.16   Solve the example problem in Sec. 8.3.3, but assume that the factor of safety against liquefaction is equal to 1.2. *Answer:* $q_{all} = 21$ kPa for the strip footing and $q_{all} = 28$ kPa for the spread footing.

Cohesive Soil Weakened by the Earthquake

8.17   Assume a tall building will be constructed at a level-ground site. The foundation will consist of a mat constructed near ground surface. The mat foundation will be 75 ft long and 50 ft wide, and the structural engineer has determined that the design vertical load (including seismic effects) is 20,000 kips located at the center of the mat. Assume that the soil located beneath the mat is a clay that has the shear strength properties shown in Fig. 8.12. Determine the factor of safety for a bearing capacity failure using the fully weakened shear strength. *Answer:* FS = 4.5.

Subsoil Profile

8.18 Assume an oil tank will be constructed at a level-ground site, and the subsurface soil conditions are shown in Fig. 8.14. The groundwater table is located at a depth of 1 m below ground surface.

The standard penetration test values shown in Fig. 8.14 are uncorrected N values. Assume a hammer efficiency E_m of 0.6 and a boring diameter of 100 mm, and the length of the drill rods is equal to the depth of the SPT below ground surface. The design earthquake conditions are a peak ground acceleration a_{max} of $0.20g$ and a magnitude of 7.5.

For the materials shown in Fig. 8.14, assume the following:

a. The surface soil layer (0 to 2.3 m) is clay having an undrained shear strength s_u of 50 kPa. The total unit weight of the soil above the groundwater table γ_t is 19.2 kN/m³, and the buoyant unit weight γ_b is equal to 9.4 kN/m³.

b. The fine sand with gravel layer (2.3 to 8 m) has a low gravel content and can be considered to be essentially a clean sand (γ_b = 9.7 kN/m³).

c. The sand layer (8 to 11.2 m) has less than 5 percent fines (γ_b = 9.6 kN/m³).

d. The silty sand layer (11.2 to 18 m) meets the requirements for a potentially liquefiable soil and has 35 percent fines (γ_b = 9.6 kN/m³).

e. The Flysh claystone (>18 m) is essentially solid rock, and it is not susceptible to earthquake-induced liquefaction or settlement.

Assume the oil tank will be constructed at ground surface and will have a diameter of 20 m and an internal storage capacity equal to a 3-m depth of oil (unit weight of oil = 9.4 kN/m³), and the actual weight of the tank can be ignored in the analysis. Determine the factor of safety against liquefaction and the amount of fill that must be placed at the site to prevent liquefaction-induced ground surface fissuring and sand boils. With the fill layer in place, determine the liquefaction-induced settlement of the tank, and calculate the factor of safety against a bearing capacity failure of the tank. Assume that the fill will be obtained from a borrow site that contains clay, and when compacted, the clay will have an undrained shear strength s_u of 50 kPa. *Answer:* Zone of liquefaction extends from 2.3 to 18 m, thickness of required fill layer at site = 0.7 m, liquefaction-induced settlement of the oil tank = 54 to 66 cm based on Figs. 7.1 and 7.2, and factor of safety against a bearing capacity failure = 1.06.

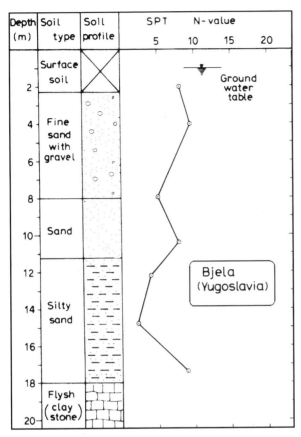

FIGURE 8.14   Subsoil profile, Bjela, Yugoslavia. (*Reproduced from Ishihara 1985.*)

CHAPTER 9
SLOPE STABILITY ANALYSES FOR EARTHQUAKES

The following notation is used in this chapter:

SYMBOL   DEFINITION

a	Acceleration
a_{max}	Maximum horizontal acceleration at ground surface (also known as peak ground acceleration)
a_y	Yield acceleration, which is defined as that acceleration that produces a pseudostatic FS = 1.0
c	Cohesion based on total stress analysis
c'	Cohesion based on effective stress analysis
d	Downslope movement caused by earthquake
D_H	Horizontal ground displacement due to lateral spreading
D_r	Relative density of soil
D_{50}	Grain size corresponding to 50 percent finer of soil
F	Fines content of soil comprising layer T
F_h	Pseudostatic lateral force
FS	Factor of safety
FS_L	Factor of safety against liquefaction
g	Acceleration of gravity
h	Depth below ground surface (for calculation of r_u)
H	Height of free face
k_h	Seismic coefficient, also known as pseudostatic coefficient
k_v	Vertical pseudostatic coefficient
K_α	Factor used to adjust factor of safety against liquefaction for sloping ground
L	Length of slip surface
L	Horizontal distance from base of free face to site location (Sec. 9.5.2)
m	Total mass of slide material
M	Magnitude of design earthquake
M_L	Local magnitude of earthquake
N	Normal force on slip surface
N'	Effective normal force on slip surface
$(N_1)_{60}$	N value corrected for field testing procedures and overburden pressure
r_u	Pore water pressure ratio
R	Distance to expected epicenter or nearest fault rupture
s_u	Undrained shear strength of soil
S	Slope gradient
T	Shear force along slip surface
T	Cumulative thickness of submerged sand layers having $(N_1)_{60} < 15$ (Sec. 9.5.2)
u	Pore water pressure

u_e	Earthquake-induced pore water pressure
u_i	Initial pore water pressure
W	Total weight of failure wedge or failure slice
W	Free face ratio (Sec. 9.5.2)
α	Slope inclination
β	Angular distortion as defined by Boscardin and Cording (1989)
ε_h	Horizontal strain of foundation
ϕ	Friction angle based on total stress analysis
ϕ'	Friction angle based on effective stress analysis
ϕ'_r	Drained residual friction angle
γ_t	Total unit weight of soil
γ_w	Unit weight of water
σ	Total stress
σ'	Effective stress
σ_n	Total normal stress
σ'_n	Effective normal stress
σ'_{v0}	Vertical effective stress
τ_f	Shear strength of soil
$\tau_{h\,\text{static}}$	Static shear stress acting on a horizontal plane

9.1 INTRODUCTION

Section 3.5 presents an introduction to slope movement. Types and examples of earthquake-induced slope movement are discussed in that section. In addition, Sec. 3.4 deals with flow slides and lateral spreading of slopes caused by the liquefaction of soil during the earthquake. Tables 3.1 and 3.2 list the different types of slope movement for rock and soil slopes.

There would appear to be a shaking threshold that is needed to produce earthquake-induced slope movement. For example, as discussed in Sec. 6.3, the threshold values needed to produce liquefaction are a peak ground acceleration a_{max} of about $0.10g$ and local magnitude M_L of about 5 (National Research Council 1985, Ishihara 1985). Thus, those sites having a peak ground acceleration a_{max} less than $0.10g$ or a local magnitude M_L less than 5 would typically not require a liquefaction-related flow slide or lateral spreading analysis. Other threshold values for different types of slope movement are summarized in Tables 9.1 and 9.2.

Tables 9.1 and 9.2 also indicate the relative abundance of earthquake-induced slope failures based on a historical study of 40 earthquakes by Keefer (1984). In general, the most abundant types of slope failures during earthquakes tend to have the lowest threshold values and can involve both small and large masses. For example, rockfalls have a low threshold value ($M_L = 4.0$) and can consist of only one or a few individual rocks, such as shown in Figs. 9.1 and 9.2. Other rockfalls during earthquakes can involve much larger masses of rock, such as shown in Fig. 9.3.

Another example of a very abundant type of earthquake-induced slope movement is a rock slide. As indicated in Table 9.1, rock slides also have a low threshold value ($M_L = 4.0$) and can involve small or large masses of rock. Figure 9.4 shows an example of a rock slide at Pacoima Dam, which was triggered by the San Fernando earthquake in California on February 9, 1971.

Those slope failures listed as *uncommon* in Tables 9.1 and 9.2 tend to have higher threshold values and also typically involve larger masses of soil and rock. Because of their large volume, they tend to be less common. For example, in comparing rock slides and rock block slides in Table 9.1, the rock block slides tend to involve massive blocks of rock that remain

FIGURE 9.1   A rockfall that struck a house located at the mouth of Pacoima Gorge. The rockfall was caused by the San Fernando earthquake (magnitude 6.6) in California on February 9, 1971. (*Photograph from the Steinbrugge Collection, EERC, University of California, Berkeley.*)

FIGURE 9.2   Large rocks from a rockfall that rolled onto the road located on the west side of Carroll Summit. The rockfall was caused by the Dixie Valley–Fairview Peaks earthquake (magnitude 7.0) in Nevada on December 16, 1954. (*Photograph from the Steinbrugge Collection, EERC, University of California, Berkeley.*)

FIGURE 9.3 A large rockfall caused by the Hebgen Lake earthquake in Montana on August 17, 1959. (*Photograph from the Steinbrugge Collection, EERC, University of California, Berkeley.*)

FIGURE 9.4 Rock slide at the Pacoima Dam caused by the San Fernando earthquake in California on February 9, 1971. (*Photograph from the Steinbrugge Collection, EERC, University of California, Berkeley.*)

relatively intact during the earthquake-induced slope movement. Another example is a rock avalanche, which by definition implies a large mass of displaced material. Figure 9.5 shows a rock avalanche caused by the Hebgen Lake earthquake, in Montana, which blocked a canyon and created a temporary lake.

As discussed in Sec. 3.5.3, the seismic evaluation of slope stability can be grouped into two general categories: inertia slope stability analysis and weakening slope stability analysis, as discussed in the following sections.

9.1.1 Inertia Slope Stability Analysis

The inertia slope stability analysis is preferred for those materials that retain their shear strength during the earthquake. Examples of these types of soil and rock are as follows:

- Massive crystalline bedrock and sedimentary rock that remain intact during the earthquake, such as earthquake-induced rock block slide (see Tables 3.1 and 9.1).

- Soils that tend to dilate during the seismic shaking, or, for example, dense to very dense granular soil and heavily overconsolidated cohesive soil such as very stiff to hard clays.

- Soils that have a stress-strain curve that does not exhibit a significant reduction in shear strength with strain. Earthquake-induced slope movement in these soils often takes the form of soil slumps or soil block slides (see Tables 3.2 and 9.2).

- Clay that has a low sensitivity.

- Soils located above the groundwater table. These soils often have negative pore water pressure due to capillary action.

- Landslides that have a distinct rupture surface, and the shear strength along the rupture surface is equal to the drained residual shear strength ϕ_r'.

FIGURE 9.5 Rock avalanche caused by the Hebgen Lake earthquake in Montana on August 17, 1959. (*Photograph from the Steinbrugge Collection, EERC, University of California, Berkeley.*)

TABLE 9.1 Earthquake-Induced Slope Movement in Rock

Main type of slope movement	Subdivisions	Material type	Minimum slope inclination	Threshold values	Relative abundance
Falls	Rockfalls	Rocks weakly cemented, intensely fractured, or weathered; contain conspicuous planes of weakness dipping out of slope or contain boulders in a weak matrix	40° (1.2 : 1)	$M_L = 4.0$	Very abundant (more than 100,000 in the 40 earthquakes)
Slides	Rock slides	Rocks weakly cemented, intensely fractured, or weathered; contain conspicuous planes of weakness dipping out of slope or contain boulders in a weak matrix	35° (1.4 : 1)	$M_L = 4.0$	Very abundant (more than 100,000 in the 40 earthquakes)
	Rock avalanches	Rocks intensely fractured and exhibiting one of the following properties: significant weathering, planes of weakness dipping out of slope, weak cementation, or evidence of previous landsliding	25° (2.1 : 1)	$M_L = 6.0$	Uncommon (100 to 1000 in the 40 earthquakes)
	Rock slumps	Intensely fractured rocks, preexisting rock slump deposits, shale, and other rocks containing layers of weakly cemented or intensely weathered material	15° (3.7 : 1)	$M_L = 5.0$	Moderately common (1000 to 10,000 in the 40 earthquakes)
	Rock block slides	Rocks having conspicuous bedding planes or similar planes of weakness dipping out of slopes	15° (3.7 :1)	$M_L = 5.0$	Uncommon (100 to 1000 in the 40 earthquakes)

Note: Also see Table 3.1 for additional comments.
Sources: Keefer (1984) and Division of Mines and Geology (1997).

TABLE 9.2 Earthquake-Induced Slope Movement in Soil

Main type of slope movement	Subdivisions	Material type	Minimum slope inclination	Threshold values	Relative abundance
Falls	Soil falls	Granular soils that are slightly cemented or contain clay binder	40° (1.2 : 1)	$M_L = 4.0$	Moderately common (1000 to 10,000 in the 40 earthquakes)
Slides	Soil avalanches	Loose, unsaturated sands	25° (2.1 : 1)	$M_L = 6.5$	Abundant (10,000 to 100,000 in the 40 earthquakes)
	Disrupted soil slides	Loose, unsaturated sands	15° (3.7 : 1)	$M_L = 4.0$	Very abundant (more than 100,000 in the 40 earthquakes)
	Soil slumps	Loose, partly to completely saturated sand or silt; uncompacted or poorly compacted artificial fill composed of sand, silt, or clay, preexisting soil slump deposits	10° (5.7 : 1)	$M_L = 4.5$	Abundant (10,000 to 100,000 in the 40 earthquakes)
	Soil block slides	Loose, partly or completely saturated sand or silt; uncompacted or slightly compacted artificial fill composed of sand or silt, bluffs containing horizontal or subhorizontal layers of loose, saturated sand or silt	5° (11 : 1)	$M_L = 4.5$	Abundant (10,000 to 100,000 in the 40 earthquakes)
Flow slides and lateral spreading	Slow earth flows	Stiff, partly to completely saturated clay and preexisting earth flow deposits	10° (5.7 : 1)	$M_L = 5.0$	Uncommon (100 to 1000 in the 40 earthquakes)
	Flow slides	Saturated, uncompacted or slightly compacted artificial fill composed of sand or sandy silt (including hydraulic fill earth dams and tailings dams); loose, saturated granular soils	2.3° (25 : 1)	$M_L = 5.0$ $a_{max} = 0.10g$	Moderately common (1000 to 10,000 in the 40 earthquakes)

TABLE 9.2 Earthquake-Induced Slope Movement in Soil (*Continued*)

Flow slides and lateral spreading (*Continued*)	Subaqueous flows	Loose, saturated granular soils	0.5° (110 : 1)	$M_L = 5.0$ $a_{max} = 0.10g$	Uncommon (100 to 1000 in the 40 earthquakes)
	Lateral spreading	Loose, partly or completely saturated silt or sand, uncompacted or slightly compacted artificial fill composed of sand	0.3° (190 : 1)	$M_L = 5.0$ $a_{max} = 0.10>$	Abundant (10,000 to 100,000 in the 40 earthquakes)

Note: Also see Table 3.2 for additional comments.
Sources: Keefer (1984) and Division of Mines and Geology (1997).

There are many different types of inertia slope stability analyses, and two of the most commonly used are the pseudostatic approach and the Newmark (1965) method. These two methods are described in Secs. 9.2 and 9.3.

9.1.2 Weakening Slope Stability Analysis

The weakening slope stability analysis is preferred for those materials that will experience a significant reduction in shear strength during the earthquake. Examples of these types of soil and rock are as follows:

1. Foliated or friable rock that fractures apart during the earthquake, resulting in rockfalls, rock slides, and rock slumps (see Tables 3.1 and 9.1).
2. Sensitive clays that lose shear strength during the earthquake. An example of a weakening landslide is the Turnagain Heights landslide as described Sec. 3.5.2.
3. Soft clays and organic soils that are overloaded and subjected to plastic flow during the earthquake. The type of slope movement involving these soils is often termed *slow earth flows* (see Tables 3.2 and 9.2).
4. Loose soils located below the groundwater table and subjected to liquefaction or a substantial increase in excess pore water pressure. There are two cases of weakening slope stability analyses involving the liquefaction of soil:

   a. Flow slide: As discussed in Sec. 3.4.4 and Tables 3.2 and 9.2, flow slides develop when the static driving forces exceed the shear strength of the soil along the slip surface, and thus the factor of safety is less than 1.0. Figures 3.38 to 3.40 show the flow slide of the Lower San Fernando Dam caused by the San Fernando earthquake on February 9, 1971.

   b. Lateral spreading: As discussed in Sec. 3.4.5 and Tables 3.2 and 9.2, there could be localized or large-scale lateral spreading of retaining walls and slopes. Examples of large-scale lateral spreading are shown in Figs. 3.41 and 3.42. The concept of cyclic mobility is used to describe large-scale lateral spreading of slopes. In this case, the static driving forces do not exceed the shear strength of the soil along the slip surface, and thus the ground is not subjected to a flow slide. Instead, the driving forces only exceed the resisting forces during those portions of the earthquake that impart net inertial forces in the downslope direction. Each cycle of net inertial forces in the downslope direction causes the driving forces to exceed the resisting forces along the slip surface, resulting in progressive and incremental lateral movement. Often the lateral movement and ground surface cracks first develop at the unconfined toe, and then the slope movement and ground cracks progressively move upslope.

Weakening slope stability analyses are discussed in Secs. 9.4 to 9.6.

9.1.3 Cross Section and Soil Properties

The first step in a slope stability analysis is to develop a cross section through the slope. It is important that cross sections be developed for the critical slope locations, such as those areas that are believed to have the lowest factors of safety. The cross section of the slope and the various soil properties needed for the analysis would be determined during the screening investigation and quantitative evaluation (see Secs. 5.2 to 5.5). Some of the additional items that may need to be addressed prior to performing a slope stability analysis are as follows (adapted from Division of Mines and Geology 1997):

- Do landslides or slope failures, that are active or inactive, exist on or adjacent (either uphill or downhill) to the project?

- Are there geologic formations or other earth materials located on or adjacent to the site that are known to be susceptible to slope movement or landslides?

- Do slope areas show surface manifestations of the presence of subsurface water (springs and seeps), or can potential pathways or sources of concentrated water infiltration be identified on or upslope of the site?

- Are susceptible landforms and vulnerable locations present? These include steep slopes, colluvium-filled swales, cliffs or banks being undercut by stream or water action, areas that have recently slid, and liquefaction-prone areas.

- Given the proposed development, could anticipated changes in the surface and subsurface hydrology (due to watering of lawns, on-site sewage disposal, concentrated runoff from impervious surfaces, etc.) increase the potential for future slope movement or landslides in some areas?

Other considerations for the development of the cross section to be used in the slope stability analysis are discussed in Sec. 9.2.6.

9.2 INERTIA SLOPE STABILITY—PSEUDOSTATIC METHOD

9.2.1 Introduction

As previously mentioned, the inertial slope stability analysis is preferred for those materials that retain their shear strength during the earthquake. The most commonly used inertial slope stability analysis is the pseudostatic approach. The advantages of this method are that it is easy to understand and apply and that the method is applicable for both total stress and effective stress slope stability analyses.

The original application of the pseudostatic method has been credited to Terzaghi (1950). This method ignores the cyclic nature of the earthquake and treats it as if it applied an additional static force upon the slope. In particular, the pseudostatic approach is to apply a lateral force acting through the centroid of the sliding mass, acting in an out-of-slope direction. The pseudostatic lateral force F_h is calculated by using Eq. (6.1), or

$$F_h = ma = \frac{Wa}{g} = \frac{Wa_{max}}{g} = k_h W \qquad (9.1)$$

where F_h = horizontal pseudostatic force acting through the centroid of sliding mass, in an out-of-slope direction, lb or kN. For slope stability analysis, slope is usually assumed to have a unit length (i.e., two-dimensional analysis).

m = total mass of slide material, lb or kg, which is equal to W/g

W = total weight of slide material, lb or kN

a = acceleration, which in this case is the maximum horizontal acceleration at ground surface caused by earthquake ($a = a_{max}$), ft/s^2 or m/s^2

a_{max} = maximum horizontal acceleration at ground surface that is induced by the earthquake, ft/s^2 or m/s^2. The maximum horizontal acceleration is also commonly referred to as the peak ground acceleration (see Sec. 5.6).

$a_{max}/g = k_h$ = seismic coefficient, also known as pseudostatic coefficient (dimensionless)

Note that an earthquake could subject the sliding mass to both vertical and horizontal pseudostatic forces. However, the vertical force is usually ignored in the standard pseudostatic analysis. This is because the vertical pseudostatic force acting on the sliding mass usually has much less effect on the stability of a slope. In addition, most earthquakes produce a peak vertical acceleration that is less than the peak horizontal acceleration, and hence k_v is smaller than k_h.

As indicated in Eq. (9.1), the only unknowns in the pseudostatic method are the weight of the sliding mass W and the seismic coefficient k_h. Based on the results of subsurface exploration and laboratory testing, the unit weight of the soil or rock can be determined, and then the weight of the sliding mass W can be readily calculated. The other unknown is the seismic coefficient k_h, which is much more difficult to determine. The next section discusses guidelines for the selection of the seismic coefficient k_h for the pseudostatic method.

9.2.2 Selection of the Seismic Coefficient

The selection of the seismic coefficient k_h takes considerable experience and judgment. Guidelines for the selection of k_h are as follows:

1. *Peak ground acceleration:* Section 5.6 presents an in-depth discussion of the determination of the peak ground acceleration a_{max} for a given site. The higher the value of the peak ground acceleration a_{max}, the higher the value of k_h that should be used in the pseudostatic analysis.
2. *Earthquake magnitude:* The higher the magnitude of the earthquake, the longer the ground will shake (see Table 2.2) and consequently the higher the value of k_h that should be used in the pseudostatic analysis.
3. *Maximum value of k_h:* When items 1 and 2 as outlined above are considered, keep in mind that the value of k_h should never be greater than the value of a_{max}/g.
4. *Minimum value of k_h:* Check to determine if there are any agency rules that require a specific seismic coefficient. For example, a common requirement by many local agencies in California is the use of a minimum seismic coefficient $k_h = 0.15$ (Division of Mines and Geology 1997).
5. *Size of the sliding mass:* Use a lower seismic coefficient as the size of the slope failure mass increases. The larger the slope failure mass, the less likely that during the earthquake the entire slope mass will be subjected to a destabilizing seismic force acting in the out-of-slope direction. Suggested guidelines are as follows:

a. Small slide mass: Use a value of $k_h = a_{max}/g$ for a small slope failure mass. Examples would include small rockfalls or surficial stability analyses.

b. Intermediate slide mass: Use a value of $k_h = 0.65a_{max}/g$ for slopes of moderate size (Krinitzsky et al. 1993, Taniguchi and Sasaki 1986). Note that this value of 0.65 was used in the liquefaction analysis [see Eq. 6.5)].

c. Large slide mass: Use the lowest values of k_h for large failure masses, such as large embankments, dams, and landslides. Seed (1979b) recommended the following:

$k_h = 0.10$ for sites near faults capable of generating magnitude 6.5 earthquakes. The acceptable pseudostatic factor of safety is 1.15 or greater.

$k_h = 0.15$ for sites near faults capable of generating magnitude 8.5 earthquakes. The acceptable pseudostatic factor of safety is 1.15 or greater.

Other guidelines for the selection of the value of k_h include the following:

- Terzaghi (1950) suggested the following values: $k_h = 0.10$ for "severe" earthquakes, $k_h = 0.20$ for "violent and destructive" earthquakes, and $k_h = 0.50$ for "catastrophic" earthquakes.
- Seed and Martin (1966) and Dakoulas and Gazetas (1986), using shear beam models, showed that the value of k_h for earth dams depends on the size of the failure mass. In particular, the value of k_h for a deep failure surface is substantially less than the value of k_h for a failure surface that does not extend far below the dam crest. This conclusion is identical to item 5 (size of sliding mass) as outlined above.
- Marcuson (1981) suggested that for dams $k_h = 0.33a_{max}/g$ to $0.50\ a_{max}/g$, and consider possible amplification or deamplification of the seismic shaking due to the dam configuration.
- Hynes-Griffin and Franklin (1984), based on a study of the earthquake records from more than 350 accelerograms, use $k_h = 0.50a_{max}/g$ for earth dams. By using this seismic coefficient and having a psuedostatic factor of safety greater than 1.0, it was concluded that earth dams will not be subjected to "dangerously large" earthquake deformations.
- Kramer (1996) states that the study on earth dams by Hynes-Griffin and Franklin (1984) would be appropriate for most slopes. Also Kramer indicates that there are no hard and fast rules for the selection of the pseudostatic coefficient for slope design, but that it should be based on the actual anticipated level of acceleration in the failure mass (including any amplification or deamplification effects).

9.2.3   Wedge Method

The simplest type of slope stability analysis is the wedge method. Figure 9.6 illustrates the free-body diagram for the wedge method. Note in this figure that the failure wedge has a planar slip surface inclined at an angle α to the horizontal. Although the failure wedge passes through the toe of the slope in Fig. 9.6, the analysis could also be performed for the case of the planar slip surface intersecting the face of the slope.

For the pseudostatic wedge analysis, there are four forces acting on the wedge:

W = weight of failure wedge, lb or kN. Usually a two-dimensional analysis is performed based on an assumed unit length of slope (i.e., length of slope = 1 ft or 1 m). Thus the weight of the wedge is calculated as the total unit weight γ_t times the cross-sectional area of the failure wedge.

$F_h = k_h W$ = horizontal psuedostatic force acting through the centroid of the sliding mass, in an out-of-slope direction, lb or kN. The value of the seismic coefficient k_h is discussed in Sec. 9.2.2.

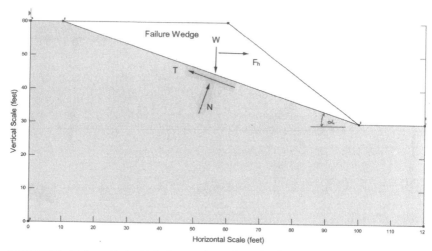

FIGURE 9.6 Wedge method, with the forces acting on the wedge shown in this diagram.

N = normal force acting on the slip surface, lb or kN

T = shear force acting along the slip surface, lb or kN. The shear force is also known as the resisting force because it resists failure of the wedge. Based on the Mohr-Coulomb failure law, the shear force is equal to the following:
For a total stress analysis:

$$T = cL + N \tan \phi$$

or

$$T = s_u L$$

For an effective stress analysis:

$$T = c'L + N' \tan \phi'$$

where L = length of the planar slip surface, ft or m
c, ϕ = shear strength parameters in terms of a total stress analysis
s_u = undrained shear strength of the soil (total stress analysis)
N = total normal force acting on the slip surface, lb or kN
c', ϕ' = shear strength parameters in terms of an effective stress analysis
N' = effective normal force acting on the slip surface, lb or kN

The assumption in this slope stability analysis is that there will be movement of the wedge in a direction that is parallel to the planar slip surface. Thus the factor of safety of the slope can be derived by summing forces parallel to the slip surface, and it is as follows:
Total stress pseudostatic analysis:

$$\text{FS} = \frac{\text{resisting force}}{\text{driving forces}} = \frac{cL + N \tan \phi}{W \sin \alpha + F_h \cos \alpha} = \frac{cL + (W \cos \alpha - F_h \sin \alpha) \tan \phi}{W \sin \alpha + F_h \cos \alpha}$$

$$(9.2a)$$

Effective stress pseudostatic analysis:

$$\text{FS} = \frac{c'L + N' \tan \phi'}{W \sin \alpha + F_h \cos \alpha} = \frac{c'L + (W \cos \alpha - F_h \sin \alpha - uL) \tan \phi'}{W \sin \alpha + F_h \cos \alpha} \qquad (9.2b)$$

where FS = factor of safety for the pseudostatic slope stability (dimensionless parameter)

u = average pore water pressure along the slip surface, kPa or lb/ft^2

Because the wedge method is a two-dimensional analysis based on a unit length of slope (i.e., length = 1 m or 1 ft), the numerator and denominator of Eq. (9.2) are in pounds (or kilonewtons). The resisting force in Eq. (9.2) is equal to the shear strength (in terms of total stress or effective stress) of the soil along the slip surface. The driving forces [Eq. (9.2)] are caused by the pull of gravity and the pseudostatic force and are equal to their components that are parallel to the slip surface.

The total stress pseudostatic analysis is performed in those cases where the total stress parameters of the soil are known. A total stress analysis could be performed by using the consolidated undrained shear strength c and ϕ or the undrained shear strength s_u of the slip surface material. When the undrained shear strength is used, $s_u = c$ and $\phi = 0$ are substituted into Eq. (9.2a). A total stress pseudostatic analysis is often performed for cohesive soil, such as silts and clays.

The effective stress pseudostatic analysis is performed in those cases where the effective stress parameters (c' and ϕ') of the soil are known. Note that in order to use an effective stress analysis [Eq. (9.2b)], the pore water pressure u along the slip surface must also be known. The effective stress analysis is often performed for cohesionless soil, such as sands and gravels.

9.2.4 Method of Slices

The most commonly used method of slope stability analysis is the *method of slices,* where the failure mass is subdivided into vertical slices and the factor of safety is calculated based on force equilibrium equations. A circular arc slip surface and rotational type of failure mode are often used for the method of slices, and for homogeneous soil, a circular arc slip surface provides a lower factor of safety than assuming a planar slip surface.

The calculations for the method of slices are similar to those for the wedge-type analysis, except that the resisting and driving forces are calculated for each slice and then summed in order to obtain the factor of safety of the slope. For the *ordinary method of slices* [also known as the *Swedish circle method* or *Fellenius method,* (Fellenius 1936)], the equation used to calculate the factor of safety is identical to Eq. (9.2), with the resisting and driving forces calculated for each slice and then summed to obtain the factor of safety.

Commonly used methods of slices to obtain the factor of safety are listed in Table 9.3. The method of slices is not an exact method because there are more unknowns than equilibrium equations. This requires that an assumption be made concerning the interslice forces. Table 9.3 presents a summary of the assumptions for the various methods. For example, for the ordinary method of slices (Fellenius 1936), it is assumed that the resultant of the interslice forces is parallel to the average inclination of the slice α. It has been determined that because of this interslice assumption for the ordinary method of slices, this method provides a factor of safety that is too low for some situations (Whitman and Bailey 1967). As a result, the other methods listed in Table 9.3 are used more often than the ordinary method of slices.

Because of the tedious nature of the calculations, computer programs are routinely used to perform the analysis. Most slope stability computer programs have the ability to perform pseudostatic slope stability analyses, and the only additional item that needs to be input is the seismic coefficient k_h. In southern California, an acceptable minimum factor of safety of the slope is 1.1 to 1.15 for a pseudostatic slope stability analysis.

TABLE 9.3 Assumptions Concerning Interslice Forces for Different Method of Slices

Type of method of slices	Assumption concerning interslice forces	Reference
Ordinary method of slices	Resultant of interslice forces is parallel to average inclination of slice	Fellenius (1936)
Bishop simplified method	Resultant of interslice forces is horizontal (no interslice shear forces)	Bishop (1955)
Janbu simplified method	Resultant of interslice forces is horizontal (a correction factor is used to account for interslice shear forces)	Janbu (1968)
Janbu generalized method	Location of interslice normal force is defined by an assumed line of thrust	Janbu (1957)
Spencer method	Resultant of interslice forces is of constant slope throughout the sliding mass	Spencer (1967, 1968)
Morgenstern-Price method	Direction of resultant interslice forces is determined by using a selected function	Morgenstern and Price (1965)

Sources: Lambe and Whitman (1969) and Geo-Slope (1991).

Duncan (1996) states that the nearly universal availability of computers and much improved understanding of the mechanics of slope stability analyses have brought about considerable change in the computational aspects of slope stability analysis. Analyses can be done much more thoroughly and, from the point of view of mechanics, more accurately than was possible previously. However, problems can develop because of a lack of understanding of soil mechanics, soil strength, and the computer programs themselves, as well as the inability to analyze the results in order to avoid mistakes and misuse (Duncan 1996).

Section 9.2.7 presents an example problem dealing with the use of the pseudostatic slope stability analysis based on the method of slices.

9.2.5 Landslide Analysis

As mentioned in Sec. 9.1.1, the pseudostatic method can be used for landslides that have a distinct rupture surface, and the shear strength along the rupture surface is equal to the drained residual shear strength ϕ_r'. The residual shear strength ϕ_r' is defined as the remaining (or residual) shear strength of cohesive soil after a considerable amount of shear deformation has occurred. In essence, ϕ_r' represents the minimum shear resistance of a cohesive soil along a fully developed failure surface. The drained residual shear strength is primarily used to evaluate slope stability when there is a preexisting shear surface. An example of a preexisting shear surface is shown in Fig. 9.7, which is the Niguel Summit landslide slip surface that was exposed during its stabilization. In addition to landslides, other conditions that can be modeled using the drained residual shear strength include slopes in overconsolidated fissured clays, slopes in fissured shales, and other types of preexisting shear surfaces, such as sheared bedding planes, joints, and faults (Bjerrum 1967, Skempton and Hutchinson 1969, Skempton 1985, Hawkins and Privett 1985, Ehlig 1992).

Skempton (1964) states that the residual shear strength ϕ_r' is independent of the original shear strength, water content, and liquidity index; and it depends only on the size, shape, and mineralogical composition of the constituent particles. The drained residual friction angle ϕ_r' of cohesive soil could be determined by using the direct shear apparatus. For example, a clay specimen could be placed in the direct shear box and then sheared back and

FIGURE 9.7   Photograph of the slide plane, which was exposed during the stabilization of the Niguel Summit landslide. Note that the direction of movement of the landslide can be inferred by the direction of striations in the slide plane.

forth several times to develop a well-defined shear failure surface. By shearing the soil specimen back and forth, the clay particles become oriented parallel to the direction of shear. Once the shear surface is developed, the drained residual shear strength can be determined by performing a final, slow shear of the specimen.

Besides the direct shear equipment, the drained residual shear strength can be determined by using the torsional ring shear apparatus (Stark and Eid 1994). Back calculations of landslide shear strength indicate that the residual shear strength from torsional ring shear tests is reasonably representative of the slip surface (Watry and Ehlig 1995). Test specifications have recently been developed, i.e., "Standard Test Method for Torsional Ring Shear Test to Determine Drained Residual Shear Strength of Cohesive Soils" (ASTM D 6467-99, 2000).

Figures 9.8 and 9.9 present an example of data obtained from torsional ring shear laboratory tests performed on slide plane material of an actual landslide (Day and Thoeny 1998). It can be seen in Fig. 9.8 that the failure envelope is nonlinear, which is a common occurrence for residual soil (Maksimovic 1989). If a linear failure envelope is assumed to pass through the origin and the shear stress at an effective normal stress of 100 kPa (2090 lb/ft^2), the residual friction angle ϕ_r' is 8.2°. If a linear failure envelope is assumed to pass through the origin and the shear stress at an effective normal stress of 700 kPa (14,600 lb/ft^2), the residual friction angle ϕ_r' is 6.2°. These drained residual friction angles are very low and are probably close to the lowest possible drained residual friction angles of soil. Also note in Fig. 9.9 that the stress-strain curve does not exhibit a reduction in shear strength with strain. This is to be expected since it is the lowest possible shear strength the soil can possess.

When the stability of a landslide is evaluated, the first step is to perform a static analysis. Since the drained residual shear strength is being utilized in the analysis (that is, ϕ_r'), an effective stress analysis must be performed. This means that the location of the groundwater table or the pore water pressures must also be known. After the static analysis is com-

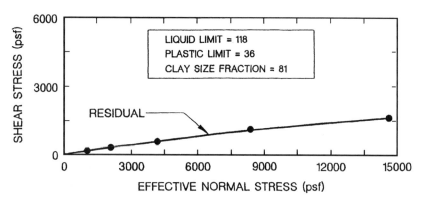

FIGURE 9.8   Drained residual shear strength envelope from torsional ring shear test on slide plane material. Also see Fig. 9.9 for the stress-strain plot.

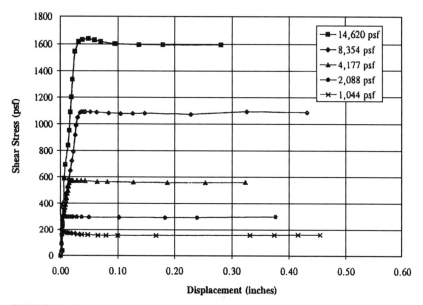

FIGURE 9.9   Shear stress versus displacement from torsional ring shear test on slide plane material. Also see Fig. 9.8 for the shear strength envelope.

pete, a pseudostatic analysis can be performed, and the only additional information that will be needed is the seismic coefficient k_h.

9.2.6   Other Slope Stability Considerations

To perform a pseudostatic slope stability analysis, a cross section must be developed that accurately models the existing or design conditions of the slope. Some of the important factors that may need to be considered in the development of the cross section to be used for the pseudostatic slope stability analysis are as follows:

Different Soil Layers. If a proposed slope or existing slope contains layers of different soil or rock types with different engineering properties, then these layers must be input into the slope stability computer program. Most slope stability computer programs have this capability. Note that for all the different soil layers, either the effective shear strength (c' and ϕ') or the shear strength in terms of total stress parameters (s_u) must be known. It is important that the horizontal pseudostatic force P_h be specified for every layer that comprises the slope cross section.

Slip Surfaces. In some cases, a planar slip surface or a composite-type slip surface may need to be used for the analysis. Most slope stability computer programs have the capability of specifying various types of failure surfaces.

Tension Cracks. It has been stated that tension cracks at the top of the slope can reduce the factor of safety of a slope by as much as 20 percent and are usually regarded as an early and important warning sign of impending failure in cohesive soil (Cernica 1995b). Slope stability programs often have the capability to model or input tension crack zones. The destabilizing effects of water in tension cracks can also be modeled by some slope stability computer programs. When the pseudostatic approach is used, these features should be included in the slope stability analysis.

Surcharge Loads. There may be surcharge loads (such as a building load) at the top of the slope or even on the slope face. Most slope stability computer programs have the capability of including surcharge loads. In some computer programs, other types of loads, such as due to tie-back anchors, can also be included in the analysis. These permanent surcharge loads should also be included in the pseudostatic method.

Nonlinear Shear Strength Envelope. In some cases, the shear strength envelope for soil or rock is nonlinear (e.g., see Fig. 9.8). If the shear strength envelope is nonlinear, then a slope stability computer program that has the capability of using a nonlinear shear strength envelope should be used in the analysis.

Plane Strain Condition. Similar to strip footings, long uniform slopes will be in a plane strain condition. As discussed in Sec. 8.2.3, the friction angle ϕ is about 10 percent higher in the plane strain condition than the friction angle ϕ measured in the triaxial apparatus (Meyerhof 1961, Perloff and Baron 1976). Since plane strain shear strength tests are not performed in practice, there will be an additional factor of safety associated with the plane strain condition. For uniform fill slopes that have a low factor of safety, it is often observed that the "end" slopes (slopes that make a 90° turn) are the first to show indications of slope movement or the first to fail during an earthquake. This is because the end slope is not subjected to a plane strain condition and the shear strength is actually lower than in the center of a long, continuous slope.

Progressive Failure. For the method of slices, the factor of safety is an average value of all the slices. Some slices, such as at the toe of the slope, may have a lower factor of safety which is balanced by other slices that have a higher factor of safety. For those slices that have a low factor of safety, the shear stress and strain may exceed the peak shear strength. For some soils, such as stiff-fissured and sensitive clays, there may be a significant drop in shear strength as the soil deforms beyond the peak values. This reduction in shear strength will then transfer the load to an adjacent slice, which will cause it to experience the same condition. Thus the movement and reduction of shear strength will progress along the slip surface, eventually leading to failure of the slope. Because of this weakening of the soil during the earthquake, it is best to use a weakening slope stability analysis (see Sec. 9.6).

Other Structures. Slope stability analysis can be used for other types of engineering structures. For example, the stability of the ground underneath a retaining wall is often analyzed by considering a slip surface beneath the foundation of the wall.

Effective Stress Analysis. The pseudostatic slope stability analysis can be performed using the effective shear strength of the soil. For this type of analysis, the effective shear strength parameters c' and ϕ' are input into the computer program. The pore water pressures must also be input into the computer program. For the pseudostatic method, it is common to assume that the same pore water pressures exist for the static case and the pseudostatic case. Several different options can be used concerning the pore water pressures:

1. *Zero pore water pressure:* A common assumption for those soil layers that are above the groundwater table is to assume zero pore water pressure. This is a conservative assumption since the soil will often have negative pore water pressures due to capillary effects.

2. *Groundwater table:* A second situation concerns those soils located below the groundwater table. If the groundwater table is horizontal, then the pore water pressures below the groundwater table are typically assumed to be hydrostatic. For the condition of seepage through the slope (i.e., a sloping groundwater table), a flow net can be drawn in order to estimate the pore water pressures below the groundwater table. Most slope stability computer programs have the ability to estimate the pore water pressures below a sloping groundwater table.

3. *Pore water pressure ratio r_u:* A third choice for dealing with pore water pressures is to use the pore water pressure ratio. The pore water pressure ratio is $r_u = u/(\gamma_t h)$, where u = pore water pressure, γ_t = total unit weight of the soil, and h = depth below the ground surface. If a value of $r_u = 0$ is selected, then the pore water pressures u are assumed to be equal to zero in the slope.

Suppose an r_u value is used for the entire slope. In many cases the total unit weight is about equal to 2 times the unit weight of water (that is, $\gamma_t = 2\,\gamma_w$), and thus a value of $r_u = 0.25$ is similar to the effect of a groundwater table at midheight of the slope. A value of $r_u = 0.5$ would be similar to the effect of a groundwater table corresponding to the ground surface.

The pore water pressure ratio r_u can be used for existing slopes where the pore water pressures have been measured in the field, or for the design of proposed slopes where it is desirable to obtain a quick estimate of the effect of pore water pressures on the stability of the slope.

In summary, the pseudostatic approach utilizes the same cross section and conditions that apply for the static slope stability case. The only additional information that most computer programs require to perform the pseudostatic method is the seismic coefficient k_h.

9.2.7 Example Problem

The purpose of this section is to present an example problem dealing with the use of the pseudostatic slope stability analysis based on the method of slices. A cross section through the slope is shown in Fig. 9.10. Specific details on the condition of the slope are as follows:

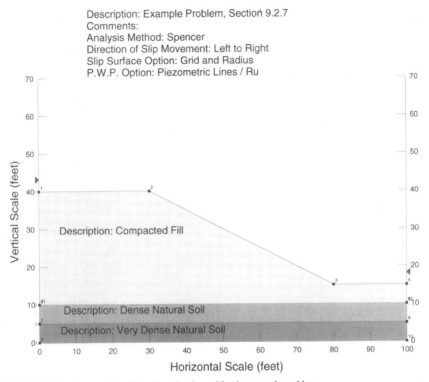

FIGURE 9.10   Cross section of the slope that is used for the example problem.

- Type of analysis: effective stress analysis
- Slope inclination: 2 : 1 (horizontal : vertical)
- Slope height = 25 ft
- Soil types:

  1. *Compacted fill:*   It consists of dense granular soil having the following shear strength parameters: $\phi' = 37°$ and $c' = 0$. The total unit weight of the soil $\gamma_t = 125$ lb/ft³.
  2. *Dense natural soil:*   Underlying the fill, there is a 5-ft-thick dense natural soil layer having the following shear strength parameters: $\phi' = 38°$ and $c' = 100$ lb/ft². The total unit weight of the soil $\gamma_t = 125$ lb/ft³.
  3. *Very dense natural soil:*   Underlying the dense natural soil, there is a very dense natural soil layer having the following shear strength parameters: $\phi' = 40°$ and $c' = 200$ lb/ft². The total unit weight of the soil $\gamma_t = 130$ lb/ft³.

- Groundwater table: The seasonal high groundwater table is located at the top of the dense natural soil layer. For the compacted fill, the pore water pressures u have been assumed to be equal to zero.

The slope stability analyses for the example problem were performed by using the SLOPE/W (Geo-Slope 1991) computer program. In particular, the slope stability analyses for the cross section shown in Fig. 9.10 were performed for two cases: the static case and the pseudostatic case, as described below:

Static Case. The first slope stability analysis was used to calculate the factor of safety of the slope for the static case (i.e., no earthquake forces). Particular details of the analysis are as follows:

• *Critical slip surface:* For this analysis, the computer program was requested to perform a trial-and-error search for the critical slip surface (i.e., the slip surface having the lowest factor of safety). Note in Fig. 9.11 that a grid of points has been produced above the slope. Each one of these points represents the center of rotation of a circular arc slip surface passing through the base of the slope. The computer program has actually performed about 1300 slope stability analyses, using the Spencer method of slices. In Fig. 9.11, the dot with the number 1.590 indicates the center of rotation of the circular arc slip surface with the lowest factor of safety (i.e., lowest factor of safety = 1.59).

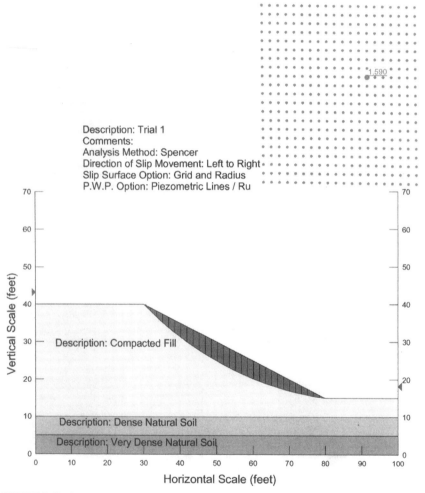

FIGURE 9.11 Slope stability analysis for the static condition using the SLOPE/W computer program (Geo-Slope 1991).

- *Check the results:*   It is always a good idea to check the final results from the computer program. For this slope, with $c' = 0$, the factor of safety can be approximated as follows:

$$FS = \frac{\tan \phi'}{\tan \alpha} = \frac{\tan 37°}{\tan 26.6°} = 1.50$$

This value of FS $= 1.50$ is close to the value calculated by the computer program— 1.59—and provides a check on the answer. A factor of safety of 1.5 is typically an acceptable condition for a permanent slope.

- *Shear strength and shear mobilized:*   The SLOPE/W (Geo-Slope 1991) computer program has the ability to print out different forces acting on the individual slices that comprise the critical failure mass. For example, Fig. 9.12 shows the shear strength and the mobilized shear along the base of each slice. The distance referred to in Fig. 9.12 is the distance measured along the slip surface, starting at the uppermost slice. Notice in Fig. 9.12 that the shear strength is always greater than the mobilized shear for each slice, which makes sense because the factor of safety is much greater than 1.0.

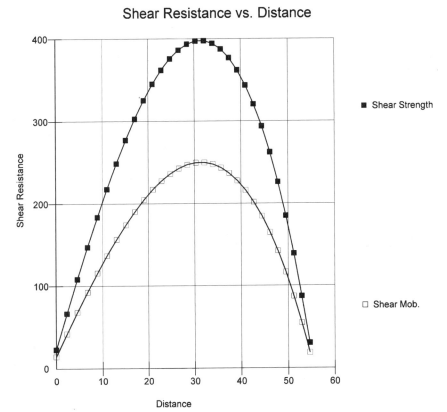

FIGURE 9.12   Shear strength and mobilized shear along the base of each slice for the static slope stability analysis. The SLOPE/W computer program was used to generate the plot (Geo-Slope 1991).

- *Seismic force divided by slice width:* Figure 9.13 shows the seismic force divided by slice width for each slice. Since the first analysis was performed for the static case, the seismic force is equal to zero for all slices.
- *Interslice forces:* Figure 9.14 shows the interslice forces (normal force and shear force). The interslice forces increase and decrease in a similar fashion to the shear forces along the base of the slices (Fig. 9.12). This is to be expected since it is the middle slices that have the greatest depth, and hence greatest shear resistance and highest interslice forces.

Pseudostatic Case. The second slope stability analysis was for the pseudostatic condition. All three soil types for this example problem are in a dense to very dense state and are not expected to lose shear strength during the seismic shaking. Therefore a pseudostatic slope stability analysis can be performed for this slope. The only additional item that needs to be input into the computer program is the seismic coefficient k_h. For this example problem, it was assumed that the seismic coefficient $k_h = 0.40$. Particular details of the analysis are as follows:

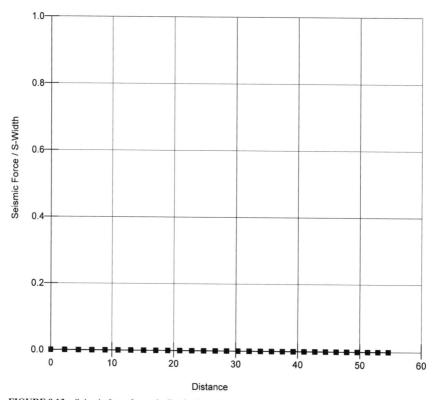

FIGURE 9.13   Seismic force for each slice in the static slope stability analysis. Note that the seismic force must equal zero for the static case. The SLOPE/W computer program was used to generate the plot (Geo-Slope 1991).

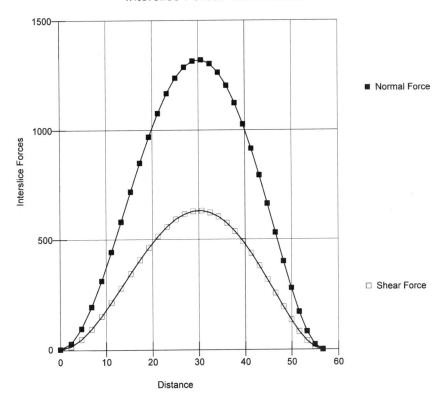

FIGURE 9.14   The shear force and normal force acting between each slice for the static slope stability analysis. These forces are also known as the interslice forces. The SLOPE/W computer program was used to generate the plot (Geo-Slope 1991).

- *Critical slip surface:*   Similar to the static analysis, the computer program was requested to perform a trial-and-error search for the critical slip surface (i.e., the slip surface having the lowest factor of safety). In Fig. 9.15, the dot with the number 0.734 indicates the center of rotation of the circular arc slip surface with the lowest factor of safety (i.e., lowest factor of safety = 0.734). Since this factor of safety is less than 1.0, it is expected that the slope will fail during the earthquake.
- *Shear strength and shear mobilized:*   Figure 9.16 shows the shear strength and the mobilized shear along the base of each slice. Notice in Fig. 9.16 that the shear strength is always less than the mobilized shear for each slice, which makes sense because the factor of safety is less than 1.0.
- *Seismic force divided by slice width:*   Figure 9.17 shows the seismic force divided by the slice width for each slice. The seismic force is higher for the middle slices because they are deeper slices and hence have a larger weight.
- *Interslice forces:*   Figure 9.18 shows the interslice forces (normal force and shear force). The interslice forces increase and decrease in a similar fashion to the shear forces along the base of the slices (Fig. 9.16). This is to be expected since it is the middle slices

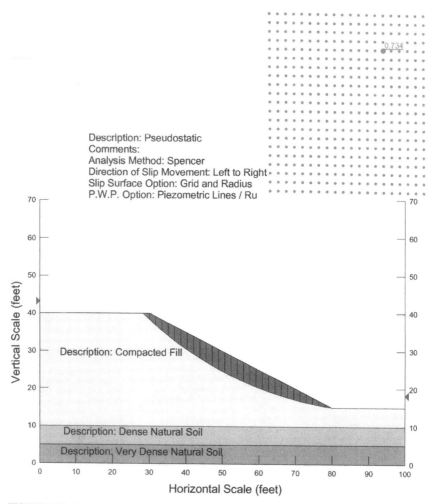

FIGURE 9.15   Slope stability analysis for the pseudostatic condition using the SLOPE/W computer program (Geo-Slope 1991).

that have the greatest depth, and hence greatest shear resistance and highest interslice forces.

- *Extent of slope failure:*   The pseudostatic method shows that only the outer face of the slope is most susceptible to failure. The slip surface could be forced farther back into the slope to evaluate the factor of safety versus distance from the top of slope. The extent of the slope that would be subjected to failure could then be determined (i.e., that slip surface that has FS = 1.0).

In summary, the factor of safety of the slope for the pseudostatic condition is less than 1.0, and failure of the slope is expected to occur during the earthquake. Acceptable values of the pseudostatic factor of safety are typically in the range of 1.1 to 1.15.

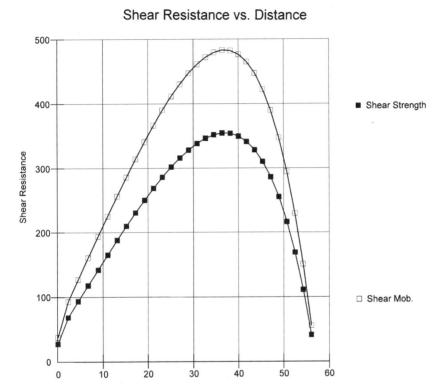

FIGURE 9.16 Shear strength and mobilized shear along the base of each slice for the pseudostatic slope stability analysis. The SLOPE/W computer program was used to generate the plot (Geo-Slope 1991).

9.3 INERTIA SLOPE STABILITY—NEWMARK METHOD

9.3.1 Introduction

The purpose of the Newmark (1965) method is to estimate the slope deformation for those cases where the pseudostatic factor of safety is less than 1.0 (i.e., the failure condition). The Newmark (1965) method assumes that the slope will deform only during those portions of the earthquake when the out-of-slope earthquake forces cause the pseudostatic factor of safety to drop below 1.0. When this occurs, the slope will no longer be stable, and it will be accelerated downslope. The longer that the slope is subjected to a pseudostatic factor of safety below 1.0, the greater the slope deformation. On the other hand, if the pseudostatic factor of safety drops below 1.0 for a mere fraction of a second, then the slope deformation will be limited.

Figure 9.19 can be used to illustrate the basic premise of the Newmark (1965) method. Figure 9.19a shows the horizontal acceleration of the slope during an earthquake. Those accelerations that plot above the zero line are considered to be out-of-slope accelerations, while

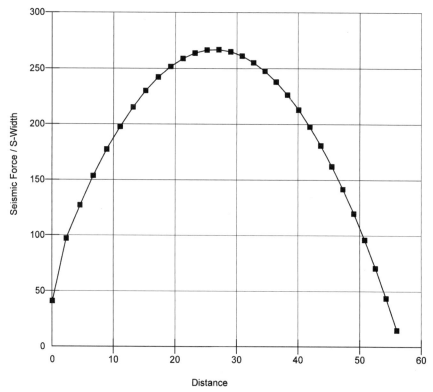

FIGURE 9.17   Seismic force for each slice in the pseudostatic slope stability analysis. The SLOPE/W computer program was used to generate the plot (Geo-Slope 1991).

those accelerations that plot below the zero line are considered to be into-the-slope accelerations. It is only the out-of-slope accelerations that cause downslope movement, and thus only the acceleration that plots above the zero line is considered in the analysis. In Fig. 9.19a, a dashed line has been drawn that corresponds to the horizontal yield acceleration, which is designated a_y. This horizontal yield acceleration a_y is considered to be the horizontal earthquake acceleration that results in a pseudostatic factor of safety that is exactly equal to 1.0. The portions of the two acceleration pulses that plot above a_y have been darkened. According to the Newmark (1965) method, it is these darkened portions of the acceleration pulses that will cause lateral movement of the slope.

Figure 9.19b and c presents the corresponding horizontal velocity and slope displacement that occur in response to the darkened portions of the two acceleration pulses. Note that the slope displacement is incremental and occurs only when the horizontal acceleration from the earthquake exceeds the horizontal yield acceleration a_y. The magnitude of the slope displacement depends on the following factors:

1. *Horizontal yield acceleration a_y:*   The higher the horizontal yield acceleration a_y, the more stable the slope is for any given earthquake.

Interslice Forces vs. Distance

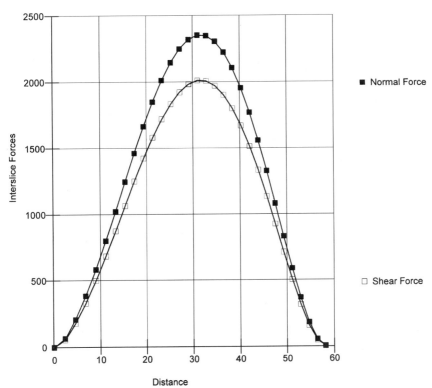

FIGURE 9.18   The shear force and normal force acting between each slice for the pseudostatic slope stability analysis. These forces are also known as the interslice forces. The SLOPE/W computer program was used to generate the plot (Geo-Slope 1991).

2. *Peak ground acceleration a_{max}:*   The peak ground acceleration a_{max} represents the highest value of the horizontal ground acceleration. In essence, this is the amplitude of the maximum acceleration pulse. The greater the difference between the peak ground acceleration a_{max} and the horizontal yield acceleration a_y, the larger the downslope movement.

3. *Length of time:*   The longer the earthquake acceleration exceeds the horizontal yield acceleration a_y, the larger the downslope deformation. Considering the combined effects of items 2 and 3, it can be concluded that the larger the shaded area shown in Fig. 9.19a, the greater the downslope movement.

4. *Number of acceleration pulses:*   The larger the number of acceleration pulses that exceed the horizontal yield acceleration a_y, the greater the cumulative downslope movement during the earthquake.

Many different equations have been developed utilizing the basic Newmark (1965) method as outlined above. One simple equation that is based on the use of two of the four main parameters discussed above is as follows (Ambraseys and Menu 1988):

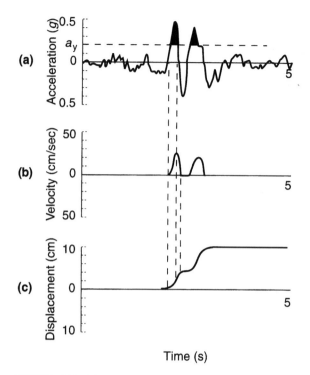

Time (s)

FIGURE 9.19 Diagram illustrating the Newmark method. (*a*) Acceleration versus time; (*b*) velocity versus time for the darkened portions of the acceleration pulses; (*c*) the corresponding downslope displacement versus time in response to the velocity pulses. (*After Wilson and Keefer 1985.*)

$$\log d = 0.90 + \log\left[\left(1 - \frac{a_y}{a_{max}}\right)^{2.53}\left(\frac{a_y}{a_{max}}\right)^{-1.09}\right] \qquad (9.3)$$

where　d = estimated downslope movement caused by the earthquake, cm
　　　　a_y = yield acceleration, defined as the horizontal earthquake acceleration that results in a pseudostatic factor of safety that is exactly equal to 1.0
　　　　a_{max} = peak ground acceleration of the design earthquake

Based on the Newmark (1965) method, Eq. (9.3) is valid only for those cases where the pseudostatic factor of safety is less than 1.0. In essence, the peak ground acceleration a_{max} must be greater then the horizontal yield acceleration a_y. To use Eq. (9.3), the first step is to determine the pseudostatic factor of safety, using the method outlined in Sec. 9.2. Provided the pseudostatic factor of safety is less than 1.0, the next step is to reduce the value of the seismic coefficient k_h until a factor of safety exactly equal to 1.0 is obtained. This can usually be quickly accomplished when using a slope stability computer program. The value of k_h that corresponds to a pseudostatic factor of safety equal to 1.0 can easily be converted to the yield acceleration [i.e., see Eq. (9.1)]. Substituting the values of the peak ground acceleration a_{max} and the yield acceleration a_y into Eq. (9.3), we can determine the slope deformation in centimeters.

Because Eq. (9.3) utilizes the peak ground acceleration a_{max} from the earthquake, the analysis tends to be more accurate for small or medium-sized failure masses where the seismic coefficient k_h is approximately equal to a_{max}/g (see Sec. 9.2.2).

9.3.2 Example Problem

Consider the example problem in Sec. 9.2.7. For this example problem, it was determined that the pseudostatic factor of safety = 0.734 for a peak ground acceleration a_{max} = 0.40g (i.e., the seismic coefficient k_h is equal to 0.40). Since the pseudostatic factor of safety is less than 1.0, the Newmark (1965) method can be used to estimate the slope deformation. Although the stability analysis is not shown, the SLOPE/W (Geo-Slope 1991) computer program was utilized to determine the value of k_h that corresponds to a pseudostatic factor of safety of 1.0. This value of k_h is equal to 0.22, and thus the yield acceleration a_y is equal to 0.22g. Substituting the ratio of a_y/a_{max} = 0.22g/0.40g = 0.55 into Eq. (9.3) yields

$$\log d = 0.90 + \log [(1 - 0.55)^{2.53} (0.55)^{-1.09}]$$

or

$$\log d = 0.90 + \log 0.254 = 0.306$$

And solving the above equation reveals the slope deformation d is equal to about 2 cm. Thus, although the pseudostatic factor of safety is well below 1.0 (i.e., pseudostatic factor of safety = 0.734), Eq. (9.3) predicts that only about 2 cm of downslope movement will occur during the earthquake.

9.3.3 Limitation of the Newmark Method

Introduction. The major assumption of the Newmark (1965) method is that the slope will deform only when the peak ground acceleration a_{max} exceeds the yield acceleration a_y. This type of analysis is most appropriate for a slope that deforms as a single massive block, such as a wedge-type failure. In fact, Newmark (1965) used the analogy of a sliding block on an inclined plane to develop the displacement equations.

A limitation of the Newmark (1965) method is that it may prove unreliable for those slopes that do not tend to deform as a single massive block. An example is a slope composed of dry and loose granular soil (i.e., sands and gravels). The individual soil grains that compose a dry and loose granular soil will tend to individually deform, rather than the entire slope deforming as one massive block.

The earthquake-induced settlement of dry and loose granular soil is discussed in Sec. 7.4 (i.e., volumetric compression). As discussed in that section, the settlement of a dry and loose granular soil is primarily dependent on three factors: (1) the relative density D_r of the soil, which can be correlated with the SPT blow count $(N_1)_{60}$ value; (2) the maximum shear strain induced by the design earthquake; and (3) the number of shear strain cycles.

The amount of lateral movement of slopes composed of dry and loose granular soils is difficult to determine. The method outlined in Sec. 7.4 will tend to underestimate the amount of settlement of a slope composed of dry and loose granular soil. This is because in a sloping environment, the individual soil particles not only will settle, but also will deform laterally in response to the unconfined slope face. In terms of initial calculations, the method outlined in Sec. 7.4 could be used to determine the minimum settlement at the top of slope. However, the actual settlement will be greater because of the unconfined slope condition. In addition, it is anticipated that the lateral movement will be the same order of magnitude as the calculated settlement. The following example problem illustrates these calculations.

Example Problem. To illustrate the analysis for dry and loose sand, assume that a slope has a height of 50 ft (15.2 m) and consists of dry and loose sand that has an $(N_1)_{60}$ value equal to 5. Further assume that the slope has a 22° slope inclination and that the slope is underlain by rock. A cross section illustrating these conditions is presented in Fig. 9.20.

To use the pseudostatic method, the earthquake must not weaken the soil. The pseudostatic slope stability methods can be used for the dry and loose sand because it will not lose shear strength during the earthquake. In fact, as the sand settles during the earthquake, there may even be a slight increase in shear strength. The friction angle ϕ' of well-graded dry and loose sand typically varies from about 30° to 34° (Table 6.12, Day 2001b). Based on an average value, a friction angle ϕ' of the dry and loose sand that is equal to 32° will be used in the slope stability analysis. In addition, for the design earthquake, a peak ground acceleration a_{max} equal to 0.20g will be used. Furthermore, the unit weight of the sand is assumed to be equal to 95 lb/ft³ (15 kN/m³).

Figure 9.21 shows the results of the pseudostatic slope stability analysis. For a peak ground acceleration a_{max} of 0.20g, the pseudostatic factor of safety is equal to 1.116. Since the pseudostatic factor of safety is greater than 1.0, there will be no slope deformation per the Newmark (1965) method. However, by using the method by Tokimatsu and Seed presented in Sec. 7.4.3, a 50-ft- (15.2-m-) thick layer of dry and loose sand having an $(N_1)_{60}$ value of 5 will experience about 2 in (5 cm) of settlement (see Prob. 7.25). Thus the minimum amount of downward movement of the top of slope will be 2 in (5 cm). Because of

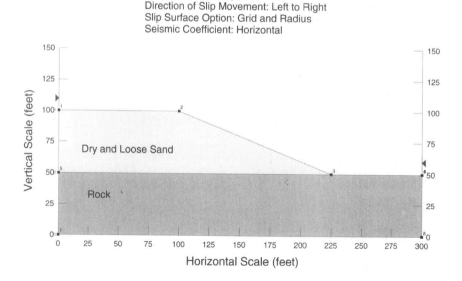

Description: Example Problem - Section 9.3.3
Analysis Method: Spencer
Direction of Slip Movement: Left to Right
Slip Surface Option: Grid and Radius
Seismic Coefficient: Horizontal

Vertical Scale (feet)

Dry and Loose Sand

Rock

Horizontal Scale (feet)

FIGURE 9.20   Cross section of the slope used for the example problem.

Description: Example Problem - Section 9.3.3
Analysis Method: Spencer
Direction of Slip Movement: Left to Right
Slip Surface Option: Grid and Radius
Seismic Coefficient: Horizontal

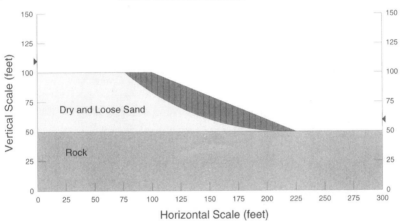

FIGURE 9.21   Slope stability analysis for the pseudostatic condition using the SLOPE/W computer program (Geo-Slope 1991).

the unconfined slope face, it is anticipated that the downward movement will exceed 2 in (5 cm). In addition, there will be lateral movement of the slope, which will also most likely exceed 2 in (5 cm).

Summary.   In summary, the Newmark (1965) method assumes no deformation of the slope during the earthquake if the pseudostatic factor of safety is greater than 1.0. However, as indicated by the above example, a slope composed of dry and loose sand could both settle and deform laterally even if the pseudostatic factor of safety is greater than 1.0. Thus the Newmark (1965) method should be used only for slopes that will deform as an intact massive block, and not for those cases of individual soil particle movement (such as a dry and loose granular soil).

9.4 WEAKENING SLOPE STABILITY—FLOW SLIDES

9.4.1 Introduction

The next three sections discuss the weakening slope stability analysis, which is the preferred method for those materials that will experience a significant reduction in shear strength during the earthquake. This section is devoted to flow slides. As discussed in Sec. 3.4.4 and Tables 3.2 and 9.2, flow slides develop when the static driving forces exceed the weakened shear strength of the soil along the slip surface, and thus the factor of safety is less than 1.0. There are three general types of flow slides:

1. *Mass liquefaction:* This type of flow slide occurs when nearly the entire sloping mass is susceptible to liquefaction. These types of failures often occur to partially or completely submerged slopes, such as shoreline embankments. For example, Fig. 9.22 shows damage to a marine facility at Redondo Beach King Harbor. This damage was caused by the California Northridge earthquake on January 17, 1994. The 5.5 m (18 ft) of horizontal displacement was due to the liquefaction of the offshore sloping fill mass that was constructed as part of the marine facility.

Another example of a mass liquefaction is shown in Fig. 9.23. This figure shows the flow slide of the Middle Niteko Dam caused by the Kobe earthquake in Japan on January 17, 1995.

For design conditions, the first step in the analysis is to determine the factor of safety against liquefaction. If it is determined that the entire sloping mass, or a significant portion

FIGURE 9.22   Damage to a marine facility caused by the California Northridge earthquake on January 17, 1994. (*From Kerwin and Stone 1997; reprinted with permission from the American Society of Civil Engineers.*)

FIGURE 9.23   Flow slide of the Middle Niteko Dam caused by the Kobe earthquake, in Japan, on January 17, 1995. (*Photograph from the Kobe Geotechnical Collection, EERC, University of California, Berkeley.*)

of the sloping mass, will be subjected to liquefaction during the design earthquake, then the slope will be susceptible to a flow slide.

2. *Zonal liquefaction:*   This second type of flow slide develops because there is a specific zone of liquefaction within the slope. For example, Figs. 3.38 to 3.40 show the flow slide of the Lower San Fernando Dam caused by the San Fernando earthquake on February 9, 1971. Figure 3.38 shows the zone of liquefaction that was believed to have caused the flow slide of the Lower San Fernando Dam. As indicated in Fig. 3.38, this flow slide was analyzed by using a circular arc slip surface that passes through the zone of liquefaction.

For design conditions, the first step is to determine the location of the zone of soil expected to liquefy during the design earthquake. Then a slope stability analysis is performed by using various circular arc slip surfaces that pass through the zone of expected liquefaction. If the factor of safety of the slope is less than 1.0, then a flow slide is likely to occur during the earthquake.

3. *Landslide movement caused by liquefaction of soil layers or seams:*   The third type of slope failure develops because of liquefaction of horizontal soil layers or seams of soil. For example, there can be liquefaction of seams of loose saturated sands within a slope. This can cause the entire slope to move laterally along the liquefied layer at the base. These types of landslides caused by liquefied seams of soil caused extensive damage during the 1964 Alaskan earthquake (Shannon and Wilson, Inc. 1964, Hansen 1965). Buildings located in the graben area are subjected to large differential settlements and are often completely destroyed by this type of liquefaction-induced landslide movement (Seed 1970).

For design conditions, it can be difficult to evaluate the possibility of landslide movement due to liquefaction of soil layers or seams. This is because the potentially liquefiable soil layers or seams can be rather thin and may be hard to discover during the subsurface exploration. In addition, when the slope stability analysis is carried out, the slip surface must pass through these horizontal layers or seams of liquefied soil. Thus a slope stability analysis is often performed using a block-type failure mode (rather than using circular arc slip surfaces).

9.4.2 Factor of Safety against Liquefaction for Slopes

Chart for Adjustment of Factor of Safety. As indicated in Sec. 9.4.1, the first step in the flow slide analysis is to determine those zones of soil that will liquefy during the design earthquake. This can be accomplished by using the liquefaction analysis presented in Chap. 6. However, this liquefaction analysis was developed for level-ground sites. For sloping ground sites, the factor of safety against liquefaction calculated from Eq. (6.8) may need to be adjusted. Figure 9.24 presents a chart that can be used to adjust the factor of safety for sloping ground conditions.

In Fig. 9.24, the horizontal axis is designated α, defined as

$$\alpha = \frac{\tau_{h\,static}}{\sigma'_{v0}} \qquad (9.4)$$

where $\tau_{h\,static}$ = static shear stress acting on a horizontal plane
σ'_{v0} = vertical effective stress

For an infinite slope, the value of $\tau_{h\,static}/\sigma'_{v0}$ is approximately equal to the slope ratio (i.e., slope ratio = vertical distance/horizontal distance). Thus a slope that has an inclination of

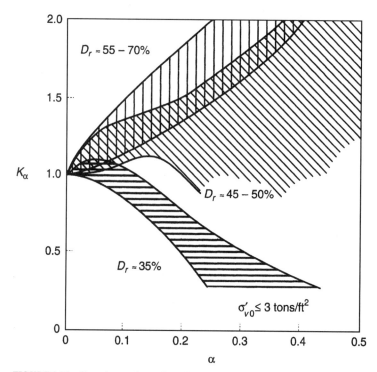

FIGURE 9.24 Chart that can be used to adjust the factor of safety against liquefaction for sloping ground. (*Original from Seed and Harder 1990, reproduced from Kramer 1996.*)

5.7° (which is equivalent to a 10 : 1 horizontal : vertical slope) will have a value of $\alpha = \frac{1}{10} = 0.10$.

In Fig. 9.24, the vertical axis is K_α. In essence, the value of K_α represents the increase (values greater than 1.0) or decrease (values less than 1.0) in the liquefaction resistance of the soil for the sloping ground site. To determine the factor of safety against liquefaction for sloping ground, FS from Eq. (6.8) is multiplied by the value of K_α from Fig. 9.24.

For sites that have a level-ground surface and geostatic soil conditions, the horizontal shear stress is equal to zero ($\tau_{h\,static} = 0$). But for sites with sloping ground or level-ground sites that support heavy structures, there will be a horizontal static shear stress $\tau_{h\,static}$ that is induced into the soil. The presence of this horizontal static shear stress makes loose soil more susceptible to liquefaction. The reason is that less earthquake-induced shear stress is required to cause contraction and hence liquefaction when the soil is already subjected to a static horizontal shear stress $\tau_{h\,static}$.

Example Problem. To use Fig. 9.24, the horizontal axis is entered with the slope ratio α. The appropriate relative density curve of the soil is intersected, and then the value of K_α is obtained from the vertical axis. The factor of safety against liquefaction for the sloping ground is determined as the value of K_α times the factor of safety from Eq. (6.8).

For example, suppose a slope is inclined at an angle of 14°. This slope inclination of 14° is equivalent to a 4 : 1 (horizontal : vertical) slope. Thus $\alpha = \frac{1}{4} = 0.25$.

For this example problem, further assume that the relative density D_r of the soil that comprises the slope is equal to 35 percent. In addition, assume that the factor of safety against liquefaction based on the analysis in Chap. 6 is equal to 1.2 [FS = 1.2 from Eq. (6.8)]. Therefore, entering Fig. 9.24 with $\alpha = 0.25$ and intersecting the middle of the hatched area designated $D_r = 35$ percent, we find the value of K_α is approximately equal to 0.50. And finally, the factor of safety against liquefaction for the sloping ground condition is equal to 1.2 times 0.50, or FS = 0.6. Thus while this soil in a level-ground condition would not have been susceptible to liquefaction (FS = 1.2), in a sloping ground condition (slope inclination = 14°), the soil would be susceptible to liquefaction (FS = 0.6).

Suggested Guidelines. To determine the factor of safety against liquefaction for sloping ground, the slope ratio α, relative density D_r, and factor of safety against liquefaction FS from Eq. (6.8) must be known. Note that Table 5.3 can be used to estimate the value of the relative density D_r for various $(N_1)_{60}$ values. The following guidelines are proposed for sloping ground conditions:

1. $\alpha \leq 0.10$, use $K_\alpha = 1.0$:   In Fig. 9.24, a value of $\alpha = 0.10$ corresponds to a slope inclination of about 6°. Note in Fig. 9.24 that the value of K_α is approximately equal to or greater than 1.0 for a wide range of relative density values. Thus for a slope that has an inclination less than or equal to 6°, ignore the effects of the sloping ground and simply use the FS calculated from Eq. (6.8).

2. $D_r \geq 45$ percent, use $K_\alpha = 1.0$:   In Fig. 9.24, the value of K_α is greater than 1.0 when the relative density of the soil is equal to or greater than 55 percent. Likewise, the value of K_α tends to be greater than 1.0 for a wide range of α values when the relative density is between 45 and 50 percent. Thus for a slope that has soil with a relative density D_r equal to or greater than 45 percent, ignore the effects of the sloping ground and simply use the FS calculated from Eq. (6.8).

3. $\alpha > 0.10$ and $D_r < 45$ percent, use $K_\alpha < 1.0$:   Whenever the slope inclination is greater than about 6° and the relative density D_r of the soil is less than 45 percent, consider a reduction in the factor of safety against liquefaction as calculated from Eq. (6.8). Unfortunately, Fig. 9.24 only shows a range in values of K_α for a relative density of 35

percent. Thus experience and judgment will be required in the selection of the value of K_α from Fig. 9.24 for sloping ground conditions.

9.4.3 Stability Analysis for Liquefied Soil

Introduction. The first step in a flow analysis is to determine the factor of safety against liquefaction for the various soil layers that comprise the slope. The factor of safety against liquefaction is based on level-ground assumptions (Chap. 6). Then the factor of safety against liquefaction [Eq. (6.8)] is adjusted for the sloping ground conditions, by using Fig. 9.24. If it is determined that the entire sloping mass, or a significant portion of the sloping mass, will be subjected to liquefaction during the design earthquake, then the slope will be susceptible to a flow slide. No further analyses will be required for the "mass liquefaction" case.

For the cases of zonal liquefaction or liquefaction of soil layers or seams, a slope stability analysis is required. To perform a slope stability analysis for soil that is anticipated to liquefy during the earthquake, there are two different approaches: (1) using a pore water pressure ratio = 1.0 or (2) using zero shear strength for the liquefiable soil.

Pore Water Pressure Ratio ($r_u = 1.0$). The first approach is to assume that the pore water pressure ratio of the liquefied soil is equal to 1.0. As previously mentioned, the pore water pressure ratio r_u is defined as $r_u = u/(\gamma_t h)$, where u = pore water pressure, γ_t = total unit weight of the soil, and h = depth below the ground surface.

As indicated in Fig. 5.15, at a factor of safety against liquefaction FS_L equal to 1.0 (i.e., liquefied soil), $r_u = 1.0$. Using a value of $r_u = 1.0$, then $r_u = 1.0 = u/(\gamma_t h)$. This means that the pore water pressure u must be equal to the total stress $\sigma = \gamma_t h$, and hence the effective stress σ' is equal to zero ($\sigma' = \sigma - u$). For a granular soil, an effective stress equal to zero means that the soil will not possess any shear strength (i.e., it has liquefied). A pore water pressure ratio r_u of 1.0 for liquefiable soil should be used only when the soil has an effective cohesion c' equal to zero, such as sands and gravels.

Note that the pore water pressure ratio determines the pore water pressures within the soil. Thus the pore water pressure ratio must only be used with an effective stress analysis. For those soil layers that do not liquefy during the earthquake, the effective shear strength parameters (c' and ϕ') and the estimated pore water pressures must be used in the slope stability analysis.

Shear Strength Equals Zero for Liquefied Soil. The second approach is to assume that the liquefied soil has zero shear strength. If a total stress analysis is used, then the liquefied soil layers are assumed to have an undrained shear strength equal to zero ($s_u = 0$). If an effective stress analysis is used, then the effective shear strength parameters are assumed to be equal to zero ($c' = 0$ and $\phi' = 0$).

Example Problem. The purpose of the remainder of this section is to present an example problem dealing with a flow slide analysis. A cross section through the slope is shown in Fig. 9.25. Specific details on the condition of the slope are as follows:

- *Type of analysis:* effective stress analysis
- *Slope inclination:* 2 : 1 (horizontal : vertical)
- *Slope height* = 25 ft
- *Soil types:*

  1. *Compacted fill:* It consists of dense granular soil having the following shear strength parameters: $\phi' = 37°$ and $c' = 0$. The total unit weight of the soil $\gamma_t = 125$ lb/ft^3.

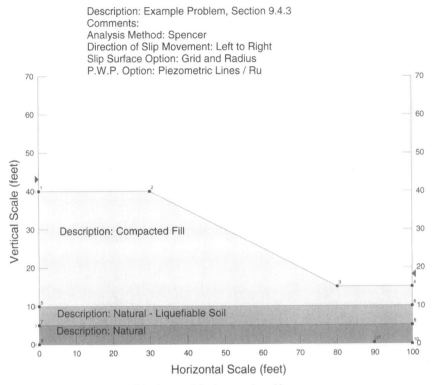

Description: Example Problem, Section 9.4.3
Comments:
Analysis Method: Spencer
Direction of Slip Movement: Left to Right
Slip Surface Option: Grid and Radius
P.W.P. Option: Piezometric Lines / Ru

Description: Compacted Fill

Description: Natural - Liquefiable Soil

Description: Natural

Vertical Scale (feet)

Horizontal Scale (feet)

FIGURE 9.25   Cross section of the slope used for the example problem.

2. *Natural liquefiable soil:*   Underlying the fill, there is a 5-ft-thick layer of loose natural sand having the following shear strength parameters: $\phi' = 30°$ and $c' = 0$ lb/ft^2. The total unit weight of the sand $\gamma_t = 125$ lb/ft^3.

   The results of the liquefaction analysis (Chap. 6) indicate that this layer has a factor of safety against liquefaction less than 1.0. When the correction factor for sloping ground (i.e., Fig. 9.24) is applied, the factor of safety against liquefaction is further reduced. It is thus anticipated that during the design earthquake, this 5-ft-thick layer will liquefy.

3. *Natural:*   Underlying the natural liquefiable soil layer, there is a denser natural soil that has the following shear strength parameters: $\phi' = 40°$ and $c' = 0$ lb/ft^2. The total unit weight of the soil $\gamma_t = 130$ lb/ft^3. This soil has a factor of safety against liquefaction that is well in excess of 1.0, and thus this layer will not liquefy during the design earthquake.

- *Groundwater table:*   The groundwater table is located at the top of the natural liquefiable soil layer. For the compacted fill layer, the pore water pressures u have been assumed to be equal to zero.

The slope stability analyses for the example problem were performed by using the SLOPE/W (Geo-Slope 1991) computer program. In particular, the slope stability analyses for the cross section shown in Figure 9.25 were performed for three cases: (1) the static case, (2) the weakening slope stability case using the pore water pressure (r_u) ratio, and (3) the weakening slope stability case using a shear strength of zero for the liquefied soil layer, as described below.

Example Problem, Static Case.   The first slope stability analysis was used to calculate the factor of safety of the slope for the static case (i.e., no earthquake forces). Particular details of the analysis are as follows:

- *Critical slip surface:*   For this analysis, the computer program was requested to perform a trial-and-error search for the critical slip surface (i.e., the slip surface having the lowest factor of safety). Although the results are not shown, the circular arc slip surface having the lowest factor of safety is equal to 1.59. The critical arc slip surface with the lowest factor of safety is located within the upper compacted fill layer, and the weaker natural soil ($\phi' = 30°$) located below the compacted fill did not have any impact on the static factor of safety results.

- *Check the results:*   It is always a good idea to check the final results from the computer program. For this slope, with $c' = 0$, the factor of safety can be approximated as follows:

$$FS = \frac{\tan \phi'}{\tan \alpha} = \frac{\tan 37°}{\tan 26.6°} = 1.50$$

   This value of FS = 1.50 is close to the value calculated by the computer program—1.59—and provides a check on the answer. A factor of safety of 1.5 would typically be an acceptable condition in terms of the static stability of a permanent slope.

Example Problem, Weakening Slope Stability Analysis ($r_u = 1.0$).   The second slope stability analysis was for the earthquake-induced weakened soil condition. An effective stress analysis was performed, utilizing the following pore water pressure and effective shear strength data:

- *Compacted fill:*   Because the fill is above the groundwater table and in a dense state, the earthquake could generate negative pore water pressures as the soil dilates. However, this effect was ignored, and both the initial (u_i) and the earthquake-induced pore water pressures (u_e) were assumed to be equal to zero ($u_i = 0$ and $u_e = 0$). The effective shear strength parameters used for this soil are the same as those in the static case: $\phi' = 37°$ and $c' = 0$.

- *Natural liquefiable soil:*   For this soil layer, the groundwater table is located at the top of the soil layer. For the slope stability computer program, a pore water pressure ratio r_u of 1.0 was specified. As discussed below, this has the effect of causing the effective stress to equal zero. The effective shear strength parameters used for this soil are the same as those for the static case: $\phi' = 30°$ and $c' = 0$ lb/ft^2.

- *Natural:*   Underlying the natural liquefiable soil layer, there is a denser natural soil that has a factor of safety against liquefaction that is well in excess of 1.0; thus this layer will not liquefy during the design earthquake. In addition, the design earthquake will produce negative pore water pressures as the dense soil dilates. Thus as a conservative approach, the pore water pressures were assumed to be equal to the hydrostatic values, and the effective shear strength parameters of $\phi' = 40°$ and $c' = 0$ lb/ft^2 were used in the analysis.

   Particular details of the weakening slope stability analysis are as follows:

- *Critical slip surface:*   Similar to the static analysis, the computer program was requested to perform a trial-and-error search for the critical slip surface (i.e., the slip surface having the lowest factor of safety). In Fig. 9.26, the dot with the number 0.899 indi-

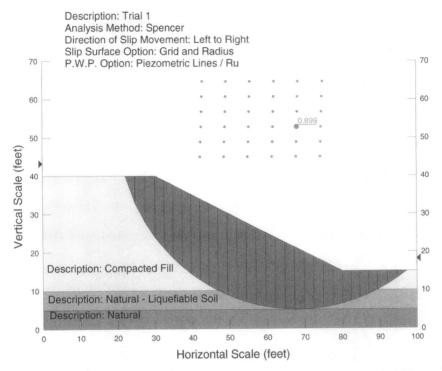

Description: Trial 1
Analysis Method: Spencer
Direction of Slip Movement: Left to Right
Slip Surface Option: Grid and Radius
P.W.P. Option: Piezometric Lines / Ru

FIGURE 9.26   Slope stability analysis for the weakening condition (r_u = 1.0) using the SLOPE/W computer program (Geo-Slope 1991).

cates the center of rotation of the circular arc slip surface with the lowest factor of safety (i.e., lowest factor of safety = 0.899). Since this factor of safety is less than 1.0, it is expected that there will be a flow slide during the earthquake.

- *Normal stress:* Figure 9.27 shows the normal stress acting on the base of each slice. The horizontal axis is the distance measured along the slip surface, starting at the uppermost slice. Each data point in Fig. 9.27 represents the normal stress for an individual slice. Notice in Fig. 9.27 that the highest normal stress occurs for those slices located near the middle part of the failure mass (see Fig. 9.26). These slices have large depths and low angles of inclination, hence high values of normal stress.

- *Pore water pressure:* Figure 9.28 shows the pore water pressure acting on the base of each slice. As previously mentioned, the compacted fill is above the groundwater table, and both the initial (u_i) and earthquake-induced pore water pressures (u_e) were assumed to be equal to zero. Thus the first 10 slices and the final 3 slices, which are those slices in the compacted fill, have zero pore water pressure.

  All those slices within the natural liquefiable soil layer have pore water pressures. These pore water pressures were calculated by the computer program, assuming (1) that the groundwater table is located at the top of the natural liquefiable soil layer and (2) that this layer has a pore water pressure ratio r_u equal to 1.0. In comparing Figs. 9.27 and 9.28, it is evident that all the slices within the natural liquefiable soil layer have nearly identical values of normal stress σ_n and pore water pressure u.

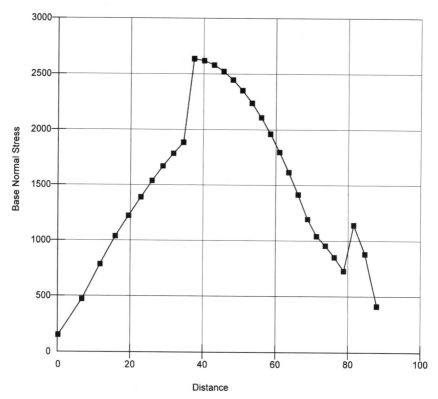

FIGURE 9.27   Normal stress along the base of each slice for the weakening slope stability analysis ($r_u = 1.0$). The SLOPE/W computer program was used to generate the plot (Geo-Slope 1991).

- *Shear strength and shear mobilized:*   Figure 9.29 shows the shear strength and the mobilized shear along the base of each slice. Notice in Fig. 9.29 that the first 10 slices and the last 3 slices have both a shear strength and a mobilized shear value. This is to be expected since all these slices are within the upper compacted fill layer. However, all the slices within the natural liquefiable soil layer have essentially zero shear strength and hence zero mobilized shear. The reason is that a pore water pressure ratio r_u equal to 1.0 was utilized for this soil layer. An $r_u = 1.0$ caused the normal stress σ_n to be equal to the pore water pressure u, which can be observed by comparing Figs. 9.27 and 9.28. If the normal stress σ_n acting on the slip surface is equal to the pore water pressure u, then the effective normal stress σ_n' will be equal to zero ($\sigma_n' = \sigma_n - u$). Using the Mohr-Coulomb failure law ($\tau_f = c' + \sigma_n' \tan \phi'$), with $c' = 0$ and $\sigma_n' = 0$, the shear strength τ_f is equal to zero, as indicated in Fig. 9.29.

Example Problem, Weakening Slope Stability Analysis (Shear Strength = 0).   The third slope stability analysis was for the weakening slope stability case using a shear strength of zero for the liquefied soil layer. The analysis was identical to that in the previous case,

Pore-Water Pressure vs. Distance

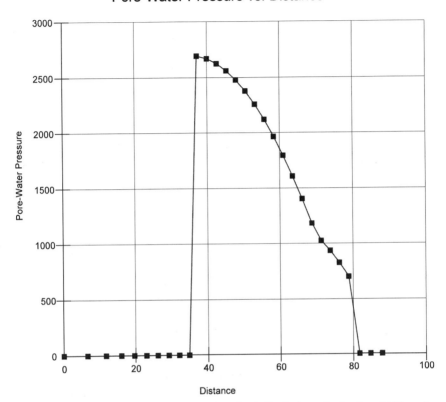

FIGURE 9.28 Pore water pressure acting on the base of each slice for the weakening slope stability analysis (r_u = 1.0). The SLOPE/W computer program was used to generate the plot (Geo-Slope 1991).

except that instead of using r_u = 1.0, the shear strength of the liquefiable soil layer is assumed to be equal to zero (c' = 0 and ϕ' = 0).

Figure 9.30 shows the results of this weakening slope stability analysis. The factor of safety, as indicted in Fig. 9.30 (FS = 0.905), is nearly identical to the factor of safety using an r_u = 1.0 (see Fig. 9.26, FS = 0.899). Thus it can be concluded that the weakening slope stability analysis for granular soils that are anticipated to liquefy during the design earthquake can be performed by using either r_u = 1.0 or ϕ' = 0 for the liquefiable soil layer.

Figure 9.31 shows the shear strength and the mobilized shear along the base of each slice. By using effective shear strength parameters that are equal to zero (c' = 0 and ϕ' = 0), the shear strengths for those slices within the liquefied soil layer are also equal to zero. Note that Fig. 9.31 is nearly identical to Fig. 9.29.

Example Problem Summary. In summary, the weakening slope stability analysis for granular soils that are anticipated to liquefy during the design earthquake can be performed by using either r_u = 1.0 or ϕ' = 0 for the liquefiable soil layer.

The results of the weakening slope stability analysis for the example problem indicate a factor of safety that is less than 1.0, and thus a flow slide is expected to occur during the

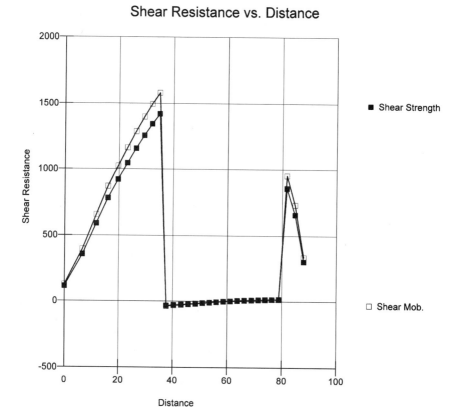

FIGURE 9.29 Shear strength and mobilized shear along the base of each slice for the weakening slope stability analysis ($r_u = 1.0$). The SLOPE/W computer program was used to generate the plot (Geo-Slope 1991).

earthquake. Figures 9.26 and 9.30 show that a significant portion of the slope will be susceptible to a flow slide during the design earthquake. This slope will thus require mitigation measures in order to prevent a flow slide from developing during the design earthquake. Mitigation measures are discussed in Sec. 9.7.

For this example problem, if the factor of safety were greater than 1.0, then a flow slide would not occur during the design earthquake. However, because of the zone of liquefaction, it is likely that there could still be lateral spreading of the slope. Section 9.5 discusses lateral spreading.

9.4.4 Liquefied Shear Strength

It has been stated that the shear strength of liquefied soil may not necessarily be equal to zero. For example, even though the soil liquefies, there may still be a small undrained shear strength caused by the individual soil particles trying to shear past one another as the flow slide develops. This undrained shear strength of liquefied soil has been termed the *liquefied shear strength* (Olson et al. 2000).

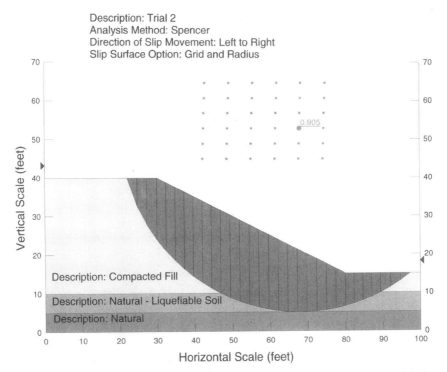

Description: Trial 2
Analysis Method: Spencer
Direction of Slip Movement: Left to Right
Slip Surface Option: Grid and Radius

FIGURE 9.30 Slope stability analysis for the weakening condition based on zero shear strength of the liquefied soil. The SLOPE/W computer program was used to perform the stability analysis (Geo-Slope 1991).

Figure 9.32 presents a plot of the normalized liquefied shear strength versus the equivalent clean sand $(N_1)_{60}$ value (R. B. Seed and Harder 1990, Stark and Mesri 1992). The vertical axis is the undrained liquefied shear strength divided by the pre-earthquake average vertical effective stress σ'_{vo}. The horizontal axis is the equivalent clean sand standard penetration test $(N_1)_{60}$ value. For those flow failures in clean sand, the horizontal axis represents the $(N_1)_{60}$ value from Eq. (5.2). For those sites having silty sands, the $(N_1)_{60}$ values have been adjusted (hence the terminology of equivalent clean sand values).

The data points shown in Fig. 9.32 are based on previous flow slide studies. Using the distance from the initiation of the flow slide to the location where it eventually stopped moving, a kinetics analysis has been performed to calculate the undrained liquefied shear strength needed to stop the movement (Olson et al. 2000). However, it has been argued that perhaps the flow slides stopped moving simply because water was able to drain out of the soil pores during the flow side, resulting in a reduction in pore water pressure and an increase in shear strength. Setting aside the issue of whether these studies actually predict the undrained liquefied shear strength, the liquefied shear strength shown in Fig. 9.32 tends to be very small, especially at low values of $(N_1)_{60}$.

The flow slide analysis presented in Sec. 9.4.3 is approximate. The greatest uncertainty lies in the determination of the factor of safety against liquefaction for sloping ground. For example, the first step is to determine which soil layers will liquefy during the design earthquake based on the analysis in Chap. 6. Then the factor of safety against liquefaction as calculated from Eq. (6.8) must be adjusted for the sloping ground condition. Figure 9.24 would be used

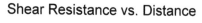

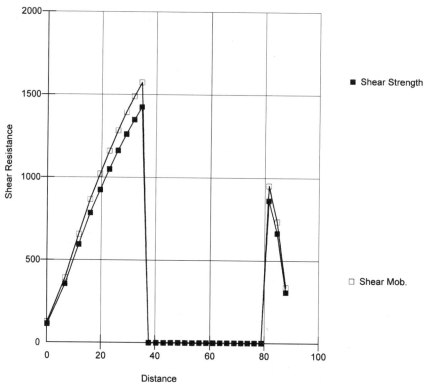

FIGURE 9.31   Shear strength and mobilized shear along the base of each slice for the weakening slope stability analysis (zero shear strength of the liquefied soil). The SLOPE/W computer program was used to generate the plot (Geo-Slope 1991).

to adjust the factor of safety against liquefaction for sloping ground. Since Fig. 9.24 only shows a range in values of K_α for a relative density of 35 percent, experience and judgment will be required in the selection of the value of K_α for different soil densities and sloping ground conditions.

   When the slope stability analysis is performed for flow slides, a total stress analysis could be performed using the undrained liquefied shear strength from Fig. 9.32. Because the flow slide analysis is approximate, a conservative approach is to perform an effective stress analysis (Sec. 9.4.3) and assume that the liquefied shear strength is equal to zero (or use $r_u = 1.0$).

9.5   WEAKENING SLOPE STABILITY—
LIQUEFACTION-INDUCED LATERAL SPREADING

9.5.1   Introduction

Lateral spreading is introduced in Sec. 3.4.5 where, as shown in Figs. 3.27 to 3.30, the principal factor in the damage at the Port of Kobe was attributed to liquefaction, which caused

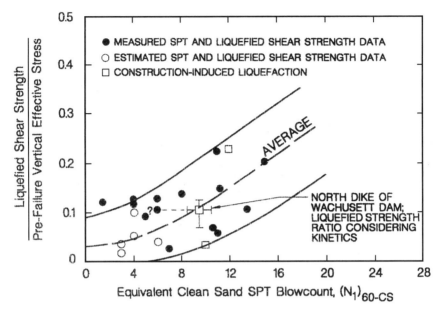

FIGURE 9.32   Normalized liquefied shear strength versus equivalent clean sand $(N_1)_{60}$ value. (*From Olson et al. 2000, reproduced with permission of the American Society of Civil Engineers.*)

lateral deformation of the retaining walls. This liquefaction-induced lateral spreading was usually restricted to the ground surface behind the retaining walls, and thus it is called *localized lateral spreading,* which is discussed in Sec. 10.3.

If the liquefaction-induced lateral spreading causes lateral movement of the ground surface over an extensive distance, then the effect is known as *large-scale lateral spreading.* Such lateral spreads often form adjacent waterways on gently sloping or even flat ground surfaces that liquefy during the earthquake. This section deals with the analysis of large-scale lateral spreading of slopes.

The concept of cyclic mobility is used to describe large-scale lateral spreading. Because the ground is gently sloping or flat, the static driving forces do not exceed the resistance of the soil along the slip surface, and thus the ground is not subjected to a flow slide (i.e., the factor of safety is greater than 1.0). Instead, the driving forces only exceed the resisting forces during those portions of the earthquake that impart net inertial forces in the downslope direction. Each cycle of net inertial forces in the downslope direction causes the driving forces to exceed the resisting forces along the slip surface, resulting in progressive and incremental lateral movement. Often the lateral movement and ground surface cracks first develop at the unconfined toe, and then the slope movement and ground cracks progressively move upslope.

Figure 3.41 shows an example of large-scale lateral spreading caused by liquefaction during the Loma Prieta earthquake on October 17, 1989. As seen in Fig. 3.41, as the displaced ground breaks up internally, it causes fissures, scarps, and depressions to form at ground surface. Notice in Fig. 3.41 that the main ground surface cracks tend to develop parallel to one another. Some of the cracks filled with water from the adjacent waterway. As the ground moves laterally, the blocks of soil between the main cracks tend to settle and break up into even smaller pieces.

Large-scale lateral spreads can damage all types of structures built on top of the lateral spreading soil. Lateral spreads can pull apart foundations of buildings built in the failure area, they can sever sewer pipelines and other utilities in the failure mass, and they can cause compression or buckling of structures, such as bridges, founded at the toe of the failure mass. Figure 3.42 shows lateral spreading caused by liquefaction during the Prince William Sound earthquake in Alaska on March 27, 1964, that damaged a paved parking area.

9.5.2 Empirical Method

A commonly used approach for predicting the amount of horizontal ground displacement resulting from liquefaction-induced lateral spreading is to use the empirical method developed by Bartlett and Youd (1995). As stated in their paper, both U.S. and Japanese case histories of lateral spreading of liquefied sand were used to develop the displacement equations. Based on the regression analysis, two different equations were developed: (1) for lateral spreading toward a free face, such as a riverbank, and (2) for lateral spreading of gently sloping ground where a free face is absent. The equations are as follows:
Lateral spreading toward a free face:

$$\log D_H = -16.366 + 1.178M - 0.927 \log R - 0.013R + 0.657 \log W$$
$$+ 0.348 \log T + 4.527 \log (100 - F) - 0.922D_{50} \qquad (9.5)$$

Lateral spreading of gently sloping ground:

$$\log D_H = -15.787 + 1.178M - 0.927 \log R - 0.013R + 0.429 \log S$$
$$+ 0.348 \log T + 4.527 \log (100 - F) - 0.922D_{50} \qquad (9.6)$$

where D_H = horizontal ground displacement due to lateral spreading, m
  M = earthquake magnitude of the design earthquake
  R = distance to the expected epicenter or nearest fault rupture of the design earthquake, km
  W = free face ratio, expressed as a percentage. The *free face ratio* is defined as $100H / L$, where H = height of the free face and L = horizontal distance from base of free face to location of the site.
  T = cumulative thickness (meters) of the submerged sand layers having $(N_1)_{60}$ < 15. In general, layer T is expected to liquefy during earthquake, and data in Chap. 6 can be used to determine factor of safety against liquefaction [Eq. (6.8)]. As discussed below, the above equations are only valid if the slope inclination is 6 percent or less (that is, $\alpha \leq 0.10$) and hence use $K_\alpha = 1.0$ (see Sec. 9.4.2).
  F = fines content of soil comprising layer T, expressed as percentage. The fines content is defined as the percent of soil particles, based on dry weight, that pass the No. 200 sieve.
  D_{50} = grain size corresponding to 50 percent fines of soil comprising layer T, mm. Both F and D_{50} typically are obtained from a grain size curve (such as shown in Fig. 6.12).
  S = slope gradient, expressed as percentage. For example, a 20 : 1 (horizontal : vertical) slope has an angle of inclination of 2.9° and a slope gradient of 5 percent (that is, $\frac{1}{20}$ = 0.05, or expressed as a percentage 5 percent).

The various terms in Eqs. (9.5) and (9.6) can be divided into three general categories:

1. *Earthquake properties:* Both Eqs. (9.5) and (9.6) have identical terms that are used to account for the earthquake properties (that is, $1.178M - 0.927 \log R - 0.013R$). Instead of using the peak ground acceleration a_{max}, the equations use a comparable earthquake effect defined by the magnitude of the design earthquake M and the distance to the epicenter or fault rupture R. Note that the higher the value of M and the smaller the value of R, the greater the amount of lateral spreading.

2. *Site conditions:*   The site conditions are accounted for by using either the free face ratio [0.657 log W, Eq. (9.5)] or the slope gradient [0.429 log S, Eq. (9.6)]. In addition, the thickness of the soil layer that is anticipated to liquefy during the design earthquake is also accounted for in both equations (that is, 0.348 log T). Note that the higher the values of W, S, and T, the greater the amount of lateral spreading.

3. *Liquefied soil conditions:*   The type of liquefied soil is accounted for by the percent of fines F and the grain size corresponding to D_{50}. Of the two parameters, the effect of the fines has much greater impact on the amount of lateral spreading [that is, 4.52 log $(100 - F) -$ $0.922D_{50}$]. The fines content could affect the liquefied shear strength, and hence greater lateral spreading is expected for those liquefied soils having the lowest fines content.

According to Bartlett and Youd (1995), those sites that will be subject to an $M \leq 8$ earthquake and have soil with $(N_1)_{60}$ values >15 are resistant to lateral spreading, and Eqs. (9.5) and (9.6) need not be applied. Figure 9.33 shows the measured displacements for U.S. and

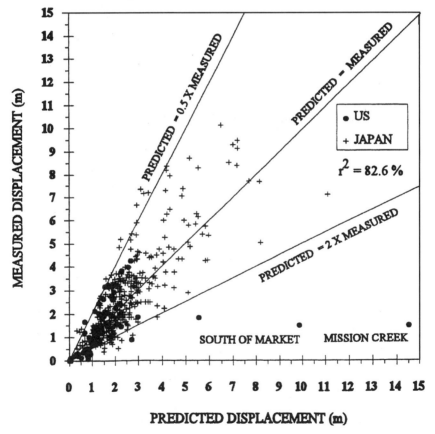

PREDICTED DISPLACEMENT (m)

FIGURE 9.33   Measured displacement plotted against displacements predicted by Eqs. (9.5) and (9.6) for U.S. and Japanese case history data. (*From Bartlett and Youd 1995, reproduced with permission of the American Society of Civil Engineers.*)

Japanese cases of lateral spreading versus the displacement predicted by using Eqs. (9.5) and (9.6). This plot shows that Eqs. (9.5) and (9.6) are generally accurate within a factor of ± 2, and thus Bartlett and Youd (1995) recommend that the value of D_H from Eqs. (9.5) and (9.6) be multiplied by 2 in order to obtain a conservative design estimate of the lateral spreading.

According to Bartlett and Youd (1995), to obtain reliable deformations, the terms in Eqs. (9.5) and (9.6) must be within the following ranges:

- Earthquake magnitude: $6 \le M \le 8$ (however, the interpolation of the model for $M < 6$ conditions appears to yield plausible predictions)
- Free face ratio: 1.0 percent $< W < 20$ percent
- Slope gradient: 0.1 percent $\le S \le 6$ percent
- Thickness of layer having $(N_1)_{60} < 15$: $1 \text{ m} \le T \le 15 \text{ m}$
- Fines content: $F \le 50$ percent
- Mid-grain size: $D_{50} \le 1$ mm

According to Bartlett and Youd (1995), other limitations of Eqs. (9.5) and (9.6) are as follows:

- *Liquefied soil layer:*   The liquefied soil layer must be within 10 m of the ground surface.
- *Gravels:*   It appears that liquefied gravels have a displacement behavior that is significantly different from that of liquefied sandy and silty sand sediments. Thus Eqs. (9.5) and (9.6) overestimate the displacement due to lateral spreading of liquefied gravels.
- *Free face condition:*   For sites very close to the free face (such as a riverbank), there could be slumping or even a flow slide. Thus caution is warranted in applying Eq. (9.5) at sites that are very close to the free face.
- *Free face calculations:*   For the free face, it is recommended that both Eqs. (9.5) and (9.6) be used to calculate the deformation due to lateral spreading. The higher value should be used for the design of the project.

9.5.3 Example Problem

To illustrate the empirical method by Bartlett and Youd (1995), assume the following conditions:

- Free face condition
- Factor of safety against a flow slide at site > 1
- Earthquake magnitude $M = 7.5$
- Distance from expected fault rupture to site $R = 50$ km
- Free face ratio W: The height of the free face $H = 5$ m, and the distance from the free face to the site $L = 50$ m. Therefore the free face ratio $W = 100(5/50) = 10$ percent.
- Slope gradient: The slope inclination is 1 : 20 (horizontal : vertical), and therefore the slope gradient $S = \frac{1}{20} = 0.05$, or expressed as a percentage, $S = 5$ percent.
- Thickness of layer T: The submerged sand having $(N_1)_{60} < 15$ is at a depth of 1 m and extends to a depth of 6 m, and therefore $T = 5$ m. This layer is also expected to liquefy during the design earthquake.
- Soil properties: Assume the soil comprising layer T has the same grain size curve as soil 4 in Fig. 6.12.

Solution.   Using the grain size curve for soil 4 in Fig. 6.12, the percent passing the No. 200 sieve $F = 6$ percent, and the grain size corresponding to 50 percent fines is $D_{50} = 0.38$ mm. Checking both the free face and gently sloping ground conditions reveals the following:

Lateral spreading toward the free face:

$$\log D_H = -16.366 + 1.178\,(7.5) - 0.927 \log 50 - 0.013\,(50) + 0.657 \log 10$$
$$+ 0.348 \log 5 + 4.527 \log (100 - 6) - 0.922\,(0.38)$$

or

$$\log D_H = -0.27 \quad \text{or} \quad D_H = 0.53 \text{ m}$$

Lateral spreading of gently sloping ground:

$$\log D_H = -15.787 + 1.178\,(7.5) - 0.927 \log 50 - 0.013\,(50) + 0.429 \log 5$$
$$+ 0.348 \log 5 + 4.527 \log (100 - 6) - 0.922\,(0.38)$$

or

$$\log D_H = -0.05 \quad \text{or} \quad D_H = 0.9 \text{ m}$$

The lateral spreading of gently sloping ground [i.e., Eq. (9.6)] predicts a larger deformation, and so use this value of $D_H = 0.9$ m. As previously mentioned, Bartlett and Youd (1995) recommend that the value of D_H from Eqs. (9.5) and (9.6) be multiplied by 2 in order to obtain a conservative design estimate of the lateral spreading. Thus the final expected horizontal deformation at the site due to lateral spreading $D_H = (0.9 \text{ m})(2) = 1.8$ m.

9.5.4 Summary

As discussed in previous sections, the liquefaction of soil can cause flow failures or lateral spreading. It is also possible that even with a factor of safety against liquefaction greater than 1.0, there could be still be significant weakening of the soil and deformation of the slope. In summary, the type of analysis should be based on the factor of safety against liquefaction FS_L as follows:

1. $FS_L \leq 1.0$:   In this case, the soil is expected to liquefy during the design earthquake, and thus a flow slide analysis (Sec. 9.4) and/or a lateral spreading analysis (Sec. 9.5.2) will be performed.

2. $FS_L > 2.0$:   If the factor of safety against liquefaction is greater than about 2.0, the pore water pressures generated by the earthquake-induced contraction of the soil are usually small enough that they can be neglected. In this case, it could be assumed that the soil is not weakened by the earthquake, and thus the inertia slope stability analyses outlined in Secs. 9.2 and 9.3 could be performed.

3. $1.0 < FS_L \leq 2.0$:   For this case, the soil is not expected to liquefy during the earthquake. However, as the loose granular soil contracts during the earthquake, there could still be a substantial increase in pore water pressure and hence weakening of the soil. Figure 5.15 can be used to estimate the pore water pressure ratio for various values of the factor of safety against liquefaction FS_L. Using the estimated pore water pressure ratio from Fig. 5.15, an effective stress slope stability analysis could be performed. If the results of the effective stress slope stability analysis indicate a factor of safety less than 1.0, then failure of the slope is expected during the earthquake.

Even with a slope stability factor of safety greater than 1.0, there could still be substantial deformation of the slope. There could be two different types of slope deformation. The first type occurs as the earthquake-induced pore water pressures dissipate and the soil contracts (i.e., see Figs. 7.1 and 7.2). The second type of deformation occurs when the earthquake imparts net inertial forces that cause the driving forces to exceed the resisting forces. Each cycle of net inertial forces in the downslope direction that cause the driving forces to exceed the resisting forces along the slip surface will result in progressive and incremental lateral movement. If the factor of safety from the slope stability analysis is only marginally in excess of 1.0, then the lateral spreading approach in Sec. 9.5.2 could be used to obtain a rough estimate of the lateral deformation of the slope.

9.6 WEAKENING SLOPE STABILITY— STRAIN-SOFTENING SOIL

Section 9.1.2 discusses the types of soil that will be susceptible to a reduction in shear strength during the earthquake. These types of soil include the following:

• Sensitive clays that lose shear strength during the earthquake. An example of a weakening landslide is the Turnagain Heights landslide, described in Sec. 3.5.2.

• Soft clays and organic soils that are overloaded and subjected to plastic flow during the earthquake. The type of slope movement involving these soils is often termed *slow earth flows*.

These types of plastic soil are often characterized as strain-softening soils, because there is a substantial reduction in shear strength once the peak shear strength is exceeded. An example of the stress-strain curve for a strain-softening soil is presented in Fig. 8.12. During the earthquake, often failure first occurs at the toe of the slope, and then the ground cracks and displacement of the slope progresses upslope. Blocks of soil are often observed to have moved laterally during the earthquake, such as illustrated in Fig. 3.46.

It is very difficult to evaluate the amount of lateral movement of slopes containing strain-softening soil. The most important factors are the level of static shear stress versus the peak shear stress of the soil and the amount of additional shear stress that will be induced into the soil by the earthquake. If the existing static shear stress is close to the peak shear stress, then only a small additional earthquake-induced shear stress will be needed to exceed the peak shear strength. Once this happens, the shear strength will significantly decrease with strain, resulting in substantial lateral movement of the slope. If it is anticipated that this will occur during the design earthquake, then one approach is to use the ultimate (i.e., softened) shear strength of the soil. For example, in Fig. 8.12, the ultimate shear strength of the soil is about 25 lb/ft^2 (170 kPa). Suppose a slope was composed of this clay and a shear strength of 25 lb/in^2 was used in the slope stability analysis. If the factor of safety is greater than 1.0, then it could be concluded that a massive shear failure of the slope during the earthquake is unlikely.

9.7 MITIGATION OF SLOPE HAZARDS

To evaluate the effect of the earthquake-induced slope movement upon the structure, the first step is to estimate the amount of lateral movement. The prior sections present different types of analyses based on differing soil and slope movement conditions. Once the amount of earthquake-induced lateral movement has been estimated, then it can be compared with the allowable lateral movement for the proposed structure. If the anticipated earthquake-induced lateral movement exceeds the allowable lateral movement, then slope stabilization options will be required.

9.7.1 Allowable Lateral Movement

To evaluate the lateral movement of buildings, a useful parameter is the *horizontal strain* ε_h, defined as the change in length divided by the original length of the foundation. Figure 9.34 shows a correlation between horizontal strain ε_h and severity of damage (Boone 1996, originally from Boscardin and Cording 1989). Assuming a 6-m- (20-ft-) wide zone of the foundation subjected to lateral movement, Fig. 9.34 indicates that a building can be damaged by as little as 3 mm (0.1 in) of lateral movement. Figure 9.34 also indicates that a lateral movement of 25 mm (1 in) would cause "severe" to "very severe" building damage.

The ability of a facility to resist lateral movement depends on its tensile strength. Those facilities that cannot resist the tensile forces imposed by lateral movement will be the most severely damaged. For example, Figs. 9.35 and 9.36 show roadway damage caused by earthquake-induced slope movement. Asphalt pavements have low tensile strength, and hence they will be simply pulled apart during the earthquake-induced slope movement, such as shown in Figs. 9.35 and 9.36.

The ability of buildings to resist lateral movement also depends on the tensile strength of the foundation. For example, Fig. 9.37 shows severe damage to a building caused by landslide movement. The foundation is too weak, and the amount of slope movement is too large for the building shown in Fig. 9.37 to be able to resist the lateral deformation. Those buildings that have isolated footings or foundations that have joints or planes of weakness are most susceptible to damage from lateral movement. Buildings having a mat foundation or a posttensioned slab are less susceptible to damage because of the high tensile resistance of these foundations.

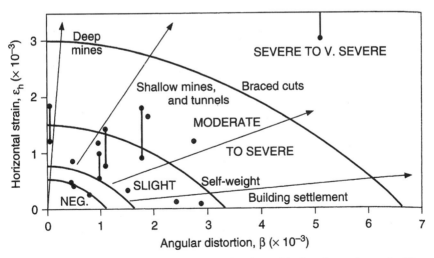

FIGURE 9.34 Relationship of damage to angular distortion and horizontal extension strain. (*From Boscardin and Cording 1989, reprinted with permission from the American Society of Civil Engineers.*)

FIGURE 9.35 Earthquake damage to a two-lane road in Santa Cruz Mountains. The slope movement was caused by the Loma Prieta earthquake in California on October 17, 1989. (*Photograph from the Loma Prieta Collection, EERC, University of California, Berkeley.*)

Figures 9.35 to 9.37 show damage caused by the pulling apart of the facilities by the earthquake-induced slope movement. In addition, the slope movement can cause compression-type damage to facilities located at the toe of the slope. For example, Fig. 9.38 shows damage to a bridge located at the toe of a landslide. The earthquake-induced landslide movement shoved the bridge sections into each other, resulting in the bridge failure shown in Fig. 9.38.

9.7.2 Mitigation Options

As discussed above, one mitigation option is to construct a foundation that has sufficient tensile strength to resist the expected earthquake-induced lateral movement. In general, mitigation options can be divided into three basic categories (Division of Mines and Geology 1997):

FIGURE 9.36 One of several earthquake-induced slides along Skyline Boulevard at Lake Merced, San Francisco. The slides were caused by the Daly City earthquake in California on March 22, 1957. (*Photograph from the Steinbrugge Collection, EERC, University of California, Berkeley.*)

FIGURE 9.37 Building damage caused by the Fourth Avenue landslide, Anchorage, Alaska. The landslide movement occurred during the Prince William Sound earthquake in Alaska on March 27, 1964. (*Photograph from the Steinbrugge Collection, EERC, University of California, Berkeley.*)

FIGURE 9.38   Earthquake-induced landslide movement that damaged the south abutment of the Shih-Wui Bridge. The landslide was caused by the Chi-chi (Taiwan) earthquake on September 21, 1999. (*Photograph from the Taiwan Collection, EERC, University of California, Berkeley.*)

1. *Avoid the failure hazard:*   Where the potential for failure is beyond the acceptable level and not preventable by practical means, as in mountainous terrain subject to massive planar slides or rock and debris avalanches, the hazard should be avoided. Developments should be built sufficiently far away from the threat that they will not be affected even if the slope does fail. Planned development areas on the slope or near its base should be avoided and relocated to areas where stabilization is feasible.

2. *Protect the site from the failure:*   While it is not always possible to prevent slope failures that occur above a project site, it is sometimes possible to protect the site from the runout of failed slope materials. This is particularly true for sites located at or near the base of steep slopes which can receive large amounts of material from shallow disaggregated landslides or debris flows. Methods include catchment and/or protective structures such as basins, embankments, diversion or barrier walls, and fences. Diversion methods should only be employed where the diverted landslide materials will not affect other sites.

3. *Reduce the hazard to an acceptable level:*   Unstable slopes affecting a project can be rendered stable (that is, by increasing the factor of safety to greater than 1.5 for static and greater than 1.1 for dynamic loads) by eliminating the slope, removing the unstable soil and rock materials, or applying one or more appropriate slope stabilization methods (such as buttress fills, subdrains, soil nailing, or crib walls). For deep-seated slope instability, strengthening the design of the structure (e.g., reinforced foundations) is generally not by itself an adequate mitigation measure.

Mitigation options for various types of earthquake-induced slope movement are as follows:

A. Mitigation options for potential rockfalls and rock slides (see Table 9.1)

    1. *Reduce the driving forces:*   This can be accomplished by flattening the slope inclination, removing the unstable or potentially unstable rocks, diverting water from the slope face, and incorporating benches into the slope (Piteau and Peckover 1978).

2. *Increase the stability:*   Retaining the individual blocks of rock on the slope face can increase the slope stability. Measures to increase the stability of the slope include the installation of anchoring systems (such as bolts, rods, or dowels), applying shotcrete to the rock slope face, and constructing retaining walls.

3. *Protect the site:*   For rockfalls, it may be possible to intercept or deflect the falling rocks around the structure. This can be accomplished by using toe-of-slope ditches, wire mesh catch fences, and catch walls (Peckover 1975). Recommendations for the width and depth of toe-of-slope ditches are given by Ritchie (1963) and Piteau and Peckover (1978).

B. Mitigation options for potential soil slides (see Table 9.2)

1. *Reduce the driving forces:*   This can be accomplished by reducing the weight of the potential slide mass by cutting off the head of the slide or totally removing the slide. Other options include flattening the slope inclination through grading, preventing water infiltration by controlling surface drainage, and lowering the groundwater table by installing subsurface drainage systems in order to dewater the slide mass.

2. *Increase the stability:*   A shear key or buttress can be constructed to increase the stability of the slope. Other options include the construction of pier walls or retaining walls, and the pinning of shallow slide masses with soil anchors or nails.

C. Mitigation options for potential flow slides and lateral spreading (see Table 9.2)

1. *Reduce the driving forces:*   This can be accomplished by removing the liquefaction-prone material from the site, using grading or excavation techniques. Other options include lowering the groundwater table by using subsurface galleries or subdrains so that the soil is no longer susceptible to liquefaction.

2. *Increase the stability:*   A shear key or buttress can be constructed in order to increase the stability of the slope. Other options include the densification of the soil by using various vibratory techniques to reduce the liquefaction potential of the soil.

3. *Protect the site:*   It may be possible to divert the flow away from the project by using diversion barriers or channels, or provide catchment structures to contain the flowing mass.

Other mitigation measures that may be appropriate for slopes are discussed in Chap. 12.

9.8   REPORT PREPARATION

The results of the slope stability investigation often need to be summarized in report form for review by the client and governing agency. The types of information that should be included in the report, per *Guidelines for Evaluating and Mitigating Seismic Hazards in California* (Division of Mines and Geology 1997), are as follows:

• Description of the shear strength test procedures (ASTM or other) and test specimens

• Shear strength plots, including identification of samples tested, whether data points reflect peak or residual values, and moisture conditions at the time of testing

• Summary table or text describing methods of analysis, shear strength values, assumed groundwater conditions, and other pertinent assumptions used in the stability calculations

- Explanation of choice of seismic coefficient and/or design strong-motion record used in slope stability analysis, including site and/or topographic amplification estimates
- Slope stability analyses of critical (least stable) cross sections which substantiate conclusions and recommendations concerning stability of natural and as-graded slopes
- Factors of safety against slope failure and/or calculated displacements for the various anticipated slope configurations (cut, fill, and/or natural slopes)
- Conclusions regarding the stability of slopes with respect to earthquake-induced landslides and their likely impact on the proposed project
- Discussion of proposed mitigation measures, if any, necessary to reduce damage from potential earthquake-initiated landslides to an acceptable level of risk (see Sec. 9.7)
- Acceptable testing criteria (e.g., pseudostatic factor of safety), if any, that will be used to demonstrate satisfactory remediation

9.9 PROBLEMS

Pseudostatic Wedge Analysis

9.1   A slope has a height of 9.1 m (30 ft), and the slope face is inclined at a 2 : 1 (horizontal : vertical) ratio. Assume a wedge-type analysis where the slip surface is planar through the toe of the slope and is inclined at a 3 : 1 (horizontal : vertical) ratio. The total unit weight of the slope material $\gamma_t = 18.1$ kN/m^3 (115 lb/ft^3). Using the undrained shear strength parameters of $c = 14.5$ kPa (2.1 lb/in^2) and $\phi = 0$, calculate the factor of safety for the static case and an earthquake condition of $k_h = 0.3$. Assume that the shear strength does not decrease with strain (i.e., not a weakening-type soil). *Answer:* Static FS $= 1.8$, pseudostatic FS $= 0.94$.

9.2   Use the data from Prob. 9.1, except assume that the slip surface has an effective shear strength of $c' = 3.4$ kPa (70 lb/ft^2) and $\phi' = 29°$. Also assume that piezometers have been installed along the slip surface, and the average measured steady-state pore water pressure $u = 2.4$ kPa (50 lb/ft^2). Calculate the factor of safety of the failure wedge based on an effective stress analysis for the static case and an earthquake condition of $k_h = 0.2$. Assume that the shear strength does not decrease with strain (i.e., not a weakening-type soil) and the pore water pressures will not increase during the earthquake. *Answer:* Static FS $= 1.95$, pseudostatic FS $= 1.15$.

9.3   A near-vertical rock slope has a continuous horizontal joint through the toe of the slope and another continuous vertical joint located 10 ft back from the top of the slope. The height of the rock slope is 20 ft, and the unit weight of the rock is 140 lb/ft^3. The shear strength parameters for the horizontal joint are $c' = 0$ and $\phi' = 40°$, and the pore water pressure u is equal to zero. For the vertical joint, assume zero shear strength. Neglecting possible rotation of the rock block and considering only a sliding failure, calculate the pseudostatic factor of safety if $k_h = 0.50$. *Answer:* Pseudostatic FS $= 1.68$.

Pseudostatic Analysis Using the Method of Slices

9.4   Use the data from the example problems in Secs. 9.2.7 and 9.3.2. Calculate the pseudostatic factor of safety if $k_h = 0.30$. *Answer:* Pseudostatic FS $= 0.88$.

Newmark Method

9.5   Use the data from Prob. 9.1, and calculate the slope deformation based on the Newmark method [i.e., Eq. (9.3)]. *Answer: d* = 0.06 cm.

9.6   Use the data from Prob. 9.2, and calculate the slope deformation based on the Newmark method [i.e., Eq. (9.3)]. *Answer:* Since pseudostatic FS > 1.0, d = 0.

9.7   Use the data from Prob. 9.2 and assume k_h = 0.5. Calculate the slope deformation based on the Newmark method [i.e., Eq. (9.3)]. *Answer: d* = 2.3 cm.

Weakening Slope Stability Analysis: Flow Slides

9.8   Use the data from Prob. 6.12. Assume the subsoil profile shown in Fig. 6.13 pertains to a level-ground site that is adjacent to a riverbank. The riverbank has a 3:1 (horizontal:vertical) slope inclination, and assume that the average level of water in the river corresponds to the depth of the groundwater table (1.5 m below ground surface). Further assume that the depth of water in the river is 9 m. Will the riverbank experience a flow failure during the design earthquake? What type of flow failure is expected? *Answer:* A mass liquefaction flow failure would develop during the earthquake.

9.9   Use the data from Prob. 6.15. Assume the subsoil profile shown in Fig. 6.15 pertains to a level-ground site that is adjacent to a riverbank. The riverbank has a 4:1 (horizontal:vertical) slope inclination, and assume that the average level of water in the river corresponds to the depth of the groundwater table (0.4 m below ground surface). Further assume that the depth of water in the river is 5 m. Will the riverbank experience a flow failure during the design earthquake? What type of flow failure is expected? *Answer:* A mass liquefaction flow failure would develop during the earthquake.

9.10   Use the data from Prob. 6.18. Assume the *N* value data shown in Fig. 6.11 pertain to a level-ground site that is adjacent to a riverbank. The riverbank has a 3:1 (horizontal:vertical) slope inclination, and assume that the average level of water in the river corresponds to the depth of the groundwater table (0.5 m below ground surface). Further assume that the depth of water in the river is 5 m. Using the "before improvement" *N* values, will the riverbank experience a flow failure during the design earthquake? If the entire riverbank and adjacent land are densified such that they have the "after improvement" *N* values, will the riverbank experience a flow failure during the design earthquake? *Answer:* Before-improvement condition: a mass liquefaction flow failure would develop during the earthquake. After-improvement condition: the riverbank is not susceptible to a flow failure.

Weakening Slope Stability Analysis: Liquefaction-Induced Lateral Spreading

9.11   Solve the example problem in Sec. 9.5.3, assuming that the slope inclination is only 1 percent. *Answer: D_H* = 0.9 m.

9.12   Solve the example problem in Sec. 9.5.3, assuming that the thickness of the liquefied layer *T* is only 1 m. *Answer: D_H* = 1.0 m.

9.13   Figure 9.39 shows the subsoil profile at Baosic, Yugoslavia. Assume the site consists of sloping ground with an inclination of 6 percent. As indicated in Fig. 9.39, the groundwater table is located 1.5 m below ground surface. The different soil types shown in Fig. 9.39 have the following properties:

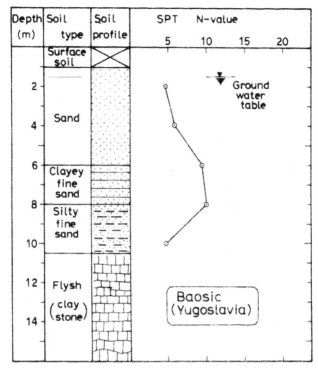

FIGURE 9.39   Subsoil profile, Baosic, Yugoslavia. (*Reproduced from Ishihara 1985.*)

1. *Surface soil:*   The surface soil is located from ground surface to a depth of 1 m and has a total unit weight γ_t equal to 18.3 kN/m³.

2. *Sand:*   The sand layer is located 1 to 6 m below ground surface. The sand has 2 percent fines and has a grain size corresponding to 50 percent fines (D_{50}) that is about 0.6 mm. The total unit weight γ_t of the sand above the groundwater table is 18.3 kN/m³, and the buoyant unit weight γ_b of the sand below the groundwater table is 9.7 kN/m³.

3. *Clayey fine sand and silty fine sand:*   These two layers are located 6 to 10.5 m below ground surface. Both soils have a liquid limit greater than 35 and thus are not susceptible to liquefaction. Assume that the SPT blow count at a depth of 6 m is located within the clayey fine sand.

4. *Flysh claystone:*   The Flysh claystone is located at a depth of 10.5 m below ground surface. This material is also not susceptible to liquefaction.

The standard penetration data shown in Fig. 9.39 are uncorrected N values. Assume a hammer efficiency E_m of 0.6 and a boring diameter of 100 mm, and the length of drill rods is equal to the depth of the SPT test below ground surface.

The design earthquake conditions are a peak ground acceleration a_{max} of 0.25g, magnitude of 7.5, and the distance from the site to the fault is 37 km. Using the standard penetration test data, determine the factor of safety against liquefaction versus depth. Also determine the amount of horizontal ground displacement caused by liquefaction-induced lateral spreading. *Answer:* The sand layer will liquefy during the design earthquake and $D_H = 2.7$ m.

Strain-Softening Soil

9.14   A site consists of a slope that has a height H of 40 ft, and the slope inclination is 2:1 (horizontal:vertical). At a depth of 20 ft below the toe of slope, there is a dense material (i.e. $D = 20$ ft). Assume that all the soil above this depth, including the slope itself, consists of uniform clay that has a total unit weight γ_t of 125 lb/ft³. Also assume that the unconfined compression test data shown in Fig. 9.40 are representative of the uniform clay.

Using the Taylor chart shown in Fig. 9.41, calculate the static factor of safety of the slope using a total stress analysis. It is anticipated that during its design life, the slope will be sub-

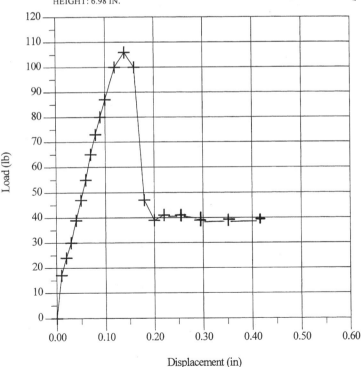

Unconfined Compression Test Results
ASTM D 2166

PROJECT NAME: n/a
PROJECT NO.: 22193.02
DATE: MAY 2000

BORING NO.: AGSB-1
SAMPLE NO.: n/a

SAMPLE TYPE: CORE
DIAMETER: 2.50 IN.
HEIGHT: 6.98 IN.

DEPTH (ft): 35-36
USCS CLASS. C H

MOISTURE CONTENT (%): 10.3
DRY DENSITY (pcf): 115.3

FIGURE 9.40   Unconfined compression test results for Prob. 9.14.

jected to strong ground shaking which would likely reduce the clay's shear strength to its strain-softened state. Calculate the factor of safety of the slope for this earthquake condition using a total stress analysis. Is a massive shear failure of the slope likely during the earthquake? *Answer:* Static factor of safety = 1.89. For earthquake conditions, factor of safety = 0.66, and yes, a massive shear failure of the slope is likely during the earthquake.

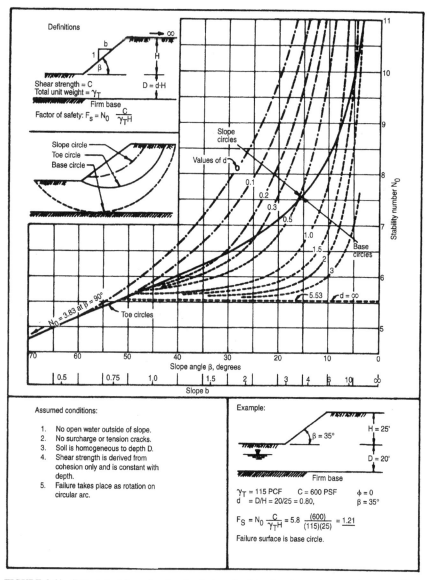

FIGURE 9.41   Taylor chart for estimating the factor of safety of a slope using a total stress analysis. (*Reproduced from NAVFAC DM-7.1, 1982.*)

CHAPTER 10
RETAINING WALL ANALYSES FOR EARTHQUAKES

The following notation is used in this chapter:

SYMBOL   DEFINITION

a	Acceleration (Sec. 10.2)
a	Horizontal distance from W to toe of footing
a_{max}	Maximum horizontal acceleration at ground surface (also known as peak ground acceleration)
A_p	Anchor pull force (sheet pile wall)
c	Cohesion based on total stress analysis
c'	Cohesion based on effective stress analysis
c_a	Adhesion between bottom of footing and underlying soil
d	Resultant location of retaining wall forces (Sec. 10.1.1)
d_1	Depth from ground surface to groundwater table
d_2	Depth from groundwater table to bottom of sheet pile wall
D	Depth of retaining wall footing
D	Portion of sheet pile wall anchored in soil (Fig. 10.9)
e	Lateral distance from P_v to toe of retaining wall
F, FS	Factor of safety
FS_L	Factor of safety against liquefaction
g	Acceleration of gravity
H	Height of retaining wall
H	Unsupported face of sheet pile wall (Fig. 10.9)
k_A	Active earth pressure coefficient
k_{AE}	Combined active plus earthquake coefficient of pressure (Mononobe-Okabe equation)
k_h	Seismic coefficient, also known as pseudostatic coefficient
k_0	Coefficient of earth pressure at rest
k_p	Passive earth pressure coefficient
k_v	Vertical pseudostatic coefficient
L	Length of active wedge at top of retaining wall
m	Total mass of active wedge
M_{max}	Maximum moment in sheet pile wall
N	Sum of wall weights W plus, if applicable, P_v
P_A	Active earth pressure resultant force
P_E	Pseudostatic horizontal force acting on retaining wall
P_{ER}	Pseudostatic horizontal force acting on restrained retaining wall
P_F	Sum of sliding resistance forces (Fig. 10.2)
P_H	Horizontal component of active earth pressure resultant force
P_L	Lateral force due to liquefied soil
P_p	Passive resultant force

P_R	Static force acting upon restrained retaining wall
P_v	Vertical component of active earth pressure resultant force
P_1	Active earth pressure resultant force ($P_1 = P_A$, Fig. 10.7)
P_2	Resultant force due to uniform surcharge
Q	Uniform vertical surcharge pressure acting on wall backfill
R	Resultant of retaining wall forces (Fig. 10.2)
s_u	Undrained shear strength of soil
W	Total weight of active wedge (Sec. 10.2)
W	Resultant of vertical retaining wall loads
β	Slope inclination behind the retaining wall
δ, ϕ_{cv}	Friction angle between bottom of wall footing and underlying soil
δ, ϕ_w	Friction angle between back face of wall and soil backfill
ϕ	Friction angle based on total stress analysis
ϕ'	Friction angle based on effective stress analysis
γ_b	Buoyant unit weight of soil
γ_{sat}	Saturated unit weight of soil
γ_t	Total unit weight of the soil
θ	Back face inclination of retaining wall
σ_{avg}	Average bearing pressure of retaining wall foundation
σ_{mom}	That portion of bearing pressure due to eccentricity of N
ψ	Equal to $\tan^{-1}(a_{max}/g)$

10.1 INTRODUCTION

A *retaining wall* is defined as a structure whose primary purpose is to provide lateral support for soil or rock. In some cases, the retaining wall may also support vertical loads. Examples include basement walls and certain types of bridge abutments. The most common types of retaining walls are shown in Fig. 10.1 and include gravity walls, cantilevered walls, counterfort walls, and crib walls. Table 10.1 lists and describes various types of retaining walls and backfill conditions.

10.1.1 Retaining Wall Analyses for Static Conditions

Figure 10.2 shows various types of retaining walls and the soil pressures acting on the walls for static (i.e., nonearthquake) conditions. There are three types of soil pressures acting on a retaining wall: (1) active earth pressure, which is exerted on the backside of the wall; (2) passive earth pressure, which acts on the front of the retaining wall footing; and (3) bearing pressure, which acts on the bottom of the retaining wall footing. These three pressures are individually discussed below.

Active Earth Pressure. To calculate the active earth pressure resultant force P_A, in kilonewtons per linear meter of wall or pounds per linear foot of wall, the following equation is used for granular backfill:

$$P_A = \frac{1}{2} k_A \gamma_t H^2 \tag{10.1}$$

where k_A = active earth pressure coefficient, γ_t = total unit weight of the granular backfill, and H = height over which the active earth pressure acts, as defined in Fig. 10.2. In its simplest form, the active earth pressure coefficient k_A is equal to

$$k_A = \tan^2(45° - \frac{1}{2}\phi) \tag{10.2}$$

where ϕ = friction angle of the granular backfill. Equation (10.2) is known as the active Rankine state, after the British engineer Rankine who in 1857 obtained this relationship. Equation (10.2) is only valid for the simple case of a retaining wall that has a vertical rear face, no friction between the rear wall face and backfill soil, and the backfill ground surface is horizontal. For retaining walls that do not meet these requirements, the active earth pressure

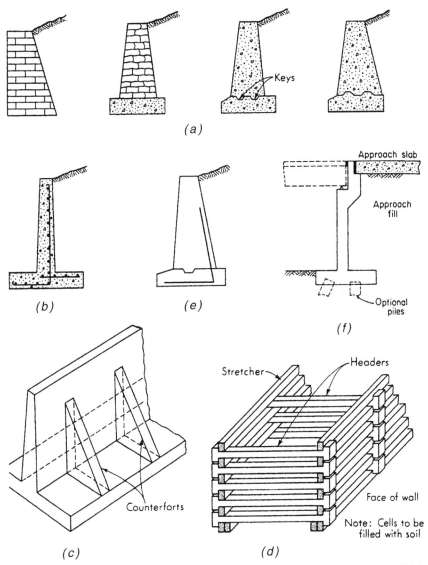

FIGURE 10.1   Common types of retaining walls. (*a*) Gravity walls of stone, brick, or plain concrete. Weight provides overturning and sliding stability. (*b*) Cantilevered wall. (*c*) Counterfort, or buttressed wall. If backfill covers counterforts, the wall is termed a *counterfort*. (*d*) Crib wall. (*e*) Semigravity wall (often steel reinforcement is used). (*f*) Bridge abutment. (*Reproduced from Bowles 1982 with permission of McGraw-Hill, Inc.*)

TABLE 10.1   Types of Retaining Walls and Backfill Conditions

Topic	Discussion
Types of retaining walls	As shown in Fig. 10.1, some of the more common types of retaining walls are gravity walls, counterfort walls, cantilevered walls, and crib walls (Cernica 1995a). Gravity retaining walls are routinely built of plain concrete or stone, and the wall depends primarily on its massive weight to resist failure from overturning and sliding. Counterfort walls consist of a footing, a wall stem, and intermittent vertical ribs (called counterforts) which tie the footing and wall stem together. Crib walls consist of interlocking concrete members that form cells which are then filled with compacted soil. Although mechanically stabilized earth retaining walls have become more popular in the past decade, cantilever retaining walls are still probably the most common type of retaining structure. There are many different types of cantilevered walls, with the common feature being a footing that supports the vertical wall stem. Typical cantilevered walls are T-shaped, L-shaped, or reverse L-shaped (Cernica 1995a).
Backfill material	Clean granular material (no silt or clay) is the standard recommendation for backfill material. There are several reasons for this recommendation: 1. *Predictable behavior:* Import granular backfill generally has a more predictable behavior in terms of earth pressure exerted on the wall. Also, expansive soil-related forces will not be generated by clean granular soil. 2. *Drainage system:* To prevent the buildup of hydrostatic water pressure on the retaining wall, a drainage system is often constructed at the heel of the wall. The drainage system will be more effective if highly permeable soil, such as clean granular soil, is used as backfill. 3. *Frost action:* In cold climates, frost action has caused many retaining walls to move so much that they have become unusable. If freezing temperatures prevail, the backfill soil can be susceptible to frost action, where ice lenses form parallel to the wall and cause horizontal movements of up to 0.6 to 0.9 m (2 to 3 ft) in a single season (Sowers and Sowers 1970). Backfill soil consisting of clean granular soil and the installation of a drainage system at the heel of the wall will help to protect the wall from frost action.
Plane strain condition	Movement of retaining walls (i.e., active condition) involves the shear failure of the wall backfill, and the analysis will naturally include the shear strength of the backfill soil. Similar to the analysis of strip footings and slope stability, for most field situations involving retaining structures, the backfill soil is in a plane strain condition (i.e., the soil is confined along the long axis of the wall). As previously mentioned, the friction angle ϕ is about 10 percent higher in the plane strain condition compared to the friction angle ϕ measured in the triaxial apparatus. In practice, plane strain shear strength tests are not performed, which often results in an additional factor of safety for retaining wall analyses.

coefficient k_A for Eq. (10.1) is often determined by using the Coulomb equation (see Fig. 10.3). Often the wall friction is neglected ($\delta = 0°$), but if it is included in the analysis, typical values are $\delta = \frac{3}{4}\phi$ for the wall friction between granular soil and wood or concrete walls and $\delta = 20°$ for the wall friction between granular soil and steel walls such as sheet pile walls. Note in Fig. 10.3 that when the wall friction angle δ is used in the analysis, the active earth pressure resultant force P_A is inclined at an angle equal to δ. Additional important details concerning the active earth pressure follow.

1. *Sufficient movement:*   There must be sufficient movement of the retaining wall in order to develop the active earth pressure of the backfill. For dense granular soil, the amount of wall translation to reach the active earth pressure state is usually very small (i.e., to reach active state, wall translation $\geq 0.0005H$, where H = height of wall).

2. *Triangular distribution:*   As shown in Figs. 10.2 and 10.3, the active earth pressure is a triangular distribution, and thus the active earth pressure resultant force P_A is located at a distance equal to $^1/_3H$ above the base of the wall.

3. *Surcharge pressure:*   If there is a uniform surcharge pressure Q acting upon the entire ground surface behind the wall, then an additional horizontal pressure is exerted upon the retaining wall equal to the product of k_A and Q. Thus the resultant force P_2, in kilonewtons per linear

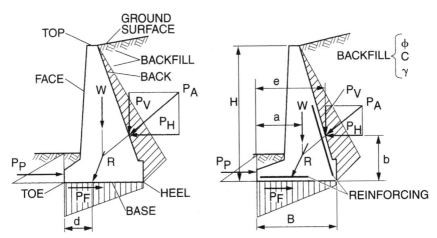

FIGURE 10.2a   Gravity and semigravity retaining walls. (*Reproduced from NAVFAC DM-7.2, 1982.*)

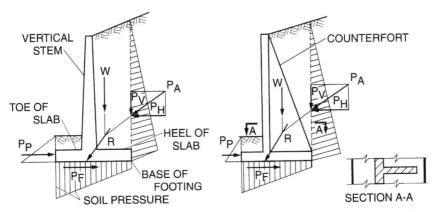

FIGURE 10.2b   Cantilever and counterfort retaining walls. (*Reproduced from NAVFAC DM-7.2, 1982.*)

LOCATION OF RESULTANT

MOMENTS ABOUT TOE:

$$d = \frac{Wa + P_Ve - P_Hb}{W + P_V}$$

ASSUMING $\quad P_P = 0$

OVERTURNING

MOMENTS ABOUT TOE:

$$F = \frac{Wa}{P_Hb - P_Ve} \geqq 1.5$$

IGNORE OVERTURNING IF R IS WITHIN MIDDLE THIRD (SOIL), MIDDLE HALF (ROCK). CHECK R AT DIFFERENT HORIZONTAL PLANES FOR GRAVITY WALLS.

RESISTANCE AGAINST SLIDING

$$F = \frac{(W + P_V) \, TAN \, \delta + C_aB}{P_H} \geqq 1.5$$

$$F = \frac{(W + P_V) \, TAN \, \delta + C_aB + P_P}{P_H} \geqq 2.0$$

$$P_F = (W + P_V) \, TAN \, \delta + C_aB$$

C_a = ADHESION BETWEEN SOIL AND BASE

TAN δ = FRICTION FACTOR BETWEEN SOIL AND BASE

W = INCLUDES WEIGHT OF WALL AND SOIL IN FRONT FOR GRAVITY AND SEMIGRAVITY WALLS. INCLUDES WEIGHT OF WALL AND SOIL ABOVE FOOTING, FOR CANTILEVER AND COUNTERFORT WALLS.

FIGURE 10.2c Design analysis for retaining walls shown in Fig. 10.2*a* and *b*. (*Reproduced from NAVFAC DM-7.2, 1982.*)

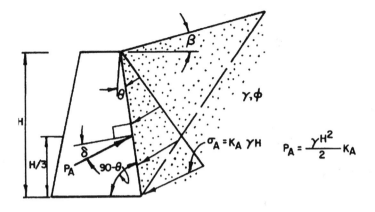

A) Coulomb's Equation (Static Condition):

$$K_A = \frac{\cos^2(\phi - \theta)}{\cos^2\theta \, \cos(\delta + \theta)\left[1 + \sqrt{\dfrac{\sin(\delta + \phi)\sin(\phi - \beta)}{\cos(\delta + \theta)\cos(\beta - \theta)}}\right]^2}$$

B) Mononobe-Okabe Equation (Earthquake Condition):

$$K_{AE} = \frac{\cos^2(\phi - \theta - \psi)}{\cos\psi \, \cos^2\theta \, \cos(\delta + \theta + \psi)\left[1 + \sqrt{\dfrac{\sin(\delta + \phi)\sin(\phi - \beta - \psi)}{\cos(\delta + \theta + \psi)\cos(\beta - \theta)}}\right]^2}$$

FIGURE 10.3   Coulomb's earth pressure (k_A) equation for static conditions. Also shown is the Mononobe-Okabe equation (k_{AE}) for earthquake conditions. (*Figure reproduced from NAVFAC DM-7.2, 1982, with equations from Kramer 1996.*)

meter of wall or pounds per linear foot of wall, acting on the retaining wall due to the surcharge Q is equal to $P_2 = QHk_A$, where Q = uniform vertical surcharge acting upon the entire ground surface behind the retaining wall, k_A = active earth pressure coefficient [Eq. (10.2) or Fig. 10.3], and H = height of the retaining wall. Because this pressure acting upon the retaining wall is uniform, the resultant force P_2 is located at midheight of the retaining wall.

   4. *Active wedge:*   The *active wedge* is defined as that zone of soil involved in the development of the active earth pressures upon the wall. This active wedge must move laterally to develop the active earth pressures. It is important that building footings or other

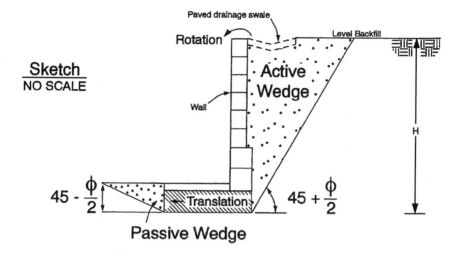

Sketch
NO SCALE

$45 - \dfrac{\phi}{2}$

Passive Wedge

$45 + \dfrac{\phi}{2}$

Note: For active and passive wedge development there must be movement of the retaining wall as illustrated above.

FIGURE 10.4   Active wedge behind retaining wall.

load-carrying members not be supported by the active wedge, or else they will be subjected to lateral movement. The active wedge is inclined at an angle of $45° + \phi/2$ from the horizontal, as indicated in Fig. 10.4.

Passive Earth Pressure.   As shown in Fig. 10.4, the passive earth pressure is developed along the front side of the footing. Passive pressure is developed when the wall footing moves laterally into the soil and a passive wedge is developed. To calculate the passive resultant force P_p, the following equation is used, assuming that there is cohesionless soil in front of the wall footing:

$$P_p = \frac{1}{2} k_p \gamma_t D^2 \tag{10.3}$$

where P_p = passive resultant force in kilonewtons per linear meter of wall or pounds per linear foot of wall, k_p = passive earth pressure coefficient, γ_t = total unit weight of the soil located in front of the wall footing, and D = depth of the wall footing (vertical distance from the ground surface in front of the retaining wall to the bottom of the footing). The passive earth pressure coefficient k_p is equal to

$$k_p = \tan^2 (45° + \frac{1}{2}\phi) \tag{10.4}$$

where ϕ = friction angle of the soil in front of the wall footing. Equation (10.4) is known as the passive Rankine state. To develop passive pressure, the wall footing must move laterally into the soil. The wall translation to reach the passive state is at least twice that required to reach the active earth pressure state. Usually it is desirable to limit the amount of wall translation by applying a reduction factor to the passive pressure. A commonly used reduction factor is 2.0. The soil engineer routinely reduces the passive pressure by one-half (reduction factor = 2.0) and then refers to the value as the allowable passive pressure.

Footing Bearing Pressure. To calculate the footing bearing pressure, the first step is to sum the vertical loads, such as the wall and footing weights. The vertical loads can be represented by a single resultant vertical force, per linear meter or foot of wall, that is offset by a distance (eccentricity) from the toe of the footing. This can then be converted to a pressure distribution by using Eq. (8.7). The largest bearing pressure is routinely at the toe of the footing, and it should not exceed the allowable bearing pressure (Sec. 8.2.5).

Retaining Wall Analyses. Once the active earth pressure resultant force P_A and the passive resultant force P_p have been calculated, the design analysis is performed as indicated in Fig. 10.2c. The retaining wall analysis includes determining the resultant location of the forces (i.e., calculate d, which should be within the middle third of the footing), the factor of safety for overturning, and the factor of safety for sliding. The adhesion c_a between the bottom of the footing and the underlying soil is often ignored for the sliding analysis.

10.1.2 Retaining Wall Analyses for Earthquake Conditions

The performance of retaining walls during earthquakes is very complex. As stated by Kramer (1996), laboratory tests and analyses of gravity walls subjected to seismic forces have indicated the following:

1. Walls can move by translation and/or rotation. The relative amounts of translation and rotation depend on the design of the wall; one or the other may predominate for some walls, and both may occur for others (Nadim and Whitman 1984, Siddharthan et al. 1992).
2. The magnitude and distribution of dynamic wall pressures are influenced by the mode of wall movement, e.g., translation, rotation about the base, or rotation about the top (Sherif et al. 1982, Sherif and Fang 1984a, b).
3. The maximum soil thrust acting on a wall generally occurs when the wall has translated or rotated toward the backfill (i.e., when the inertial force on the wall is directed toward the backfill). The minimum soil thrust occurs when the wall has translated or rotated away from the backfill.
4. The shape of the earthquake pressure distribution on the back of the wall changes as the wall moves. The point of application of the soil thrust therefore moves up and down along the back of the wall. The position of the soil thrust is highest when the wall has moved toward the soil and lowest when the wall moves outward.
5. Dynamic wall pressures are influenced by the dynamic response of the wall and backfill and can increase significantly near the natural frequency of the wall-backfill system (Steedman and Zeng 1990). Permanent wall displacements also increase at frequencies near the natural frequency of the wall-backfill system (Nadim 1982). Dynamic response effects can also cause deflections of different parts of the wall to be out of phase. This effect can be particularly significant for walls that penetrate into the foundation soils when the backfill soils move out of phase with the foundation soils.
6. Increased residual pressures may remain on the wall after an episode of strong shaking has ended (Whitman 1990).

Because of the complex soil-structure interaction during the earthquake, the most commonly used method for the design of retaining walls is the pseudostatic method, which is discussed in Sec. 10.2.

10.1.3 One-Third Increase in Soil Properties for Seismic Conditions

When the recommendations for the allowable soil pressures at a site are presented, it is common practice for the geotechnical engineer to recommend that the allowable bearing pressure and the allowable passive pressure be increased by a factor of one-third when performing seismic analyses. For example, in soil reports, it is commonly stated: "For the analysis of earthquake loading, the allowable bearing pressure and passive resistance may be increased by a factor of one-third." The rationale behind this recommendation is that the allowable bearing pressure and allowable passive pressure have an ample factor of safety, and thus for seismic analyses, a lower factor of safety would be acceptable.

Usually the above recommendation is appropriate if the retaining wall bearing material and the soil in front of the wall (i.e., passive wedge area) consist of the following:

- Massive crystalline bedrock and sedimentary rock that remains intact during the earthquake.
- Soils that tend to dilate during the seismic shaking or, e.g., dense to very dense granular soil and heavily overconsolidated cohesive soil such as very stiff to hard clays.
- Soils that have a stress-strain curve that does not exhibit a significant reduction in shear strength with strain.
- Clay that has a low sensitivity.
- Soils located above the groundwater table. These soils often have negative pore water pressure due to capillary action.

These materials do not lose shear strength during the seismic shaking, and therefore an increase in bearing pressure and passive resistance is appropriate.

A one-third increase in allowable bearing pressure and allowable passive pressure should not be recommended if the bearing material and/or the soil in front of the wall (i.e., passive wedge area) consists of the following:

- Foliated or friable rock that fractures apart during the earthquake, resulting in a reduction in shear strength of the rock.
- Loose soil located below the groundwater table and subjected to liquefaction or a substantial increase in pore water pressure.
- Sensitive clays that lose shear strength during the earthquake.
- Soft clays and organic soils that are overloaded and subjected to plastic flow.

These materials have a reduction in shear strength during the earthquake. Since the materials are weakened by the seismic shaking, the static values of allowable bearing pressures and allowable passive resistance should not be increased for the earthquake analyses. In fact, the allowable bearing pressure and the allowable passive pressure may actually have to be reduced to account for the weakening of the soil during the earthquake. Sections 10.3 and 10.4 discuss retaining wall analyses for the case where the soil is weakened during the earthquake.

10.2 PSEUDOSTATIC METHOD

10.2.1 Introduction

The most commonly used method of retaining wall analyses for earthquake conditions is the pseudostatic method. The pseudostatic method is also applicable for earthquake slope stability analyses (see Sec. 9.2). As previously mentioned, the advantages of this method are that it is easy to understand and apply.

Similar to earthquake slope stability analyses, this method ignores the cyclic nature of the earthquake and treats it as if it applied an additional static force upon the retaining wall. In particular, the pseudostatic approach is to apply a lateral force upon the retaining wall. To derive the lateral force, it can be assumed that the force acts through the centroid of the active wedge. The pseudostatic lateral force P_E is calculated by using Eq. (6.1), or

$$P_E = ma = \frac{W}{g} a = W \frac{a_{max}}{g} = k_h W \qquad (10.5)$$

where
P_E = horizontal pseudostatic force acting upon the retaining wall, lb or kN. This force can be assumed to act through the centroid of the active wedge. For retaining wall analyses, the wall is usually assumed to have a unit length (i.e., two-dimensional analysis)

m = total mass of active wedge, lb or kg, which is equal to W/g

W = total weight of active wedge, lb or kN

a = acceleration, which in this case is maximum horizontal acceleration atground surface caused by the earthquake ($a = a_{max}$), ft/s² or m/s²

a_{max} = maximum horizontal acceleration at ground surface that is induced by the earthquake, ft/s² or m/s². The maximum horizontal acceleration is also commonly referred to as the peak ground acceleration (see Sec. 5.6)

$a_{max}/g = k_h$ = seismic coefficient, also known as pseudostatic coefficient (dimensionless)

Note that an earthquake could subject the active wedge to both vertical and horizontal pseudostatic forces. However, the vertical force is usually ignored in the standard pseudostatic analysis. This is because the vertical pseudostatic force acting on the active wedge usually has much less effect on the design of the retaining wall. In addition, most earthquakes produce a peak vertical acceleration that is less than the peak horizontal acceleration, and hence k_v is smaller than k_h.

As indicated in Eq. (10.5), the only unknowns in the pseudostatic method are the weight of the active wedge W and the seismic coefficient k_h. Because of the usual relatively small size of the active wedge, the seismic coefficient k_h can be assumed to be equal to a_{max}/g. Using Fig. 10.4, the weight of the active wedge can be calculated as follows:

$$W = \tfrac{1}{2} H L \gamma_t = \tfrac{1}{2} H [H \tan (45° - \tfrac{1}{2}\phi)] \gamma_t = \tfrac{1}{2} k_A^{1/2} H^2 \gamma_t \qquad (10.6)$$

where W = weight of the active wedge, lb or kN per unit length of wall

H = height of the retaining wall, ft or m

L = length of active wedge at top of retaining wall. Note in Fig. 10.4 that the active wedge is inclined at an angle equal to $45° + \tfrac{1}{2}\phi$. Therefore the internal angle of the active wedge is equal to $90° - (45° + \tfrac{1}{2}\phi) = 45° - \tfrac{1}{2}\phi$. The length L can then be calculated as $L = H \tan (45° - \tfrac{1}{2}\phi) = H k_A^{1/2}$

γ_t = total unit weight of the backfill soil (i.e., unit weight of soil comprising active wedge), lb/ft³ or kN/m³

Substituting Eq. (10.6) into Eq. (10.5), we get for the final result:

$$P_E = k_h W = \tfrac{1}{2} k_h k_A^{1/2} H^2 \gamma_t = \tfrac{1}{2} k_A^{1/2} \left(\frac{a_{max}}{g} \right) (H^2 \gamma_t) \qquad (10.7)$$

Note that since the pseudostatic force is applied to the centroid of the active wedge, the location of the force P_E is at a distance of $\tfrac{2}{3}H$ above the base of the retaining wall.

10.2.2 Method by Seed and Whitman

Seed and Whitman (1970) developed an equation that can be used to determine the horizontal pseudostatic force acting on the retaining wall:

$$P_E = \frac{3}{8} \frac{a_{max}}{g} H^2 \gamma_t \tag{10.8}$$

Note that the terms in Eq. (10.8) have the same definitions as the terms in Eq. (10.7). Comparing Eqs. (10.7) and (10.8), we see the two equations are identical for the case where $\frac{1}{2} k_A^{1/2} = \frac{3}{8}$. According to Seed and Whitman (1970), the location of the pseudostatic force from Eq. (10.8) can be assumed to act at a distance of $0.6H$ above the base of the wall.

10.2.3 Method by Mononobe and Okabe

Mononobe and Matsuo (1929) and Okabe (1926) also developed an equation that can be used to determine the horizontal pseudostatic force acting on the retaining wall. This method is often referred to as the Mononobe-Okabe method. The equation is an extension of the Coulomb approach and is

$$P_{AE} = P_A + P_E = \frac{1}{2} k_{AE} H^2 \gamma_t \tag{10.9}$$

where P_{AE} = the sum of the static (P_A) and the pseudostatic earthquake force (P_E). The equation for k_{AE} is shown in Fig. 10.3. Note that in Fig. 10.3, the term ψ is defined as

$$\psi = \tan^{-1} k_h = \tan^{-1} \frac{a_{max}}{g} \tag{10.10}$$

The original approach by Mononobe and Okabe was to assume that the force P_{AE} from Eq. (10.9) acts at a distance of $\frac{1}{3}H$ above the base of the wall.

10.2.4 Example Problem

Figure 10.5 (from Lambe and Whitman 1969) presents an example of a proposed concrete retaining wall that will have a height of 20 ft (6.1 m) and a base width of 7 ft (2.1 m). The wall will be backfilled with sand that has a total unit weight γ_t of 110 lb/ft^3 (17.3 kN/m^3), friction angle ϕ of 30°, and an assumed wall friction $\delta = \phi_w$ of 30°. Although $\phi_w = 30°$ is used for this example problem, more typical values of wall friction are $\phi_w = \frac{3}{4}\phi$ for the wall friction between granular soil and wood or concrete walls, and $\phi_w = 20°$ for the wall friction between granular soil and steel walls such as sheet pile walls. The retaining wall is analyzed for the static case and for the earthquake condition assuming $k_h = 0.2$. It is also assumed that the backfill soil, bearing soil, and soil located in the passive wedge are not weakened by the earthquake.

Static Analysis

Active Earth Pressure. For the example problem shown in Fig. 10.5, the value of the active earth pressure coefficient k_A can be calculated by using Coulomb's equation (Fig. 10.3) and inserting the following values:

- Slope inclination: $\beta = 0$ (no slope inclination)
- Back face of the retaining wall: $\theta = 0$ (vertical back face of the wall)

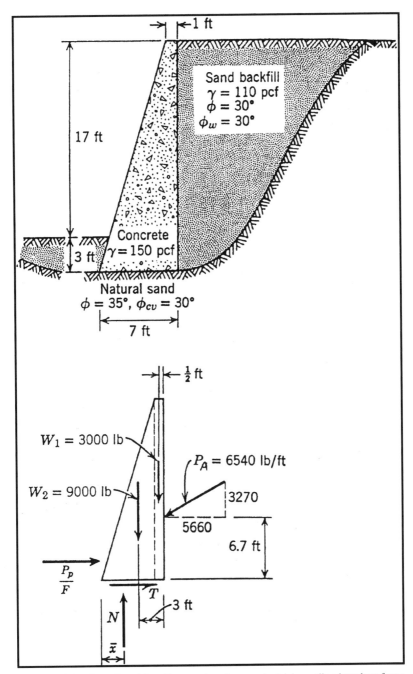

FIGURE 10.5a Example problem. Cross section of proposed retaining wall and resultant forces acting on the retaining wall. (*From Lambe and Whitman 1969; reproduced with permission of John Wiley & Sons.*)

Find. Adequacy of wall.
Solution. The first step is to determine the active thrust;
The next step is to compute the weights:

$$W_1 = (1)(20)(150) = 3000 \text{ lb/ft}$$

$$W_2 = \tfrac{1}{2}(6)(20)(150) = 9000 \text{ lb/ft}$$

Next N and $\bar{x}$ are computed:

$$N = 9000 + 3000 + 3270 = 15{,}270 \text{ lb/ft}$$

Overturning moment $= 5660(6.67) - 3270(7) = 37{,}800 - 22{,}900 = 14{,}900$

Moment of weight $= (6.5)(3000) + (4)(9000) = 19{,}500 + 36{,}000 = 55{,}500$

Ratio $= 3.73$ <u>OK</u>

$$\bar{x} = \frac{55{,}500 - 14{,}900}{15{,}270} = \frac{40{,}600}{15{,}270} = 2.66 \text{ ft} \quad \underline{\text{OK}}$$

The location of N

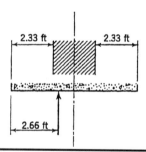

FIGURE 10.5b   Example problem (*continued*). Calculation of the factor of safety for overturning and the location of the resultant force N. (*From Lambe and Whitman 1969; reproduced with permission of John Wiley & Sons.*)

- Friction between the back face of the wall and the soil backfill: $\delta = \phi_w = 30°$
- Friction angle of backfill sand: $\phi = 30°$

Inputting the above values into Coulomb's equation (Fig. 10.3), the value of the active earth pressure coefficient $k_A = 0.297$.

By using Eq. (10.1) with $k_A = 0.297$, total unit weight $\gamma_t = 110 \text{ lb/ft}^3$ (17.3 kN/m³), and the height of the retaining wall $H = 20$ ft (see Fig. 10.5a), the active earth pressure resultant force $P_A = 6540$ lb per linear foot of wall (95.4 kN per linear meter of wall). As indicated in Fig. 10.5a, the active earth pressure resultant force $P_A = 6540$ lb/ft is inclined at an angle of 30° due to the wall friction assumptions. The vertical ($P_v = 3270$ lb/ft) and horizontal ($P_H = 5660$ lb/ft) resultants of P_A are also shown in Fig. 10.5a. Note in Fig. 10.3 that even

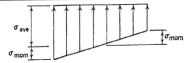

Next the bearing stress is computed. The average bearing stress is $15,270/7 = 2180$ psf. Assuming that the bearing stress is distributed linearly, the maximum stress can be found

$$\sigma_{mom} = \frac{M}{S}$$

where

$$M = \text{moment about } \mathcal{L} = 15,270(3.5 - 2.66) = 12,820 \text{ lb-ft/ft}$$

$$S = \text{section modulus} = \tfrac{1}{6}B^2 = \tfrac{1}{6}(7)^2 = 8.17 \text{ ft}^2$$

where B is width of base

$$\sigma_{mom} = \frac{12,820}{8.17} = 1570 \text{ psf}$$

$$\text{Maximum stress} = 2180 + 1570 = 3750 \text{ psf}$$

Finally, the resistance to horizontal sliding is checked. Assuming passive resistance without wall friction,

$$K_p = 3$$

$$P_p = \tfrac{1}{2}(110)(3^2)(3) = 1500 \text{ lb/ft}$$

With reduction factor of 2,

$$\frac{P_p}{F} = 750 \text{ lb/ft}$$

$$T = 5660 - 750 = 4910 \text{ lb/ft}$$

$$N \tan 30° = 8810 \text{ lb/ft}$$

$$\frac{N \tan \phi_{cv}}{T} = 1.79 < 2 \qquad \underline{\underline{\text{not OK}}}$$

Ignoring passive resistance

$$T = 5660 \text{ lb/ft}$$

$$\frac{N \tan \phi_{cv}}{T} = 1.55 > 1.5 \qquad \underline{\underline{\text{OK}}}$$

FIGURE 10.5c   Example problem (*continued*). Calculation of the maximum bearing stress and the factor of safety for sliding. (*From Lambe and Whitman 1969, reproduced with permission of John Wiley & Sons.*)

with wall friction, the active earth pressure is still a triangular distribution acting upon the retaining wall, and thus the location of the active earth pressure resultant force P_A is at a distance of $\frac{1}{3}H$ above the base of the wall, or 6.7 feet (2.0 m).

 Passive Earth Pressure. As shown in Fig. 10.5a, the passive earth pressure is developed by the soil located at the front of the retaining wall. Usually wall friction is ignored for the passive earth pressure calculations. For the example problem shown in Fig. 10.5, the passive resultant force P_p was calculated by using Eqs. (10.3) and (10.4) and neglecting wall friction and the slight slope of the front of the retaining wall (see Fig. 10.5c for passive earth pressure calculations).

Footing Bearing Pressure. The procedure for the calculation of the footing bearing pressure is as follows:

1. *Calculate N:* As indicated in Fig. 10.5b, the first step is to calculate N (15,270 lb/ft), which equals the sum of the weight of the wall, footing, and vertical component of the active earth pressure resultant force (that is, $N = W + P_A \sin \phi_w$).

2. *Determine resultant location of N:* The resultant location of N from the toe of the retaining wall (that is, 2.66 ft) is calculated as shown in Fig. 10.5b. The moments are determined about the toe of the retaining wall. Then the location of N is equal to the difference in the opposing moments divided by N.

3. *Determine average bearing pressure:* The average bearing pressure (2180 lb/ft²) is calculated in Fig. 10.5c as N divided by the width of the footing (7 ft).

4. *Calculate moment about the centerline of the footing:* The moment about the centerline of the footing is calculated as N times the eccentricity (0.84 ft).

5. *Section modulus:* The section modulus of the footing is calculated as shown in Fig. 10.5c.

6. *Portion of bearing stress due to moment:* The portion of the bearing stress due to the moment (σ_{mom}) is determined as the moment divided by the section modulus.

7. *Maximum bearing stress:* The maximum bearing stress is then calculated as the sum of the average stress ($\sigma_{avg} = 2180$ lb/ft²) plus the bearing stress due to the moment ($\sigma_{mom} = 1570$ lb/ft²).

As indicated in Fig. 10.5c, the maximum bearing stress is 3750 lb/ft² (180 kPa). This maximum bearing stress must be less than the allowable bearing pressure (Chap. 8). It is also a standard requirement that the resultant normal force N be located within the middle third of the footing, such as illustrated in Fig. 10.5b. As an alternative to the above procedure, Eq. (8.7) can be used to calculate the maximum and minimum bearing stress.

Sliding Analysis. The factor of safety (FS) for sliding of the retaining wall is often defined as the resisting forces divided by the driving force. The forces are per linear meter or foot of wall, or

$$\text{FS} = \frac{N \tan \delta + P_p}{P_H} \qquad (10.11)$$

where $\delta = \phi_{cv}$ = friction angle between the bottom of the concrete foundation and bearing soil; N = sum of the weight of the wall, footing, and vertical component of the active earth pressure resultant force (or $N = W + P_A \sin \phi_w$); P_p = allowable passive resultant force [P_p from Eq. (10.3) divided by a reduction factor]; and P_H = horizontal component of the active earth pressure resultant force ($P_H = P_A \cos \phi_w$).

There are variations of Eq. (10.11) that are used in practice. For example, as illustrated in Fig. 10.5c, the value of P_p is subtracted from P_H in the denominator of Eq. (10.11), instead of P_p being used in the numerator. For the example problem shown in Fig. 10.5, the factor of safety for sliding is FS = 1.79 when the passive pressure is included and FS = 1.55 when the passive pressure is excluded. For static conditions, the typical recommendations for minimum factor of safety for sliding are 1.5 to 2.0 (Cernica 1995b).

Overturning Analysis. The factor of safety for overturning of the retaining wall is calculated by taking moments about the toe of the footing and is

$$\text{FS} = \frac{Wa}{\frac{1}{3}P_H H - P_v e} \qquad (10.12)$$

where a = lateral distance from the resultant weight W of the wall and footing to the toe of the footing, P_H = horizontal component of the active earth pressure resultant force, P_v = vertical component of the active earth pressure resultant force, and e = lateral distance from the location of P_v to the toe of the wall. In Fig. 10.5b, the factor of safety (ratio) for overturning is calculated to be 3.73. For static conditions, the typical recommendations for minimum factor of safety for overturning are 1.5 to 2.0 (Cernica 1995b).

Settlement and Stability Analysis. Although not shown in Fig. 10.5, the settlement and stability of the ground supporting the retaining wall footing should also be determined. To calculate the settlement and evaluate the stability for static conditions, standard settlement and slope stability analyses can be utilized (see chaps. 9 and 13, Day 2000).

Earthquake Analysis. The pseudostatic analysis is performed for the three methods outlined in Secs. 10.2.1 to 10.2.3.

Equation (10.7). Using Eq. (10.2) and neglecting the wall friction, we find

$$k_A = \tan^2 (45° - \tfrac{1}{2}\phi) = \tan^2 (45° - \tfrac{1}{2}30°) = 0.333$$

Substituting into Eq. (10.7) gives

$$P_E = \tfrac{1}{2}\, k_A^{1/2} \left(\frac{a_{max}}{g} \right) (H^2 \gamma_t)$$

$$= \tfrac{1}{2}\, (0.333)^{1/2}\, (0.2)\, (20 \text{ ft})^2\, (110 \text{ lb/ft}^3) = 2540 \text{ lb per linear foot of wall length}$$

This pseudostatic force acts at a distance of $\tfrac{2}{3}H$ above the base of the wall, or $\tfrac{2}{3}H = \tfrac{2}{3}(20 \text{ ft}) = 13.3$ ft. Similar to Eq. (10.11), the factor of safety for sliding is

$$\text{FS} = \frac{N \tan \delta + P_p}{P_H + P_E} \tag{10.13}$$

Substituting values into Eq. (10.13) gives

$$\text{FS} = \frac{15{,}270 \tan 30° + 750}{5660 + 2540} = 1.17$$

Based on Eq. (10.12), the factor of safety for overturning is

$$\text{FS} = \frac{Wa}{\tfrac{1}{3}P_H H - P_v e + \tfrac{2}{3}H P_E} \tag{10.14}$$

Inserting values into Eq. (10.14) yields

$$\text{FS} = \frac{55{,}500}{\tfrac{1}{3}(5660)(20) - 3270(7) + \tfrac{2}{3}(20)(2540)} = 1.14$$

Method by Seed and Whitman (1970). Using Eq. (10.8) and neglecting the wall friction, we get

$$P_E = \frac{3}{8} \left(\frac{a_{max}}{g} \right) H^2 \gamma_t$$

$$= \tfrac{3}{8}\, (0.2)\, (20 \text{ ft})^2\, (110 \text{ lb/ft}^3) = 3300 \text{ lb per linear foot of wall length}$$

This pseudostatic force acts at a distance of $0.6H$ above the base of the wall, or $0.6H = (0.6)(20 \text{ ft}) = 12$ ft. Using Eq. (10.13) gives

$$FS = \frac{N \tan \delta + P_p}{P_H + P_E} = \frac{15{,}270 \tan 30° + 750}{5660 + 3300} = 1.07$$

Similar to Eq. (10.14), the factor of safety for overturning is

$$FS = \frac{Wa}{\frac{1}{3}P_H H - P_v e + 0.6HP_E} \tag{10.15}$$

Substituting values into Eq. (10.15) gives

$$FS = \frac{55{,}500}{\frac{1}{3}(5660)(20) - 3270(7) + 0.6(20)(3300)} = 1.02$$

Mononobe-Okabe Method.   We use the following values:

θ (wall inclination) $= 0°$

ϕ (friction angle of backfill soil) $= 30°$

β (backfill slope inclination) $= 0°$

$\delta = \phi_w$ (friction angle between the backfill and wall) $= 30°$

$\psi = \tan^{-1} k_h = \tan^{-1} \dfrac{a_{max}}{g} = \tan^{-1} 0.2 = 11.3°$

Inserting the above values into the K_{AE} equation in Fig. 10.3, we get $K_{AE} = 0.471$. Therefore, using Eq. (10.9) yields

$$P_{AE} = P_A + P_E = \frac{1}{2} k_{AE} H^2 \gamma_t$$

$$= \frac{1}{2}(0.471)(20)^2(110) = 10{,}400 \text{ lb per linear foot of wall length}$$

This force P_{AE} is inclined at an angle of $30°$ and acts at a distance of $0.33H$ above the base of the wall, or $0.33H = (0.33)(20 \text{ ft}) = 6.67$ ft. The factor of safety for sliding is

$$FS = \frac{N \tan \delta + P_p}{P_H} = \frac{(W + P_{AE} \sin \phi_w) \tan \delta + P_p}{P_{AE} \cos \phi_w} \tag{10.16}$$

Substituting values into Eq. (10.16) gives

$$FS = \frac{(3000 + 9000 + 10{,}400 \sin 30°)(\tan 30°) + 750}{10{,}400 \cos 30°} = 1.19$$

The factor of safety for overturning is

$$FS = \frac{Wa}{\frac{1}{3} H P_{AE} \cos \phi_w - P_{AE} \sin \phi_w e} \tag{10.17}$$

Substituting values into Eq. (10.17) produces

$$FS = \frac{55{,}500}{\frac{1}{3}(20)(10{,}400)(\cos 30°) - (10{,}400)(\sin 30°)(7)} = 2.35$$

Summary of Values. The values from the static and earthquake analyses using $k_h = a_{max}/g$ = 0.2 are summarized below:

Type of condition		P_E or P_{AE}, lb/ft	Location of P_E or P_{AE} above base of wall, ft	Factor of safety for sliding	Factor of safety for overturning
Static		$P_E = 0$	—	1.69[*]	3.73
Earthquake ($k_h = 0.2$)	Equation (10.7)	$P_E = 2,540$	$\frac{2}{3}H = 13.3$	1.17	1.14
	Seed and Whitman	$P_E = 3,300$	$0.6H = 12$	1.07	1.02
	Mononobe-Okabe	$P_{AE} = 10,400$	$\frac{1}{3}H = 6.7$	1.19	2.35

[*]Factor of safety for sliding using Eq. (10.11).

For the analysis of sliding and overturning of the retaining wall, it is common to accept a lower factor of safety (1.1 to 1.2) under the combined static and earthquake loads. Thus the retaining wall would be considered marginally stable for the earthquake sliding and overturning conditions.

Note in the above table that the factor of safety for overturning is equal to 2.35 based on the Mononobe-Okabe method. This factor of safety is much larger than that for the other two methods. This is because the force P_{AE} is assumed to be located at a distance of $\frac{1}{3}H$ above the base of the wall. Kramer (1996) suggests that it is more appropriate to assume that P_E is located at a distance of $0.6H$ above the base of the wall [that is, $P_E = P_{AE} - P_A$, see Eq. (10.9)].

Although the calculations are not shown, it can be demonstrated that the resultant location of N for the earthquake condition is outside the middle third of the footing. Depending on the type of material beneath the footing, this condition could cause a bearing capacity failure or excess settlement at the toe of the footing during the earthquake.

10.2.5 Mechanically Stabilized Earth Retaining Walls

Introduction. Mechanically stabilized earth (MSE) retaining walls are typically composed of strip- or grid-type (geosynthetic) reinforcement. Because they are often more economical to construct than conventional concrete retaining walls, mechanically stabilized earth retaining walls have become very popular in the past decade.

A mechanically stabilized earth retaining wall is composed of three elements: (1) wall facing material, (2) soil reinforcement, such as strip- or grid-type reinforcement, and (3) compacted fill between the soil reinforcement. Figure 10.6 shows the construction of a mechanically stabilized earth retaining wall.

The design analyses for a mechanically stabilized earth retaining wall are more complex than those for a cantilevered retaining wall. For a mechanically stabilized earth retaining wall, both the internal and external stability must be checked, as discussed below.

External Stability—Static Conditions. The analysis for the external stability is similar to that for a gravity retaining wall. For example, Figs. 10.7 and 10.8 present the design analysis for external stability for a level backfill condition and a sloping backfill condition. In both

FIGURE 10.6 Installation of a mechanically stabilized earth retaining wall. The arrow points to the wall facing elements, which are in the process of being installed.

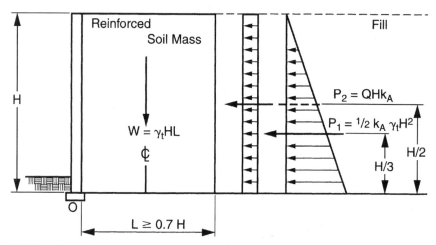

FIGURE 10.7 Static design analysis for mechanically stabilized earth retaining wall having horizontal backfill. (*Adapted from Standard Specifications for Highway Bridges, AASHTO 1996.*)

Figs. 10.7 and 10.8, the zone of mechanically stabilized earth mass is treated in a similar fashion as a massive gravity retaining wall. For static conditions, the following analyses must be performed:

1. *Allowable bearing pressure:* The bearing pressure due to the reinforced soil mass must not exceed the allowable bearing pressure.

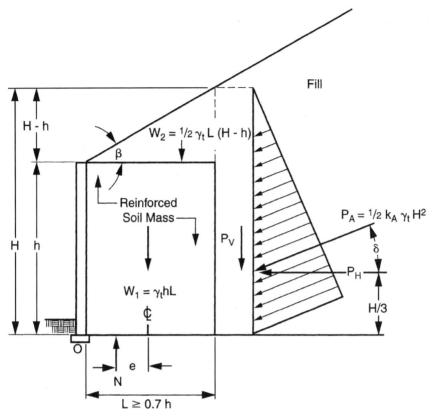

FIGURE 10.8   Static design analysis for mechanically stabilized earth retaining wall having sloping backfill. (*Adapted from Standard Specifications for Highway Bridges, AASHTO 1996.*)

2. *Factor of safety for sliding:*   The reinforced soil mass must have an adequate factor of safety for sliding.

3. *Factor of safety for overturning:*   The reinforced soil mass must have an adequate factor of safety for overturning about point O.

4. *Resultant of vertical forces:*   The resultant of the vertical forces N must be within the middle one-third of the base of the reinforced soil mass.

5. *Stability of reinforced soil mass.*   The stability of the entire reinforced soil mass (i.e., shear failure below the bottom of the wall) should be checked.

Note in Fig. 10.7 that two forces P_1 and P_2 are shown acting on the reinforced soil mass. The first force P_1 is determined from the standard active earth pressure resultant equation [Eq. (10.1)]. The second force P_2 is due to a uniform surcharge Q applied to the entire ground surface behind the mechanically stabilized earth retaining wall. If the wall does not have a surcharge, then P_2 is equal to zero.

Figure 10.8 presents the active earth pressure force for an inclined slope behind the retaining wall. As shown in Fig. 10.8, the friction δ of the soil along the backside of the reinforced soil mass has been included in the analysis. The value of k_A would be obtained from Coulomb's earth pressure equation (Fig. 10.3). As a conservative approach, the friction angle δ can be assumed to be equal to zero, and then $P_H = P_A$. As indicated in both Figs. 10.7 and 10.8, the minimum width of the reinforced soil mass must be at least $^7\!/_{10}$ times the height of the reinforced soil mass.

External Stability—Earthquake Conditions.   For earthquake conditions, the most commonly used approach is the pseudostatic method. The pseudostatic force can be calculated from Eqs. (10.7), (10.8), or (10.9). Once the pseudostatic force and location are known, then the five items listed in "External Stability—Static Conditions" would need to be checked. Acceptable values of the factors of safety for sliding and overturning are typically in the range of 1.1 to 1.2 for earthquake conditions.

Internal Stability.   To check the static stability of the mechanically stabilized zone, a slope stability analysis can be performed in which the soil reinforcement is modeled as horizontal forces equivalent to its allowable tensile resistance. For earthquake conditions, the slope stability analysis could incorporate a pseudostatic force (i.e., Sec. 9.2.4). In addition to calculating the factor of safety for both the static and earthquake conditions, the pullout resistance of the reinforcement along the slip surface should be checked.

Example Problem.   Using the mechanically stabilized earth retaining wall shown in Fig. 10.7, let $H = 20$ ft, the width of the mechanically stabilized earth retaining wall $= 14$ ft, the depth of embedment at the front of the mechanically stabilized zone $= 3$ ft, and there is a level backfill with no surcharge pressures (that is, $P_2 = 0$). Assume that the soil behind and in front of the mechanically stabilized zone is a clean sand having a friction angle $\phi = 30°$, a total unit weight of $\gamma_t = 110$ lb/ft^3, and there will be no shear stress (that is, $\delta = 0°$) along the vertical back and front sides of the mechanically stabilized zone. For the mechanically stabilized zone, assume the soil will have a total unit weight $\gamma_t = 120$ lb/ft^3 and $\delta = 23°$ along the bottom of the mechanically stabilized zone. For earthquake design conditions, use $a_{\max} = 0.20g$. Calculate the factor of safety for sliding and for overturning for both the static and earthquake conditions.

   Solution: Static Analysis

$$k_A = \tan^2 (45° - \tfrac{1}{2}\phi) = \tan^2 [45° - \tfrac{1}{2}(30°)] = 0.333$$

$$k_p = \tan^2 (45° + \tfrac{1}{2}\phi) = \tan^2 [45° + \tfrac{1}{2}(30°)] = 3.0$$

$$P_A = \tfrac{1}{2}k_A\gamma_t H^2 = \tfrac{1}{2}(0.333)(110)(20)^2 = 7330 \text{ lb/ft}$$

$$P_p = \tfrac{1}{2}k_p\gamma_t D^2 = \tfrac{1}{2}(3.0)(110)(3)^2 = 1490 \text{ lb/ft}$$

With reduction factor $= 2$,

$$\text{Allowable } P_p = 740 \text{ lb/ft}$$

For sliding analysis:

$$\text{FS} = \frac{N \tan \delta + P_p}{P_A} \qquad \text{Eq. (10.11) , where } P_A = P_H$$

$$W = N = HL\gamma_t = (20)(14)(120 \text{ lb/ft}^3) = 33{,}600 \text{ lb per linear foot of wall length}$$

$$FS = \frac{33{,}600 \tan 23° + 740}{7330} = 2.05$$

For overturning analysis: Taking moments about the toe of the wall gives

$$\text{Overturning moment} = P_A \frac{H}{3} = 7330 \frac{20}{3} = 48{,}900$$

$$\text{Moment of weight} = 33{,}600 \frac{14}{2} = 235{,}000$$

$$FS = \frac{235{,}000}{48{,}900} = 4.81$$

Solution: Earthquake Analysis. Using Eq. (10.7), we get

$$P_E = \tfrac{1}{2} k_A^{1/2} \, (a_{max}/g) \, (H^2 \gamma_t) = \tfrac{1}{2} \, (0.333)^{1/2} \, (0.20)(20)^2 (110) = 2540 \text{ lb/ft}$$

For sliding analysis, use Eq. (10.13):

$$FS = \frac{N \tan \delta + P_p}{P_H + P_E} = \frac{33{,}600 \tan 23° + 740}{7330 + 2540} = 1.52$$

For overturning analysis, use Eq. (10.14) with $P_v = 0$.

$$FS = \frac{Wa}{\tfrac{1}{3} P_H H + \tfrac{2}{3} H P_E} = \frac{33{,}600(7)}{\tfrac{1}{3}(7330)(20) + \tfrac{2}{3}(20)(2540)} = 2.84$$

In summary,

Static conditions:

$$FS \text{ sliding} = 2.05$$

$$FS \text{ overturning} = 4.81$$

Earthquake conditions ($a_{max} = 0.20g$):

$$FS \text{ sliding} = 1.52$$

$$FS \text{ overturning} = 2.84$$

10.3 RETAINING WALL ANALYSES FOR LIQUEFIED SOIL

10.3.1 Introduction

Retaining walls are commonly used for port and wharf facilities, which are often located in areas susceptible to liquefaction. Many of these facilities have been damaged by earthquake-induced liquefaction. The ports and wharves often contain major retaining structures, such as seawalls, anchored bulkheads, gravity and cantilever walls, and sheet pile cofferdams, that allow large ships to moor adjacent to the retaining walls and then load or unload cargo. Examples of liquefaction-induced damage to retaining walls are presented in Sec. 3.4.3.

There are often three different types of liquefaction effects that can damage the retaining wall:

1. *Passive wedge liquefaction:* The first is liquefaction of soil in front of the retaining wall. In this case, the passive resistance in front of the retaining wall is reduced.

2. *Active wedge liquefaction:* In the second case, the soil behind the retaining wall liquefies, and the pressure exerted on the wall is greatly increased. Cases 1 and 2 can act individually or together, and they can initiate an overturning failure of the retaining wall or cause the wall to progressively slide outward (localized lateral spreading) or tilt toward the water. Another possibility is that the increased pressure exerted on the wall could exceed the strength of the wall, resulting in a structural failure of the wall.

Liquefaction of the soil behind the retaining wall can also affect tieback anchors. For example, the increased pressure due to liquefaction of the soil behind the wall could break the tieback anchors or reduce their passive resistance.

3. *Liquefaction below base of wall:* The third case is liquefaction below the bottom of the wall. Many waterfront retaining walls consist of massive structures, such as the concrete box caissons shown in Fig. 3.31. In this case, the bearing capacity or slide resistance of the wall is reduced, resulting in a bearing capacity failure or promoting lateral spreading of the wall.

10.3.2 Design Pressures

The first step in the analysis is to determine the factor of safety against liquefaction for the soil behind the retaining wall, in front of the retaining wall, and below the bottom of the wall. The analysis presented in Chap. 6 can be used to determine the factor of safety against liquefaction. The retaining wall may exert significant shear stress into the underlying soil, which can decrease the factor of safety against liquefaction for loose soils (i.e., see Fig. 9.24). Likewise, there could be sloping ground in front of the wall or behind the wall, in which case the factor of safety against liquefaction may need to be adjusted (see Sec. 9.4.2).

After the factor of safety against liquefaction has been calculated, the next step is to determine the design pressures that act on the retaining wall:

1. *Passive pressure:* For those soils that will be subjected to liquefaction in the passive zone, one approach is to assume that the liquefied soil has zero shear strength. In essence, the liquefied zones no longer provide sliding or overturning resistance.

2. *Active pressure:* For those soils that will be subjected to liquefaction in the active zone, the pressure exerted on the face of the wall will increase. One approach is to assume zero shear strength of the liquefied soil (that is, $\phi' = 0$). There are two possible conditions:

    a. *Water level located only behind the retaining wall:* In this case, the wall and the ground beneath the bottom of the wall are relatively impermeable. In addition, there is a groundwater table behind the wall with dry conditions in front of the wall. The thrust on the wall due to liquefaction of the backfill can be calculated by using Eq. (10.1) with $k_A = 1$ [i.e., for $\phi' = 0$, $k_A = 1$, see Eq. (10.2)] and $\gamma_t = \gamma_{sat}$ (i.e., γ_{sat} = saturated unit weight of the soil).

    b. *Water levels are approximately the same on both sides of the retaining wall:* The more common situation is that the elevation of the groundwater table behind the wall is approximately the same as the water level in front of the wall. The thrust on the wall due to liquefaction of the soil can be calculated by using Eq. (10.1) with $k_A = 1$ [i.e., for $\phi' = 0$, $k_A = 1$, see Eq. (10.2)] and using γ_b (buoyant unit weight) in place of γ_t.

The only difference between the two cases is that the first case includes the unit weight of water ($\gamma_{sat} = \gamma_b + \gamma_w$), while the second case does not include γ_w because it is located on both sides of the wall and hence its effect is canceled out.

In addition to the increased pressure acting on the retaining wall due to liquefaction, consider a reduction in support and/or resistance of the tieback anchors.

3. *Bearing soil:* For the liquefaction of the bearing soil, use the analysis in Sec. 8.2.

10.3.3 Sheet Pile Walls

Introduction. Sheet pile retaining walls are widely used for waterfront construction and consist of interlocking members that are driven into place. Individual sheet piles come in many different sizes and shapes. Sheet piles have an interlocking joint that enables the individual segments to be connected together to form a solid wall.

Static Design. Many different types of static design methods are used for sheet pile walls. Figure 10.9 shows the most common type of static design method. In Fig. 10.9, the term H represents the unsupported face of the sheet pile wall. As indicated in Fig. 10.9, this sheet pile wall is being used as a waterfront retaining structure, and the elevation of the water in front of the wall is the same as that of the groundwater table behind the wall. For highly permeable soil, such as clean sand and gravel, this often occurs because the water can quickly flow underneath the wall in order to equalize the water levels.

In Fig. 10.9, the term D represents that portion of the sheet pile wall that is anchored in soil. Also shown in Fig. 10.9 is a force designated as A_p. This represents a restraining force on the sheet pile wall due to the construction of a tieback, such as by using a rod that has a

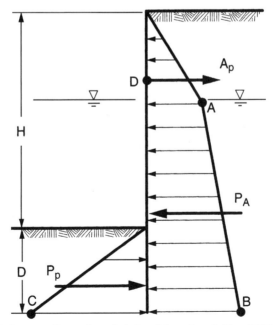

FIGURE 10.9   Earth pressure diagram for static design of sheet pile wall. (*From NAVFAC DM-7.2, 1982.*)

grouted end or is attached to an anchor block. Tieback anchors are often used in sheet pile wall construction to reduce the bending moments in the sheet pile. When tieback anchors are used, the sheet pile wall is typically referred to as an anchored bulkhead, while if no tiebacks are utilized, the wall is called a cantilevered sheet pile wall.

Sheet pile walls tend to be relatively flexible. Thus, as indicated in Fig. 10.9, the design is based on active and passive earth pressures. The soil behind the wall is assumed to exert an active earth pressure on the sheet pile wall. At the groundwater table (point A), the active earth pressure is equal to

$$\text{Active earth pressure at point } A, \text{ kPa or lb/ft}^2 = k_A \gamma_t d_1 \qquad (10.18)$$

where k_A = active earth pressure coefficient from Eq. (10.2) (dimensionless parameter). Friction between sheet pile wall and soil is usually neglected in design analysis
γ_t = total unit weight of the soil above the groundwater table, kN/m^3 or lb/ft^3
d_1 = depth from the ground surface to the groundwater table, m or ft

In using Eq. (10.18), a unit length (1 m or 1 ft) of sheet pile wall is assumed. At point B in Fig. 10.9, the active earth pressure equals

$$\text{Active earth pressure at point } B, \text{ kPa or lb/ft}^2 = k_A \gamma_t d_1 + k_A \gamma_b d_2 \qquad (10.19)$$

where γ_b = buoyant unit weight of the soil below the groundwater table and d_2 = depth from the groundwater table to the bottom of the sheet pile wall. For a sheet pile wall having assumed values of H and D (see Fig. 10.9) and using the calculated values of active earth pressure at points A and B, the active earth pressure resultant force P_A, in kilonewtons per linear meter of wall or pounds per linear foot of wall, can be calculated.

The soil in front of the wall is assumed to exert a passive earth pressure on the sheet pile wall. The passive earth pressure at point C in Fig. 10.9 is

$$\text{Passive earth pressure at point } C, \text{ kPa or lb/ft}^2 = k_p \gamma_b D \qquad (10.20)$$

where the passive earth pressure coefficient k_p can be calculated from Eq. (10.4). Similar to the analysis of cantilever retaining walls, if it is desirable to limit the amount of sheet pile wall translation, then a reduction factor can be applied to the passive pressure. Once the allowable passive pressure is known at point C, the passive resultant force P_p can be readily calculated.

As an alternative solution for the passive pressure, Eq. (10.3) can be used to calculate P_p with the buoyant unit weight γ_b substituted for the total unit weight γ_t and the depth D as shown in Fig. 10.9.

Note that a water pressure has not been included in the analysis. This is because the water level is the same on both sides of the wall, and water pressure cancels and thus should not be included in the analysis.

The static design of sheet pile walls requires the following analyses: (1) evaluation of the earth pressures that act on the wall, such as shown in Fig. 10.9; (2) determination of the required depth D of piling penetration; (3) calculation of the maximum bending moment M_{max} which is used to determine the maximum stress in the sheet pile; and (4) selection of the appropriate piling type, size, and construction details.

A typical design process is to assume a depth D (Fig. 10.9) and then calculate the factor of safety for toe failure (i.e., toe kick-out) by the summation of moments at the tieback anchor (point D). The factor of safety is defined as the moment due to the passive force divided by the moment due to the active force. Values of acceptable FS for toe failure are 2 to 3. An alternative solution is to first select the factor of safety and then develop the active and passive resultant forces and moment arms in terms of D. By solving the equation, the value of D for a specific factor of safety can be directly calculated.

Once the depth D of the sheet pile wall is known, the anchor pull A_p must be calculated. The anchor pull is determined by the summation of forces in the horizontal direction, or

$$A_p = P_A - \frac{P_p}{FS} \qquad (10.21)$$

where P_A and P_p are the resultant active and passive forces (see Fig. 10.9) and FS is the factor of safety that was obtained from the toe failure analysis. Based on the earth pressure diagram (Fig. 10.9) and the calculated value of A_p, elementary structural mechanics can be used to determine the maximum moment in the sheet pile wall. The maximum moment divided by the section modulus can then be compared with the allowable design stresses of the sheet piling.

Some other important design considerations for the static design of sheet pile walls include the following:

1. *Soil layers:* The active and passive earth pressures should be adjusted for soil layers having different engineering properties.

2. *Penetration depth:* The penetration depth D of the sheet pile wall should be increased by at least an additional 20 percent to allow for the possibility of dredging and scour. Deeper penetration depths may be required based on a scour analysis.

3. *Surcharge loads:* The ground surface behind the sheet pile wall is often subjected to surcharge loads. The equation $P_2 = QHk_A$ can be used to determine the active earth pressure resultant force due to a uniform surcharge pressure applied to the ground surface behind the wall. Note in this equation that the entire height of the sheet pile wall (that is, $H + D$, see Fig. 10.9) must be used in place of H. Typical surcharge pressures exerted on sheet pile walls are caused by railroads, highways, dock loading facilities and merchandise, ore piles, and cranes.

4. *Unbalanced hydrostatic and seepage forces:* The previous discussion has assumed that the water levels on both sides of the sheet pile wall are at the same elevation. Depending on factors such as the watertightness of the sheet pile wall and the backfill permeability, it is possible that the groundwater level could be higher than the water level in front of the wall, in which case the wall would be subjected to water pressures. This condition could develop when there is a receding tide or a heavy rainstorm that causes a high groundwater table. A flow net can be used to determine the unbalanced hydrostatic and upward seepage forces in the soil in front of the sheet pile wall.

5. *Other loading conditions:* The sheet pile wall may have to be designed to resist the lateral loads due to ice thrust, wave forces, ship impact, mooring pull, and earthquake forces. If granular soil behind or in front of the sheet pile wall is in a loose state, it could be susceptible to liquefaction during an earthquake.

Earthquake Analysis. In the case of liquefaction of soil, the earthquake design pressures must be modified. As indicated in Sec. 10.3.2, higher pressures will be exerted on the back face of the wall if this soil should liquefy. Likewise, there will be less passive resistance if the soil in front of the sheet pile wall will liquefy during the design earthquake. Section 10.3.2 should be used as a guide in the selection of the pressures exerted on the sheet pile wall during the earthquake. Once these earthquake-induced pressures behind and in front of the wall are known, then the factor of safety for toe failure and the anchor pull force can be calculated in the same manner as outlined in the previous section.

Example Problems. Using the sheet pile wall diagram shown in Fig. 10.9, assume that the soil behind and in front of the sheet wall is uniform sand with a friction angle $\phi' = 33°$, buoyant unit weight $\gamma_b = 64$ lb/ft^3, and above the groundwater table, the total unit weight

$\gamma_t = 120$ lb/ft^3. Also assume that the sheet pile wall has $H = 30$ ft and $D = 20$ ft, the water level in front of the wall is at the same elevation as the groundwater table which is located 5 ft below the ground surface, and the tieback anchor is located 4 ft below the ground surface. In the analysis, neglect wall friction.

Static Design. Calculate the factor of safety for toe kick-out and the tieback anchor force. Equation (10.2):

$$k_A = \tan^2 (45° - \tfrac{1}{2}\phi) = \tan^2 [45° - \tfrac{1}{2}(33°)] = 0.295$$

Equation (10.4):

$$k_p = \tan^2 (45° + \tfrac{1}{2}\phi) = \tan^2 [45° + \tfrac{1}{2}(33°)] = 3.39$$

From 0 to 5 ft:

$$P_{1A} = \tfrac{1}{2}k_A\gamma_t(5)^2 = \tfrac{1}{2}(0.295)(120)(5)^2 = 400 \text{ lb/ft}$$

From 5 to 50 ft:

$$P_{2A} = k_A\gamma_t(5)(45) + \tfrac{1}{2}k_A\gamma_b(45)^2 = 0.295(120)(5)(45) + \tfrac{1}{2}(0.295)(64)(45)^2$$

$$= 8000 + 19,100 = 27,100$$

$$P_A = P_{1A} + P_{2A} = 400 + 27,100 = 27,500 \text{ lb/ft}$$

Equation (10.3) with γ_b:

$$P_p = \tfrac{1}{2}k_p\gamma_b D^2 = \tfrac{1}{2}(3.39)(64)(20)^2 = 43,400 \text{ lb/ft}$$

Moment due to passive force $= 43,400(26 + \tfrac{2}{3}20) = 1.71 \times 10^6$

Neglecting P_{1A},

Moment due to active force (at tieback anchor)

$$= 8000\left(1 + \frac{45}{2}\right) + 19,100[1 + \tfrac{2}{3}(45)] = 7.8 \times 10^5$$

$$FS = \frac{\text{resisting moment}}{\text{destabilizing moment}} = \frac{1.71 \times 10^6}{7.8 \times 10^5}$$

$$= 2.19$$

$$A_p = P_A - \frac{P_p}{FS} = 27,500 - \frac{43,400}{2.19} = 7680 \text{ lb/ft}$$

For a 10-ft spacing, therefore,

$$A_p = 10(7680) = 76,800 \text{ lb} = 76.8 \text{ kips}$$

Earthquake Analysis, Pseudostatic Method. For the first earthquake analysis, assume that the sand behind, beneath, and in front of the wall has a factor of safety against liquefaction that is greater than 2.0. The design earthquake condition is $a_{max} = 0.20g$. Using the pseudostatic approach [i.e., Eq. (10.7)], calculate the factor of safety for toe kick-out and the tieback anchor force.

Since the effect of the water pressure tends to cancel on both sides of the wall, use Eq. (10.7) and estimate P_E based on the buoyant unit weight $\gamma_b = 64$ lb/ft^3, or

$$P_E = \frac{1}{2} k_A^{1/2} \left(\frac{a_{max}}{g} \right)(H^2 \gamma_b) = \frac{1}{2}(0.295)^{1/2}(0.20)(50)^2(64) = 8690 \text{ lb/ft}$$

And P_E acts at a distance of $\frac{2}{3}(H + D)$ above the bottom of the sheet pile wall.

$$\text{Moment due to } P_E = 8690[\frac{1}{3}(50) - 4] = 1.10 \times 10^5$$

$$\text{Total destabilizing moment} = 7.80 \times 10^5 + 1.10 \times 10^5 = 8.90 \times 10^5$$

$$\text{Moment due to passive force} = 1.71 \times 10^6$$

$$\text{FS} = \frac{\text{resisting moment}}{\text{destabilizing moment}} = \frac{1.71 \times 10^6}{8.90 \times 10^5} = 1.92$$

$$A_p = P_A + P_E - \frac{P_p}{\text{FS}} = 27,500 + 8690 - \frac{43,400}{1.92} = 13,600 \text{ lb/ft}$$

For a 10-ft spacing, therefore

$$A_p = 10 (13,600) = 136,000 \text{ lb} = 136 \text{ kips}$$

Earthquake Analysis, Liquefaction of Passive Wedge. For the second earthquake analysis, assume that the sand located behind the retaining wall has a factor of safety against liquefaction greater than 2.0. Also assume that the upper 10 ft of sand located in front of the retaining wall will liquefy during the design earthquake, while the sand located below a depth of 10 ft has a factor of safety greater than 2.0. Calculate the factor of safety for toe kick-out and the tieback anchor force.

For the passive wedge:

- 0 to 10 ft: Passive resistance = 0
- At 10-ft depth: Passive resistance = $k_p \gamma_b d = 3.39(64)(10) = 2170$ lb/ft^2
- At 20-ft depth: Passive resistance = $k_p \gamma_b d = 3.39(64)(20) = 4340$ lb/ft^2

$$\text{Passive force} = \frac{(2170 + 4340)}{2}(10) = 32,600 \text{ lb/ft}$$

$$\text{Moment due to passive force} = 2170(10)(45 - 4) + \frac{4340 - 2170}{2}(10)[40 + \frac{2}{3}(10) - 4]$$

$$= 890,000 + 463,000 = 1.35 \times 10^6$$

Including a pseudostatic force in the analysis gives these results:

$$P_E = \frac{1}{2} k_A^{1/2} \left(\frac{a_{max}}{g} \right)(H^2 \gamma_b) = \frac{1}{2}(0.295)^{1/2}(0.20)(50)^2(64) = 8690 \text{ lb/ft}$$

And P_E acts at a distance of $\frac{2}{3}(H + D)$ above the bottom of the sheet pile wall.

$$\text{Moment due to } P_E = 8690[\frac{1}{3}(50) - 4] = 1.10 \times 10^5$$

$$\text{Total destabilizing moment} = 7.80 \times 10^5 + 1.10 \times 10^5 = 8.90 \times 10^5$$

Moment due to passive force $= 1.35 \times 10^6$

$$FS = \frac{\text{resisting moment}}{\text{destabilizing moment}} = \frac{1.35 \times 10^6}{8.90 \times 10^5}$$

$$= 1.52$$

$$A_p = P_A + P_E - \frac{P_p}{FS} = 27{,}500 + 8690 - \frac{32{,}600}{1.52} = 14{,}700 \text{ lb/ft}$$

For a 10-ft spacing, therefore,

$$A_p = 10 \,(14{,}700) = 147{,}000 \text{ lb} = 147 \text{ kips}$$

Earthquake Analysis, Liquefaction of Active Wedge. For the third earthquake analysis, assume that the sand located in front of the retaining wall has a factor of safety against lique-faction greater than 2.0. However, assume that the submerged sand located behind the retaining will liquefy during the earthquake. Further assume that the tieback anchor will be unaffected by the liquefaction. Calculate the factor of safety for toe kick-out.

As indicated in Sec. 10.3.2, when the water levels are approximately the same on both sides of the retaining wall, use Eq. (10.1) with $k_A = 1$ [i.e., for $\phi' = 0$, $k_A = 1$, see Eq. (10.2)] and use γ_b (buoyant unit weight) in place of γ_t.

As an approximation, assume that the entire 50 ft of soil behind the sheet pile wall will liquefy during the earthquake. Using Eq. (10.1), with $k_A = 1$ and $\gamma_b = 64$ lb/ft^3,

$$P_L = \tfrac{1}{2} k_A \gamma_b (H + D)^2 = \tfrac{1}{2}(1.0)(64)(50)^2 = 80{,}000 \text{ lb/ft}$$

Moment due to liquefied soil $= 80{,}000[\tfrac{2}{3}(50) - 4] = 2.35 \times 10^6$

Moment due to passive force $= 1.71 \times 10^6$

$$FS = \frac{\text{resisting moment}}{\text{destabilizing moment}} = \frac{1.71 \times 10^6}{2.35 \times 10^6}$$

$$= 0.73$$

Summary of Values

Example problem		Factor of safety for toe kick-out	A_p, kips
Static analysis		2.19	76.8
Earthquake	Pseudostatic method [Eq. (10.7)]	1.92	136
	Partial passive wedge liquefaction[*]	1.52	147
	Liquefaction of soil behind wall	0.73	—

[*]Pseudostatic force included for the active wedge.

As indicated by the values in this summary table, the sheet pile wall would not fail for partial liquefaction of the passive wedge. However, liquefaction of the soil behind the retaining wall would cause failure of the wall.

10.3.4 Summary

As discussed in the previous sections, the liquefaction of soil can affect the retaining wall in many different ways. It is also possible that even with a factor of safety against liquefaction greater than 1.0, there could be still be significant weakening of the soil, leading to a retaining wall failure. In summary, the type of analysis should be based on the factor of safety against liquefaction FS_L as follows:

1. **$FS_L \le 1.0$:**   In this case, the soil is expected to liquefy during the design earthquake, and thus the design pressures acting on the retaining wall must be adjusted (see Sec. 10.3.2).

2. **$FS_L > 2.0$:**   If the factor of safety against liquefaction is greater than about 2.0, the pore water pressures generated by the earthquake-induced contraction of the soil are usually small enough that they can be neglected. In this case, it could be assumed that the earthquake does not weaken the soil, and the pseudostatic analyses outlined in Sec. 10.2 could be performed.

3. **$1.0 < FS_L \le 2.0$:**   For this case, the soil is not anticipated to liquefy during the earthquake. However, as the loose granular soil contracts during the earthquake, there could still be a substantial increase in pore water pressure and hence weakening of the soil. Figure 5.15 can be used to estimate the pore water pressure ratio r_u for various values of the factor of safety against liquefaction FS_L. The analysis would vary depending on the location of the increase in pore water pressure as follows:

   - *Passive wedge:*   If the soil in the passive wedge has a factor of safety against liquefaction greater than 1.0 but less than 2.0, then the increase in pore water pressure would decrease the effective shear strength and the passive resisting force would be reduced [i.e., passive resistance = $P_p(1 - r_u)$].
   - *Bearing soil:*   For an increase in the pore water pressure in the bearing soil, use the analysis in Sec. 8.3.
   - *Active wedge:*   In addition to the pseudostatic force P_E and the active earth pressure resultant force P_A, include a force that is equivalent to the anticipated earthquake-induced pore water pressure.

10.4 RETAINING WALL ANALYSES FOR WEAKENED SOIL

Besides the liquefaction of soil, many other types of soil can be weakened during the earthquake. In general, there are three cases:

1. *Weakening of backfill soil:*   In this case, only the backfill soil is weakened during the earthquake. An example would be backfill soil that is susceptible to strain softening during the earthquake. As the backfill soil weakens during the earthquake, the force exerted on the back face of the wall increases. One design approach would be to estimate the shear strength corresponding to the weakened condition of the backfill soil and then use this strength to calculate the force exerted on the wall. The bearing pressure, factor of safety for sliding, factor of safety for overturning, and location of the resultant vertical force could then be calculated for this weakened backfill soil condition.

2. *Reduction in the soil resistance:*   In this case, the soil beneath the bottom of the wall or the soil in the passive wedge is weakened during the earthquake. For example, the bearing soil could be susceptible to strain softening during the earthquake. As the bearing soil weakens during the earthquake, the wall foundation could experience additional settlement, a bearing capacity failure, sliding failure, or overturning failure. In addition, the weakening of

the ground beneath or in front of the wall could result in a shear failure beneath the retaining wall. One design approach would be to reduce the shear strength of the bearing soil or passive wedge soil to account for its weakened state during the earthquake. The settlement, bearing capacity, factor of safety for sliding, factor of safety for overturning, and factor of safety for a shear failure beneath the bottom of the wall would then be calculated for this weakened soil condition.

3. *Weakening of the backfill soil and reduction in the soil resistance:* This is the most complicated case and would require combined analyses of both items 1 and 2 as outlined above.

10.5 RESTRAINED RETAINING WALLS

10.5.1 Introduction

As mentioned in Sec. 10.1.1, in order for the active wedge to be developed, there must be sufficient movement of the retaining wall. In many cases movement of the retaining wall is restricted. Examples include massive bridge abutments, rigid basement walls, and retaining walls that are anchored in nonyielding rock. These cases are often described as *restrained retaining walls*.

10.5.2 Method of Analysis

To determine the static earth pressure acting on a restrained retaining wall, Eq. (10.1) can be utilized where the coefficient of earth pressure at rest k_0 is substituted for k_A. For static design conditions of restrained retaining walls that have granular backfill, a commonly used value of k_0 is 0.5. Restrained retaining walls are especially susceptible to higher earth pressures induced by heavy compaction equipment, and extra care must be taken during the compaction of backfill for restrained retaining walls.

For earthquake conditions, restrained retaining walls will usually be subjected to larger forces compared to those retaining walls that have the ability to develop the active wedge. One approach is to use the pseudostatic method to calculate the earthquake force, with an increase to compensate for the unyielding wall conditions, or

$$P_{ER} = \frac{P_E k_0}{k_A} \qquad (10.22)$$

where P_{ER} = pseudostatic force acting upon a restrained retaining wall, lb or kN
   P_E = pseudostatic force assuming wall has the ability to develop the active wedge, i.e., use Eq. (10.7), (10.8), or (10.9), lb or kN
   k_0 = coefficient of earth pressure at rest
   k_A = active earth pressure coefficient, calculated from Eq. (10.2) or using the k_A equation in Fig. 10.3

10.5.3 Example Problem

Use the example problem from Sec. 10.2.4 (i.e., Fig. 10.5), but assume that it is an unyielding bridge abutment. Determine the static and earthquake resultant forces acting on the restrained retaining wall. Neglect friction between the wall and backfill ($\delta = \phi_w = 0$).

Static Analysis.   Using a value of $k_0 = 0.5$ and substituting k_0 for k_A in Eq. (10.1), we see the static earth pressure resultant force exerted on the restrained retaining wall is

$$P_R = \tfrac{1}{2}k_0\gamma_t H^2 = \tfrac{1}{2}(0.5)(110)(20)^2 = 11{,}000 \text{ lb per linear foot of wall}$$

The location of this static force is at a distance of $\tfrac{1}{3}H = 6.7$ ft above the base of the wall.

Earthquake Analysis.   Using the method outlined in Sec. 10.2.1, we find the value of $k_A = 0.333$ and $P_E = 2540$ lb per linear foot of wall length (see Sec. 10.2.4). Therefore, using Eq. (10.22), we have

$$P_{ER} = P_E\frac{k_0}{k_A} = 2540\,\frac{0.5}{0.333} = 3800 \text{ lb per linear foot of wall}$$

The location of this pseudostatic force is assumed to act at a distance of $\tfrac{2}{3}H = 13.3$ ft above the base of the wall.

In summary, the resultant earth pressure forces acting on the retaining wall are static $P_R = 11{,}000$ lb/ft acting at a distance of 6.7 ft above the base of the wall and earthquake $P_{ER} = 3800$ lb/ft acting at a distance of 13.3 ft above the base of the wall.

10.6   TEMPORARY RETAINING WALLS

10.6.1   Static Design

Temporary retaining walls are often used during construction, such as for the support of the sides of an excavation that is made below grade to construct the building foundation. If the temporary retaining wall has the ability to develop the active wedge, then the basic active earth pressure principles described in Sec. 10.1.1 can be used for the design of the temporary retaining walls.

Especially in urban areas, movement of the temporary retaining wall may have to be restricted to prevent damage to adjacent property. If movement of the retaining wall is restricted, the earth pressures will typically be between the active (k_A) and at-rest (k_0) values.

For some projects, temporary retaining walls may be constructed of sheeting (such as sheet piles) that are supported by horizontal braces, also known as *struts*. Near or at the top of the temporary retaining wall, the struts restrict movement of the retaining wall and prevent the development of the active wedge. Because of this inability of the retaining wall to deform at the top, earth pressures near the top of the wall are in excess of the active (k_A) pressures. At the bottom of the wall, the soil is usually able to deform into the excavation, which results in a reduction in earth pressure. Thus the earth pressures at the bottom of the excavation tend to be constant or even decrease, as shown in Fig. 10.10.

The earth pressure distributions shown in Fig. 10.10 were developed from actual measurements of the forces in struts during the construction of braced excavations. In Fig. 10.10, case *a* shows the earth pressure distribution for braced excavations in sand and cases *b* and *c* show the earth pressure distribution for clays. In Fig. 10.10, the distance H represents the depth of the excavation (i.e., the height of the exposed wall surface). The earth pressure distribution is applied over the exposed height H of the wall surface with the earth pressures transferred from the wall sheeting to the struts (the struts are labeled with forces F_1, F_2, etc.).

Any surcharge pressures, such as surcharge pressures on the ground surface adjacent to the excavation, must be added to the pressure distributions shown in Fig. 10.10. In addition, if the sand deposit has a groundwater table that is above the level of the bottom of the excavation, then water pressures must be added to the case *a* pressure distribution shown in Fig. 10.10.

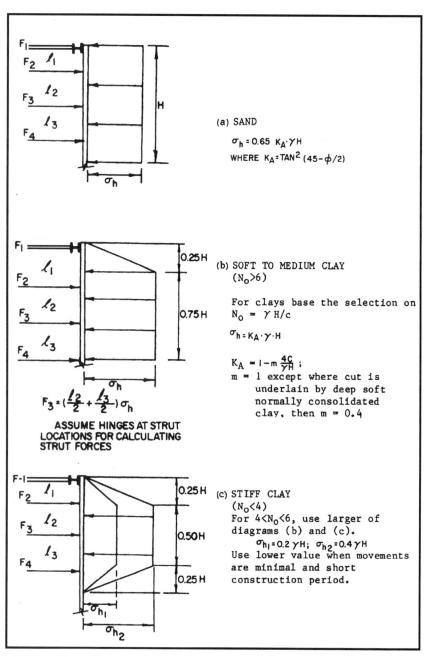

(a) SAND

$$\sigma_h = 0.65 \; K_A \cdot \gamma H$$

WHERE $K_A = TAN^2 (45 - \phi/2)$

(b) SOFT TO MEDIUM CLAY
$(N_o > 6)$

For clays base the selection on
$N_o \; = \; \gamma H/c$

$$\sigma_h = K_A \cdot \gamma \cdot H$$

$K_A \; = \; 1 - m \frac{4c}{\gamma H}$;
m = 1 except where cut is
    underlain by deep soft
    normally consolidated
    clay, then m = 0.4

$$F_3 = \left(\frac{\ell_2}{2} + \frac{\ell_3}{2}\right)\sigma_h$$

ASSUME HINGES AT STRUT
LOCATIONS FOR CALCULATING
STRUT FORCES

(c) STIFF CLAY
$(N_o < 4)$
For $4 < N_o < 6$, use larger of
diagrams (b) and (c).
$\sigma_{h_1} = 0.2 \; \gamma H; \; \sigma_{h_2} = 0.4 \; \gamma H$
Use lower value when movements
are minimal and short
construction period.

FIGURE 10.10   Earth pressure distribution on temporary braced walls. (*From NAVFAC DM-7.2 1982, originally developed by Terzaghi and Peck 1967.*)

Because the excavations are temporary (i.e., short-term condition), the undrained shear strength (s_u = c) is used for the analysis of the earth pressure distributions for clay. The earth pressure distributions for clay (i.e., cases b and c) are not valid for permanent walls or for walls where the groundwater table is above the bottom of the excavation.

10.6.2 Earthquake Analysis

Since temporary retaining walls are usually only in service for a short time, the possibility of earthquake effects is typically ignored. However, in active seismic zones or if the consequence of failure could be catastrophic, it may be prudent to perform an earthquake analysis. Depending on whether the wall is considered to be yielding or restrained, the analysis would be based on the data in Sec. 10.2 or Sec. 10.5. Weakening of the soil during the design earthquake and its effects on the temporary retaining wall should also be included in the analysis.

10.7 PROBLEMS

The problems have been divided into basic categories as indicated below.

Pseudostatic Method

10.1   Using the retaining wall shown in Fig. 10.4, assume H = 4 m, the thickness of the reinforced concrete wall stem = 0.4 m, the reinforced concrete wall footing is 3 m wide by 0.5 m thick, the ground surface in front of the wall is level with the top of the wall footing, and the unit weight of concrete = 23.5 kN/m^3. The wall backfill will consist of sand having ϕ = 32° and γ_t = 20 kN/m^3. Also assume that there is sand in front of the footing with these same soil properties. The friction angle between the bottom of the footing and the bearing soil δ = 38°. For the condition of a level backfill and neglecting the wall friction on the backside of the wall and the front side of the footing, determine the resultant normal force N and the distance of N from the toe of the footing, the maximum bearing pressure q' and the minimum bearing pressure q'' exerted by the retaining wall foundation, factor of safety for sliding, and factor of safety for overturning for static conditions and earthquake conditions [using Eq. (10.7)] if a_{max} = 0.20g. *Answer:* Static conditions: N = 68.2 kN/m and location = 1.16 m from toe, q' = 37.9 kPa and q'' = 7.5 kPa, FS for sliding = 1.17, and FS for overturning = 2.2. Earthquake conditions: P_E = 17.7 kN/m, N is not within the middle third of the footing, FS for sliding = 0.86, FS for overturning = 1.29.

10.2   Solve Prob. 10.1, using Eq. (10.8). *Answer:* Static values are the same. Earthquake conditions: P_E = 24 kN/m, N is not within the middle third of the footing, FS for sliding = 0.78, FS for overturning = 1.18.

10.3   Solve Prob. 10.1, but include wall friction in the analysis (use Coulomb's earth pressure equation, Fig. 10.3). Assume the friction angle between the backside of the retaining wall and the backfill is equal to $^3/_4$ of ϕ (that is, ϕ_w = $^3/_4\phi$ = 24°). Use Eq. (10.9) for the earthquake analysis. *Answer:* Static condition: N = 86.1 kN/m and location = 1.69 m from toe, q' = 39.6 kPa and q'' = 17.8 kPa, FS for sliding = 1.78, and FS for overturning = ∞. Earthquake conditions: P_{AE} = 68.5 kN/m, N is 1.51 m from the toe of the footing, q' = q'' = 32.0 kPa, FS for sliding = 1.26, FS for overturning = ∞.

10.4   Using the retaining wall shown at the top of Fig. 10.2b (i.e., a cantilevered retaining wall), assume H = 4 m, the thickness of the reinforced concrete wall stem = 0.4 m and the wall stem is located at the centerline of the footing, the reinforced concrete wall footing is 2 m wide by 0.5 m thick, the ground surface in front of the wall is level with the top of the wall footing, and the unit weight of concrete = 23.5 kN/m³. The wall backfill will consist of sand having ϕ = 32° and γ_t = 20 kN/m³. Also assume that there is sand in front of the footing with these same soil properties. The friction angle between the bottom of the footing and the bearing soil δ = 24°. For the condition of a level backfill and assuming total mobilization of the shear strength along the vertical plane at the heel of the wall, calculate the resultant normal force N and the distance of N from the toe of the footing, the maximum bearing pressure q' and the minimum bearing pressure q'' exerted by the retaining wall foundation, factor of safety for sliding, and factor of safety for overturning for static conditions and earthquake conditions [using Eq. (10.9)] if a_{max} = 0.20g. *Answer:* Static conditions: N = 136 kN/m and location = 1.05 m from toe, q' = 78.1 kPa and q'' = 57.8 kPa, FS for sliding = 1.72, and FS for overturning = 47. Earthquake conditions: P_{AE} = 71.2 kN/m, N is 0.94 m from the toe of the footing, q' = 88.6 kPa and q'' = 61.5 kPa, FS for sliding = 1.17, FS for overturning = 29.

10.5   For the example problem shown in Fig. 10.5, assume that there is a vertical surcharge pressure of 200 lb/ft² located at ground surface behind the retaining wall. Calculate the factor of safety for sliding and the factor of safety for overturning, and determine if N is within the middle third of the retaining wall foundation for the static conditions and earthquake conditions [using Eq. (10.9)] if a_{max} = 0.20g. *Answer:* Static conditions: FS for sliding = 1.48, FS for overturning = 2.64, and N is not within the middle third of the retaining wall foundation. Earthquake conditions: FS for sliding = 1.02, FS for overturning = 0.91, and N is not within the middle third of the retaining wall foundation.

10.6   For the example problem shown in Fig. 10.5, assume that the ground surface behind the retaining wall slopes upward at a 3:1 (horizontal:vertical) slope inclination. Calculate the factor of safety for sliding and factor of safety for overturning, and determine if N is within the middle third of the retaining wall foundation for the static conditions and earthquake conditions [using Eq. (10.9)] if a_{max} = 0.20g. *Answer:* Static conditions: FS for sliding = 1.32, FS for overturning = 2.73, and N is not within the middle third of the retaining wall foundation. Earthquake conditions: FS for sliding = 0.72, FS for overturning = 1.06, and N is not within the middle third of the retaining wall foundation.

10.7   Use the data from the example problem in Sec. 10.2.5, and assume that there is a vertical surcharge pressure of 200 lb/ft² located at ground surface behind the mechanically stabilized earth retaining wall. Calculate the factor of safety for sliding, factor of safety for overturning, and maximum pressure exerted by the base of the mechanically stabilized earth retaining wall for static and earthquake conditions. *Answer:* Static conditions: FS for sliding = 1.73, FS for overturning = 3.78, and maximum pressure q' = 4300 lb/ft². Earthquake conditions: FS for sliding = 1.29, FS for overturning = 2.3, and N is not within the middle third of the base of the wall.

10.8   Use the data from the example problem in Sec. 10.2.5, and assume that the ground surface behind the mechanically stabilized earth retaining wall slopes upward at a 3:1 (horizontal:vertical) slope inclination. Also assume that the 3:1 slope does not start at the upper front corner of the rectangular reinforced soil mass (such as shown in Fig. 10.8), but instead the 3:1 slope starts at the upper back corner of the rectangular reinforced soil mass. Calculate the factor of safety for sliding, factor of safety for overturning, and maximum pressure exerted by the retaining wall foundation for the static and earthquake conditions, using the equations in Fig. 10.3. *Answer:* Static conditions: FS for sliding = 1.60, FS for overturning = 3.76, and maximum pressure q' = 4310 lb/ft². Earthquake conditions: FS for sliding = 0.81, FS for overturning = 1.91, and N is not within the middle third of the base of the wall.

10.9   For the example problem in Sec. 10.2.5, the internal stability of the mechanically stabilized zone is to be checked by using wedge analysis. Assume a planar slip surface that is inclined at an angle of 61° (that is, $\alpha = 61°$) and passes through the toe of the mechanically stabilized zone. Also assume that the mechanically stabilized zone contains 40 horizontal layers of Tensar SS2 geogrid which has an allowable tensile strength = 300 lb/ft of wall length for each geogrid. In the wedge analysis, these 40 layers of geogrid can be represented as an allowable horizontal resistance force = 12,000 lb/ft of wall length (that is, 40 layers times 300 lb). If the friction angle ϕ of the sand = 32° in the mechanically stabilized zone, calculate the factor of safety for internal stability of the mechanically stabilized zone, using the wedge analysis for static and earthquake conditions. *Answer:* Static conditions: $F = 1.82$; earthquake conditions: FS = 1.29.

Sheet Pile Wall Analyses for Liquefied Soil

10.10 For the example problem in Sec. 10.3.3, assume that there is a uniform vertical surcharge pressure = 200 lb/ft^2 applied to the ground surface behind the sheet pile wall. Calculate the factor of safety for toe kick-out and the anchor pull force for the static condition and the earthquake conditions, using the pseudostatic method, and for partial liquefaction of the passive wedge. *Answer:* See App. E for solution.

10.11 For the example problem in Sec. 10.3.3, assume that the ground surface slopes upward at a 3:1 (horizontal:vertical) slope ratio behind the sheet pile wall. Calculate the factor of safety for toe kick-out and the anchor pull force for the static condition and the earthquake conditions, using the pseudostatic method, and for partial liquefaction of the passive wedge. *Answer:* See App. E for solution.

10.12 For the example problem in Sec. 10.3.3, assume that the ground in front of the sheet pile wall (i.e., the passive earth zone) slopes downward at a 3:1 (horizontal:vertical) slope ratio. Calculate the factor of safety for toe kick-out for the static condition and the earthquake conditions, using the pseudostatic method. *Answer:* Static condition: FS for toe kick-out = 1.18; earthquake condition: FS for toe kick-out = 1.04.

10.13 For the example problem in Sec. 10.3.3, assume that the anchor block is far enough back from the face of the sheet pile wall that it is not in the active zone. Also assume that the anchor block is located at a depth of 3 to 5 ft below ground surface, it is 5 ft by 5 ft in plan dimensions, and it consists of concrete that has a unit weight of 150 lb/ft^3. Further assume that the tieback rod is located at the center of gravity of the anchor block. For friction on the top and bottom of the anchor block, use a friction coefficient = $^2/_3\phi$, where ϕ = friction angle of the sand. Determine the lateral resistance of the anchor block for static conditions and for earthquake conditions, assuming that all the soil behind the retaining wall will liquefy during the earthquake. *Answer:* Static condition: lateral resistance = 26.6 kips; earthquake conditions: lateral resistance = 0.

Braced Excavations

10.14 A braced excavation will be used to support the vertical sides of a 20-ft-deep excavation (that is, $H = 20$ ft in Fig. 10.10). If the site consists of a sand with a friction angle $\phi = 32°$ and a total unit weight $\gamma_t = 120$ lb/ft^3, calculate the earth pressure σ_h and the resultant earth pressure force acting on the braced excavation for the static condition and the earthquake condition [using Eq. (10.7)] if $a_{max} = 0.20g$. Assume the groundwater table is well below the bottom of the excavation. *Answer:* Static condition: $\sigma_h = 480$ lb/ft^2 and the resultant force = 9600 lb per linear foot of wall length. Earthquake condition: $P_E = 2700$ lb per linear foot of wall length.

10.15 Solve Prob. 10.14, but assume the site consists of a soft clay having an undrained shear strength $s_u = 300$ lb/ft^2 (that is, $c = s_u = 300$ lb/ft^2) and use Eq. (10.8).

Answer: Static condition: $\sigma_h = 1200$ lb/ft^2, and the resultant force $= 21,000$ lb per linear foot of wall length. Earthquake condition: $P_E = 3600$ lb per linear foot of wall length.

10.16 Solve Prob. 10.15, but assume the site consists of a stiff clay having an undrained shear strength $s_u = 1200$ lb/ft^2 and use the higher earth pressure condition (that is, σ_{h2}). *Answer:* Static condition: $\sigma_{h2} = 960$ lb/ft^2, and resultant force $= 14,400$ lb per linear foot of wall length. Earthquake condition: $P_E = 3600$ lb per linear foot of wall length.

Subsoil Profiles

10.17 Use the data from Prob. 6.15 and Fig. 6.15 (i.e., sewage site at Niigata). Assume the subsoil profile represents conditions behind a retaining wall. Also assume that the type of retaining wall installed at the site is a concrete box structure, having height $= 8$ m, width $= 5$ m, and total weight of the concrete box structure $= 823$ kilonewtons per linear meter of wall length. The soil behind the retaining wall is flush with the top of the concrete box structure. The water level in front of the retaining wall is at the same elevation as the groundwater table behind the wall. The effective friction angle ϕ' of the soil can be assumed to be equal to $30°$, wall friction along the back face of the wall can be neglected, and the coefficient of friction along the bottom of the wall $= ^2/_3 \phi'$. In addition, the ground in front of the wall is located 1 m above the bottom of the wall, and the subsoil profile in Fig. 6.15 starting at a depth of 7 m can be assumed to be applicable for the soil in front of the wall. For the static conditions and earthquake conditions, determine the resultant normal force N and the distance of N from the toe of the wall, the maximum bearing pressure q' and the minimum bearing pressure q'' exerted by the retaining wall foundation, factor of safety for sliding, and factor of safety for overturning. *Answer:* Static conditions: $N = 450$ kN/m and location $= 1.89$ m from toe, $q' = 156$ kPa and $q'' = 24$ kPa, FS for sliding $= 1.66$, and FS for overturning $= 4.1$. Earthquake conditions: $N = 450$ kN/m, N is not within the middle third of the footings, FS for sliding $= 0.55$, FS for overturning $= 1.36$.

Submerged Backfill Condition

10.18 A cantilevered retaining wall (3 m in height) has a granular backfill with $\phi = 30°$ and $\gamma_t = 20$ kN/m^3. Neglect wall friction, and assume the drainage system fails and the water level rises 3 m above the bottom of the retaining wall (i.e., the water table rises to the top of the retaining wall). Determine the initial active earth pressure resultant force P_A and the resultant force (due to earth plus water pressure) on the wall due to the rise in water level. For the failed drainage system condition, also calculate the total force on the wall if the soil behind the retaining wall should liquefy during the earthquake. For both the static and earthquake conditions, assume that there is no water in front of the retaining wall (i.e., only a groundwater table behind the retaining wall). *Answer:* Static condition: $P_A = 30$ kN/m (initial condition). With a rise in water level the force acting on the wall $= 59.4$ kN/m. Earthquake condition: $P_L = 90$ kN/m.

CHAPTER 11
OTHER GEOTECHNICAL EARTHQUAKE ENGINEERING ANALYSES

The following notation is used in this chapter:

SYMBOL	DEFINITION
a	Acceleration
a_{max}	Maximum horizontal acceleration at ground surface (also known as peak ground acceleration)
A, B, C	Seismic source types
B	Width of pipeline (for trench conditions B = width of trench at top of pipeline)
C_a, C_v	Seismic coefficients needed for development of a response spectrum
C_w	Coefficient used to calculate load on a pipeline for trench or jacked condition
D	Diameter of pipeline
E'	Modulus of soil resistance
F_v	Vertical pseudostatic force (pipeline design)
g	Acceleration of gravity
H	Height of soil above top of pipeline
k_h	Horizontal seismic coefficient
k_v	Vertical seismic coefficient
K_b	Bedding coefficient
m	Total mass of soil bearing on pipeline
N_a, N_v	Near-source factors
$(N_1)_{60}$	N value corrected for field testing procedures and overburden pressure
s_u	Undrained shear strength
S_A, S_B, etc.	Soil profile types
T	Period of vibration
T_0, T_s	Periods needed for determination of response spectrum
V_{s1}	Corrected shear wave velocity [Equation (6.9)]
W	Total weight of soil bearing on top of pipeline
W_{min}	Minimum vertical load on rigid pipeline
γ_t	Total unit weight of soil

11.1 INTRODUCTION

The prior chapters in Part 2 have described field investigation, liquefaction analyses, earthquake-induced settlement, bearing capacity, slope stability, and retaining wall analyses. There are many other types of earthquake analyses that may be required by the geotechnical engineer. This final chapter of Part 2 describes some of these analyses. Items included in this chapter are:

- Surface rupture zone
- Groundwater
- Pavement design
- Pipeline design
- Response spectrum

11.2 SURFACE RUPTURE ZONE

11.2.1 Introduction

Section 3.2 presents an introduction into surface rupture. Examples of damage caused by surface rupture are shown in Figs. 3.3 to 3.13.

The best individual to determine the location and width of the surface rupture zone is the engineering geologist. Seismic study maps, such as the *State of California Special Studies Zones Maps* (1982), which were developed as part of the Alquist-Priolo Special Studies Zones Act, delineate the approximate location of active fault zones that require special geologic studies. These maps also indicate the approximate locations of historic fault offsets, which are indicated by year of earthquake-associated event, as well as the locations of ongoing surface rupture due to fault creep. There are many other geologic references, such as the cross section shown in Fig. 5.2, that can be used to identify active fault zones. Trenches, such as shown in Fig. 5.8, can be excavated across the fault zone to more accurately identify the width of the surface rupture zone.

11.2.2 Design Approach

Since most structures will be unable to resist the shear movement associated with surface rupture, one design approach is to simply restrict construction in the fault shear zone. Often the local building code will restrict the construction in fault zones. For example, the *Southern Nevada Building Code Amendments* (1997) state the following:

> Minimum Distances to Ground Faulting:
>
> 1. No portion of the foundation system of any habitable space shall be located less than five feet to a fault.
> 2. When the geotechnical report establishes that neither a fault nor a fault zone exists on the project, no fault zone set back requirements shall be imposed.
> 3. If through exploration, the fault location is defined, the fault and/or the no-build zone shall be clearly shown to scale on grading and plot plan(s).
> 4. When the fault location is not fully defined by explorations but a no build zone of potential fault impact is established by the geotechnical report, no portion of the foundation system of any habitable space shall be constructed to allow any portion of the foundation system to be located within that zone. The no build zone shall be clearly shown to scale on grading and plot plan(s).
> 5. For single lot, single family residences, the fault location may be approximated by historical research as indicated in the geotechnical report. A no build zone of at least 50 feet each side of the historically approximated fault edge shall be established. The no build zone shall be clearly shown to scale on grading and plot plan(s).

FIGURE 11.1 Offset of a road north of Zacapa caused by the Guatemala (Gualan) earthquake (magnitude 7.5) on February 4, 1976. (*Photograph from the Steinbrugge Collection, EERC, University of California, Berkeley.*)

In many cases, structures will have to be constructed in the surface rupture zone. For example, transportation routes may need to cross the active shear fault zones. One approach is to construct the roads such that they cross the fault in a perpendicular direction. In addition, it is desirable to cross the surface rupture zone at a level ground location so that bridges or overpasses need not be constructed in the surface rupture zone. Probably the best type of pavement material to be used in the fault zone is asphalt concrete, which is relatively flexible and easy to repair. For example, Fig. 11.1 shows an asphalt concrete road that crosses a surface rupture zone. The damage shown in Fig. 11.1 was caused by the surface rupture associated with the Guatemala (Gulan) earthquake. This damage will be relatively easy to repair. In fact, the road was still usable even in its sheared condition.

Pipelines also must often pass through surface rupture zones. Similar to pavements, it is best to cross the fault rupture zone in a perpendicular direction and at a level ground site. There are many different types of design alternatives for pipelines that cross the rupture zone. For example, a large tunnel can be constructed with the pipeline suspended within the center of the tunnel. The amount of open space between the tunnel wall and the pipeline would be based on the expected amount of surface rupture. Another option is to install automatic shutoff valves that will close the pipeline if there is a drop in pressure. With additional segments of the pipeline stored nearby, the pipe can then be quickly repaired.

11.2.3 Groundwater

The fault plane often contains of a thin layer of *fault gouge,* which is a clayey seam that has formed during the slipping or shearing of the fault and often contains numerous striations. For example, Fig. 11.2 shows surface rupture associated with the August 31, 1968, Kakh earthquake in Iran. Figure 11.3 shows a close-up view of the fault gouge. The cracks in the fault gouge are due to the drying out of the clay upon exposure. The fault gouge tends to act as a barrier to the migration of water, and it can have a strong influence on the regional groundwater table.

Earthquakes can also change the quality of the groundwater. For example, after the Gujarat earthquake (magnitude 7.9) in India on January 23, 2001, it was reported that black saline water was oozing from cracks in the ground and that farm animals were dying of thirst because they refused to drink the black water. It was also reported that near the Indian cities of Bhuj and Bhachau, which were among the worst hit by the tremor, the normally saline well water now tastes better. According to the M. S. Patel, Irrigation Secretary (*Earthweek* 2001), "Sweet water is coming from wells, and traces are seeping from the ground in several places. In some villages, where we could only find salty water at around 100–150 meters deep, we are now finding sweet water at 20 meters." This change in quality of the groundwater is usually attributed to fracturing of the ground during the earthquake which can alter the groundwater flow paths.

FIGURE 11.2   Surface rupture caused by the Kakh earthquake (magnitude 7.3) in Iran on August 31, 1968. The view is to the east along the Dasht-i-bayaz fault, located east of Baskobad. There was about 6 ft of lateral slip and about 2 ft of vertical movement. (*Photograph from the Steinbrugge Collection, EERC, University of California, Berkeley.*)

FIGURE 11.3   Close-up view of the fault plane. The striations indicate predominantly horizontal movement with some vertical movement. (*Photograph from the Steinbrugge Collection, EERC, University of California, Berkeley.*)

11.3   PAVEMENT DESIGN

11.3.1   Introduction

In terms of pavement design, one of the main objectives is to provide an adequate pavement thickness in order to prevent a bearing capacity failure. For example, unpaved roads and roads with a weak subgrade can be susceptible to bearing capacity failures caused by heavy wheel loads. The heavy wheel loads can cause a general bearing capacity failure or a punching-type shear failure. These bearing capacity failures are commonly known as *rutting*, and they develop when the unpaved road or weak pavement section is unable to support the heavy wheel load.

Because the thickness of the pavement design is governed by the shear strength of the soil supporting the road, usually the geotechnical engineer tests the soil and determines the pavement design thickness. The transportation engineer often provides design data to the geotechnical engineer, such as the estimated traffic loading, required width of pavement, and design life of the pavement.

Pavements are usually classified as either *rigid* or *flexible* depending on how the surface loads are distributed. A rigid pavement consists of Portland cement concrete slabs, which tend to distribute the loads over a fairly wide area. Flexible pavements are discussed in the next section.

11.3.2   Flexible Pavements

A *flexible pavement* is defined as a pavement having a sufficiently low bending resistance, yet having the required stability to support the traffic loads, e.g., macadam, crushed stone, gravel, and asphalt (California Division of Highways 1973).

The most common type of flexible pavement consists of the following:

- *Asphalt concrete:* The uppermost layer (surface course) is typically *asphalt concrete* that distributes the vehicle load in a cone-shaped area under the wheel and acts as the wearing surface. The ingredients in asphalt concrete are asphalt (the cementing agent), coarse and fine aggregates, mineral filler (i.e., fines such as limestone dust), and air. Asphalt concrete is usually hot-mixed in an asphalt plant and then hot-laid and compacted by using smooth-wheeled rollers. Other common names for asphalt concrete are *black-top, hot mix,* or simply *asphalt* (Atkins 1983).

- *Base:* Although not always a requirement, in many cases there is a base material that supports the asphalt concrete. The base typically consists of aggregates that are well graded, hard, and resistant to degradation from traffic loads. The base material is compacted into a dense layer that has a high frictional resistance and good load distribution qualities. The base can be mixed with up to 6 percent Portland cement to give it greater strength, and this is termed a *cement-treated base* (CTB).

- *Subbase:* In some cases, a subbase is used to support the base and asphalt concrete layers. The subbase often consists of a lesser-quality aggregate that is lower-priced than the base material.

- *Subgrade:* The subgrade supports the pavement section (i.e., the overlying subbase, base, and asphalt concrete). The subgrade could be native soil or rock, a compacted fill, or soil that has been strengthened by the addition of lime or other cementing agents. Instead of strengthening the subgrade, a geotextile could be placed on top of the subgrade to improve its load-carrying capacity.

Many different types of methods can be used for the design of the pavements. For example, empirical equations and charts have been developed based on the performance of pavements in actual service. For the design of flexible pavements in California, an empirical equation is utilized that relates the required pavement thickness to the anticipated traffic loads, shear strength of the materials (R value), and gravel equivalent factor (California Division of Highways 1973; ASTM Standard No. D 2844-94, 2000). Instead of using the R value, some methods utilize the *California bearing ratio* (CBR) as a measure of the shear strength of the base and subgrade. Numerous charts have also been developed that relate the shear strength of the subgrade and the traffic loads to a recommended pavement thickness (e.g., Asphalt Institute 1984). When designing pavements, the geotechnical engineer should always check with the local transportation authority for design requirements as well as the local building department or governing agency for possible specifications on the type of method that must be used for the design.

11.3.3 Earthquake Design

The design of an asphalt concrete road typically does not include any factors to account for earthquake conditions. The reason is that usually the surface course, base, and subbase are in a compacted state and are not affected by the ground shaking. In addition, the cumulative impact and vibration effect of cars and trucks tends to have greater impact than the shaking due to earthquakes.

Concrete pavement and concrete median barriers are often damaged at their joints, or they are literally buckled upward. This damage frequently develops because the concrete sections are so rigid and there are insufficient joint openings to allow for lateral movement during the earthquake. For example, Fig. 11.4 shows compressional damage to the roadway and at the median barrier caused by the Northridge earthquake, in California, on January 17, 1994. In additional to rigid pavements, flexible pavements can be damaged by localized compression, such as shown in Fig. 11.5.

FIGURE 11.4   View to the north along northbound Interstate 405, 250 yards south of Rinaldi Overcrossing. The compressional damage to the roadway and at the median barrier was caused by the Northridge earthquake, in California, on January 17, 1994. (*Photograph from the Northridge Collection, EERC, University of California, Berkeley.*)

FIGURE 11.5   Localized compression feature and pavement damage caused by the Northridge earthquake, in California, on January 17, 1994. (*Photograph from the Northridge Collection, EERC, University of California, Berkeley.*)

Other common causes of damage to roadways are the following:

- Surface rupture, such as shown in Fig. 11.1.
- Slope instability, such as shown in Figs. 3.54, 9.35, and 9.36.
- Liquefaction flow slides or lateral spreading, such as shown in Fig. 3.42.
- Settlement of soft soils. For example, Fig. 11.6 shows the failure of a concrete surface highway during the Chile earthquake in May 1960. The highway was constructed on top of a marshy region.
- Collapse of underlying structures. For example, Fig. 11.7 shows street damage caused by the collapse of the Daikai subway station during the Kobe earthquake.

In summary, the pavement design typically is not based on seismic conditions or modified for earthquake effects. Common causes of damage are due to localized compression and movement of the underlying ground, such as earthquake-induced slope instability, settlement, or collapse of underlying structures.

11.4 PIPELINE DESIGN

11.4.1 Introduction

Similar to pavements, pipelines are often damaged due to surface rupture or movement of the underlying soil caused by earthquake-induced slope movement, liquefaction flow slides or lateral spreading, and earthquake-induced settlement of soft soils.

FIGURE 11.6 This picture shows the failure of a concrete surfaced highway due to an earthquake-induced foundation failure. This area was observed to be a marshy region. This main highway is located 6 km north of Perto Montt. The May 1960 Chile earthquake (moment magnitude = 9.5) caused the highway damage. (*Photograph from the Steinbrugge Collection, EERC, University of California, Berkeley.*)

FIGURE 11.7   This picture shows street damage caused by the underlying collapse of the Daikai subway station. The January 17, 1995, Kobe earthquake (moment magnitude = 6.9) in Japan caused this damage. (*Photograph from the Kobe Geotechnical Collection, EERC, University of California, Berkeley.*)

The pipeline can also be crushed by the dynamic soil forces exerted upon the pipeline. The pseudostatic approach is often utilized in the design of the pipeline. As previously mentioned, this method ignores the cyclic nature of the earthquake and treats it as if it applied an additional static force upon the pipeline. In particular, the pseudostatic approach is to apply a vertical force acting through the centroid of the mass of soil bearing on the top of the pipeline. The pseudostatic vertical force F_v is calculated by using Eq. (6.1), or

$$F_v = ma = \frac{W}{g} a = W \frac{a}{g} = k_v W \qquad (11.1)$$

where F_v = vertical pseudostatic force acting through the centroid of the mass of soil bearing on top of the pipeline, lb or kN. For pipeline analysis, the pipe is usually assumed to have a unit length (i.e., two-dimensional analysis)

m = total mass of soil bearing on top of the pipeline, lb or kg, which is equal to W/g

W = total weight of soil bearing on top of the pipeline, lb or kN

a = acceleration, which in this case is the vertical acceleration at ground surface caused by the earthquake, ft/s^2 or m/s^2

$a/g = k_v$ = vertical seismic coefficient (dimensionless). The vertical seismic coefficient k_v is often assumed to be equal to $\frac{2}{3}k_h$. As previously mentioned, $k_h = a_{max}/g$, where a_{max} is the maximum horizontal acceleration at ground surface that is induced by the earthquake, ft/s^2 or m/s^2. The maximum horizontal acceleration is also commonly referred to as peak ground acceleration (see Sec. 5.6).

Note that an earthquake could subject the soil to both vertical and horizontal pseudostatic forces. However, the horizontal force is usually ignored in the standard pipeline pseudostatic analysis. This is because the vertical pseudostatic force acting on the soil mass supported by the pipeline will usually cause a more critical design condition than the addition of a horizontal pseudostatic force acting on the sides of the pipeline.

As indicated in Eq. (11.1), the only unknowns in the pseudostatic method are the weight of the soil mass bearing on the top of the pipeline W and the seismic coefficient k_v. As previously mentioned, the seismic coefficient k_v can be assumed to be equal to $\frac{2}{3}(a_{max}/g)$. The determination of W is described in the next section.

11.4.2 Static Design

For static design, the external load on a pipeline depends on many different factors. One important factor is the type of pipeline (rigid versus flexible). Another important factor is the placement conditions, i.e., whether the pipeline is constructed under an embankment, in a trench, or is pushed or jacked into place. Figure 11.8 illustrates the three placement conditions of trench, embankment, and tunnel (or pushed or jacked condition).

Other factors that affect the external load on a pipeline for the static design include the unit weight and thickness of overburden soil, the surface loads such as applied by traffic, compaction procedures, and the presence of groundwater (i.e., buoyant conditions on an empty pipeline).

Rigid Pipeline Design for Static Conditions.   Examples of rigid pipelines include precast concrete, cast-in-place concrete, and cast iron. Design pressures due to the overlying soil pressure are as follows:

Minimum Design Load.   In general, the minimum vertical load W on a rigid pipeline is equal to the unit weight of soil γ_t times the height H of soil above the top of the pipeline times the diameter of the pipe D, or

$$W_{min} = \gamma_t H D \tag{11.2}$$

As an example, suppose the pipeline has a diameter D of 24 in (2 ft) and a depth of overburden H of 10 ft, and the backfill soil has a total unit weight γ_t of 125 lb/ft^3. Therefore, the minimum vertical load W_{min} acting on the pipeline is

$$W_{min} = (125 \text{ lb/ft}^3) (10 \text{ ft}) (2 \text{ ft}) = 2500 \text{ lb per linear foot of pipe length}$$

Embankment Condition.   Different types of embankment conditions are shown in Fig. 11.8. In many cases, compaction of fill or placement conditions will impose vertical loads greater than the minimum values calculated above. Also, because the pipe is rigid, the arching effect of soil adjacent to the pipe will tend to transfer load to the rigid pipe.

Figure 11.9a shows the recommendations for a pipeline to be constructed beneath a fill embankment. In Fig. 11.9a, W = vertical dead load on the pipeline, D = diameter of the pipeline, and B = width of the pipeline (that is, $B = D$). Note that Fig. 11.9a was developed for an embankment fill having a total unit weight $\gamma_t = 100$ lb/ft^3 and an adjustment is required for conditions having different unit weights.

As an example, use the same conditions as before ($B = D = 2$ ft, $H = 10$ ft, and $\gamma_t = 125$ lb/ft^3). Figure 11.9a is entered with $H = 10$ ft, the curve marked 24 in (2 ft) is intersected, and the value of W read from the vertical axis is about 3800 pounds. Therefore,

$$W = 3800 \, \frac{125}{100} = 4750 \text{ lb per linear foot of pipeline length}$$

Note that this value of 4750 pounds is greater than the minimum dead load (2500 lb), and the above value (4750 lb) would be used for the embankment condition.

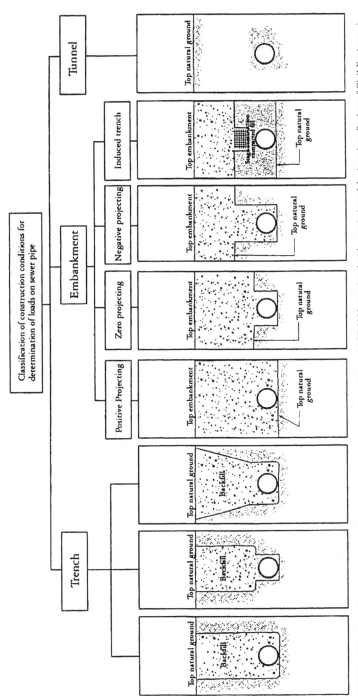

FIGURE 11.8 Classification of construction conditions for buried pipelines. (*From ASCE 1982, reprinted with permission of the American Society of Civil Engineers.*)

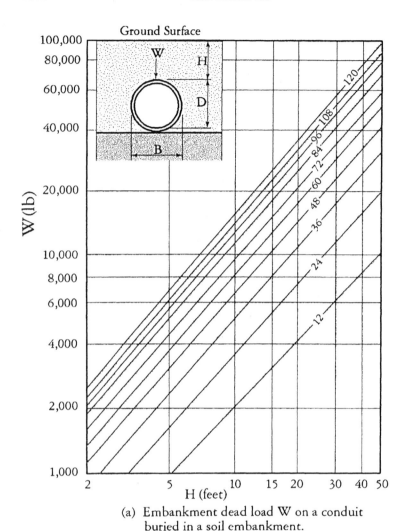

(a) Embankment dead load W on a conduit
buried in a soil embankment.

FIGURE 11.9 Embankment load W for rigid pipelines buried in a soil embankment.
(*Reproduced from NAVFAC DM-7.1, 1982.*)

Trench Condition.   Different types of trench conditions are shown in Fig. 11.8. Figure
11.9b shows the recommendations for a pipeline to be constructed in a trench. Note that in
Fig. 11.9b the dimension B is *not* the diameter of the pipeline, but rather is the width of the
trench at the top of the pipeline. This is because studies have shown that if the pipeline is
rigid, it will carry practically all the load on the plane defined by B (Marston 1930, ASCE
1982). Curves are shown for both sand and clay backfill in Fig. 11.9b. The procedure is to
enter the chart with the H/B ratio, intersect the "sands" or "clays" curve, and then determine
C_w. Once C_w is obtained, the vertical load W on the pipeline is calculated from

$$W = C_w \gamma_t B^2 \tag{11.3}$$

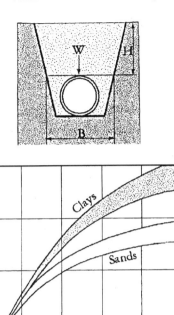

(b) C_w for conduit in a trench

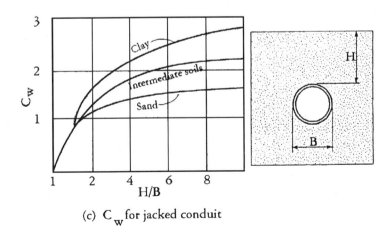

(c) C_w for jacked conduit

FIGURE 11.9 (*Continued*) Embankment load W and backfill coefficient C_w for rigid pipelines in a trench or for a jacked condition. (*Reproduced from NAVFAC DM-7.1, 1982.*)

As an example, use the same conditions as before ($D = 2$ ft, $H = 10$ ft, and $\gamma_t = 125$ lb/ft³). Also assume that the trench width at the top of the pipeline will be 4 ft (that is, $B = 4$ ft) and the trench will be backfilled with sand. Figure 11.9*b* is entered with $H/B = 10/4 = 2.5$, the curve marked "sands" is intersected, and the value of C_w of about 1.6 is obtained from the vertical axis. Therefore,

$$W = C_w \gamma_t B^2 = (1.6)(125)(4)^2 = 3200 \text{ lb per linear foot}$$

Note this value of 3200 lb is greater than the minimum value (2500 lb), and thus 3200 lb would be used for the trench condition.

It should be mentioned that as the width of the trench increases, the values from this section may exceed the embankment values. If this occurs, the embankment condition should be considered to be the governing loading condition.

Jacked or Driven Pipelines. The jacked or driven pipeline condition (i.e., tunnel condition) is shown in Fig. 11.8. Figure 11.9*c* shows the recommendations for a jacked or driven pipeline. Note in Fig. 11.9*c* that the dimension B is equal to the diameter of the pipeline ($B = D$). The curves shown in Fig. 11.9*c* are for pipelines jacked or driven through sand, clay, or intermediate soils. The procedure is to enter the chart with the H/B ratio, intersect the appropriate curve, and then determine C_w. Once C_w is obtained, the vertical load W on the pipeline is calculated from

$$W = C_w \gamma_t B^2 \tag{11.4}$$

As an example, use the same conditions as before ($D = B = 2$ ft, $H = 10$ ft, and $\gamma_t = 125$ lb/ft³), and the pipeline will be jacked through a sand deposit. Figure 11.9*c* is entered with $H/B = 10/2 = 5$, the curve marked "sand" is intersected, and the value of C_w of about 1.5 is obtained from the vertical axis. Therefore,

$$W = C_w \gamma_t B^2 = (1.5)(125)(2)^2 = 750 \text{ lb per linear foot}$$

Note this value of 750 lb is less than the minimum load value (2500 lb), and thus the value of 2500 lb would be used for the jacked or driven pipe condition. Basic soil mechanics indicates that the long-term load for rigid pipelines will be at least equal to the overburden soil pressure (i.e., the minimum design load).

Factor of Safety. A factor of safety should be applied to the static design dead load W calculated above. The above values also consider only the vertical load W on the pipeline due to soil pressure. Other loads, such as traffic or seismic loads, may need to be included in the static design of the pipeline. For pressurized pipes, rather than the exterior soil load W, the interior fluid pressure may govern the design.

Flexible Pipeline Design for Static Conditions. Flexible pipelines under embankments or in trenches derive their ability to support loads from their inherent strength plus the passive resistance of the soil as the pipe deflects and the sides of the flexible pipe move outward against the soil. Examples of flexible pipes are ductile iron pipe, ABS pipe, polyvinyl chloride (PVC) pipe, and corrugated metal pipe (CMP). Proper compaction of the soil adjacent to the sides of the flexible pipe is essential in its long-term performance. Flexible pipes often fail by excessive deflection and by collapse, buckling, and cracking, rather than by rupture, as in the case of rigid pipes.

The design of flexible pipelines depends on the amount of deflection considered permissible, which in turn depends on the physical properties of the pipe material and the project use. Because flexible pipe can deform, the dead load on the pipe W is usually less than that calculated for rigid pipes. Thus as a conservative approach, the value of the design dead load W calculated from the rigid pipe section can be used for the flexible pipeline design.

To complete the static design of flexible pipelines, the designer will need to calculate the deflection of the pipeline. The deflection depends on the applied vertical dead load W as well as other factors, such as the modulus of elasticity of the pipe, pipe diameter and thickness, modulus of soil resistance (E', see ASCE 1982, Table 9-10), and bedding constant K_b. Per ASCE (1982, Table 9-11), the values of the bedding constant K_b vary from 0.110 (bedding angle = 0°) to about 0.083 (bedding angle = 180°). The bedding angle may vary along the trench, and thus a conservative value of 0.10 is often recommended.

11.4.3 Earthquake Design

Once the weight W of the soil bearing on top of the pipeline is known [i.e., Eqs. (11.2), (11.3), and (11.4)], the pseudostatic force can be calculated by using Eq. (11.1)]. As an example, use the same data from Sec. 11.4.2 ($B = 2$ ft, $H = 10$ ft, and $\gamma_t = 125$ lb/ft³), and assume that for the design earthquake, the peak ground acceleration $a_{max} = 0.30g$. Using $k_v = \frac{2}{3}k_h = \frac{2}{3}(0.30) = 0.20$, the pseudostatic forces are as follows [Eq. (11.1)]:

Minimum pseudostatic force:

$$F_v = k_v W_{min} = 0.20\,(2500) = 500 \text{ lb per linear foot}$$

Embankment condition:

$$F_v = k_v W = 0.20\,(4750) = 950 \text{ lb per linear foot}$$

Trench condition:

$$F_v = k_v W = 0.20\,(3200) = 640 \text{ lb per linear foot}$$

Jacked or driven pipeline:

$$F_v = k_v W = 0.20\,(750) = 150 \text{ lb per linear foot}$$

For jacked or driven pipeline, use the minimum value of $F_v = 500$ lb per linear foot.
In summary, for the example problem of a 2-ft-diameter pipeline having 10 ft of overburden soil with a total unit weight of 125 lb/ft³, the soil loads are as follows:

Pipeline design	Minimum design load, lb/ft	Embankment condition, lb/ft	Trench condition, lb/ft	Jacked or driven pipeline, lb/ft
Static load W	2500	4750	3200	2500*
Pseudostatic load F_v	500	950	640	500*

*Using minimum design values.

11.5 RESPONSE SPECTRUM

11.5.1 Introduction

As discussed in Sec. 4.6, a response spectrum can be used to directly assess the nature of the earthquake ground motion on the structure. A response spectrum is basically a plot of the maximum displacement, velocity, or acceleration versus the natural period of a single-degree-of-freedom system. Different values of system damping can be used, and thus a family of such curves could be obtained. The structural engineer can then use this information for the design of the building.

The geotechnical engineer may be required to provide a response spectrum to the structural engineer. The response spectrum could be based on site-specific geology, tectonic activity, seismology, and soil characteristics. As an alternative, a simplified response spectrum can be developed based on the seismic zone and the site soil profile. This method is described in the following section.

11.5.2 Response Spectrum per the *Uniform Building Code*

One easy approach for the preparation of a response spectrum is to use the method outlined in the *Uniform Building Code* (1997). Figure 11.10 shows the elastic response spectrum in terms of the spectral acceleration g versus the period of vibration (in seconds) for 5 percent system damping. To prepare the response spectra shown in Fig. 11.10, only two parameters are needed: the seismic coefficients C_a and C_v. The steps in determining C_a and C_v are as follows:

1. *Determine seismic zone:*   Figure 5.17 presents the seismic zone map for the United States. By using Fig. 5.17, the seismic zone (i.e., 0, 1, 2A, 2B, 3, or 4) can be determined for the

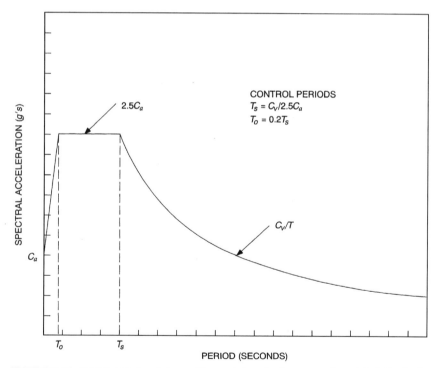

FIGURE 11.10   Response spectrum in terms of the spectral acceleration g versus the period of vibration (in seconds) for 5 percent system damping. (*Reproduced from the Uniform Building Code, 1997, with permission from the International Conference of Building Officials.*)

TABLE 11.1 Soil Profile Types

		Average soil properties from ground surface to a depth of 100 ft (30 m)		
Soil profile type	Material descriptions	Shear wave velocity V_{s1} [see Eq. (6.9)], ft/s (m/s)	Granular soil: $(N_1)_{60}$ value [see Eq. (5.2)], blows per foot	Cohesive soil: undrained shear strength s_u, lb/ft^2 (kPa)
S_A	Hard rock	>5000 (>1500)	—	—
S_B	Rock	2500–5000 (760–1500)	—	—
S_C	Soft rock, very dense granular soil, very stiff to hard cohesive soil	1200–2500 (360–760)	>50	>2000 (>100)
S_D	Dense granular soil, stiff cohesive soil	600–1200 (180–360)	15–50	1000–2000 (50–100)
S_E*	Granular soil having a loose to medium density; cohesive soil having a soft to medium consistency	<600 (<180)	<15	<1000 (<50)
S_F		Soil requiring a site-specific evaluation (see Sec. 11.5.2)		

*Soil profile S_E also includes any subsoil profile having more than 10 ft (3 m) of soft clay, defined as a soil with a plasticity index > 20, water content $\geq$ 40 percent, and s_u < 500 lb/ft^2 (24 kPa).
Data obtained from the *Uniform Building Code* (1997).

site. The *Uniform Building Code* (1997) also provides the seismic zone values for other countries. A response spectrum would usually not be needed for sites that have a seismic zone = 0.

2. *Soil profile type:*   Using Table 11.1, the next step is to determine the soil type profile (i.e., S_A, S_B, S_C, S_D, S_E, or S_F), as follows:

   a. *Soil profile types S_A, S_B, S_C, and S_D:*   For the first four soil profile types, the classification is based on the average condition of the material that exists at the site from ground surface to a depth of 100 ft (30 m). If the ground surface will be raised or lowered by grading operations, then the analysis should be based on the final as-built conditions. As indicated in Table 11.1, the selection of the first four soil profile types is based on the material type and engineering properties, such as shear wave velocity, standard penetration test $(N_1)_{60}$ values, and the undrained shear strength.

   b. *Soil profile type S_E:*   Similar to the first four soil profiles, the classification for S_E is based on material type and engineering properties, such as shear wave velocity, standard penetration test $(N_1)_{60}$ values, and the undrained shear strength. In addition, any site that contains a clay layer that is thicker than 10 ft and has a plasticity index > 20, water content $\geq$ 40 percent, and undrained shear strength s_u < 500 lb/ft^2 (24 kPa) would be considered to be an S_E soil profile.

   c. *Soil profile type S_F:*   The definition of this last soil profile is as follows:

   • Soils vulnerable to potential failure or collapse under seismic loading such as liquefiable soil, quick and highly sensitive clays, and collapsible weakly cemented soils

TABLE 11.2 Seismic Coefficient C_a

| | Seismic zone (see Fig. 5.17) | | | | |
Soil profile type	Zone 1	Zone 2A	Zone 2B	Zone 3	Zone 4
S_A	0.06	0.12	0.16	0.24	$0.32\ N_a$
S_B	0.08	0.15	0.20	0.30	$0.40\ N_a$
S_C	0.09	0.18	0.24	0.33	$0.40\ N_a$
S_D	0.12	0.22	0.28	0.36	$0.44\ N_a$
S_E	0.19	0.30	0.34	0.36	$0.44\ N_a$
S_F	Soil requiring a site-specific evaluation (see Sec. 11.5.2)				

Data obtained from the *Uniform Building Code* (1997). Data for soil profile type S_E at zone 4 adjusted to be more consistent with published data. Obtain N_a from Table 11.3.

TABLE 11.3 Near-Source Factor N_a

| Seismic source type (see Table 11.4) | Closest distance to known seismic source | | |
	≤ 1.2 mi (≤ 2 km)	3 mi (5 km)	≥ 6 mi (≥ 10 km)
A	1.5	1.2	1.0
B	1.3	1.0	1.0
C	1.0	1.0	1.0

Notes: Data obtained from the *Uniform Building Code* (1997). Near-source factor N_a is only needed if the seismic zone = 4 (see Table 11.2). The near-source factor may be based on the linear interpolation of values for distances other than those shown in the table. The location and type of seismic sources to be used for design can be based on geologic data, such as recent mapping of active faults by the U.S. Geological Survey or the California Division of Mines and Geology. The closest distance to the known seismic source can be calculated as the minimum distance between the site and the surface location of the fault plane (or the surface projection of the fault plane). If there are several sources of seismic activity, then the closest one to the site should be considered to be the governing case.

- Greater than 10-ft (3-m) thickness of peats and/or highly organic clays
- Greater than 25-ft (8-m) thickness of very highly plastic clays having a plasticity index >75
- Greater than 120-ft (37-m) thickness of soft, medium, or stiff clays. If the soil at a site meets any one of these criteria, then a site-specific analysis is required and the method outlined in this section is not applicable.

3. *Seismic coefficient C_a:* Given the seismic zone and the soil profile type, the seismic coefficient C_a can be obtained from Table 11.2. If the seismic zone is equal to 4, then the near-source factor N_a must be known. To calculate the near-source factor N_a, use Table 11.3 as follows:

   a. Closest distance to known seismic source: The location and type of seismic sources to be used for design can be based on geologic data, such as recent mapping of active faults by the U.S. Geological Survey or the California Division of Mines and Geology. The closest distance to the known seismic source can be calculated as the minimum distance between the site and the surface location of the fault plane (or the

TABLE 11.4 Seismic Source Type

	Seismic source definition		
Seismic source type	Seismic source description	Maximum moment magnitude M_w	Slip rate, mm/yr
A	Faults that are capable of producing large-magnitude events and that have a high rate of seismic activity	$M_w \geq 7$	SR ≥ 5
B	All faults other than types A and C	$M_w \geq 7$ $M_w < 7$ $M_w \geq 6.5$	SR < 5 SR > 2 SR < 2
C	Faults that are not capable of producing large-magnitude earthquakes and that have a relatively low rate of seismic activity	$M_w < 6.5$	SR ≤ 2

Notes: Data obtained from the *Uniform Building Code* (1997). Seismic source type is only needed if the seismic zone = 4 (see Table 11.2). Subduction sources shall be evaluated on a site-specific basis. For the seismic source definition, both the maximum moment magnitude M_w and slip rate (SR) conditions must be satisfied concurrently when determining the seismic source type.

TABLE 11.5 Seismic Coefficient C_v

	Seismic zone (see Fig. 5.17)				
Soil profile type	Zone 1	Zone 2A	Zone 2B	Zone 3	Zone 4
S_A	0.06	0.12	0.16	0.24	$0.32N_v$
S_B	0.08	0.15	0.20	0.30	$0.40N_v$
S_C	0.13	0.25	0.32	0.45	$0.56N_v$
S_D	0.18	0.32	0.40	0.54	$0.64N_v$
S_E	0.26	0.50	0.64	0.84	$0.96N_v$
S_F	Soil requiring a site-specific evaluation (see Sec. 11.5.2)				

Data obtained from the *Uniform Building Code* (1997). Obtain N_v from Table 11.6.

surface projection of the fault plane). If there are several sources of seismic activity, then the closest one to the site should be considered to be the governing case.

   b. Seismic source type: The data in Table 11.4 can be used to determine the seismic source type (A, B, or C).

   c. Near-source factor N_a: Given the seismic source type and the closest distance to a known seismic source, the near-source factor N_a can be determined from Table 11.3.

4. *Seismic coefficient C_v:* Given the seismic zone and the soil profile type, the seismic coefficient C_v can be obtained from Table 11.5. If the seismic zone is equal to 4, then the near-source factor N_v must be known. To calculate the near-source factor N_v, the following steps are performed:

   a. Closest distance to known seismic source: As outlined in step 3a, the closest distance to the known seismic source must be determined.

TABLE 11.6 Near-Source Factor N_v

Seismic source type (see Table 11.4)	Closest distance to known seismic source			
	≤1.2 mi (≤2 km)	3 mi (5 km)	6 mi (10 km)	≥9 mi (≥15 km)
A	2.0	1.6	1.2	1.0
B	1.6	1.2	1.0	1.0
C	1.0	1.0	1.0	1.0

Notes: Data obtained from the *Uniform Building Code* (1997). Near-source factor N_v is only needed if the seismic zone = 4 (see Table 11.5). The near-source factor may be based on the linear interpolation of values for distances other than those shown in the table. The location and type of seismic sources to be used for design can be based on geologic data, such as recent mapping of active faults by the U.S. Geological Survey or the California Division of Mines and Geology. The closest distance to the known seismic source can be calculated as the minimum distance between the site and the surface location of the fault plane (or the surface projection of the fault plane). If there are several sources of seismic activity, then the closest one to the site should be considered to be the governing case.

   b. Seismic source type: As indicated in step 3*b*, the data in Table 11.4 can be used to determine the seismic source type (A, B, or C).

   c. Near-source factor N_v: Given the seismic source type and the closest distance to a known seismic source, the near-source factor N_v can be determined from Table 11.6.

   Once the seismic coefficients C_a and C_v are known, then the response spectrum shown in Fig. 11.10 can be developed. The first step is to determine the periods T_s and T_0, defined as follows:

$$T_s = \frac{C_v}{2.5C_a} \tag{11.5}$$

$$T_0 = 0.2T_s \tag{11.6}$$

   At a period of 0 s, the spectral acceleration is equal to C_a. The spectral acceleration then linearly increases to a value of $2.5C_a$ at a period of T_0. As shown in Fig. 11.10, the spectral acceleration is constant until a period equal to T_s has been reached. For any period greater than T_s, the spectral acceleration is equal to C_v/T, where T = period of vibration, in seconds, corresponding to the horizontal axis in Fig. 11.10.

11.5.3 Alternate Method

In Fig. 11.10, the seismic coefficient C_a determines the highest value of the spectral acceleration. It is expected that the spectral acceleration will increase when (1) the intensity of the earthquake increases and (2) as the ground becomes softer (see Sec. 4.6.1). This is why the values of the seismic coefficient C_a in Table 11.2 increase as the seismic zone increases. In addition, the values of C_a in Table 11.2 increase for softer ground conditions.

   In Fig. 11.10, a period of zero would correspond to a completely rigid structure. Thus when $T = 0$, the spectral acceleration is equal to the peak ground acceleration (that is, $C_a = a_{max}$). The geotechnical engineer will often need to determine a_{max} in order to perform liquefaction, settlement, slope stability, and retaining wall analyses. Once the peak ground acceleration a_{max} has been determined, it can be used in place of C_a to construct the response spectrum (i.e., in Fig. 11.10, use $C_a = a_{max}$). Since a_{max} is based on site-specific conditions (see Sec. 5.6), the use of $C_a = a_{max}$ would seem to be an appropriate revision to the method outlined in Sec. 11.5.2.

11.5.4 Example Problem

For this example problem, assume the following:

- The subsurface exploration revealed that the site is underlain by soft sedimentary rock that has an average shear wave velocity V_{s1} = 2300 ft/s.
- Seismic zone = 4.
- Design earthquake conditions: maximum moment magnitude M_w = 7, SR = 5 mm/yr, and distance to seismic source = 3 mi.

To develop the response spectrum, the following data are utilized:

1. *Soil profile type (Table 11.1)*: For soft sedimentary rock that has an average shear wave velocity V_{s1} = 2300 ft/s, the soil profile type is S_C (see Table 11.1).
2. *Seismic source type (Table 11.4)*: Since the maximum moment magnitude M_w = 7 and SR = 5 mm/yr, the seismic source type is A (see Table 11.4).
3. *Seismic coefficient C_a (Table 11.2)*: Entering Table 11.2 with soil profile type = S_C and zone 4, the value of C_a = 0.40N_a. Entering Table 11.3 with seismic source type = A and distance to the seismic source = 3 mi, the value of N_a = 1.2. Therefore, the value of the seismic coefficient C_a = 0.40N_a = 0.40(1.2) = 0.48.
4. *Seismic coefficient C_v (Table 11.5)*: Entering Table 11.5 with soil profile type = S_C and zone 4, the value of C_v = 0.56N_v. Entering Table 11.6 with seismic source type = A and distance to the seismic source = 3 mi, the value of N_v = 1.6. Therefore, the value of the seismic coefficient C_v = 0.56N_v = 0.56(1.6) = 0.90.
5. *Values of T_s and T_0 [Eqs. (11.5) and (11.6)]*: The values of T_s and T_0 can be calculated as follows:

$$T_s = \frac{C_v}{2.5 C_a} = \frac{0.90}{2.5 \, (0.48)} = 0.75 \text{ s}$$

$$T_0 = 0.2 T_s = 0.20 \, (0.75) = 0.15 \text{ s}$$

By using Fig. 11.10 and the values of C_a = 0.48, C_v = 0.90, T_s = 0.75 s, and T_0 = 0.15 s, the response spectrum can be developed such as shown in Fig. 11.11.

11.6 PROBLEMS

11.1 Solve the example problem in Secs. 11.4.2 and 11.4.3, but assume that the pipe is located 20 ft below ground surface. *Answer:* See App. E for the solution.

11.2 Solve the example problem in Sec. 11.5.4, but assume that the seismic zone = 1. Compare the results with the solution to the example problem in Sec. 11.5.4. *Answer:* See App. E for the solution and Fig. 11.12 for the response spectrum.

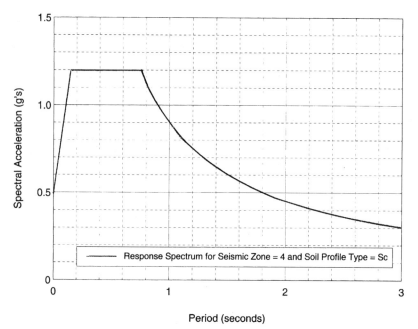

FIGURE 11.11   Response spectrum in terms of the spectral acceleration g versus the period of vibration (in seconds) for 5 percent system damping using the data from the example problem in Sec. 11.5.4.

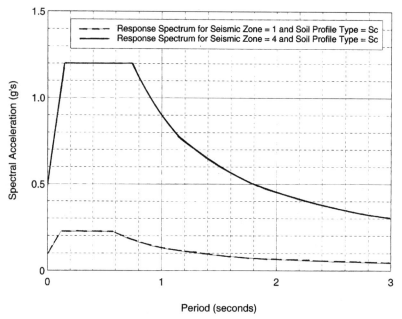

FIGURE 11.12   Answer to Prob. 11.2. Response spectrum in terms of the spectral acceleration g versus the period of vibration (in seconds) for 5 percent system damping using the data from the example problem in Sec. 11.5.4 and seismic zones 1 and 4.

SITE IMPROVEMENT METHODS TO MITIGATE EARTHQUAKE EFFECTS

CHAPTER 12
GRADING AND OTHER SOIL IMPROVEMENT METHODS

12.1 INTRODUCTION

Part 3 of the book (Chaps. 12 and 13) discusses the various methods that can be used to mitigate the effects of the earthquake on the structure. The next two chapters deal with site mitigation methods such as grading and soil improvement (Chap. 12) and foundation alternatives to resist the earthquake effects (Chap. 13).

The mitigation of slope hazards has already been discussed in Sec. 9.7.2. Options include avoiding the slope failure, protecting the site from the failure mass, and reducing the risk to an acceptable level by increasing the factor of safety of the slope. For slope hazards dealing with liquefaction-induced flow slides and lateral spreading, Seed (1987) states:

> It is suggested that, at the present time, the must prudent method of minimizing the hazards associated with liquefaction-induced sliding and deformations is to plan new construction or devise remedial measures in such a way that either high pore water pressures cannot build up in the potentially liquefiable soil, and thus liquefaction cannot be triggered, or, alternatively, to confine the liquefiable soils by means of stable zones so that no significant deformations can occur; by this means, the difficult problems associated with evaluating the consequences of liquefaction (sliding or deformations) are avoided.

Types of stable zones that can be used to confine the liquefiable soils include robust edge containment structures and shear keys (i.e., compacted soil zones). Examples of robust edge containment structures that are capable of resisting failure or excessive displacement under the seismic loading include compacted berms and dikes as well as massive seawalls or retaining structures (R. B. Seed 1991). The construction of stable zones may need to be used in conjunction with other methods that mitigate liquefaction-induced settlement, bearing capacity, and ground damage (surface cracking and sand boils).

Other options for dealing with liquefaction hazards are as follows (Federal Emergency Management Agency 1994):

> Four general approaches apply to the mitigation of liquefaction hazards (avoidance, prevention, engineered design, and post earthquake repairs). A prime way to limit the damage due to liquefaction is to avoid areas susceptible to liquefaction. This approach is not always possible

because some facilities such as transportation routes, irrigation canals, pipelines, etc., commonly must cross susceptible areas. In some instances ground can be stabilized by compaction, dewatering or replacement of soil. In other cases, structures can be designed to resist liquefaction by attachment to the soil strata below all liquefiable layers.

These general approaches for the mitigation of liquefaction hazards when designing or constructing new buildings or other structures such as bridges, tunnels, and roads can be summarized as follows:

1. *Avoid liquefaction-susceptible soils:*   The first option is to avoid construction on liquefaction-susceptible soils. Those sites that have thick deposits of soils that have a low factor of safety against liquefaction can be set aside as parks or open-space areas. Buildings and other facilities would be constructed in those areas that have more favorable subsurface conditions.

2. *Remove or improve the soil:*   The second option involves mitigation of the liquefaction hazards by removing or improving the soil. For example, the factor of safety against liquefaction can be increased by densifying the soil and/or by improving the drainage characteristics of the soil. This can be done using a variety of soil improvement techniques; such as removal and replacement of liquefiable soil; in situ stabilization by grouting, densification, and dewatering; and buttressing of lateral spread zones. These various options are discussed in Secs. 12.2 to 12.4.

3. *Build liquefaction-resistant structures:*   For various reasons, such as the lack of available land, a structure may need to be constructed on liquefaction-prone soils. It may be possible to make the structure liquefaction-resistant by using mat or deep foundation systems. This is discussed further in Chap. 13.

12.2 GRADING

Since most building sites start out as raw land, the first step in site construction work usually involves the grading of the site. *Grading* is defined as any operation consisting of excavation, filling, or a combination thereof. The glossary (App. A, Glossary 4) presents a list of common construction and grading terms and their definitions. Most projects involve grading, and it is an important part of geotechnical engineering.

The geotechnical engineer often prepares a set of grading specifications for the project. These specifications are then used to develop the grading plans, which are basically a series of maps that indicate the type and extent of grading work to be performed at the site. Often the grading specifications will be included as an appendix in the preliminary or feasibility report prepared by the geotechnical engineer and engineering geologist (see App. B of Day 2000 for an example of grading specifications).

An important part of the grading of the site often includes the compaction of fill. *Compaction* is defined as the densification of a fill by mechanical means. This physical process of getting the soil into a dense state can increase the shear strength, decrease the compressibility, and decrease the permeability of the soil.

Some examples of activities that can be performed during grading to mitigate earthquake effects include the following:

1. *Slope stabilization:*   Examples are the flattening of the slope, decreasing the height of the slope, or increasing the factor of safety of the slope by constructing a fill buttress or shear key.

2. *Liquefaction-prone soils:*   If the liquefaction-prone soils are shallow and the ground-water table can be temporarily lowered, then these soils can be removed and replaced with different soil during the grading operations. Another option is to remove the potentially liquefiable soil, stockpile the soil and allow it to dry out (if needed), and then recompact the soil as structural fill.

3. *Earthquake-induced settlement:*   As discussed in Sec. 7.3, one approach for level-ground sites that can be used to reduce the potential for liquefaction-induced ground damage, such as surface fissuring and sand boils, is to add a fill layer to the site. This operation could be performed during the grading of the site. It should be mentioned that this method will provide relatively little benefit for sloping ground since it will not prevent structural damage and surface fissuring due to lateral spreading.

4. *Volumetric settlement and rocking settlement:*   Loose soils and those types of soils that are susceptible to plastic flow or strain softening can be removed and replaced during the grading operations. Another option is to remove the soil, stockpile the soil and allow it to dry out, and then recompact the soil as structural fill.

Instead of removing and recompacting the soil during grading, another approach is to use precompression, which is often an effective method of soil improvement for soft clays and organic soils. The process consists of temporarily surcharging the soils during the grading operations in order to allow the soils to consolidate, which will reduce their compressibility and increase their shear strength.

5. *Earthquake-induced bearing capacity:*   Similar to the options for settlement, poor bearing soils can be removed and replaced or surcharged during the grading operations.

6. *Drainage and dewatering systems:*   Drainage systems could be installed during the grading operations. Drainage and dewatering are discussed in Sec. 12.4.

12.3   OTHER SITE IMPROVEMENT METHODS

12.3.1   Soil Replacement

As discussed in the previous section, soil replacement typically occurs during grading. As indicated in Table 12.1, there are basically two types of soil replacement methods: (1) removal and replacement and (2) displacement. The first method is the most common approach, and it consists of the removal of the compressible soil layer and replacement with structural fill during the grading operations. Usually the remove-and-replace grading option is only economical if the compressible soil layer is near the ground surface and the groundwater table is below the compressible soil layer, or the groundwater table can be economically lowered.

12.3.2   Water Removal

Table 12.1 lists several different types of water removal site improvement techniques. If the site contains an underlying compressible cohesive soil layer, the site can be surcharged with a fill layer placed at ground surface. Vertical drains (such as wick drains or sand drains) can be installed in the compressible soil layer to reduce the drainage path and to speed up the consolidation process. Once the compressible cohesive soil layer has had sufficient consolidation, the fill surcharge layer is removed and the building is constructed.

TABLE 12.1 Site Improvement Methods

Method	Technique	Principles	Suitable soils	Remarks
Soil replacement methods	Remove and replace	Excavate weak or undesirable material and replace with better soils	Any	Limited depth and area where cost-effective; generally ≤ 30 ft
	Displacement	Overload weak soils so that they shear and are displaced by stronger fill	Very soft	Problems with mud waves and trapped compressible soil under the embankment; highly dependent on specific site
Water removal methods	Trenching	Allows water drainage	Soft, fine-grained soils and hydraulic fills	Effective depth up to 10 ft; speed dependent on soil and trench spacing; resulting desiccated crust can improve site mobility
	Precompression	Loads applied prior to construction to allow soil consolidation	Normally consolidated fine-grained soil, organic soil, fills	Generally economical; long time may be needed to obtain consolidation; effective depth only limited by ability to achieve needed stresses
	Precompression with vertical drains	Shortens drainage path to speed consolidation	Same as above	More costly; effective depth usually limited to ≤ 100 ft
	Electroosmosis	Electric current causes water to flow to cathode	Normally consolidated silts and silty clays	Expensive; relatively fast; usable in confined area; not usable in conductive soils; best for small areas
Site strengthening methods	Dynamic compaction	Large impact loads applied by repeated dropping of a 5- to 35-ton weight; larger weights have been used	Cohesionless best; possible use for soils with fines; cohesive soils below groundwater table give poorest results	Simple and rapid; usable above and below the groundwater table; effective depths up to 60 ft; moderate cost; potential vibration damage to adjacent structures

Method	Description	Applicable soils	Comments
Vibrocompaction	Vibrating equipment densifies soils	Cohesionless soils with <20 percent fines	Can be effective up to 100-ft depth; can achieve good density and uniformity; grid spacing of holes critical, relatively expensive
Vibroreplacement	Jetting and vibration used to penetrate and remove soil; compacted granular fill then placed in hole to form support columns surrounded by undisturbed soil	Soft cohesive soils (s_u = 15 to 50 kPa)	Relatively expensive
Vibrodisplacement	Similar to vibroreplacement except soil is displaced laterally rather than removed from the hole	Stiffer cohesive soils (s_u = 30 to 60 kPa)	Relatively expensive
Grouting — Injection of grout	Fill soil voids with cementing agents to strengthen and reduce permeability	Wide spectrum of coarse- and fine-grained soils	Expensive; more expensive grouts needed for finer-grained soils; may use pressure injection, soil fracturing, or compaction techniques
Deep mixing	Jetting or augers used to physically mix stabilizer and soil	Wide spectrum of coarse- and fine-grained soils	Jetting poor for highly cohesive clays and some gravelly soils; deep mixing best for soft soils up to 165 ft deep
Thermal — Heat	Heat used to achieve irreversible strength gain and reduced water susceptibility	Cohesive soils	High energy requirements; cost limits practicality
Freezing	Moisture in soil frozen to hold particles together and increase shear strength and reduce permeability	All soils below the groundwater table; cohesive soils above the groundwater table	Expensive; highly effective for excavations and tunneling; high groundwater flows troublesome; slow process
Geosynthetics — Geogrids, geotextiles, geonets, and geomembranes	Use geosynthetic materials for filters, erosion control, water barriers, drains, or soil reinforcing	Effective filters for all soils; reinforcement often used for soft soils	Widely used to accomplish a variety of tasks; commonly used in conjunction with other methods (e.g., strip drain with surcharge or to build a construction platform for site access)

Source: Rollings and Rollings (1996).

12.7

12.3.3 Site Strengthening

Many different methods can be used to strengthen the on-site soil (see Table 12.1). Examples are as follows:

- *Dynamic compaction methods:* For example, heavy tamping consists of using a crane that repeatedly lifts and drops a large weight onto the ground surface in order to vibrate the ground and increase the density of near-surface granular soils. Although this method can increase the density of soil to a depth of 60 ft (18 m), it is usually only effective to depths of approximately 20 to 30 ft (6 to 9 m). In addition, this method requires the filling of impact craters and releveling of the ground surface.

- *Compaction piles:* Large-displacement piles, such as precast concrete piles or hollow steel piles with a closed end, can be driven into the ground to increase the density of the soil. The soil is densified by both the actual displacement of the soil and the vibration of the ground that occurs during the driving process. The piles are typically left in place, which makes this method more expensive than the other methods. In addition, there must be relatively close spacing of the piles in order to provide meaningful densification of soil between the piles.

- *Blasting:* Deep densification of the soil can be accomplished by blasting. This method has a higher risk of injury and damage to adjacent structures. There may be local restrictions on the use of such a method.

- *Compaction with vibratory probes:* Deep vibratory techniques, such as illustrated in Fig. 12.1, are often used to increase the density of loose sand deposits. This method is considered to be one of the most reliable and comprehensive methods for the mitigation of liquefaction hazard when liquefiable soils occur at depth (R. B. Seed 1991). Some techniques can be used to construct vertical gravel drains (discussed below).

- *Vertical gravel drains:* Vibroflotation or other methods are used to make a cylindrical vertical hole, which is filled with compacted gravel or crushed rock. These columns of gravel or crushed rock have a very high permeability and can quickly dissipate the earthquake-induced pore water pressures in the surrounding soil. This method can be effective in reducing the loss of shear strength, but it will not prevent overall site settlements. In addition, the method can be effective in relatively free-draining soils, but the vertical columns must be closely spaced to provide meaningful pore pressure dissipation. If the drain capacity is exceeded by the rate of pore pressure increase, there will be no partial mitigation (R. B. Seed 1991).

12.3.4 Grouting

There are many types of grouting methods that can be used to strengthen the on-site soil (see Table 12.1). For example, to stabilize the ground, fluid grout can be injected into the ground to fill in joints, fractures, or underground voids (Graf 1969, Mitchell 1970). For the releveling of existing structures, one option is *mudjacking,* which has been defined as a process whereby a water and soil-cement or soil-lime cement grout is pumped beneath the slab, under pressure, to produce a lifting force which literally floats the slab to the desired position (Brown 1992). Other site improvement grouting methods are as follows:

- *Compaction grouting:* A commonly used site improvement technique is compaction grouting, which consists of intruding a mass of very thick consistency grout into the soil, which both displaces and compacts the loose soil (Brown and Warner 1973; Warner 1978, 1982). Compaction grouting has proved successful in increasing the density of

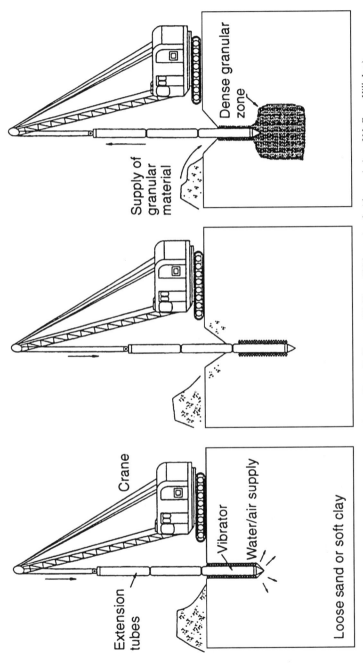

FIGURE 12.1 Equipment used for deep vibratory techniques. (*From Rollings and Rollings 1996, reprinted with permission of McGraw-Hill, Inc.*)

12.9

poorly compacted fill, alluvium, and compressible or collapsible soil. The advantages of compaction grouting are less expense and disturbance to the structure than foundation underpinning, and it can be used to relevel the structure. The disadvantages of compaction grouting are that it is difficult to analyze the results, it is usually ineffective near slopes or for near-surface soils because of the lack of confining pressure, and there is the danger of filling underground pipes with grout (Brown and Warner 1973).

- *Jet grouting (columnar):* This process is used to create columns of grouted soil. The grouted columns are often brittle and may provide little or no resistance to lateral movements and may be broken by lateral ground movements (R. B. Seed 1991).

- *Deep mixing:* Jetting or augers are used to physically mix the stabilizer and soil. There can be overlapping of treated columns in order to create a more resistant treated zone.

12.3.5 Thermal

As indicated in Table 12.1, the thermal site improvement method consists of either heating or freezing the soil in order to improve its shear strength and reduce its permeability. These types of soil improvement methods are usually very expensive and thus have limited uses.

12.3.6 Summary

Figure 12.2 presents a summary of site improvement methods as a function of soil grain size. Whatever method of soil improvement is selected, the final step should be to check the results in the field, using such methods as the cone penetration test (CPT) or standard penetration test (SPT). For example, Fig. 6.11 shows actual field test data, where standard penetration tests were performed before and after soil improvement. If the soil improvement is unsatisfactory, then it should be repeated until the desired properties are attained.

12.4 GROUNDWATER CONTROL

12.4.1 Introduction

The groundwater table (also known as the phreatic surface) is the top surface of underground water, the location of which is often determined from piezometers, such as an open standpipe. A perched groundwater table refers to groundwater occurring in an upper zone separated from the main body of groundwater by underlying unsaturated rock or soil.

Groundwater can affect all types of civil engineering projects. Probably more failures in geotechnical earthquake engineering are either directly or indirectly related to groundwater than to any other factor. Groundwater can cause or contribute to failure because of excess saturation, seepage pressures, uplift forces, and loss of shear strength due to liquefaction. It has been stated that uncontrolled saturation and seepage cause many billions of dollars yearly in damage. Examples of geotechnical and foundation problems due to groundwater are as follows (Cedergren 1989):

- Piping failures of dams, levees, and reservoirs
- Seepage pressures that cause or contribute to slope failures and landslides
- Deterioration and failure of roads due to the presence of groundwater in the base or subgrade
- Highway and other fill foundation failures caused by perched groundwater

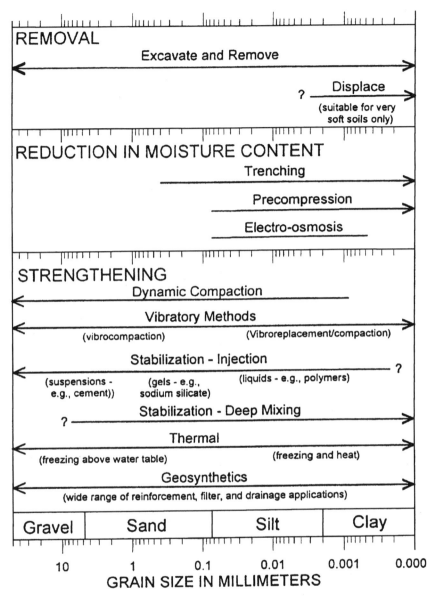

FIGURE 12.2 Site improvement methods as a function of soil grain size. (*From Rollings and Rollings 1996, reprinted with permission of McGraw-Hill, Inc.*)

- Earth embankment and foundation failures caused by excess pore water pressures
- Retaining wall failures caused by hydrostatic water pressures
- Canal linings, dry docks, and basement or spillway slabs uplifted by groundwater pressures
- Soil liquefaction, caused by earthquake shocks, because of the presence of loose granular soil that is below the groundwater table

Proper drainage design and construction of drainage facilities can mitigate many of these groundwater problems. For example, for canyon and drainage channels where fill is to be placed, a canyon subdrain system should be installed to prevent the buildup of groundwater in the canyon fill. The drain consists of a perforated pipe (perforations on the underside of the pipe), an open graded gravel around the pipe, with the gravel wrapped in a geofabric that is used to prevent the gravel and pipe from being clogged with soil particles.

12.4.2 Methods of Groundwater Control

For sites that have highly permeable soil and that are adjacent to a large body of water, such as coastal areas, it is usually not economical to permanently lower the groundwater table. However, for other sites, it may be possible to use groundwater control to mitigate earthquake effects. Table 12.2 lists various methods of groundwater control.

One commonly used method of lowering the groundwater table is to install a well point system with suction pumps. The purpose of this method is to lower the groundwater table by installing a system of perimeter wells. As illustrated in Fig. 12.3, this method is often utilized for temporary excavations, but it can also be used as a permanent groundwater control system. The well points are small-diameter pipes having perforations at the bottom ends. Pumps are used to extract water from the pipes, which then lowers the groundwater table, as illustrated in Fig. 12.3. It is important to consider the possible damage to adjacent structures caused by the lowering of the groundwater table at the site. For example, lowering of the groundwater table could lead to consolidation of soft clay layers or rotting of wood piling.

Another type of system that can be installed for groundwater control consists of a sump. Figure 12.4 illustrates the basic elements of this system.

12.4.3 Groundwater Control for Slopes

Groundwater can affect slopes in many different ways. Table 12.3 presents common examples and the influence of groundwater on slope failures. The main destabilizing factors of groundwater on slope stability are as follows (Cedergren 1989):

1. Reducing or eliminating cohesive strength
2. Producing pore water pressures which reduce effective stresses, thereby lowering shear strength
3. Causing horizontally inclined seepage forces which increase the driving forces and reduce the factor of safety of the slope
4. Providing for the lubrication of slip surfaces
5. Trapping of groundwater in soil pores during earthquakes or other severe shocks, which leads to liquefaction failures

There are many different construction methods that can be used to mitigate the effects of groundwater on slopes. During construction of slopes, built-in drainage systems can be installed. For existing slopes, drainage devices such as trenches or galleries, relief wells, or horizontal drains can be installed. Another common slope stabilization method is the construction of a drainage buttress at the toe of a slope. In its simplest form, a drainage buttress can consist of cobbles or crushed rock placed at the toe of a slope. The objective of the drainage buttress is to be as heavy as possible to stabilize the toe of the slope and also have a high permeability so that seepage is not trapped in the underlying soil.

TABLE 12.2 Methods of Groundwater Control

Method	Soils suitable for treatment	Uses	Comments
Sump pumping	Clean gravels and coarse sands	Open shallow excavations	Simplest pumping equipment. Fines easily removed from the ground. Encourages instability of formation. See Fig. 12.4
Well-point system with suction pump	Sandy gravels down to fine sands (with proper control can also be used in silty sands)	Open excavations including utility trench excavations	Quick and easy to install in suitable soils. Suction lift limited to about 18 ft (5.5 m). If greater lift needed, multistage installation is necessary. See Fig. 12.3
Deep wells with electric submersible pumps	Gravels to silty fine sands, and water-bearing rocks	Deep excavation in, through, or above water-bearing formations	No limitation on depth of drawdown. Wells can be designed to draw water from several layers throughout its depth. Wells can be sited clear of working area
Jetting system	Sands, silty sand, and sandy silts	Deep excavations in confined space where multistage well points cannot be used	Jetting system uses high-pressure water to create vacuum as well as to lift the water. No limitation on depth of drawdown
Sheet piling cutoff wall	All types of soil (except boulder beds)	Practically unrestricted use	Tongue-and-groove wood sheeting utilized for shallow excavations in soft and medium soils. Well-understood method and can be rapidly installed. Steel sheet piling for other cases. Well-understood method and can be rapidly installed. Steel sheet piling can be incorporated into permanent works or recovered. Interlock leakage can be reduced by filling interlock with bentonite, cement, grout, or similar materials
Slurry trench cutoff wall	Silts, sands, gravels, and cobbles	Practically unrestricted use; extensive curtain walls around open excavations	Rapidly installed. Can be keyed into impermeable strata such as clays or soft shales. May be impractical to key into hard or irregular bedrock surfaces, or into open gravels

12.14

TABLE 12.2 Methods of Groundwater Control (*Continued*)

Method	Soils suitable for treatment	Uses	Comments
Freezing: ammonium and brine refrigerant	All types of saturated soils and rock	Formation of ice in void spaces stops groundwater flow	Treatment is effective from a working surface outward. Better for large applications of long duration. Treatment takes longer time to develop
Freezing: liquid nitrogen refrigerant	All types of saturated soils and rock	Formation of ice in void spaces stops groundwater flow	Better for small applications of short duration where quick freezing is required. Liquid nitrogen is expensive and requires strict site control. Some ground heave could occur
Diaphragm structural walls: structural concrete	All soil types including those containing boulders	Deep basements, underground construction, and shafts	Can be designed to form a part of the permanent foundation. Particularly efficient for circular excavations. Can be keyed into rock. Minimum vibration and noise. Can be used in restricted space. Also can be installed very close to the existing foundation
Diaphragm structural walls: bored piles or mixed-in-place piles	All soil types, but penetration through boulders may be difficult and costly	Deep basements, underground construction, and shafts	A type of diaphragm wall that is rapidly installed. Can be keyed into impermeable strata such as clays or soft shales

Sources: NAVFAC DM-7.2 (1982); based on the work by Cashman and Harris (1970).

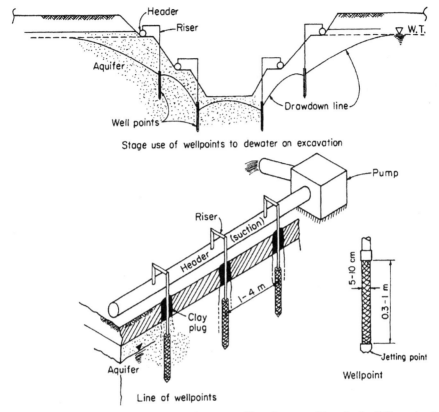

FIGURE 12.3   Groundwater control: well point system with suction pump. (*From Bowles 1982, reprinted with permission of McGraw-Hill, Inc.*)

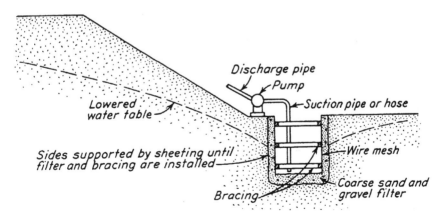

FIGURE 12.4   Groundwater control: example of a sump being used to lower the groundwater table. (*From Peck, Hanson, and Thornburn 1974, reprinted with permission of John Wiley & Sons.*)

TABLE 12.3 Common Groundwater Conditions Causing Slope Failures

Kind of slope	Conditions leading to failure	Type of failure and its consequences
Natural earth slopes above developed land areas (homes, industrial)	Earthquake shocks, heavy rains, snow, freezing and thawing, undercutting at toe, mining excavations	Mud flows, avalanches, landslides; destroying property, burying villages, damming rivers
Natural earth slopes within developed land areas	Undercutting of slopes, heaping fill on unstable slopes, leaky sewers and water lines, lawn sprinkling	Usually slow creep type of failure; breaking water mains, sewers, destroying buildings, roads
Reservoir slopes	Increased soil and rock saturation, raised water table, increased buoyancy, rapid drawdown	Rapid or slow landslides, damaging highways, railways, blocking spillways, leading to over-topping of dams, causing flood damage with serious loss of life
Highway or railway cut or fill slopes	Excessive rain, snow, freezing, thawing, heaping fill on unstable slopes, undercutting, trapping groundwater	Cut slope failures blocking roadways, foundation slipouts removing roadbeds or tracks, property damage, some loss of life
Earth dams and levees, reservoir ridges	High seepage levels, earthquake shocks; poor drainage	Sudden slumps leading to total failure and floods downstream, much loss of life, property damage
Excavations	High groundwater level, insufficient groundwater control, breakdown of dewatering systems	Slope failures or heave of bottoms of excavations; largely delays in construction, equipment loss, property damage

Source: Cedergren 1989.

CHAPTER 13
FOUNDATION ALTERNATIVES TO MITIGATE EARTHQUAKE EFFECTS

13.1 INTRODUCTION

If the expected settlement or lateral movement for a proposed structure is too large, then different foundation support or soil stabilization options must be evaluated. One alternative is soil improvement methods, such as discussed in Chap. 12. Instead of soil improvement, the foundation can be designed to resist the anticipated soil movement caused by the earthquake. For example, mat foundations or post-tensioned slabs may enable the building to remain intact, even with substantial movements. Another option is a deep foundation system that transfers the structural loads to adequate bearing material in order to bypass a compressible or liquefiable soil layer. A third option is to construct a floating foundation, which is a special type of deep foundation in which the weight of the structure is balanced by the removal of soil and construction of an underground basement. A floating foundation could help reduce the amount of rocking settlement caused by the earthquake. Typical factors that govern the selection of a particular type of foundation are presented in Table 13.1.

13.2 SHALLOW FOUNDATIONS

A shallow foundation is often selected when the structural load and the effects of the earthquake will not cause excessive settlement or lateral movement of the underlying soil layers. In general, shallow foundations are more economical to construct than deep foundations. Common types of shallow foundations are described in Table 13.2 and shown in Figs. 13.1 and 13.2.

If it is anticipated that the earthquake will cause excessive settlement or lateral movement, then isolated footings are generally not desirable. This is because the foundation can be pulled apart during the earthquake, causing collapse of the structure. Instead, a mat foundation (Fig. 13.2) or a post-tensioned slab is more desirable. This is because such foundations may enable the building to remain intact, even with substantial movements.

TABLE 13.1 Selection of Foundation Type

Topic	Discussion
Selection of foundation type	Based on an analysis of the factors listed below, a specific type of foundation (i.e., shallow versus deep) would be recommended by the geotechnical engineer.
Adequate depth	The foundation must have an adequate depth to prevent frost damage. For such foundations as bridge piers, the depth of the foundation must be sufficient to prevent undermining by scour.
Bearing capacity failure	The foundation must be safe against a bearing capacity failure.
Settlement	The foundation must not settle to such an extent that it damages the structure.
Quality	The foundation must be of adequate quality that it is not subjected to deterioration, such as from sulfate attack.
Adequate strength	The foundation must be designed with sufficient strength that it does not fracture or break apart under the applied superstructure loads. The foundation must also be properly constructed in conformance with the design specifications.
Adverse soil changes	The foundation must be able to resist long-term adverse soil changes. An example is expansive soil, which could expand or shrink, causing movement of the foundation and damage to the structure.
Seismic forces	The foundation must be able to support the structure during an earthquake without excessive settlement or lateral movement.
Required specifications	The foundation may also have to meet special requirements or specifications required by the local building department or governing agency.

13.3 DEEP FOUNDATIONS

13.3.1 Introduction

Common types of deep foundations are described in Table 13.3. Typical pile characteristics and uses are presented in Table 13.4. Figures 13.3 and 13.4 show common types of cast-in-place concrete piles and examples of pile configurations.

Deep foundations are one of the most effective means of mitigating foundation movement during an earthquake. For example, as discussed in Sec. 3.4.2, the Niigata earthquake resulted in dramatic damage due to liquefaction of the sand deposits in the low-lying areas of Niigata City. At the time of the Niigata earthquake, there were approximately 1500 reinforced concrete buildings in Niigata City, and about 310 of these buildings were damaged, of which approximately 200 settled or tilted rigidly without appreciable damage to the superstructure (see Fig. 3.20). As noted in Sec. 3.4.2, the damaged concrete buildings were built on very shallow foundations or friction piles in loose soil. Similar concrete buildings founded on piles bearing on firm strata at a depth of 20 m (66 ft) did not suffer damage.

Besides buildings, deep foundations can be used for almost any type of structure. For example, Fig. 13.5 shows concrete piles that were used to support a storage tank. The soil

TABLE 13.2 Common Types of Shallow Foundations

Topic	Discussion
Spread footings	Spread footings are often square in plan view, are of uniform reinforced concrete thickness, and are used to support a single load directly in the center of the footing.
Strip footings	Strip footings, also known as wall footings, are often used to support load-bearing walls. They are usually long, reinforced concrete members of uniform width and shallow depth.
Combined footings	Reinforced concrete combined footings are often rectangular or trapezoidal in plan view and carry more than one column load (see Fig. 13.1).
Other types of footings	Figure 13.1 shows other types of footings, such as the cantilever (also known as strap) footing, an octagonal footing, and an eccentric loaded footing with the resultant coincident with area so that the soil pressure is uniform.
Mat foundation	If at mat foundation is constructed at or near ground surface, then it is considered to be a shallow foundation. Figure 13.2 shows different types of mat foundations. Based on economic considerations, mat foundations are often constructed for the following reasons (NAVFAC DM-7.2, 1982): 1. *Large individual footings:* A mat foundation is often constructed when the sum of individual footing areas exceeds about one-half of the total foundation area. 2. *Cavities or compressible lenses:* A mat foundation can be used when the subsurface exploration indicates that there will be unequal settlement caused by small cavities or compressible lenses below the foundation. A mat foundation would tend to span the small cavities or weak lenses and create a more uniform settlement condition. 3. *Shallow settlements:* A mat foundation can be recommended when shallow settlements predominate and the mat foundation would minimize differential settlements. 4. *Unequal distribution of loads:* For some structures, there can be a large difference in building loads acting on different areas of the foundation. Conventional spread footings could be subjected to excessive differential settlement, but a mat foundation would tend to distribute the unequal building loads and reduce the differential settlements. 5. *Hydrostatic uplift:* When the foundation will be subjected to hydrostatic uplift due to a high groundwater table, a mat foundation could be used to resist the uplift forces.
Conventional slab-on-grade	A continuous reinforced concrete foundation consists of bearing wall footings and a slab-on-grade. Concrete reinforcement often consists of steel rebar in the footings and wire mesh in the concrete slab.

beneath the storage tank liquefied during the Kobe earthquake. Concerning the performance of this deep foundation during the earthquake, it was stated (EERC 1995):

The tank, reportedly supported by 33 piles extending to depths of approximately 33 meters, was undamaged. The piles consist of reinforced-concrete sections with diameters of approximately

TABLE 13.2 Common Types of Shallow Foundations (*Continued*)

Topic	Discussion
Posttensioned slab-on-grade	Post-tensioned slab-on-grade is common in southern California and other parts of the United States. It is an economical foundation type when there is no ground freezing or the depth of frost penetration is low. The most common uses of post-tensioned slab-on-grade are to resist expansive soil forces or when the projected differential settlement exceeds the tolerable value for a conventional (lightly reinforced) slab-on-grade. For example, a posttensioned slab-on-grade is frequently recommended if the projected differential settlement is expected to exceed 2 cm (0.75 in).
	Installation and field inspection procedures for post-tensioned slab-on-grade have been prepared by the Post-Tensioning Institute (1996). Post-tensioned slab-on-grade consists of concrete with embedded steel tendons that are encased in thick plastic sheaths. The plastic sheath prevents the tendon from coming in contact with the concrete and permits the tendon to slide within the hardened concrete during the tensioning operations. Usually tendons have a dead end (anchoring plate) in the perimeter (edge) beam and a stressing end at the opposite perimeter beam to enable the tendons to be stressed from one end. However, the Post-Tensioning Institute (1996) does recommend that the tendons in excess of 30 m (100 ft) be stressed from both ends. The Post-Tensioning Institute (1996) also provides typical anchorage details for the tendons.
	Because post-tensioned slab-on-grade performs better (i.e., less shrinkage-related concrete cracking) than conventional slab-on-grade, it is more popular even for situations where low levels of settlement are expected.
Raised wood floor	Perimeter footings support wood beams and a floor system. Interior support is provided by pad or strip footings. There is a crawl space below the wood floor.
Shallow foundation alternatives	If the expected settlement or lateral movement for a proposed shallow foundation is too large, then other options for foundation support or soil stabilization must be evaluated. Commonly used alternatives include deep foundations, grading options, or other site improvement techniques. Deep foundations are discussed in this chapter, and grading and other site improvement techniques are discussed in Chap. 12.

35 centimeters. Twelve of the 33 piles were arranged in an outer ring near the perimeter of the tank; the rest are situated closer to its center [see Fig. 13.5]. Beneath the tank, the ground had liquefied and settled 28 centimeters. Damage to the exposed portions of the piles appeared to be relatively light. Several piles contained hairline cracks in the upper meter or two. At least one pile contained intersecting cracks that could allow large pieces of concrete to spall out. The most seriously damaged piles were located along the northwestern part of the perimeter. The piles appeared repairable and thus were not classified as having failed.

For earthquake conditions, two of the most commonly used types of deep foundations are the pier and grade beam system and prestressed concrete piles. These two foundation types are described individually in the next two sections.

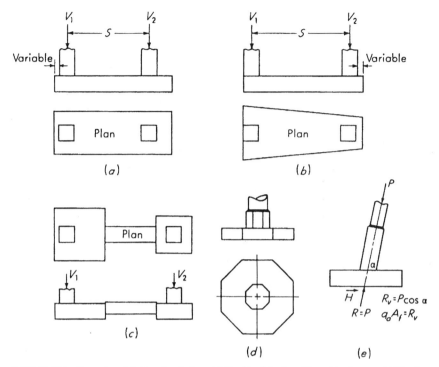

FIGURE 13.1   Examples of shallow foundations. (*a*) Combined footing; (*b*) combined trapezoidal footing; (*c*) cantilever or strap footing; (*d*) octagonal footing; (*e*) eccentric loaded footing with resultant coincident with area so soil pressure is uniform. (*Reproduced from Bowles 1982 with permission of McGraw-Hill, Inc.*)

13.3.2 Pier and Grade Beam Support

The typical steps in the construction of a foundation consisting of piers and grade beams are as follows:

1. *Excavation of piers:*   Figures 13.6 to 13.8 show the excavation of the piers using a truck-mounted auger drill rig. This type of equipment can quickly and economically excavate the piers to the desired depth. In Figs. 13.6 to 13.8, an auger with a 30-in (0.76-m) diameter is being used to excavate the pier holes.

2. *Cleaning of the bottom of the excavation:*   Piers are often designed as end-bearing members. For example, there may be a loose or compressible upper soil zone with the piers excavated through this material and into competent material. The ideal situation is to have the groundwater table below the bottom of the piers. This will then allow for a visual inspection of the bottom of the pier excavation. Often an experienced driller will be able to clean out most of the bottom of the pier by quickly spinning the auger. A light can then be lowered into the pier hole to observe the embedment conditions (i.e., see Fig. 13.9). A worker should not descend into the hole to clean out the bottom; rather, any loose material at the bottom of the pier should be pushed to one side and then scraped into a bucket lowered into the pier hole. If it is simply not possible to clean out the bottom of the pier, then the pier resistance could be based solely on skin friction in the bearing strata with the end-bearing resistance assumed to be equal to zero.

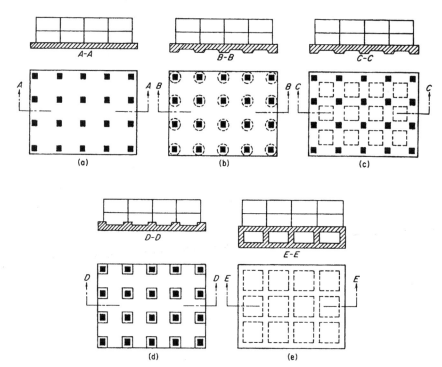

FIGURE 13.2 Examples of mat foundations. (*a*) Flat plate; (*b*) plate thickened under columns; (*c*) beam-and-slab; (*d*) plate with pedestals; (*e*) basement walls as part of mat. (*Reproduced from Bowles 1982 with permission of McGraw-Hill, Inc.*)

TABLE 13.3 Common Types of Deep Foundations

Topic	Discussion
Pile foundations	Probably the most common type of deep foundation is the pile foundation. Piles can consist of wood (timber), steel H-sections, precast concrete, cast-in-place concrete, pressure-injected concrete, concrete-filled steel pipe piles, and composite-type piles (also see Table 13.4). Piles are either driven into place or installed in predrilled holes. Piles that are driven into place are generally considered to be low displacement or high displacement depending on the amount of soil that must be pushed out of the way as the pile is driven. Examples of low-displacement piles are steel H-sections and open-ended steel pipe piles that do not form a soil plug at the end. Examples of high-displacement piles are solid section piles, such as round timber piles or square precast concrete piles, and steel pipe piles with a closed end. Various types of piles are as follows: • *Batter pile:* A pile driven in at an angle inclined to the vertical that provides high resistance to lateral loads.

TABLE 13.3   Common Types of Deep Foundations (*Continued*)

Topic	Discussion
Pile foundations (*continued*)	• *End-bearing pile:* This pile's support capacity is derived principally from the resistance of the foundation material on which the pile tip rests. End-bearing piles are often used when a soft upper layer is underlain by dense or hard strata. If the upper soft layer should settle, the pile could be subjected to down-drag forces, and the pile must be designed to resist these soil-induced forces. • *Friction pile:* This pile's support capacity is derived principally from the resistance of the soil friction and/or adhesion mobilized along the side of the pile. Friction piles are often used in soft clays where the end-bearing resistance is small because of punching shear at the pile tip. A pile that resists upward loads (i.e., tension forces) would also be considered to be a friction pile. • *Combined end-bearing and friction pile:* This pile derives its support capacity from combined end-bearing resistance developed at the pile tip and frictional and/or adhesion resistance on the pile perimeter. Piles are usually driven into specific arrangements and are used to support reinforced concrete pile caps or a mat foundation. For example, the building load from a steel column may be supported by a concrete pile cap that is in turn supported by four piles located near the corners of the concrete pile cap.
Concrete-filled steel pipe piles	Another option is a concrete-filled steel pipe pile. In this case, the steel pipe pile is driven into place. The pipe pile can be driven with either an open or a closed end. If the end is open, the soil within the pipe pile is removed (by jetting) prior to placement of the steel reinforcement and concrete. Table 13.4 provides additional details on the concrete-filled steel pipe piles.
Prestressed concrete piles	Table 13.4 presents details on typical prestressed concrete piles that are delivered to the job site and then driven into place.
Other types of piles	Table 13.4 provides additional details on various types of piles.
Piers	A pier is defined as a deep foundation system, similar to a cast-in-place pile, that consists of a columnlike reinforced concrete member. Piers are often of large enough diameter to enable downhole inspection. Piers are also commonly referred to as drilled shafts, bored piles, or drilled caissons.
Caissons	Large piers are sometimes referred to as caissons. A caisson can also be a watertight underground structure within which work is carried on.
Mat or raft foundation	If a mat or raft foundation is constructed below ground surface or if the mat or raft is supported by piles or piers, then it should be considered to be a deep foundation system.
Floating foundation	A floating foundation is a special type of deep foundation where the weight of the structure is balanced by the removal of soil and construction of an underground basement.

TABLE 13.4 Typical Pile Characteristics and Uses

Pile type	Timber	Steel	Cast-in-place concrete piles (shells driven without mandrel)	Cast-in-place concrete piles (shells withdrawn)
Maximum length	35 m	Practically unlimited	45 m	36 m
Optimum length	9–20 m	12–50 m	9–25 m	8–12 m
Applicable material specifications	ASTM-D25 for piles; P1-54 for quality of creosote; C1-60 for creosote treatment (standards of American Wood Preservers Assoc.)	ASTM-A36 for structural sections; ASTM-A1 for rail sections	ACI	ACI*
Recommended maximum stresses	Measured at midpoint of length: 4–6 MPa for cedar, western hemlock, Norway pine, spruce, and depending on code 5–8 MPa for southern pine, Douglas fir, oak, cypress, hickory	$f_s = 65$ to 140 MPa $f_s = 0.35f_y-0.5f_y$	$0.33f_c'; 0.4f_c'$ if shell gage ≤ 14; shell stress $= 0.35f_y$, if thickness of shell ≥ 3 mm	$0.25f_c'-0.33f_c'$
Maximum load for usual conditions	270 kN	Maximum allowable stress × cross section	900 kN	1300 kN
Optimum load range	130–225 kN	350–1050 kN	450–700 kN	350–900 kN
Disadvantages	Difficult to splice Vulnerable to damage in hard driving Vulnerable to decay unless treated, when piles are intermittently submerged	Vulnerable to corrosion HP section may be damaged or deflected by major obstructions	Hard to splice after concreting Considerable displacement	Concrete should be placed in dry hole More than average dependence on quality of workmanship

13.8

Advantages	Comparatively low initial cost Permanently submerged piles are resistant to decay Easy to handle	Easy to splice High capacity Small displacement Able to penetrate through light obstructions	Can be redriven Shell not easily damaged	Initial economy
Remarks	Best suited for friction pile in granular material	Best suited for end bearing on rock Reduce allowable capacity for corrosive locations	Best suited for friction piles of medium length	Allowable load on pedestal pile is controlled by bearing capacity of stratum immediately below pile
Typical illustrations				

Illustration 1 labels: Grade; Butt diameter 300-500 mm; Pile may be treated with wood preservative; Cross section; Tip diameter 150-250 mm

Illustration 2 labels: Grade; Typical cross section; Rails or sheet pile sections can be used as shown below:; Welded Rail; Welded Sheet pile

Illustration 3 labels: Grade; Sides straight or tapered; 300-450 mm diameter, Shell thickness 3-8 mm, Typical cross section (Fluted shell); 250-900 mm dia, Shell thickness 3-8 mm, Typical cross section (Spiral welded shell); Minimum tip diameter 200 mm

Illustration 4 labels: Grade; Pedestal may be omitted; 350-500 mm diameter, Typical cross section

Notes: Stresses given for steel piles and shells are for noncorrosive locations. For corrosive locations estimate possible reduction in steel cross section or provide protection from corrosion.

TABLE 13.4 Typical Pile Characteristics and Uses (*Continued*)

Pile type	Concrete-filled steel pipe piles	Composite piles	Precast concrete (including prestressed)	Cast in place (thin shell driven with mandrels)	Auger-placed pressure-injected concrete (grout) piles
Maximum length	Practically unlimited	55 m	30 m for precast 60 m for prestressed	30 m for straight sections 12 m for tapered sections	9–25 m
Optimum length	12–36 m	18–36 m	12–15 m for precast 18–30 m for prestressed	12–18 m for straight 5–12 m for tapered	12–18 m
Applicable material specifications	ASTM A36 for core ASTM A252 for pipe ACI Code 318 for concrete	ACI Code 318 for concrete ASTM A36 for structural section ASTM A252 for steel pipe ASTM D25 for timber	ASTM A15 for reinforcing steel ASTM A82 for cold-drawn wire ACI Code 318 for concrete	ACI	See ACI*
Recommended maximum stresses	$0.40f_y$ reinforcement <205 MPa $0.50f_y$ for core <175 MPa $0.33f'_c$ for concrete	Same as concrete in other piles Same as steel in other piles Same as timber piles for wood composite	$0.33f'_c$ unless local building code is less; $0.4f_y$ for reinforced unless prestressed	$0.33f'_c$; $f_s = 0.4f_y$ if shell gauge is ≤14; use $f_s = 0.35f_y$ if shell thickness ≥ 3 mm	$0.225f'_c$–$0.40f'_c$
Maximum load for usual conditions	1800 kN without cores 18,000 kN for large sections with steel cores	1800 kN	8500 kN for prestressed 900 kN for precast	675 kN	700 kN
Optimum load range	700–1100 kN without cores 4500–14,000 kN with cores	250–725 kN	350–3500 kN	250–550 kN	350–550 kN
Disadvantages	High initial cost Displacement for closed-end pipe	Difficult to attain good joint between two materials	Difficult to handle unless prestressed High initial cost Considerable displacement Prestressed difficult to splice	Difficult to splice after concreting Redriving not recommended Thin shell vulnerable during driving Considerable displacement	Dependence on workmanship Not suitable in compressible soil

Advantages	Best control during installation No displacement for open-end installation Open-end pipe best against obstructions High load capacities Easy to splice	High load capacities Corrosion resistance can be attained Hard driving possible	Initial economy Taped sections provide higher bearing resistance in granular stratum	Freedom from noise and vibration Economy High skin friction No splicing
Remarks	Provides high bending resistance where unsupported length is loaded laterally	The weakest of any material used shall govern allowable stresses and capacity	Cylinder piles in particular are suited for bending resistance	Best suited for medium-load friction piles in granular materials	Patented method
Typical illustrations				

*ACI Committee 543, "Recommendations for Design, Manufacture, and Installation of Concrete Piles," *JACI*, August 1973, October 1974.
Sources: NAVFAC DM-7.2 (1982) and Bowles (1982).

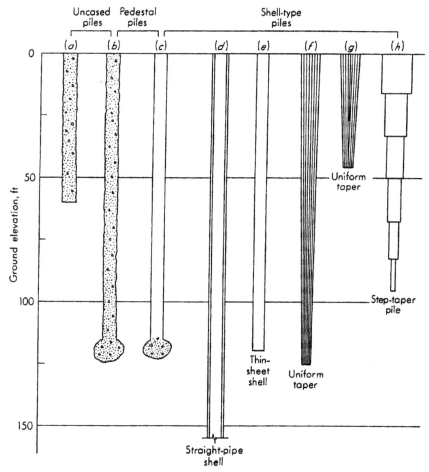

FIGURE 13.3   Common types of cast-in-place concrete piles. (*a*) Uncased pile; (*b*) Franki uncased-pedestal pile; (c) Franki cased-pedestal pile; (*d*) welded or seamless pipe pile; (*e*) cased pile using a thin sheet shell; (*f*) monotube pile; (*g*) uniform tapered pile; (*h*) step-tapered pile. (*Reproduced from Bowles 1982 with permission of McGraw-Hill, Inc.*)

3. *Steel cage and concrete:*   Once the bottom of the pier hole has been cleaned, a steel reinforcement cage is lowered into the pier hole. Small concrete blocks can be used to position the steel cage within the hole. Care should be used when inserting the steel cage so that soil is not knocked off the sides of the hole. Once the steel cage is in place, the hole is filled with concrete. Figure 13.10 shows the completion of the pier with the steel reinforcement extending out the top of the pier.

4. *Grade beam construction:*   The next step is to construct the grade beams that span between the piers. Figure 13.11 shows the excavation of a grade beam between two piers. Figure 13.12 shows the installation of steel for the grade beam. Similar to the piers, small concrete blocks are used to position the steel reinforcement within the grade beam. A visqueen moisture barrier is visible on the left side of Fig. 13.12.

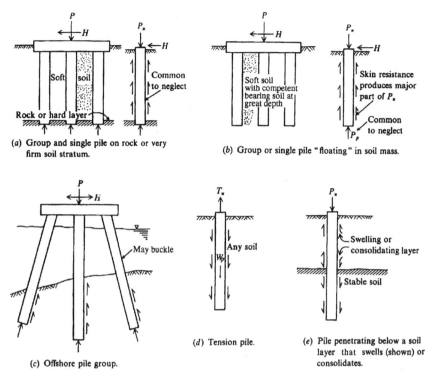

FIGURE 13.4 Typical pile configurations. (*Reproduced from Bowles 1982 with permission of McGraw-Hill, Inc.*)

Figure 13.13 shows a pier located at the corner of the building. The steel reinforcement from the grade beams is attached to the steel reinforcement from the piers. Once the steel reinforcement is in place, the final step is to place the concrete for the grade beams. Figure 13.14 shows the finished grade beams. The steel reinforcement protruding out of the grade beams will be attached to the steel reinforcement in the floor slab.

5. *Floor slab:* Prior to placement of the floor slab, a visqueen moisture barrier and a gravel capillary break should be installed. Then the steel reinforcement for the floor slab is laid out, such as shown in Fig. 13.15. Although not shown in Fig. 13.15, small concrete blocks will be used to elevate the steel reinforcement off the subgrade, and the steel will be attached to the steel from the grade beams. The final step is to place the concrete for the floor slab. Figure 13.16 shows the completed floor slab.

6. *Columns:* When the building is being designed, the steel columns that support the superstructure can be positioned directly over the center of the piers. For example, Fig. 13.17 shows the location where the bottom of a steel column is aligned with the top of a pier. A steel column having an attached baseplate will be bolted to the concrete. Then the steel reinforcement from the pier (see Fig. 13.17) will be positioned around the bottom of the steel column. Once filled with concrete, the final product will be essentially a fixed-end column condition having a high lateral resistance to earthquake shaking.

A main advantage of this type of foundation is that there are no open joints or planes of weakness that can be exploited by the seismic shaking. The strength of the foundation is

FIGURE 13.5 Storage tank supported by concrete piles. The soil underneath the tank liquefied during the Kobe earthquake on January 17, 1995. The soil around the piles was removed in order to observe the condition of the piles. (*Photograph from the Kobe Geotechnical Collection, EERC, University of California, Berkeley.*)

FIGURE 13.6 Truck-mounted auger drill rig used to excavate piers.

FIGURE 13.7   Close-up of auger being pushed into the soil.

FIGURE 13.8   Close-up of auger being extracted from the ground with soil lodged within its grooves.

due to its monolithic construction, with the floor slab attached and supported by the grade beams, which are in turn anchored to the piers. In addition, the steel columns of the superstructure can be constructed so that they bear directly on top of the piers and have fixed end connections. This monolithic foundation and the solid connection between the steel columns and piers will enable the structure to resist the seismic shaking.

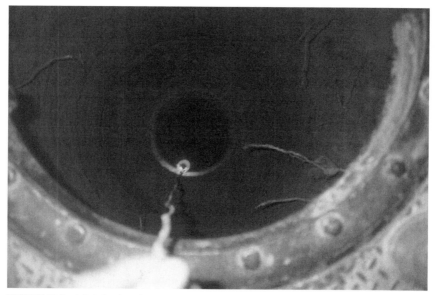

FIGURE 13.9   A light has been lowered to the bottom of the pier to observe embedment conditions.

FIGURE 13.10   The pier hole has been filled with concrete. The steel reinforcement from the pier will be attached to the steel reinforcement in the grade beam.

FIGURE 13.11 Excavation for the grade beam that will span between the two piers.

Usually this foundation system is designed by the structural engineer. The geotechnical engineer provides various design parameters, such as the estimated depth of the bearing strata, the allowable end-bearing resistance, allowable skin friction in the bearing material, allowable passive resistance of the bearing material, and any anticipated down-drag loads that could be induced on the piers if the upper loose or compressible soil should settle under its own weight or during the anticipated earthquake. The geotechnical engineer also needs to inspect the foundation during construction in order to confirm the embedment conditions of the piers.

13.3.3 Prestressed Concrete Piles

Introduction.   Common types of prestressed concrete piles are shown in Fig. 13.18. Prestressed piles are typically produced at a manufacturing plant. The first step is to set up the form, which contains the prestressed strands that are surrounded by wire spirals. The

FIGURE 13.12 Steel reinforcement is being installed within the grade beam excavation.

concrete is then placed within the form and allowed to cure. Once the concrete has reached an adequate strength, the tensioning force is released, which induces a compressive stress into the pile. The prestressed piles are then loaded onto trucks, transported to the site, and stockpiled such as shown in Fig. 13.19.

Solid square concrete piles, such as shown in Fig. 13.19, are the most commonly used type of prestressed piles. As shown in Fig. 13.19, the end of the pile that will be driven into the ground is flush, while at the opposite end the strands protrude from the concrete. A main advantage of prestressed concrete piles is that they can be manufactured to meet site conditions. For example, the prestressed concrete piles shown in Fig. 13.19 were manufactured to meet the following specifications:

- 12-in (0.3-m) square piles
- Design load = 70 tons (620 kN) per pile
- Required prestress = 700 psi (5 MPa)
- 28-Day compressive stress = 6000 psi (40 MPa)
- Maximum water-cement ratio = 0.38
- Portland cement type V (i.e., high sulfate content in the soil)

FIGURE 13.13   Corner of the building where the steel reinforcement from the two grade beams has been attached to the steel reinforcement from the pier.

FIGURE 13.14   The concrete for the grade beams has been placed. The steel reinforcement from the grade beams will be attached to the steel reinforcement in the floor slab.

FIGURE 13.15 Positioning of the steel reinforcement for the floor slab.

FIGURE 13.16 Concrete for the floor slab has been placed.

Pile Driving. Large pile-driving equipment, such as shown in Fig. 13.20, is required to drive the piles into place. If the piles are to be used as end-bearing piles and the depth to the bearing strata is variable, then the first step is to drive indicator piles. An indicator pile is essentially a prestressed pile that is manufactured so that it is longer than deemed necessary. For example,

FIGURE 13.17   Location where a steel column will be attached to the top of a pier.

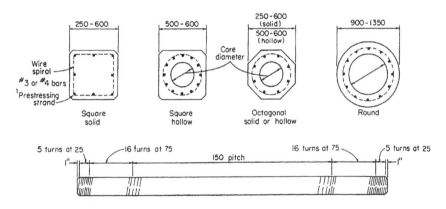

FIGURE 13.18   Typical prestressed concrete piles; dimensions in millimeters. (*Reproduced from Bowles 1982 with permission of McGraw-Hill, Inc.*)

if the depth to adequate bearing material is believed to be 30 ft (9 m), then an indicator pile could be manufactured 35 ft (11 m) long. Usually about 10 to 20 percent of the piles will be indicator piles. The indicator piles are used to confirm embedment conditions, and thus some indicator piles may be driven near the locations of prior borings, while other indicator piles are driven in areas where there is uncertainty about the depth of the bearing strata. Once the indicator piles have been driven, the remaining prestressed piles are manufactured with the lengths of the piles based on the depths to bearing strata as determined from the indicator piles.

FIGURE 13.19   Prestressed concrete piles stockpiled at the job site.

It is always desirable for the geotechnical engineer to observe the driving conditions for the prestressed piles. Prior to driving the piles, basic pile-driving information should be recorded (see Table 13.5). In addition, during the actual driving of the piles, the number of blows per foot of penetration should be recorded. The pile-driving contractor typically marks the pile in 1-ft increments so that the number of blows per foot can be easily counted.

Table 13.6 presents actual data during the driving of a prestressed pile. At this site, soft and liquefiable soil was encountered at a depth of about 15 to 30 ft (4.6 to 9.2 m) below ground surface. Although the blows per foot at this depth were reduced to about 1 per foot, the driving contractor actually allowed the hammer to free-fall, and thus the energy supplied to the top of the pile was significantly less than at the other depths. For the data in Table 13.6, the very high blow counts recorded at a depth of 31 ft (9.5 m) are due to the presence of hard bedrock that underlies the soft and loose soil. Figure 13.21 shows the completed installation of the prestressed concrete pile. The wood block shown on top of the concrete pile in Fig. 13.21 was used as a cushion to protect the pile top from being crushed during the driving operation.

A major disadvantage of prestressed concrete piles is that they can break during the driving process. The most common reason for the breakage of a prestressed concrete pile is that it strikes an underground obstruction, such as a boulder or large piece of debris, which causes the pile to deflect laterally and break. For example, Fig. 13.22 shows the lateral deflection of a prestressed concrete pile as it was driven into the ground. In some cases, the fact that the pile has broken will be obvious. In Fig. 13.23, the prestressed concrete pile hit an underground obstruction, displaced laterally, and then broke near ground surface. In other cases where the pile breaks well below ground surface, the telltale signs will be a continued lateral drifting of the piles and low blow counts at the bearing strata. If a pile should break during installation, standard procedure is to install another pile adjacent to the broken pile. Often the new pile will be offset a distance of 5 ft (1.5 m) from the broken pile. Grade beams are often used to tie together the piles, and thus the location of the new pile

FIGURE 13.20   Pile-driving equipment. A prestressed concrete pile is in the process of being hoisted into position.

should be in line with the proposed grade beam location. The structural engineer will need to redesign the grade beam for its longer span.

Pile Load Tests.   The best method to evaluate the load capacity of a pile is to use a pile load test. A pile load test takes a considerable amount of time and effort to properly set up. Thus only one or two load tests are usually recommended for a particular site. The pile load tests should be located at the most critical area of the site, such as where the bearing strata are deepest or weakest. The first step is to install the piles. In Fig. 13.24, the small arrows point to the prestressed concrete piles, which have been installed and are founded on the bearing strata. The next step is to install the anchor piles, which are used to hold the reaction frame in place and provide resistance to the load applied to the test piles. The most common type of pile load test is the simple compression load test (i.e., see "Standard Test Method for Piles under Static Axial Compressive Load," ASTM D 1143-94, 2000). A schematic setup for this test is shown in Fig. 13.25 and includes the test pile, anchor piles, test beam, hydraulic jack, load cell, and dial gauges. Figure 13.26 shows an actual load test where the reaction frame has been installed on top of the anchor piles and the hydraulic loading jack is in place. A load cell is used to measure the force applied to the top of the pile. Dial gauges, such as shown in Fig. 13.27, are used to record the vertical displacement of the piles during testing.

TABLE 13.5 Example of Pile-Driving Information that Should Be Recorded for the Project

Pile-Driving Record

- Date: March 7, 2001
- Project name and number: Grossmont Healthcare, F.N. 22132.06
- Name of contractor: Foundation Pile Inc.
- Type of pile and date of casting: Precast concrete, cast 2/6/01
- Pile location: See pile-driving records (Table 13.6)
- Sequence of driving in pile group: Not applicable
- Pile dimensions: 12 in by 12 in cross section, lengths vary
- Ground elevation: Varies
- Elevation of tip after driving: See total depth on the driving record
- Final tip and cutoff elevation of pile after driving pile group: Not applicable
- Records of redriving: No redriving
- Elevation of splices: No splices
- Type, make, model, and rated energy of hammer: D30 DELMAG
- Weight and stroke of hammer: Piston weight = 6615 lb. Double-action hammer, maximum stroke = 9 ft
- Type of pile-driving cap used: Wood blocks.
- Cushion material and thickness: Wood blocks approximately 1 ft thick
- Actual stroke and blow rate of hammer: Varies, but stroke did not exceed 9 ft
- Pile-driving start and finish times; and total driving time: See driving record (Table 13.6)
- Time, pile tip elevation, and reason for interruptions: No interruptions
- Record of number of blows per foot: See driving record (Table 13.6)
- Pile deviations from location and plumb: No deviations
- Record preboring, jetting, or special procedures used: No preboring, jetting, or special procedures
- Record of unusual occurrences during pile driving: None

TABLE 13.6 Actual Blow Count Record Obtained during Driving of a Prestressed Concrete Pile

	Blow count record	
Location: M–14.5		
Start time:	8:45 a.m.	
End time:	8:58 a.m.	
Blows per foot:	0 to 5 ft	= 1, 2, 3, 5, 9
	5 to 10 ft	= 9, 9, 11, 10, 9
	10 to 15 ft	= 7, 5, 4, 3, 2
	15 to 20 ft	= 2, 2, 1, 1, 1
	20 to 25 ft	= 1, 1, 1, 1, 1
	25 to 30 ft	= 1, 1 for 2 ft, 1, 2
	>30 ft	= 8, 50 for 10 in
Total depth		= 31.8 ft

FIGURE 13.21   A prestressed concrete pile has been successfully driven to the bearing strata. The wood block shown on top of the concrete pile was used as a cushion to protect the pile top from being crushed during the driving operation.

FIGURE 13.22   Lateral displacement of a prestressed concrete pile during the driving operations.

FIGURE 13.23 This prestressed concrete pile struck an underground obstruction, displaced laterally, and broke near ground surface. The arrow points to the location of the breakage.

The pile is often subjected to a vertical load that is at least 2 times the design value. In most cases, the objective is not to break the pile or load the pile until a bearing capacity failure occurs, but rather to confirm that the design end-bearing parameters used for the design of the piles are adequate. The advantage of this type of approach is that the piles that are load-tested can be left in place and used as part of the foundation. Figure 13.28 presents the actual load test data for the pile load test shown in Figs. 13.26 and 13.27. For this project, the prestressed concrete piles were founded on solid bedrock, and thus the data in Fig. 13.28 show very little compression of the pile. In fact, the recorded displacement of the pile was almost entirely due to elastic compression of the pile itself, instead of deformation of the bearing strata.

Pile Cap, Grade Beams, and Floor Slab. After the piles have been successfully installed, the next step is to construct the remainder of the foundation:

1. *Cut-off top of piles:* Especially for the indicator piles, the portion of the pile extending above ground surface may be much longer than needed. In this case, the pile can be cut off or the concrete chipped off by using a jackhammer, such as shown in Fig. 13.29.

2. *Grade beam excavation:* The next step is to excavate the ground for the grade beams that span between the piles. Figure 13.29 shows the excavation of a grade beam between two piles. For the foundation shown in Fig. 13.29, there is only one pile per cap; thus the pile caps are relatively small compared to the size of the grade beams.

Those prestressed piles that broke during installation should also be incorporated into the foundation. For example, in Fig. 13.30, the pile located at the bottom of the picture is the same broken pile shown in Fig. 13.23. The replacement pile, which was successfully installed to the bearing strata, is located at a distance of 5 ft (1.5 m) from the broken pile (i.e., the pile near the center of Fig. 13.30). As previously mentioned, replacement piles

FIGURE 13.24   Pile load test. The small arrows point to the prestressed concrete piles, which will be subjected to a load test. The large arrow points to one of the six anchor piles.

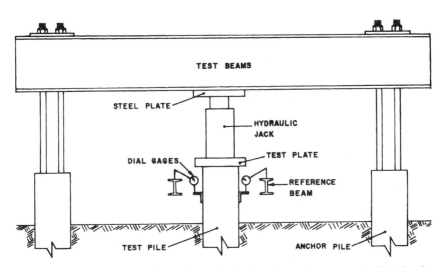

FIGURE 13.25   Schematic setup for applying vertical load to the test pile using a hydraulic jack acting against an anchored reaction frame. (*Reproduced from ASTM D 1143-94, 2000, with permission from the American Society for Testing and Materials.*)

should be installed in line with the grade beam. As shown in Fig. 13.30, both the broken pile and the replacement pile will be attached to the grade beam; however, the broken pile will be assumed to have no support capacity.

FIGURE 13.26 Pile load tests. The reaction frame has been set up, and the hydraulic jack and load cell are in place.

FIGURE 13.27 Pile load tests. This photograph shows one of the dial gauges that are used to record the vertical displacement of the top of the pile during testing.

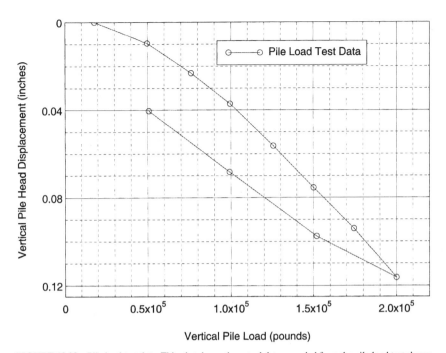

Vertical Pile Load (pounds)

FIGURE 13.28   Pile load test data. This plot shows the actual data recorded from the pile load test shown in Figs. 13.26 and 13.27. The vertical deformation is the average displacement recorded by the dial gauges. The axial load is determined from a load cell.

Once the grade beams have been excavated, the next step is to trim the top of the pre-stressed piles such that they are relatively flush, such has shown in Figs. 13.31 and 13.32. The strands at the top of the pile are not cut off because they will be tied to the steel reinforcement in the grade beam in order to make a solid connection at the top of the pile.

3. *Installation of steel in grade beams:*   After the pile caps and grade beams have been excavated, the next step is to install the steel reinforcement. Figure 13.33 shows a close-up view of the top of a prestressed concrete pile with the steel reinforcement from the grade beam positioned on top of the pile. Note in Fig. 13.33 that the strands from the prestressed pile are attached to the reinforcement steel in the grade beams. This will provide for a solid connection between the pile and the grade beam. Fig. 13.34 presents an overview of the grade beam with the steel reinforcement in place and the grade beam ready for the placement of concrete.

4. *Floor slab:*   Prior to placement of the floor slab, the visqueen moisture barrier and a gravel capillary break should be installed. Then the steel reinforcement for the floor slab is laid out, such as shown in Fig. 13.15. Although not shown, the final step is to place the concrete for the floor slab.

5. *Columns:*   When the building is designed, the steel columns that support the super-structure can be positioned directly over the center of the pile caps.

Similar to the pier and grade beam foundation, a main advantage of the prestressed pile foundation is that there are no open joints or planes of weakness that can be exploited by

FIGURE 13.29   Prestressed concrete piles have been installed, and the excavations for the pile caps and grade beams are complete. The strands at the top of the pile will be connected to the steel reinforcement in the pile cap and grade beam.

the seismic shaking. The strength of the foundation is due to its monolithic construction, with the floor slab attached and supported by the grade beams, which are in turn anchored by the pile caps and the prestressed piles. In addition, the steel columns of the superstructure can be constructed so that they bear directly on top of the pile caps and have fixed end connections. This monolithic foundation and the solid connection between the steel columns and piles will enable the structure to resist the seismic shaking.

Usually this foundation system is designed by the structural engineer. The geotechnical engineer provides various design parameters, such as the estimated depth of the bearing strata, the allowable end-bearing resistance, allowable skin friction in the bearing material, allowable passive resistance of the bearing material, and any anticipated down-drag loads that could be induced on the piles if the upper loose or compressible soil should settle under its own weight or during the anticipated earthquake. The geotechnical engineer should also perform pile load tests and inspect the foundation during construction in order to confirm the design recommendations.

Design Considerations.   There are several important earthquake design considerations for using piles, as follows:

FIGURE 13.30 The prestressed concrete pile at the bottom of the picture is the same pile shown in Figure 13.23. The pile near the center of the photograph is the replacement pile. The broken pile and the replacement pile will be attached to the grade beam.

1. *Connection between pile and cap:* It is important to have an adequate connection between the top of the pile and the pile cap. As shown in Fig. 13.33, this can be accomplished by connecting the strands from the prestressed pile to the steel reinforcement in the pile cap and grade beam. Without this reinforced connection, the pile will be susceptible to separation at the pile cap during the earthquake. For example, Figs. 13.35 and 13.36 show two examples where the tops of the piles separated from the pile cap during the Kobe earthquake.

2. *Down-drag loads due to soil liquefaction:* The pile-supported structure may remain relatively stationary, but the ground around the piles may settle as the pore pressures dissipate in the liquefied soil. The settlement of the ground relative to the pile will induce down-drag loads onto the pile. The piles should have an adequate capacity to resist the down-drag loads.

The relative movement between the relatively stationary structure and the settling soil can also damage utilities. To mitigate damage to utilities, flexible connections can be provided at the location where the utilities enter the building.

3. *Passive resistance for liquefiable soil:* A common assumption is to assume that the liquefied soil will be unable to provide any lateral resistance. If a level-ground site contains an upper layer of nonliquefiable soil that is of sufficient thickness to prevent ground

FIGURE 13.31 The excavation for the grade beams is complete, and the tops of the prestressed piles are trimmed so that they are relatively flush.

fissuring and sand boils, then this layer may provide passive resistance for the piles, caps, and grade beams.

4. *Liquefaction of sloping ground:* For liquefaction of sloping ground, there will often be lateral spreading of the ground, which could shear off the piles. One mitigation measure consists of the installation of compaction piles (see Sec. 12.3.3), in order to create a zone of nonliquefiable soil around and beneath the foundation.

13.4 FOUNDATIONS FOR SINGLE-FAMILY HOUSES

In southern California, the type of foundation for single-family houses often consists of either a raised wood floor foundation or a concrete slab-on-grade.

FIGURE 13.32 Close-up view of one of the prestressed piles showing a trimmed top surface with the strands extending out the top of the pile.

FIGURE 13.33 Close-up view of the top of a prestressed pile with the steel reinforcement from the grade beam positioned on top of the pile. The strands from the pile are attached to the steel reinforcement in the grade beam.

FIGURE 13.34 Overview of the steel reinforcement positioned within the grade beam excavation.

13.4.1 Raised Wood Floor Foundation

The typical raised wood floor foundation consists of continuous concrete perimeter footings and interior (isolated) concrete pads. The floor beams span between the continuous perimeter footings and the isolated interior pads. The continuous concrete perimeter footings are typically constructed so that they protrude about 0.3 to 0.6 m (1 to 2 ft) above the adjacent pad grade. The interior concrete pad footings are not as high as the perimeter footings, and short wood posts are used to support the floor beams. The perimeter footings and interior posts elevate the wood floor and provide for a crawl space below the floor.

In southern California, the raised wood floor foundation having isolated interior pads is common for houses 30 years or older. Most newer houses are not constructed with this foundation type. In general, damages caused by southern California earthquakes have been more severe to houses having this type of raised wood floor foundation. There may be several different reasons for this behavior:

1. *Lack of shear resistance of wood posts:* As previously mentioned, in the interior, the raised wood floor beams are supported by short wood posts bearing on interior concrete pads. During the earthquake, these short posts are vulnerable to collapse or tilting.

FIGURE 13.35 The top of the steel pile separated from the concrete pile cap during the Kobe earthquake on January 17, 1995. (*Photograph from the Kobe Geotechnical Collection, EERC, University of California, Berkeley.*)

2. *No bolts or inadequate bolted condition:* Because in many cases the house is not adequately bolted to the foundation, it can slide or even fall off the foundation during the earthquake. In other cases the bolts are spaced too far apart, and the wood sill plate splits, allowing the house to slide off the foundation.

3. *Age of residence:* The houses having this type of raised wood floor foundation are older. The wood is more brittle and in some cases weakened due to rot or termite damage. In some cases, the concrete perimeter footings are nonreinforced or have been weakened due to prior soil movement, making them more susceptible to cracking during the earthquake.

4. *Crawl-space vents:* To provide ventilation to the crawl space, long vents are often constructed just above the concrete foundation, such as shown in Fig. 13.37. These vents provide areas of weakness just above the foundation.

All these factors can contribute to the detachment of the house from the foundation. For example, Fig. 13.37 shows the sliding of the house off the foundation caused by the Fernando earthquake.

FIGURE 13.36   The top of the concrete pile separated from the concrete pile cap during the Kobe earthquake on January 17, 1995. (*Photograph from the Kobe Geotechnical Collection, EERC, University of California, Berkeley.*)

Besides determining the type of foundation to resist earthquake-related effects, the geotechnical engineer could also be involved with the retrofitting of existing structures. As previously mentioned, the raised wood floor with isolated posts is rarely used for new construction. But there are numerous older houses that have this foundation type, and in many cases, the wood sill plate is inadequately bolted to the foundation. Bolts or tie-down anchors could be installed to securely attach the wood framing to the concrete foundation. Wood bracing or plywood could be added to the open areas between posts to give the foundation greater shear resistance and prevent the house from sliding off the foundations, such as shown in Fig. 13.37.

13.4.2 Slab-on-Grade

In southern California, the concrete slab-on-grade is the most common type of foundation for houses constructed within the past 20 years. It consists of perimeter and interior continuous footings, interconnected by a slab-on-grade. Construction of the slab-on-grade begins with the excavation of the interior and perimeter continuous footings. Steel reinforcing bars are commonly centered in the footing excavations, and wire mesh or steel bars are used as reinforcement for the slab. The concrete for both the footings and the slab is usually placed at the same time, to create a monolithic foundation. Unlike the raised wood floor foundation, the slab-on-grade does not have a crawl space.

In general, for those houses with a slab-on-grade, the wood sill plate is securely bolted to the concrete foundation. In many cases, an earthquake can cause the development of an exterior crack in the stucco at the location where the sill plate meets the concrete founda-

FIGURE 13.37   Sliding of house off the foundation caused by the San Fernando earthquake in California on February 9, 1971. The house is located in the city of San Fernando, near Knox and Grove Streets. (*Photograph from the Steinbrugge Collection, EERC, University of California, Berkeley.*)

tion. In some cases, the crack can be found on all four sides of the house. The crack develops when the house framing bends back and forth during the seismic shaking.

For raised wood floor foundations and the slab-on-grade foundations subjected to similar earthquake intensity and duration, those houses having a slab-on-grade generally have the best performance. This is because the slab-on-grade is typically stronger due to steel reinforcement and monolithic construction, the houses are newer (less wood rot and concrete deterioration), there is greater frame resistance because of the construction of shear walls, and the wood sill plate is in continuous contact with the concrete foundation.

Note that although the slab-on-grade generally has the best performance, these houses can be severely damaged. In many cases, these houses do not have adequate shear walls, there are numerous wall openings, or there is poor construction. The construction of a slab-on-grade by itself is not enough to protect a structure from collapse if the structural frame above the slab does not have adequate shear resistance.

13.4.3   California Northridge Earthquake

The Northridge earthquake, which occurred in California on January 17, 1994, struck an urban area that primarily contained single-family dwellings. The type of foundation for the single-family houses was a major factor in the damage caused by the Northridge earthquake. Particulars concerning the Northridge earthquake are as follows (Day 1999, USGS 1994):

- The Northridge earthquake had a magnitude of 6.7 and occurred beneath the San Fernando Valley on a deeply buried blind thrust fault that may be an eastern extension of the Oak Ridge fault system. The fault plane ruptured from a depth of about 11 mi (17.5 km) upward to about 3 mi (5 km) beneath the surface. For 8 s following the initial break, the rupture propagated upward and northwestward along the fault plane at a rate of about 2 mi/s (3 km/s). Fortuitously, the strongest seismic energy was directed along the fault plane toward sparsely populated areas north of the San Fernando Valley.

- The earthquake deformed the earth's crust over an area of 1500 mi^2 (4000 km^2), forcing the land surface upward in the shape of an asymmetric dome. The dome manifests features and consequences of blind thrust faulting that might lead scientists to the discovery of similar faults elsewhere. The lack of clear surface rupture in 1994 may be explained by fault movement terminating at depth against another fault that moved in the 1971 San Fernando event.

- Studies of more than 250 ground-motion records showed that peak accelerations during the earthquake generally exceeded those predicted. At several locations, horizontal peaks were close to or exceeded 1g, and at one station, vertical acceleration exceeded 1g. Ground motions both near and far from the fault contained consistent, high-energy pulses of relatively long duration. Midrise to high-rise steel structures designed for lesser motions were particularly vulnerable to these pulses. In general, the ratio of horizontal to vertical shaking was similar to that of past earthquakes, and the motions, although strong, were not unusual.

- There was collapse of specially designed structures such as multistory buildings, parking garages, and freeways. In some areas, the most severe damage would indicate a modified Mercalli intensity of IX, although VII to VIII was more widespread. Because the Northridge earthquake occurred in a suburban community, damage to single-family houses was common.

- Numerous structural failures throughout the region were evidence of significant deficiencies in design or construction methods. Steel frames of buildings intended for seismic resistance were cracked, and reinforced concrete columns were crushed. Most highway structures performed well, but freeways collapsed at seven sites, and 170 bridges sustained varying degrees of damage.

- Damage estimates varied considerably. For both public and private facilities, the total cost of the Northridge earthquake was on the order of $20 to $25 billion. This makes the Northridge earthquake California's most expensive natural disaster. Given the significant damage caused by this earthquake, the number of deaths was relatively low. This was partly because most people were asleep at home at the time of the earthquake (4:31 a.m.).

The observed foundation damage caused by the California Northridge earthquake indicated the importance of tying together the various foundation elements. To resist damage during the earthquake, the foundation should be monolithic with no gaps in the footings or planes of weakness due to free-floating slabs. For new construction in southern California, many single-family houses are being constructed with post-tensioned slab-on-grade (see Fig. 13.38). This type of foundation has an induced compressive stress due to the tensioning of the steel tendons embedded in the foundation concrete. Because of the compression stress and lack of free-floating slab elements, this type of foundation will probably perform even better during an earthquake than the conventional slab-on-grade.

13.5 PROBLEMS

13.1   Use the data from Prob. 9.13 and Fig. 9.39 and assume a level-ground site. A proposed building will have a deep foundation system consisting of piles that are driven into the Flysh claystone. Assuming that the piles are widely spaced and do not increase the liquefaction resistance of the soil, calculate the differential movement between the building and adjacent ground. Answer: Using Fig. 7.1, differential movement = 20 cm. Using Fig. 7.2, differential movement = 14 cm.

13.2   Use the data from Prob. 13.1 and an effective friction angle ϕ' between the pile surface and the surface soil layer and sand layer of 28°. Assume that $k_0 = 0.5$ and that the last location for the earthquake-induced pore water pressures to dissipate will be just above the clayey fine sand layer. Further assume that the clayey fine sand layer and the silty fine sand layer are not anticipated to settle during the earthquake. If the piles are 0.3 m in diameter, calculate the down-drag load on each pile due to liquefaction at the site. Answer: Down-drag load = 61 kN.

13.3   Use the data from Prob. 6.12 and Fig. 6.13. To prevent liquefaction-induced settlement of the building, what is the minimum length of piles that should be installed at the site? Answer: 20-m-long piles.

FIGURE 13.38   Construction of a post-tensioned foundation for a single-family residence.

P · A · R · T · 4

BUILDING CODES

CHAPTER 14
EARTHQUAKE PROVISIONS IN BUILDING CODES

14.1 INTRODUCTION

This last chapter presents a discussion of the role of building codes in geotechnical earthquake engineering. The geotechnical engineer should always review local building codes and other regulatory specifications that may govern the seismic design of the project. Types of information that could be included in the building code are as follows:

1. *Earthquake potential:* The building code may specify the earthquake potential for a given site. For example, Fig. 5.17 presents the seismic zone map for the United States (*Uniform Building Code* 1997). The seismic potential often changes as new earthquake data are evaluated. For example, as discussed in Sec. 3.4.3, one of the main factors that contributed to the damage at the Port of Kobe during the Kobe earthquake was that the area had been previously considered to have a relatively low seismic risk, hence the earthquake design criteria were less stringent than in other areas of Japan.

2. *General requirements:* The building code could also specify general requirements that must be fulfilled by the geotechnical engineer. An example is presented in Sec. 1.1, where the *Uniform Building Code* (1997) requires an analysis of the potential for soil liquefaction and soil strength loss during an earthquake. This code provision also requires that the geotechnical engineer evaluate the potential consequences of any liquefaction and soil strength loss, including the estimation of differential settlement, lateral movement, and reduction in foundation bearing capacity, and discuss mitigating measures.

3. *Detailed analyses:* The building code could also provide detailed seismic analyses. For example, Sec. 11.5 outlines the method that can be used to determine the response spectrum per the *Uniform Building Code* (1997).

14.2 CODE DEVELOPMENT

One of the most important ways to develope code is to observe the performance of structures during earthquakes. There must be a desire to improve conditions and not simply accept the death and destruction from earthquakes as inevitable. Two examples of the impact of earthquakes on codes and regulations are as follows:

1. *March 10, 1933, Long Beach earthquake in California:* This earthquake brought an end to the practice of laying brick masonry without reinforcing steel. Prior to this earthquake,

14.3

the exterior walls of buildings were often of brick, or in some cases hollow clay tile. Wood was used to construct the roofs and floors, which were supported by the brick walls. This type of construction was used for schools, and the destruction to these schools was some of the most spectacular damage during the 1933 Long Beach earthquake. Fortunately, the earthquake occurred after school hours, and a catastrophic loss of life was averted. However, the destruction was so extensive and had such dire consequences that the California legislature passed the Field Law on April 10, 1933. This law required that all new public schools be constructed so that they are highly resistant to earthquakes. The Field Law also required that there be field supervision during the construction of schools.

2. *February 9, 1971, San Fernando earthquake in California:* Because of the damage caused by this earthquake, building codes were strengthened, and the California legislature passed the Alquist Priolo Special Studies Zone Act in 1972. The purpose of this act is to prohibit the construction of structures for human occupancy across the traces of active faults. The goal of this legislation is to mitigate the hazards caused by fault rupture.

There has also been a considerable amount of federal legislation in response to earthquake damage. For example, the Federal Emergency Management Agency (1994) states:

> At the federal level, there are two important pieces of legislation relating to local seismic hazard assessment. These are Public Law 93-288, amended in 1988 as the Stafford Act, which establishes basic rules for federal disaster assistance and relief, and the Earthquake Hazards Reduction Act of 1977, amended in 1990, which establishes the National Earthquake Hazards Reduction Program (NEHRP).
>
> The Stafford Act briefly mentions "construction and land use" as possible mitigation measures to be used after a disaster to forestall repetition of damage and destruction in subsequent events. However, the final rules promulgated by the Federal Emergency Management Agency (FEMA) to implement the Stafford Act (44 CFR Part 206, Subparts M and N) require post-disaster state-local hazard mitigation plans to be prepared as a prerequisite for local governments to receive disaster assistance funds to repair and restore damaged or destroyed public facilities. Under the regulations implementing Sec. 409 of the Stafford Act, a city or county must adopt a hazard mitigation plan acceptable to FEMA if it is to receive facilities restoration assistance authorized under Sec. 406.
>
> The overall purpose of the National Earthquake Hazards Reduction Act is to reduce risks to life and property from earthquakes. This is to be carried out through activities such as: hazard identification and vulnerability studies; development and dissemination of seismic design and construction standards; development of an earthquake prediction capability; preparation of national, state, and local plans for mitigation, preparedness, and response; conduct basic and applied research into causes and implications of earthquake hazards; and, education of the public about earthquakes. While this bears less directly on earthquake preparation for a particular local government, much of the growing body of earthquake-related scientific and engineering knowledge has been developed through NEHRP funded research, including this study.

14.3 LIMITATIONS OF BUILDING CODES

Common limitations of building codes are that they may not be up to date or may underestimate the potential for earthquake shaking at a particular area. In addition, the building codes may not be technically sound, or they may contain loopholes that can be exploited by developers. For example, in terms of the collapse of structures caused by the Chi-chi earthquake in Taiwan on September 21, 1999, Hands (1999) states:

> Why then were so many of these collapses occurring in 12-story buildings? Was it, as the local media suggested, a result of seismic waves hitting just the right resonant frequency to take

them out? Professor Chern dismisses this as bordering on superstition. "Basically Taiwan has a lot of 12-story buildings, especially central Taiwan. You hardly see any 20-story high-rises in those areas hit by the quake. The reason for this is that buildings under 50 meters in height don't have to go to a special engineering committee to be approved, so 12 stories is just right." Approval of a structure by qualified structural engineers, and correct enforcement of the building codes, is the crux of the problem, Chern believes.

Another example is the Kobe earthquake in Japan on January 17, 1995. It was observed that a large number of 20-year-old and older high-rise buildings collapsed at the fifth floor. The cause of these building collapses was apparently an older version of the building code that allowed a weaker superstructure beginning at the fifth floor.

Even with a technically sound building code without loopholes, there could be many other factors that are needed to produce earthquake-resistant structures:

1. *Qualified engineers:* There must be qualified structural and geotechnical engineers who can prepare seismic designs and building plans. However, the availability of a professional engineering group will not ensure adequate designs. For example, concerning the collapse of structures caused by the Chi-chi earthquake in Taiwan on September 21, 1999, Hands (1999) states:

> Professor Chern is particularly damning of some of his fellow engineers, and the professional associations to which they belong. "In 1997 we had 6,300 registered civil engineers. Three hundred of them are working in their own consultancies, and 2,800 are employed by building contractors. That means that the other 3,300, or more than half, are possibly renting their licenses." Asked to explain further, Chern said that it was common practice for an engineer to rent his engineer's license to a building contractor, so that the contractor could then claim the architectural drawings had been approved by a qualified engineer, without the engineer even having seen the blueprints. Chern sees the problem as stemming from the way the engineers' professional associations are run. "When they elect a president of the association, the candidate who favors license-renting will get all the votes from those people and win the election, and then he won't be willing to do anything about the problem."

2. *Permit process:* After the engineers have prepared the structural plans and specifications, the plans must be reviewed and approved by the governing agency. The local jurisdiction should have qualified engineers who review the designs to ensure that proper actions are taken to mitigate the impact of seismic hazards, to evaluate structural and nonstructural seismic design and construction practices so that they minimize earthquake damage in critical facilities, and to prevent the total collapse of any structure designed for human occupancy. An important aspect of the permit process is that the governing agency has the power to deny construction of the project if it is deemed to be below the standard of practice.

3. *Inspection during construction:* Similar to the permit process, there must be adequate inspection during the construction of the project to ensure that the approved building plans and specifications are being followed. Any proposed changes to the approved building plans and specifications would have to be reviewed by the governing agency. The project engineers should issue final reports to certify that the structure was built in conformance with the approved building plans.

4. *Construction industry:* An experienced workforce that will follow the approved plans and specifications is needed during construction. In addition, there must be available materials that meet project requirements in terms of quality, strength, etc. An example of lax construction follows (Hands 1999):

Professor Chern said the construction industry is riddled with problems from top to bottom. Even the concrete has problems. "In Taiwan we have quite narrow columns with a lot of rebar in them. This makes it difficult to pour the concrete and get it through and into all the spaces between the bars. Just imagine it—you usually have a small contractor doing the pouring, maybe five men with one pumping car, with two doing the vibrating. They pour 400 cubic meters in one day, and only make NT$5,000 for one morning's work."

It's also a manpower quality problem, he said. "You have low quality workers on low pay, so everything is done quickly. Very good concrete is viscous, so they add water to ready-mixed concrete to make it flow better. But then you get segregation of the cement and aggregate, and the bonding of the concrete and rebar is poor. We've seen that in a lot of the collapsed buildings. Adding water is the usual practice," Chern said. "They even bring along a water tank for the purpose." And although structural engineers are wont to criticize architects for designing pretty buildings that fall down in quakes, perhaps the opposite extreme should also be avoided. "If I had my way all buildings would be squat concrete cubes with no windows," joked Vincent Borov, an engineer with the EQE team.

APPENDIX A
GLOSSARIES

The following is a list of commonly used geotechnical engineering and engineering geology terms and definitions. The glossary has been divided into five main categories:

Glossary 1   Field Testing Terminology
Glossary 2   Laboratory Testing Terminology
Glossary 3   Terminology for Engineering Analysis and Computations
Glossary 4   Compaction, Grading, and Construction Terminology
Glossary 5   Earthquake Terminology

Basic Terms

Civil Engineer   A professional engineer who is registered to practice in the field of civil works.

Civil Engineering   The application of the knowledge of the forces of nature, principles of mechanics, and the properties of materials for the evaluation, design, and construction of civil works for the beneficial uses of humankind.

Earthquake Engineering   Study of the design of structures to resist the forces exerted on the structure by the seismic energy of the earthquake.

Engineering Geologist   A geologist who is experienced and knowledgeable in the field of engineering geology.

Engineering Geology   The application of geologic knowledge and principles in the investigation and evaluation of naturally occurring rock and soil for use in the design of civil works.

Geologist   An individual educated and trained in the field of geology.

Geotechnical Engineer   A licensed individual who performs an engineering evaluation of earth materials including soil, rock, groundwater, and artificial materials and their interaction with earth retention systems, structural foundations, and other civil engineering works.

Geotechnical Engineering   A subdiscipline of civil engineering. Geotechnical engineering requires a knowledge of engineering laws, formulas, construction techniques, and the performance of civil engineering works influenced by earth materials. Geotechnical engineering encompasses many of the engineering aspects of soil mechanics, rock mechanics, foundation engineering, geology, geophysics, hydrology, and related sciences.

Rock Mechanics   The application of the knowledge of the mechanical behavior of rock to engineering problems dealing with rock. Rock mechanics overlaps with engineering geology.

Soil Mechanics   The application of the laws and principles of mechanics and hydraulics to engineering problems dealing with soil as an engineering material.

Soils Engineer   Synonymous with geotechnical engineer (see *Geotechnical Engineer*).

Soils Engineering   Synonymous with geotechnical engineering (see *Geotechnical Engineering*).

GLOSSARY 1   FIELD TESTING TERMINOLOGY

Adobe   Sun-dried brick composed of mud and straw. Abode is commonly used for construction in the southwestern United States and in Mexico.

Aeolian (or eolian)   Particles of soil that have been deposited by the wind. Aeolian deposits include dune sands and loess.

Alluvium   Detrital deposits resulting from the flow of water, including sediments deposited in riverbeds, canyons, floodplains, lakes, fans at the foot of slopes, and estuaries.

Aquiclude   A relatively impervious rock or soil stratum that will not transmit groundwater fast enough to furnish an appreciable supply of water to a well or spring.

Aquifer   A relatively pervious rock or soil stratum that will transmit groundwater fast enough to furnish an appreciable supply of water to a well or spring.

Artesian   Groundwater that is under pressure and is confined by impervious material. If the trapped pressurized water is released, such as by drilling a well, the water will rise above the groundwater table and may even rise above the ground surface.

Ash   Fine fragments of rock, between 4 and 0.25 mm in size, that originated as airborne debris from explosive volcanic eruptions.

Badlands   An area, large or small, characterized by extremely intricate and sharp erosional sculpture. Badlands occur chiefly in arid or semiarid climates where the rainfall is concentrated in sudden heavy showers. They may, however, occur in humid regions where vegetation has been destroyed, or where soil and coarse detritus are lacking.

Bedding   The arrangement of rock in layers, strata, or beds.

Bedrock   A more or less solid, relatively undisturbed rock in place either at the surface or beneath deposits of soil.

Bentonite   A soil or formational material that has a high concentration of the clay mineral montmorillonite. Bentonite is usually characterized by high swelling upon wetting. The term *bentonite* also refers to manufactured products that have a high concentration of montmorillonite, e.g., bentonite pellets.

Bit   A device that is attached to the end of the drill stem and is used as a cutting tool to bore into soil and rock.

Bog   A peat-covered area with a high groundwater table. The surface is often covered with moss, and it tends to be nutrient-poor and acidic.

Boring   A method of investigating subsurface conditions by drilling a hole into the earth materials. Usually soil and rock samples are extracted from the boring. Field tests, such as the standard penetration test (SPT) and the vane shear test (VST), can also be performed in the boring.

Boring Log   A written record of the materials penetrated during the subsurface exploration.

California Bearing Ratio (CBR)   The CBR can be determined for soil in the field or soil compacted in the laboratory. The CBR is frequently used for the design of roads and airfields.

Casing   A steel pipe that is temporarily inserted into a boring or drilled shaft to prevent the adjacent soil from caving.

Cohesionless Soil   A soil, such as a clean gravel or sand, that when unconfined, falls apart in either a wet or dry state.

Cohesive Soil   A soil, such as a silt or clay, that when unconfined, has considerable shear strength when dried and will not fall apart in a saturated state. Cohesive soil is also known as a plastic soil, or a soil that has a plasticity index.

Colluvium   Generally loose deposits usually found near the base of slopes and brought there chiefly by gravity through slow, continuous downhill creep.

Cone Penetration Test (CPT) A field test used to identify and determine the in situ properties of soil deposits and soft rock.

Electric Cone A cone penetrometer that uses electric-force transducers built into the apparatus for measuring cone resistance and friction resistance.

Mechanical Cone A cone penetrometer that uses a set of inner rods to operate a telescoping penetrometer tip and to transmit the resistance force to the surface for measurement.

Mechanical-Friction Cone A cone penetrometer with the additional capability of measuring the local side friction component of penetration resistance.

Piezocone A cone penetrometer with the additional capability of measuring pore water pressure generated during the penetration of the cone.

Core Drilling Also known as diamond drilling, the process of cutting out cylindrical rock samples in the field.

Core Recovery (RQD) The Rock Quality Designation (RQD) is computed by summing the lengths of all pieces of the rock core (NX size) equal to or longer than 10 cm (4 in) and dividing by the total length of the core run. The RQD is multiplied by 100 to express it as a percentage.

Deposition The geologic process of laying down or accumulating natural material into beds, veins, or irregular masses. Deposition includes mechanical settling (such as sedimentation in lakes), precipitation (such as the evaporation of surface water to form halite), and the accumulation of dead plants (such as in a peat bog).

Detritus Any material worn or broken down from rocks by mechanical means.

Diatomaceous Earth Usually fine, white, siliceous powder, composed mainly of diatoms and their remains.

Erosion The wearing away of the ground surface as a result of the movement of wind, water, and/or ice.

Fold Bending or flexure of a layer or layers of rock. Examples of folded rock include anticlines and synclines. Usually folds are created by the massive compression of rock layers.

Fracture A visible break in a rock mass. Examples includes joints, faults, and fissures.

Geophysical Techniques Various methods of determining subsurface soil and rock conditions without performing subsurface exploration. A common geophysical technique is to induce a shock wave into the earth and then measure the seismic velocity of the wave's travel through the earth material. The seismic velocity has been correlated with the rippability of the earth material.

Groundwater Table (also known as Phreatic Surface) The top surface of underground water, the location of which is often determined from piezometers, such as an open standpipe. A perched groundwater table refers to groundwater occurring in an upper zone separated from the main body of groundwater by underlying unsaturated rock or soil.

Horizon One of the layers of a soil profile that can be distinguished by its texture, color, and structure.

A Horizon The uppermost layer of a soil profile which often contains remnants of organic life. Inorganic colloids and soluble materials are often leached from this horizon.

B Horizon The layer of a soil profile in which material leached from the overlying A horizon is accumulated.

C Horizon Undisturbed parent material from which the overlying soil profile has been developed.

Inclinometer An instrument that records the horizontal movement preceding or during the movement of slopes. The slope movement can be investigated by successive surveys of the shape and position of flexible vertical casings installed in the ground. The surveys are performed by lowering an inclinometer probe into the flexible vertical casing.

In Situ Used in reference to the original in-place (or in situ) condition of the soil or rock.

Iowa Borehole Shear Test (BST) A field test in which the device is lowered into an uncased borehole and then expanded against the sidewalls. The force required to pull the device toward ground surface is measured, and much like a direct shear test, the shear strength properties of the in situ soil can then be determined.

Karst Topography   A type of landform developed in a region of easily soluble limestone. It is characterized by vast numbers of depressions of all sizes; sometimes by great outcrops of limestone ledges, sinks and other solution passages; an almost total lack of surface streams; and large springs in the deeper valleys.

Kelly   A heavy tube or pipe, usually square or rectangular in cross section, that is used to provide a downward load when an auger borehole is excavated.

Landslide   Mass movement of soil or rock that involves shear displacement along one or several rupture surfaces, which are either visible or may be reasonably inferred.

Landslide Debris   Material, generally porous and of low density, produced from instability of natural or artificial slopes.

Leaching   The removal of soluble materials in soil or rock caused by percolating or moving groundwater.

Loess   A wind-deposited silt often having a high porosity and low density which is often susceptible to collapse of its soil structure upon wetting.

Mineral   An inorganic substance that has a definite chemical composition and distinctive physical properties. Most minerals are crystalline solids.

Overburden   The soil that overlies bedrock. In other cases, it refers to all material overlying a point of interest in the ground, such as the overburden pressure exerted on a clay layer.

Peat   A naturally occurring, highly organic deposit derived primarily from plant materials.

Penetration Resistance   See *Standard Penetration Test.*

Percussion Drilling   A drilling process in which a borehole is advanced by using a series of impacts to the drill rods and attached bit.

Permafrost   Perennially frozen soil. Also defined as ground that remains below freezing temperatures for 2 or more years. The bottom of permafrost lies at depths ranging from a few feet to over a thousand feet. The *active layer* is defined as the upper few inches to several feet of ground that is frozen in winter but thawed in summer.

Piezometer   A device installed for measuring the pore water pressure (or pressure head) at a specific point within the soil mass.

Pit (or Test Pit)   An excavation made for the purpose of observing subsurface conditions, performing field tests, and obtaining soil samples. A pit also refers to an excavation in the surface of the earth from which ore is extracted, such as an open-pit mine.

Pressuremeter Test (PMT)   A field test that involves the expansion of a cylindrical probe within an uncased borehole.

Refusal   During subsurface exploration, an inability to excavate any deeper with the boring equipment. Refusal could be due to many different factors, such as hard rock, boulders, or a layer of cobbles.

Residual Soil   Soil derived by in-place weathering of the underlying material.

Rock   A relatively solid mass that has permanent and strong bonds between the minerals. Rock can be classified as sedimentary, igneous, or metamorphic.

Rotary Drilling   A drilling process in which a borehole is advanced by rotation of a drill bit under constant pressure without impact.

Rubble   Rough stones of irregular shape and size that are naturally or artificially broken from larger masses of rock. Rubble is often created during quarrying, stone cutting, and blasting.

Screw Plate Compressometer (SPC)   A field test that involves a plate that is screwed down to the desired depth, and then as pressure is applied, the settlement of the plate is measured.

Seep   A small area where water oozes from the soil or rock.

Slaking   The crumbling and disintegration of earth materials when exposed to air or moisture. Slaking can also refer to the breaking up of dried clay when submerged in water, due either to compression of entrapped air by inwardly migrating water or to the progressive swelling and sloughing off of the outer layers.

Slickensides   Surfaces within a soil mass which have been smoothed and striated by shear movements on these surfaces.

Slope Wash   Soil and/or rock material that has been transported down a slope by mass wasting assisted by runoff water not confined by channels (also see *Colluvium*).

Soil   Sediments or other accumulations of mineral particles produced by the physical and chemical disintegration of rocks. Inorganic soil does not contain organic matter, while organic soil contains organic matter.

Soil Sampler   A device used to obtain soil samples during subsurface exploration. Based on the inside clearance ratio and the area ratio, soil samples can be either disturbed or undisturbed.

Standard Penetration Test (SPT)   A field test that consists of driving a thick-walled sampler (inner diameter = 1.5 in, outer diameter = 2 in) into the soil by using a 140-lb hammer falling 30 in. The number of blows to drive the sampler 18 in is recorded. The N value (penetration resistance) is defined as the number of blows required to drive the sampler from a depth interval of 6 to 18 in.

Strike and Dip   Strike and dip refer to a planar structure, such as a shear surface, fault, or bed. The strike is the compass direction of a level line drawn on the planar structure. The dip angle is measured between the planar structure and a horizontal surface.

Subgrade Modulus (also known as Modulus of Subgrade Reaction)   This value is often obtained from field plate load tests and is used in the design of pavements and airfields.

Subsoil Profile   Developed from subsurface exploration, a cross section of the ground that shows the soil and rock layers. A summary of field and laboratory tests could also be added to the subsoil profile.

Till   Material created directly by glaciers, without transportation or sorting by water. Till often consists of a wide range in particle sizes, including boulders, gravel, sand, and clay.

Topsoil   The fertile upper zone of soil which contains organic matter and is usually darker in color and loose.

Vane Shear Test (VST)   An in situ field test that consists of inserting a four-bladed vane into the borehole and then pushing the vane into the clay deposit located at the bottom of the borehole. Once it is inserted into the clay, the maximum torque required to rotate the vane and shear the clay is measured. Based on the dimensions of the vane and the maximum torque, the undrained shear strength s_u of the clay can be calculated.

Varved Silt or Varved Clay   A lake deposit with alternating thin layers of sand and silt (varved silt) or sand and clay (varved clay). It is formed by the process of sedimentation from the summer to winter months. The sand is deposited during the summer, and the silt or clay is deposited in the winter when the lake surface is covered with ice and the water is tranquil.

Weathering   The chemical and/or physical processes by which materials (such as rock) at or near the earth's surface are broken apart and disintegrated. The material can experience a change in color, texture, composition, density, and form due to the processes of weathering.

Wetland   Land which has a groundwater table at or near the ground surface, or land that is periodically under water, and supports various types of vegetation that are adapted to a wet environment.

GLOSSARY 2   LABORATORY TESTING TERMINOLOGY

Absorption   The mass of water in the aggregate divided by the dry mass of the aggregate. Absorption is used in soil mechanics for the study of oversize particles or in concrete mix design.

Activity of Clay   The ratio of plasticity index to percent dry mass of the total sample that is smaller than 0.002 mm in grain size. This property is related to the types of clay minerals in the soil.

Angle of Internal Friction   See *Friction Angle.*

Atterberg Limits   Water contents corresponding to different behavior conditions of plastic soil.

Liquid Limit   The water content corresponding to the behavior change between the liquid and plastic states of a soil. The liquid limit is arbitrarily defined as the water content at which a pat of soil, cut by a groove of standard dimensions, will flow together for a distance of 12.7 mm (0.5 in) under the impact of 25 blows in a standard liquid limit device.

Plastic Limit   The water content corresponding to the behavior change between the plastic and semisolid states of a soil. The plastic limit is arbitrarily defined as the water content at which the soil will just begin to crumble when rolled into a thread approximately 3.2 mm (^{1}/8 in) in diameter.

Shrinkage Limit   The water content corresponding to the behavior change between the semi-solid and solid states of a soil. The shrinkage limit is also defined as the water content at which any further reduction in water content will not result in a decrease in volume of the soil mass.

Average Degree of Consolidation   The ratio, expressed as a percentage, of the settlement at any given time to the primary consolidation.

Binder (Soil Binder)   Typically clay-size particles that can bind together or provide cohesion between soil particles. Organic matter and precipitation of cementing minerals can also bind together soil particles.

Boulder   A large detached rock fragment with an average dimension greater than 300 mm (12 in).

Capillarity   Also known as capillary action and capillary rise, the rise of water through a soil due to the fluid property known as surface tension. Due to capillarity, the pore water pressures are less than atmospheric because of the surface tension of pore water acting on the meniscus formed in void spaces between the soil particles. The height of capillary rise is inversely proportional to the pore size of the soil.

Cation Exchange Capacity   The capacity of clay-size particles to exchange cations with the double layer. Also see *Double Layer.*

Clay Minerals   The three most common clay minerals are listed below, with their respective activity *A* values:

Illite (A 5 0.5 to 1.3):   Clay mineral whose structure is similar to that of montmorillonite, but the layers are more strongly bonded together. In terms of cation exchange capacity, in ability to absorb and retain water, and in physical characteristics such as plasticity index, illite is intermediate in activity between clays of the kaolin and montmorillonite groups. Illite often plots just above the *A* line in the plasticity chart.

Kaolinite (A 5 0.3 to 0.5):   A group of clay minerals consisting of hydrous aluminum silicates. A common kaolin mineral is kaolinite, having the general formula $Al_2Si_2(OH)_4$. Kaolinite is usually formed by alteration of feldspars and other aluminum-bearing minerals. Kaolinite is usually a large clay mineral of low activity and often plots below the *A* line in the plasticity chart. Kaolinite is a relatively inactive clay mineral, and even though it is technically a clay, it behaves more as a silt material. Kaolinite has many industrial uses including the production of china, medicines, and cosmetics.

Montmorillonite (Na-montmorillonite, A 5 4 to 7, and Ca-montmorillonite, A 5 1.5):   A group of clay minerals that are characterized by weakly bonded layers. Each layer consists of two silica sheets with an aluminum (gibbsite) sheet in the middle. Water and exchangeable cations (Na, Ca, etc.) can enter and separate the layers, creating a very small crystal that has a strong attraction for water. Montmorillonite has the highest activity, and it can have the highest water content, greatest compressibility, and lowest shear strength of all the clay minerals. Montmorillonite plots just below the *U* line in the plasticity chart. Montmorillonite often forms as the result of the weathering of ferro-magnesian minerals, calcic feldspars, and volcanic materials. For example, sodium montmorillonite is often formed from the weathering of volcanic ash. Other environments that are likely to form montmorillonite are alkaline conditions with a supply of magnesium ions and a lack of leaching. Such conditions are often present in semiarid regions.

Clay-Size Particles   Clay-size particles are finer than 0.002 mm. Most clay particles are flat or platelike in shape, and as such they have a large surface area. The most common clay minerals belong to the kaolin, montmorillonite, and illite groups.

Coarse-Grained Soil   According to the Unified Soil Classification System, coarse-grained soils have more than 50 percent soil particles (by dry mass) retained on the No. 200 U.S. standard sieve.

Cobble   A rock fragment, usually rounded or semirounded, with an average dimension between 75 and 300 mm (3 and 12 in).

Coefficient of Compressibility   The change in void ratio divided by the corresponding change in vertical effective stress.

Coefficient of Consolidation   A coefficient used in the theory of consolidation. It is obtained from laboratory consolidation tests and is used to predict the time-settlement behavior of field loading of fine-grained soil.

Coefficient of Curvature and Coefficient of Uniformity   These two parameters are used for the classification of coarse-grained soils (USCS) and nonplastic soils (ISBP). These two parameters are used to distinguish a well-graded soil from a uniformly graded soil.

Coefficient of Permeability   See *Hydraulic Conductivity.*

Cohesion   There are two types of cohesion: (1) cohesion in terms of total stress and (2) cohesion in terms of effective stress. For total cohesion c, the soil particles are predominantly held together by capillary tension. For effective stress cohesion c', there must be actual bonding or attraction forces between the soil particles.

Cohesionless Soil   See *Nonplastic Soil.*

Cohesive Soil   See *Plastic Soil.*

Colloidal Soil Particles   Generally clay-size particles (finer than 0.002 mm) where the surface activity of the particle has an appreciable influence on the properties of the soil.

Compaction (Laboratory)

   Compaction Curve   A curve showing the relationship between the dry density and the water content of a soil for a given compaction energy.
   Compaction Test   A laboratory compaction procedure whereby a soil at a known water content is compacted into a mold of specific dimensions. The procedure is repeated for various water contents to establish the compaction curve. The most common testing procedures (compaction energy, number of soil layers in the mold, etc.) are the modified Proctor (ASTM D 1557) or standard Proctor (ASTM 698). The objective of the laboratory compaction test is to obtain the laboratory maximum dry density and the optimum moisture content for the tested soil.
   Relative Compaction   The degree of compaction (expressed as a percentage) defined as the field dry density divided by the laboratory maximum dry density.

Compression Index   For a consolidation test, the slope of the linear portion of the vertical pressure versus void ratio curve on a semilog plot. The compression index is calculated for the virgin consolidation curve.

Compressive Strength   See *Unconfined Compressive Strength.*

Consistency of Clay   Generally the firmness of a cohesive soil. For example, a cohesive soil can have a consistency that varies from *very soft* up to *hard.*

Consolidated Drained Triaxial Compression Test   See *Triaxial Test.*

Consolidated Undrained Triaxial Compression Test   See *Triaxial Test.*

Consolidation Test   A laboratory test used to measure the consolidation properties of saturated cohesive soil. The specimen is laterally confined in a ring and is compressed between porous plates (oedometer apparatus). Also see *Consolidation* in Glossary 3.

Contraction (during Shear)   During the shearing of soil, the tendency of loose soil to decrease in volume (or contract).

Controlled Strain Test   A laboratory test where the load is applied so as to control the rate of strain. The shear portions of triaxial compression tests are often performed by subjecting the soil specimen to a specific rate of axial strain.

Controlled Stress Test   A laboratory test in which the load is applied in increments. The consolidation test is often performed by subjecting the soil specimen to an incremental increase in load, with the soil specimen subjected to each load for a period of 24 h.

Creep   For laboratory tests, drained creep occurs when a plastic soil experiences continued deformation under constant effective stress. For example, secondary compression is often referred to as drained creep.

Deflocculating Agent   Used during the hydrometer test, a compound such as sodium hexametaphosphate that prevents clay-size particles from coalescing into flocs.

Density   Mass per unit volume. In the International System of units (SI), typical units for the density of soil are megagrams per cubic meter.

Deviator Stress   Difference between the major and minor principal stresses in a triaxial test.

Dilation (during Shear)   During the shearing of soil, the tendency of dense soil to increase in volume (or dilate).

Direct Shear Test   A laboratory test used to obtain the effective shear strength properties (c' and ϕ') of the soil. The test consists of applying a vertical pressure to the laterally confined soil specimen, submerging the soil specimen in distilled water, allowing the soil to consolidate, and then shearing the soil specimen by moving the top of the shear box relative to the fixed bottom. The soil specimen must be sheared slowly enough that excess pore water pressures do not develop.

Dispersing Agent   See *Deflocculating Agent.*

Double Layer   A grossly simplified interpretation of the positively charged water layer, together with the negatively charged surface of the particle itself. Two reasons for the attraction of water to the clay particle are that (1) the dipolar structure of the water molecule causes it to be electrostatically attracted to the surface of the clay particle and (2) the clay particles attract cations which contribute to the attraction of water by the hydration process. The *absorbed water layer* consists of water molecules that are tightly held to the clay particle face, such as by the process of hydrogen bonding.

Exchange Capacity   See *Cation Exchange Capacity.*

Fabric (of Soil)   Definitions vary, but in general the geometric arrangement of the soil particles. In contrast, *soil structure* refers to both the geometric arrangement of soil particles and the interparticle forces that may act between them.

Fine-grained Soil   Per the Unified Soil Classification System, a soil that contains more than 50 percent (by dry mass) of particles finer than the No. 200 sieve.

Fines   The silt and clay-size particles in the soil; i.e., soil particles that are finer than the No. 200 U.S. standard sieve.

Flocculation   When in suspension in water, the process of fines attracting one another to form a larger particle or floc. In the hydrometer test, a dispersing agent is added to prevent flocculation of fines.

Friction Angle   A relative measure of a soil's frictional shear strength. In terms of effective shear stress, the soil friction is usually considered to be due to the interlocking of the soil or rock grains and the resistance to sliding between the grains.

Grain Size Distribution   See *Particle Size Distribution.*

Gravel-Size Fragments   Rock fragments and soil particles that will pass the 3-in (76-mm) sieve and be retained on a No. 4 (4.75-mm) U.S. standard sieve.

Hydraulic Conductivity (or Coefficient of Permeability)   A measure of the soil's ability to allow water to flow through its soil pores. For laminar flow of water in soil, both terms are synonymous. The hydraulic conductivity is often measured in a constant head or falling head permeameter.

Illite   See *Clay Minerals.*

Kaolinite   See *Clay Minerals.*

Laboratory Maximum Dry Density   The peak point of the compaction curve (see *Compaction*).

Liquidity Index Index used to distinguish quick clays (liquidity index usually greater than 1.0) from highly desiccated clays (negative liquidity index).

Liquid Limit See *Atterberg Limits.*

Log-of-Time Method Using data from the laboratory consolidation test, a plot of the vertical deformation versus time on a semilog graph. The log-of-time method is used to determine the coefficient of consolidation. Also see *Square-Root-of-Time Method.*

Moisture Content (or Water Content) The ratio of the mass of water in the soil divided by the dry mass of the soil, usually expressed as a percentage. Moisture content and water content are synonymous.

Montmorillonite See *Clay Minerals.*

Nonplastic Soil A granular soil that cannot be rolled or molded at any water content. A nonplastic soil has a plasticity index equal to zero, or the plastic limit is greater than the liquid limit. A nonplastic soil is known as a cohesionless soil.

Optimum Moisture Content The moisture content, determined from a laboratory compaction test, at which the maximum dry density of a soil is obtained using a specific compaction energy. Also see *Compaction.*

Organic Soil Soil that partly or predominately consists of organic matter.

Overconsolidation Ratio (OCR) The ratio of the preconsolidation vertical effective stress to the current vertical effective stress.

Oversize Particles For fill compaction, the gravel and cobble-size particles retained on the $3/4$-in or No. 4 (4.75-mm) U.S. standard sieve. Also see *Soil Matrix.*

Particle Size Distribution The distribution of particles sizes in the soil based on dry mass. Also known as grain size distribution or gradation.

Peak Shear Strength The maximum shear strength along a shear failure surface.

Permeability The ability of water (or other fluid) to flow through a soil by traveling through the void spaces. A high permeability indicates that flow occurs rapidly, and vice versa. A measure of the soil's permeability is the hydraulic conductivity, also known as the coefficient of permeability.

Plasticity Term applied to silt and clay, to indicate the soil's ability to be rolled and molded without breaking apart. A measure of a soil's plasticity is the plasticity index.

Plasticity Index The liquid limit minus the plastic limit, often expressed as a whole number (also see *Atterberg Limits*).

Plastic Limit See *Atterberg Limits.*

Plastic Soil A soil that exhibits plasticity; i.e., the ability to be rolled and molded without breaking apart. A measure of a soil's plasticity is the plasticity index. A plastic soil is also known as a cohesive soil.

Pore Water Pressure See *Pore Water Pressure* in Glossary 3.

Principal Planes and Principal Stresses See Glossary 3.

Sand Equivalent (SE) A measure of the amount of silt or clay contamination in fine aggregate as determined by ASTM D 2419 test procedures.

Sand-Size Particles Soil particles that will pass the No. 4 (4.75-mm) sieve and be retained on the No. 200 (0.075-mm) U.S. standard sieve.

Secant Modulus On a stress-strain plot, the slope of the line from the origin to a given point on the curve. The data for the stress-strain plot are often obtained from a laboratory triaxial compression test.

Shear Strength The maximum shear stress that a soil or rock can sustain. Shear strength of soil is based on total stresses (i.e., undrained shear strength) or effective stresses (i.e., effective shear strength).

Effective Shear Strength   Shear strength of soil based on effective stresses. The effective shear strength of soil could be expressed in terms of the failure envelope, which is defined by effective cohesion c' and effective friction angle ϕ'.

Shear Strength in Terms of Total Stress   Shear strength of soil based on total stresses. The undrained shear strength of soil could be expressed in terms of the undrained shear strength s_u, or by using the failure envelope that is defined by total cohesion c and total friction angle ϕ.

Shear Strength Tests (Laboratory)   There are many types of shear strength tests that can be performed in the laboratory. The objective is to obtain the shear strength of the soil. Laboratory tests can generally be divided into two categories:

Shear Strength Tests Based on Effective Stress   The purpose of these laboratory tests is to obtain the effective shear strength of the soil based on the failure envelope in terms of effective stress. An example is a direct shear test where the saturated, submerged, and consolidated soil specimen is sheared slowly enough that excess pore water pressures do not develop (this test is known as a consolidated-drained test).

Shear Strength Tests Based on Total Stress   The purpose of these laboratory tests is to obtain the undrained shear strength of the soil or the failure envelope in terms of total stresses. An example is the unconfined compression test, which is also known as an unconsolidated-undrained test.

Shrinkage Limit   See *Atterberg Limits*.

Sieve   Laboratory equipment consisting of a pan with a screen at the bottom. U.S. standard sieves are used to separate particles of a soil sample into their various sizes.

Silt-Size Particles   That portion of a soil that is finer than the No. 200 sieve (0.075 mm) and coarser than 0.002 mm. Silt and clay size particles are considered to be *fines*.

Soil Matrix   For fill compaction, that portion of the soil that is finer than the $3/4$-in or No. 4 (4.75-mm) U.S. standard sieve. Also see *Oversize Particles*.

Soil Structure   Definitions vary, but in general both the geometric arrangement of the soil particles and the interparticle forces which may act between them. Common soil structures are as follows:

Cluster Structure   Soil grains that consist of densely packed silt or clay size particles.

Dispersed Structure   Structure in which the clay size particles are oriented parallel to one another.

Flocculated (or Cardhouse) Structure   Structure in which the clay size particles are oriented in edge-to-face arrangements.

Honeycomb Structure   Loosely arranged bundles of soil particles having a structure that resembles a honeycomb.

Single-Grained Structure   An arrangement composed of individual soil particles. This is a common structure of sands.

Skeleton Structure   An arrangement in which coarser soil grains form a skeleton with the void spaces partly filled by a relatively loose arrangement of soil fines.

Specific Gravity   The ratio of the density of the soil particles to the density of water. The specific gravity of soil or oversize particles can be determined in the laboratory.

Square-Root-of-Time Method   Using data from the laboratory consolidation test, a plot of the vertical deformation versus square root of time. The square-root-of-time method is used to determine the coefficient of consolidation. Also see *Log-of-Time-Method*.

Tangent Modulus   On a stress-strain plot, the slope of the line tangent to the stress-strain curve at a given stress value. The stress value used to obtain the tangent modulus is often the stress value that is equal to one-half of the compressive strength. The data for the stress-strain plot can be obtained from a laboratory triaxial compression test.

Tensile Test   For a geosynthetic, a laboratory test in which the geosynthetic is stretched in one direction to determine the force-elongation characteristics, breaking force, and breaking elongation.

Texture (of Soil)   The degree of fineness of the soil, such as smooth, gritty, or sharp, when the soil is rubbed between the fingers.

Thixotropy   The property of a remolded clay that enables it to stiffen (gain shear strength) in a relatively short time.

Torsional Ring Shear Test   A laboratory test in which a relatively thin soil specimen of circular or annular cross section is consolidated and then sheared at a slow rate to obtain the drained residual friction angle.

Triaxial Test   A laboratory test in which a cylindrical specimen of soil or rock encased in an impervious membrane is subjected to a confining pressure and then is loaded axially to failure. Different types of commonly used triaxial tests are as follows:

   Consolidated Drained Triaxial Compression Test   A triaxial test in which the cylindrical soil specimen first is saturated and consolidated by the effective confining pressure. Then the soil specimen is sheared by increasing the axial load. During shearing, drainage is provided to the soil specimen, and it is sheared slowly enough that the shear-induced pore water pressures can dissipate.

   Consolidated Undrained Triaxial Compression Test   A triaxial test in which the cylindrical soil specimen first is saturated and consolidated by the effective confining pressure. Then the soil specimen is sheared by increasing the axial load. During shearing, drainage is not provided to the soil specimen, hence it is an undrained test. The shear-induced pore water pressures can be measured during the shearing process.

   Unconsolidated Undrained Triaxial Compression Test   A triaxial test in which the cylindrical soil specimen retains its initial water content throughout the test (i.e., the water content remains unchanged both during the application of the confining pressure and during shearing). Since drainage is not provided during both the application of the confining pressure and during shearing, the soil specimen is unconsolidated and undrained during shearing.

Unconfined Compressive Strength   The vertical stress which causes the shear failure of a cylindrical specimen of a plastic soil or rock in a simple compression test. For the simple compression test, the undrained shear strength s_u of the plastic soil is defined as one-half of the unconfined compressive strength.

Unconsolidated Undrained Triaxial Compression Test   See *Triaxial Test.*

Unit Weight   Weight per unit volume. In the International System of units (SI), unit weight has units of kilonewtons per cubic meter. In the U.S. Customary System, unit weight has units of pounds-force per cubic foot.

Water Content (or Moisture Content)   See *Moisture Content.*

Zero Air Voids Curve   The relationship between water content and dry density for a condition of saturation ($S = 100$ percent) for a specified specific gravity. On the laboratory compaction curve, the zero air voids curve is often included.

GLOSSARY 3   TERMINOLOGY FOR ENGINEERING ANALYSIS AND COMPUTATIONS

Adhesion   Shearing resistance between two different materials. For example, for piles driven into clay deposits, there is adhesion between the surface of the pile and the surrounding clay.

Allowable Bearing Pressure   The maximum pressure that can be imposed by a foundation onto soil or rock supporting the foundation. It is derived from experience and general usage, and it provides an adequate factor of safety against shear failure and excessive settlement.

Anisotropic Soil   A soil mass having different properties in different directions at any given point, referring primarily to stress-strain or permeability characteristics.

Arching   The transfer of stress from an unconfined area to a less yielding or restrained structure. Arching is important in the design of pile or pier walls that have open gaps between the members.

Bearing Capacity

Allowable Bearing Capacity   The maximum allowable bearing pressure for the design of foundations.

Ultimate Bearing Capacity   The bearing pressure that causes failure of the soil or rock supporting the foundation.

Bearing Capacity Failure   A foundation failure that occurs when the shear stresses in the adjacent soil exceed the shear strength.

Bell   The enlarged portion of the bottom of a drilled shaft foundation. A bell is used to increase the end-bearing resistance. Not all drilled shafts have bells.

Collapsible Formations   For example, limestone formations and deep mining of coal beds. Limestone can form underground caves and caverns which can gradually enlarge, resulting in a collapse of the ground surface and the formation of a sinkhole. Sites that are underlain by coal or salt mines could also experience ground surface settlement when the underground mine collapses.

Collapsible Soil   Soil that is susceptible to a large and sudden reduction in volume upon wetting. Collapsible soil usually has a low dry density and low moisture content. Such soil can withstand a large applied vertical stress with a small compression, but then experience much larger settlements after wetting, with no increase in vertical pressure. Collapsible soil can include fill compacted dry of optimum and natural collapsible soil, such as alluvium, colluvium, or loess.

Compressibility   A decrease in volume that occurs in the soil mass when it is subjected to an increase in loading. Some highly compressible soils are loose sands, organic clays, sensitive clays, highly plastic and soft clays, uncompacted fills, municipal landfills, and permafrost soils.

Consolidation   The consolidation of a saturated clay deposit is generally divided into three separate categories:

    Initial or Immediate Settlement   The initial settlement of the structure caused by undrained shear deformations, or in some cases contained plastic flow, due to two- or three-dimensional loading.

    Primary Consolidation   The compression of clays under load that occurs as excess pore water pressures slowly dissipate with time.

    Secondary Compression   The final component of settlement, which is that part of the settlement that occurs after essentially all the excess pore water pressures have dissipated.

Creep   An imperceptibly slow and more or less continuous movement of slope-forming soil or rock debris.

Critical Height   The maximum height at which a vertical excavation or slope will stand unsupported.

Critical Slope   The maximum angle at which a sloped bank of soil or rock of given height will stand unsupported.

Crown   Generally, the highest point. For tunnels, the crown is the arched roof. For landslides, the crown is the area above the main scarp of the landslide.

Dead Load   Structural loads due to the weight of beams, columns, floors, roofs, and other fixed members. It does not include nonstructural items such as furniture, snow, occupants, or inventory.

Debris Flow   An initial shear failure of a soil mass which then transforms itself into a fluid mass that can move rapidly over the ground surface.

Depth of Seasonal Moisture Change   Also known as the active zone; the layer of expansive soil subjected to shrinkage during the dry season and swelling during the wet season. This zone extends from ground surface to the depth of significant moisture fluctuation.

Desiccation   The process of shrinkage of clays. The process involves a reduction in volume of the grain skeleton and subsequent cracking of the clay caused by the development of capillary stresses in the pore water as the soil dries.

Design Load   All forces and moments that are used to proportion a foundation. The design load includes the deadweight of a structure and, in some cases, can include live loads. Considerable judgment and experience are required to determine the design load that is to be used to proportion a foundation.

Downdrag   Force induced on deep foundation resulting from downward movement of adjacent soil relative to foundation element. Also referred to as negative skin friction.

Earth Pressure   Usually used in reference to the lateral pressure imposed by a soil mass against an earth-supporting structure such as a retaining wall or basement wall:

> **Active Earth Pressure** k_A Horizontal pressure for a condition where the retaining wall has yielded sufficiently to allow the backfill to mobilize its shear strength.
> **At-Rest Earth Pressure** k_0 Horizontal pressure for a condition where the retaining wall has not yielded or compressed into the soil. This would also be applicable to a soil mass in its natural state.
> **Passive Earth Pressure** k_p Horizontal pressure for a condition such as a retaining wall footing that has moved into and compressed the soil sufficiently to develop its maximum lateral resistance.

Effective Stress   The total stress minus the pore water pressure.

Equipotential Line   A line connecting points of equal total head.

Equivalent Fluid Pressure   Horizontal pressures of soil, or soil and water in combination, which increase linearly with depth and are equivalent to those that would be produced by a soil of a given density. Equivalent fluid pressure is often used in the design of retaining walls.

Excess Pore Water Pressure   See *Pore Water Pressure.*

Exit Gradient   The hydraulic gradient near the toe of a dam or the bottom of an excavation through which groundwater seepage is exiting the ground surface.

Finite Element   A soil and structure profile subdivided into regular geometric shapes for the purpose of numerical stress analysis.

Flow Line   The path of travel traced by moving groundwater as it flows through a soil mass.

Flow Net   A graphical representation used to study the flow of groundwater through a soil. A flow net is composed of flow lines and equipotential lines.

Head   From Bernoulli's energy equation, the sum of the velocity head, pressure head, and elevation head. Head has units of length. For seepage problems in soil, the velocity head is usually small enough to be neglected and thus for laminar flow in soil, the total head h is equal to the sum of the pressure head h_p and elevation head h_e.

Heave   The upward movement of foundations or other structures caused by frost heave or expansive soil and rock. Frost heave refers to the development of ice layers or lenses within the soil that causes the ground surface to heave upward. Heave due to expansive soil and rock is caused by an increase in the water content of clays or rocks, such as shale or slate.

Homogeneous Soil   Soil that exhibits essentially the same physical properties at every point throughout the soil mass.

Hydraulic Gradient   Difference in total head at two points divided by the distance between them. Hydraulic gradient is used in seepage analyses.

Hydrostatic Pore Water Pressure   See *Pore Water Pressure.*

Isotropic Soil   A soil mass having essentially the same properties in all directions at any given point, referring primarily to stress-strain or permeability characteristics.

Laminar Flow   Groundwater seepage in which the total head loss is proportional to the velocity.

Live Load   Structural load due to nonstructural members, such as furniture, occupants, inventory, and snow.

Mohr Circle   A graphical representation of the stresses acting on the various planes at a given point in the soil.

Negative Pore Water Pressure   See *Pore Water Pressure.*

Normally Consolidated   The condition that exists if a soil deposit has never been subjected to an effective stress greater than the existing overburden pressure and if the deposit is completely consolidated under the existing overburden pressure.

Overconsolidated   The condition that exists if a soil deposit has been subjected to an effective stress greater than the existing overburden pressure.

Piping   The movement of soil particles as a result of unbalanced seepage forces produced by percolating water, leading to the development of ground surface boils or underground erosion voids and channels.

Plastic Equilibrium   The state of stress of a soil mass that has been loaded and deformed to such an extent that its ultimate shearing resistance is mobilized at one or more points.

Pore Water Pressure   The water pressure that exists in the soil void spaces:

Excess Pore Water Pressure   The increment of pore water pressures greater than hydrostatic values, produced by consolidation stress in compressible materials or by shear strain.

Hydrostatic Pore Water Pressure   Pore water pressure or groundwater pressures exerted under conditions of no flow where the magnitudes of pore pressures increase linearly with depth below the groundwater table.

Negative Pore Water Pressure   Pore water pressure that is less than atmospheric. An example is capillary rise, which can induce a negative pore water pressure in the soil. Another example is the undrained shearing of dense or highly overconsolidated soils, where the soil wants to dilate during shear, resulting in negative pore water pressures.

Porosity   The ratio, usually expressed as a percentage, of the volume of voids divided by the total volume of the soil or rock.

Preconsolidation Pressure   The greatest vertical effective stress to which a soil, such as a clay layer, has been subjected. Also known as the maximum past pressure.

Pressure (or Stress)   The load divided by the area over which it acts.

Principal Planes   Each of three mutually perpendicular planes through a point in the soil mass on which the shearing stress is zero. For soil mechanics, compressive stresses are positive.

Intermediate Principal Plane   The plane normal to the direction of the intermediate principal stress.

Major Principal Plane   The plane normal to the direction of the major principal stress (highest stress in the soil).

Minor Principal Plane   The plane normal to the direction of the minor principal stress (lowest stress in the soil).

Principal Stresses   The stresses that occur on the principal planes. Also see *Mohr Circle.*

Progressive Failure   Formation and development of localized stresses which lead to fracturing of the soil, which spreads and eventually forms a continuous rupture surface and a failure condition. Stiff fissured clay slopes are especially susceptible to progressive failure.

Quick Clay   A clay that has a sensitivity greater than 16. Upon remolding, such clays can exhibit a fluid (or quick) condition.

Quick Condition (or Quicksand)   A condition in which groundwater is flowing upward with a sufficient hydraulic gradient to produce a zero effective stress condition in the sand deposit.

Relative Density   Term applied to a sand deposit to indicate its relative density state, defined as the ratio of (1) the difference between the void ratio in the loosest state and the in situ void ratio to (2) the difference between the void ratios in the loosest and in the densest states.

Saturation (Degree of)   The volume of water in the void space divided by the total volume of voids. It is usually expressed as a percentage. A completely dry soil has a degree of saturation of 0 percent, and a saturated soil has a degree of saturation of 100 percent.

Seepage   The infiltration or percolation of water through soil and rock.

Seepage Analysis   An analysis to determine the quantity of groundwater flowing through a soil deposit. For example, by using a flow net, the quantity of groundwater flowing through or underneath a earth dam can be determined.

Seepage Force   The frictional drag of water flowing through the soil voids.

Seepage Velocity   The velocity of flow of water in the soil, while the superficial velocity is the velocity of flow into or out of the soil.

Sensitivity   The ratio of the undrained shear strength of the undisturbed plastic soil to the remolded shear strength of the same plastic soil.

Settlement   The permanent downward vertical movement experienced by structures as the underlying soil consolidates, compresses, or collapses due to the structural load or secondary influences.

Differential Settlement   The difference in settlement between two foundation elements or between two points on a single foundation.

Total Settlement   The absolute vertical movement of the foundation.

Shear Failure   A failure in a soil or rock mass caused by shearing strain along one or more slip (rupture) surfaces.

General Shear Failure   Failure in which the shear strength of the soil or rock is mobilized along the entire slip surface.

Local Shear Failure   Failure in which the shear strength of the soil or rock is mobilized only locally along the slip surface.

Progressive Shear Failure   See *Progressive Failure.*

Punching Shear Failure   Shear failure where the foundation pushes (or punches) into the soil due to the compression of soil directly below the footing as well as vertical shearing around the footing perimeter.

Shear Plane (or Slip Surface)   A plane along which failure of soil or rock occurs by shearing.

Shear Stress   Stress that acts parallel to the surface element.

Slope Stability Analyses

Gross Slope Stability   The stability of slope material below a plane approximately 0.9 to 1.2 m (3 to 4 ft) deep, measured from and perpendicular to the slope face.

Surficial Slope Stability   The stability of the outer 0.9 to 1.2 m (3 to 4 ft) of slope material measured from and perpendicular to the slope face.

Strain   The change in shape of soil when it is acted upon by stress:

Normal Strain   A measure of compressive or tensile deformations, defined as the change in length divided by the initial length. In geotechnical engineering, strain is positive when it results in compression of the soil.

Shear Strain   A measure of the shear deformation of soil.

Subsidence   Settlement of the ground surface over a very large area, such as caused by the extraction of oil from the ground or the pumping of groundwater from wells.

Swell   Increase in soil volume, typically referring to volumetric expansion of clay due to an increase in water content.

Time Factor T   A dimensionless factor, used in the Terzaghi theory of consolidation or swelling of cohesive soil.

Total Stress   The effective stress plus the pore water pressure. The vertical total stress for uniform soil and a level ground surface can be calculated by multiplying the total unit weight of the soil by the depth below ground surface.

Underconsolidation   The condition that exists if a soil deposit is not fully consolidated under the existing overburden pressure and excess pore water pressures exist within the soil. Underconsolidation occurs in areas where a cohesive soil is being deposited very rapidly and not enough time has elapsed for the soil to consolidate under its own weight.

Void Ratio   The volume of voids divided by the volume of soil solids.

GLOSSARY 4   COMPACTION, GRADING, AND CONSTRUCTION TERMINOLOGY

Aggregate   A granular material used for a pavement base, wall backfill, etc.

   Coarse Aggregate   Gravel or crushed rock that is retained on the No. 4 sieve (4.75 mm).
   Fine Aggregate   Often sand (passes the No. 4 sieve and is retained on the No. 200 U.S. standard sieve).
   Open-Graded Aggregate   Generally a gravel that does not contain any soil particles finer than the No. 4 sieve.

Apparent Opening Size   For a geotextile, a property which indicates the approximate largest particle that would effectively pass through the geotextile.

Approval   A written engineering or geologic opinion by the responsible engineer, geologist of record, or responsible principal of the engineering company concerning the process and completion of the work unless it specifically refers to the building official.

Approved Plans   The current grading plans which bear the stamp of approval of the building official.

Approved Testing Agency   A facility whose testing operations are controlled and monitored by a registered civil engineer and which is equipped to perform and certify the tests as required by the local building code or building official.

As-Graded (or As-Built)   The surface conditions at the completion of grading.

Asphalt   A dark brown to black cementitious material whose main ingredient is bitumen (high molecular hydrocarbons) that occurs in nature or is obtained from petroleum processing.

Asphalt Concrete (AC)   A mixture of asphalt and aggregate that is compacted into a dense pavement surface. Asphalt concrete is often prepared in a batch plant.

Backdrain   Generally a pipe and gravel or similar drainage system placed behind earth retaining structures such as buttresses, stabilization fills, and retaining walls.

Backfill   Soil material placed behind or on top of an area that has been excavated. For example, backfill is placed behind retaining walls and in utility trench excavations.

Base Course or Base   A layer of specified or selected material of planned thickness constructed on the subgrade or subbase for the purpose of providing support to the overlying concrete or asphalt concrete surface of roads and airfields.

Bench   A relatively level step excavated into earth material on which fill is to be placed.

Berm   A raised bank or path of soil. For example, a berm is often constructed at the top of slopes to prevent water from flowing over the top of the slope.

Borrow   Earth material acquired from an off-site location for use in grading on a site.

Brooming   The crushing or separation of wood fibers at the butt (top of the pile) of a timber pile while it is being driven.

Building Official   The city engineer, director of the local building department or a duly delegated representative.

Bulking   The increase in volume of soil or rock caused by its excavation. For example, rock or dense soil will increase in volume upon excavation or by being dumped into a truck for transportation.

Buttress Fill A fill mass, the configuration of which is designed by engineering calculations to stabilize a slope exhibiting adverse geologic features. A buttress is generally specified by minimum key width and depth and by maximum backcut angle. A buttress normally contains a back drainage system.

Caisson Sometimes a large-diameter pier. Another definition is a large structural chamber utilized to keep soil and water from entering into a deep excavation or construction area. Caissons may be installed by being sunk in place or by excavating the bottom of the unit as it slowly sinks to the desired depth.

Cat Slang for Caterpillar grading or construction equipment.

Clearing, Brushing, and Grubbing The removal of vegetation (grass, brush, trees, and similar plant types) by mechanical means.

Clogging For a geotextile, a decrease in permeability due to soil particles that either have lodged in the geotextile openings or have built up a restrictive layer on the surface of the geotextile.

Compaction The densification of a fill by mechanical means. Also see *Compaction* in Glossary 2.

Compaction Equipment Equipment grouped generally into five different types or classifications: sheepsfoot, vibratory, pneumatic, high-speed tamping foot, and chopper wheels (for municipal landfill). Combinations of these types are also available.

Compaction Production Production expressed in compacted cubic meters (m^3) or compacted cubic yards (yd^3) per hour.

Concrete A mixture of aggregates (sand and gravel) and paste (Portland cement and water). The paste binds the aggregates together into a rocklike mass as the paste hardens because of the chemical reactions between the cement and the water.

Contractor A person or company under contract or otherwise retained by the client to perform demolition, grading, and other site improvements.

Cut-Fill Transition The location in a building pad where on one side the pad has been cut down, exposing natural or rock material, while on the other side, fill has been placed.

Dam A structure built to impound water or other fluid products such as tailing waste and wastewater effluent.

   Homogeneous Earth Dam An earth dam whose embankment is formed of one soil type without a systematic zoning of fill materials.

   Zoned Earth Dam An earth dam embankment zoned by the systematic distribution of soil types according to their strength and permeability characteristics, usually with a central impervious core and shells of coarser materials.

Debris All products of clearing, grubbing, demolition, or contaminated soil material that are unsuitable for reuse as compacted fill and/or any other material so designated by the geotechnical engineer or building official.

Dewatering The process used to remove water from a construction site, such as pumping from wells to lower the groundwater table during a foundation excavation.

Dozer Slang for bulldozer construction equipment.

Drainage The removal of surface water from the site. See App. C, Standard Detail No. 9 (Day 1999) for typical lot drainage specifications.

Drawdown The lowering of the groundwater table that occurs in the vicinity of a well that is in the process of being pumped.

Earth Material Any rock, natural soil, or fill, or any combination thereof.

Electroosmosis A method of dewatering, applicable for silts and clays, in which an electric field is established in the soil mass to cause the movement by electroosmotic forces of pore water to well point cathodes.

Erosion Control Devices (Temporary) Devices which are removable and can rarely be salvaged for subsequent reuse. In most cases they will last no longer than one rainy season. They include sandbags, gravel bags, plastic sheeting (visqueen), silt fencing, straw bales, and similar items.

Erosion Control System   A combination of desilting facilities and erosion protection, including effective planting to protect adjacent private property, watercourses, public facilities, and receiving waters from any abnormal deposition of sediment or dust.

Excavation   The mechanical removal of earth material.

Fill   A deposit of earth material placed by artificial means. An engineered (or structural) fill refers to a fill in which the geotechnical engineer has, during grading, made sufficient tests to enable the conclusion that the fill has been placed in substantial compliance with the recommendations of the geotechnical engineer and the governing agency requirements. See App. C, Standard Detail No. 5 (Day 1999) for typical canyon fill placement specifications.

  Hydraulic Fill   A fill placed by transporting soils through a pipe using large quantities of water. These fills are generally loose because they have little or no mechanical compaction during construction.

Footing   A structural member typically installed at a shallow depth that is used to transmit structural loads to the soil or rock strata. Common types of footings include combined footings, spread (or pad) footings, and strip (or wall) footings.

Forms   Structures, usually made of wood, used during the placement of concrete. Forms confine and support the fluid concrete as it hardens.

Foundation   That part of the structure that supports the weight of the structure and transmits the load to underlying soil or rock.

  Deep Foundation   A foundation that derives its support by transferring loads to soil or rock at some depth below the structure.

  Shallow Foundation   A foundation that derives its support by transferring load directly to soil or rock at a shallow depth.

Freeze   Also known as setup; an increase in the load capacity of a pile after it has been driven. Freeze is caused primarily by the dissipation of excess pore water pressures.

Geosynthetic   A planar product manufactured from polymeric material and typically placed in soil to form an integral part of a drainage, reinforcement, or stabilization system. Types include geotextiles, geogrids, geonets, and geomembranes.

Geotextile   A permeable geosynthetic composed solely of textiles.

Grade   The vertical location of the ground surface.

  Existing Grade   The ground surface prior to grading.
  Finished Grade   The final grade of the site which conforms to the approved plan.
  Lowest Adjacent Grade   Adjacent to the structure, the lowest point of elevation of the finished surface of the ground, paving, or sidewalk.
  Natural Grade   The ground surface unaltered by artificial means.
  Rough Grade   The stage at which the grade approximately conforms to the approved plan.

Grading   Any operation consisting of excavation, filling, or a combination thereof.

Grading Contractor   A contractor licensed and regulated who specializes in grading work or is otherwise licensed to do grading work.

Grading Permit   An official document or certificate issued by the building official authorizing grading activity as specified by approved plans and specifications.

Grouting   The process of injecting grout into soil or rock formations to change their physical characteristics. Common examples include grouting to decrease the permeability of a soil or rock stratum, or compaction grouting to densify loose soil or fill.

Hillside Site   A site that entails cut and/or fill grading of a slope which may be adversely affected by drainage and/or stability conditions within or outside the site, or which may cause an adverse effect on adjacent property.

Jetting   The use of a water jet to facilitate the installation of a pile. It can also refer to the fluid placement of soil, such as jetting in the soil for a utility trench.

Key   A designed compacted fill placed in a trench excavated in earth material beneath the toe of a proposed fill slope.

Keyway   An excavated trench into competent earth material beneath the toe of a proposed fill slope.

Lift   During compaction operations, a layer of soil that is dumped by the construction equipment and then subsequently compacted as structural fill.

Necking   A reduction in cross-sectional area of a drilled shaft as a result of the inward movement of the adjacent soils.

Owner   Any person, agency, firm, or corporation having a legal or equitable interest in a given real property.

Permanent Erosion Control Devices   Improvements which remain throughout the life of the development. They include terrace drains, down-drains, slope landscaping, channels, and storm drains.

Permit   An official document or certificate issued by the building official authorizing performance of a specified activity.

Pier   A deep foundation system, similar to a cast-in-place pile, that consists of columnlike reinforced concrete members. Piers are often of large enough diameter to enable down-hole inspection. Piers are also commonly referred to as drilled shafts, bored piles, or drilled caissons.

Pile   A deep foundation system, consisting of relatively long, slender, columnlike members that are often driven into the ground.

   Batter Pile   A pile driven in at an angle inclined to the vertical to provide higher resistance to lateral loads.

   Combination End-Bearing and Friction Pile   A pile that derives its capacity from combined end-bearing resistance developed at the pile tip and frictional and/or adhesion resistance on the pile perimeter.

   End-Bearing Pile   A pile whose support capacity is derived principally from the resistance of the foundation material on which the pile tip rests.

   Friction Pile   A pile whose support capacity is derived principally from the resistance of the soil friction and/or adhesion mobilized along the side of the embedded pile.

Pozzolan   For concrete mix design, a siliceous or siliceous and aluminous material which will chemically react with calcium hydroxide within the cement paste to form compounds having cementitious properties.

Precise Grading Permit   A permit that is issued on the basis of approved plans which show the precise structure location, finish elevations, and all on-site improvements.

Relative Compaction   The degree of compaction (expressed as a percentage) defined as the field dry density divided by the laboratory maximum dry density.

Ripping or Rippability   The characteristic of rock or dense and rocky soils that can be excavated without blasting. Ripping is accomplished by using equipment such as a Caterpillar ripper, ripper-scarifiers, tractor-ripper, or impact ripper. Ripper performance has been correlated with the seismic wave velocity of the soil or rock (see *Caterpillar Performance Handbook* 1997).

Riprap   Rocks that are generally less than 1800 kg (2 tons) in mass that are placed on the ground surface, on slopes or at the toe of slopes, or on top of structures to prevent erosion by wave action or strong currents.

Running Soil or Running Ground   In tunneling or trench excavations, a granular material that tends to flow or "run" into the excavation.

Sand Boil   Also known as sand blow, sand volcano, or silt volcano. The ejection of sand at ground surface, usually forming a cone shape, caused by underground piping. Sand boils can also form at ground surface when there has been liquefaction of underlying soil during an earthquake.

Shear Key   Similar to a buttress; however, generally constructed by excavating a slot within a natural slope in order to stabilize the upper portion of the slope without grading encroachment into the lower portion of the slope. A shear key is also often used to increase the factor of safety of an ancient landslide.

Shotcrete   Mortar or concrete pumped through a hose and projected at high velocity onto a surface. Shotcrete can be applied by a wet or dry mix method.

Shrinkage Factor (SF)   When the loose material is worked into a compacted state, the ratio of the volume of compacted material to the volume of borrow material.

Site   The particular lot or parcel of land where grading or other development is performed.

Slope   An inclined ground surface. For graded slopes, the steepness is generally specified as a ratio of horizontal:vertical (for example, 2:1 slope). Common types of slopes include natural (unaltered) slopes, cut slopes, false slopes (temporary slopes generated during fill compaction operations), and fill slopes.

Slough   Loose, noncompacted fill material generated during grading operations. Slough can also refer to a shallow slope failure, such as sloughing of the slope face.

Slump   In the placement of concrete, the slump is a measure of the consistency of freshly mixed concrete as measured by the slump test. In geotechnical engineering, a slump could also refer to a slope failure.

Slurry Seal   In the construction of asphalt pavements, a fluid mixture of bituminous emulsion, fine aggregate, mineral filler, and water. A slurry seal is applied to the top surface of an asphalt pavement to seal its surface and prolong its wearing life.

Soil Stabilization   The treatment of soil to improve its properties. There are many methods of soil stabilization such as adding gravel, cement, or lime to the soil. The soil could also be stabilized by using geotextiles, by drainage, or through the use of compaction.

Specification   A precise statement in the form of specific requirements. The requirements could be applicable to a material, product, system, or engineering service.

Stabilization Fill   Similar to a buttress fill, whose configuration is typically related to slope height and is specified by the standards of practice for enhancing the stability of locally adverse conditions. A stabilization fill is normally specified by minimum key width and depth and by maximum backcut angle. A stabilization fill usually has a back drainage system.

Staking   During grading, staking is the process where a land surveyor places wood stakes that indicate the elevation of existing ground surface and the final proposed elevation per the grading plans.

Structure   That which is built or constructed, an edifice or building of any kind, or any piece of work artificially built up or composed of parts joined together in some definite manner.

Subdrain (for Canyons)   A pipe and gravel or similar drainage system placed in the alignment of canyons or former drainage channels. After placement of the subdrain, structural fill is placed on top of the subdrain.

Subgrade   For roads and airfields, the underlying soil or rock that supports the pavement section (subbase, base, and wearing surface). The subgrade is also referred to as the basement soil or foundation soil.

Substructure   The foundation.

Sulfate (SO$_4$)   A chemical compound occurring in some soils which, at above certain levels of concentration, has a corrosive effect on ordinary Portland cement concrete and some metals.

Sump   A small pit excavated in the ground or through the basement floor to serve as a collection basin for surface runoff or groundwater. A sump pump is used to periodically drain the pit when it fills with water.

Superstructure   The portion of the structure located above the foundation (includes beams, columns, floors, and other structural and architectural members).

Tack Coat   In the construction of asphalt pavements, a bituminous material that is applied to an existing surface to provide a bond between different layers of the asphalt concrete.

Tailings   In terms of grading, nonengineered fill which accumulates on or adjacent to equipment haul roads. Tailings could also be the waste products generated during a mining operation.

Terrace   A relatively level step constructed in the face of a graded slope surface for drainage control and maintenance purposes.

Underpinning   Piles or other types of foundations built to provide new support for an existing foundation. Underpinning is often used as a remedial measure.

Vibrodensification   The densification or compaction of cohesionless soils by imparting vibrations into the soil mass so as to rearrange soil particles, resulting in less voids in the overall mass.

Walls

   Bearing Wall   Any metal or wood stud wall that supports more than 100 lb per linear foot of superimposed load. Any masonry or concrete wall that supports more than 200 lb per linear foot of superimposed load or is more than one story (*Uniform Building Code* 1997).
   Cutoff Wall   The construction of tight sheeting or a barrier of impervious material extending downward to an essentially impervious lower boundary to intercept and block the path of groundwater seepage. Cutoff walls are often used in dam construction.
   Retaining Wall   A wall designed to resist the lateral displacement of soil or other materials.

Water–Cementitious Materials Ratio   Similar to the water-cement ratio, the ratio of the mass of water (exclusive of that part absorbed by the aggregates) to the mass of cementitious materials in the concrete mix. Commonly used cementitious materials for the concrete mix include Portland cement, fly ash, pozzolan, slag, and silica fume.

Water-Cement Ratio   For concrete mix design, the ratio of the mass of water (exclusive of that part absorbed by the aggregates) to the mass of cement.

Well Point   During the pumping of groundwater, the perforated end section of a well pipe where the groundwater is drawn into the pipe.

Windrow   A string of large rock buried within engineered fill in accordance with guidelines set forth by the geotechnical engineer or governing agency requirements.

Workability of Concrete   The ability to manipulate a freshly mixed quantity of concrete with a minimum loss of homogeneity.

GLOSSARY 5   EARTHQUAKE TERMINOLOGY

Active Fault   See *Fault.*

Aftershock   An earthquake which follows a larger earthquake or main shock and originates in or near the rupture zone of the larger earthquake. Generally, major earthquakes are followed by a large number of aftershocks, usually decreasing in frequency with time.

Amplitude   The maximum height of a wave crest or depth of a trough.

Anticline   Layers of rock that have been folded in a generally convex upward direction. The core of an anticline contains the older rocks.

Array   An arrangement of seismometers or geophones that feed data into a central receiver.

Arrival   The appearance of seismic energy on a seismic record.

Arrival Time   The time at which a particular wave phase arrives at a detector.

Aseismic   A term that indicates the event is not due to an earthquake. An example is an aseismic zone, which indicates an area that has no record of earthquake activity.

Asthenosphere   The layer of shell of the earth below the lithosphere. Magma can be generated within the asthenosphere.

Attenuation Relationship   A relationship that is used to estimate the peak horizontal ground acceleration at a specified distance from the earthquake. Numerous attenuation relationships have been developed. Many attenuation relationships relate the peak horizontal ground acceleration to the earthquake magnitude and closest distance between the site and the focus of the earthquake. Attenuation relationships have also been developed assuming *soft soil* or *hard rock* sites.

Base Shear   The earthquake-induced total design lateral force or shear assumed to act on the base of the structure.

Body Wave   A seismic wave that travels through the interior of the earth. P waves and S waves are body waves.

Body Wave Magnitude Scales (m_b and M_B)   Scales based on the amplitude of the first few P waves to arrive at the seismograph.

Continental Drift   The theory, first advanced by Alfred Wegener, that the earth's continents were originally one land mass. Pieces of the land mass split off and migrated to form the continents.

Core (of the Earth)   The innermost layers of the earth. The inner core is solid and has a radius of about 1300 km. The outer core is fluid and is about 2300 km thick. S waves cannot travel through the outer core.

Crust   The thin outer layer of the earth's surface, averaging about 10 km thick under the oceans and up to about 50 km thick on the continents.

Cyclic Mobility   Concept used to describe large-scale lateral spreading of slopes. In this case, the static driving forces do not exceed the shear strength of the soil along the slip surface, and thus the ground is not subjected to a flow slide. Instead, the driving forces only exceed the resisting forces during those portions of the earthquake that impart net inertial forces in the downslope direction. Each cycle of net inertial forces in the downslope direction causes the driving forces to exceed the resisting forces along the slip surface, resulting in progressive and incremental lateral movement. Often the lateral movement and ground surface cracks first develop at the unconfined toe, and then the slope movement and ground cracks progressively move upslope.

Design-Basis Ground Motion   According to the *Uniform Building Code* (1997), ground motion that has a 10 percent chance of being exceeded in 50 years. The ground motion can be determined by a site-specific hazard analysis, or it may be determined from a hazard map.

Design Response Spectrum   For the design of structures, an elastic response spectrum for 5 percent equivalent viscous damping used to represent the dynamic effects of the design-basis ground motion (*Uniform Building Code* 1997). The response spectrum could be a site-specific spectrum based on a study of the geologic, tectonic, seismological, and soil characteristics of the site.

Dip   See *Strike and Dip.*

Earthquake   Shaking of the earth caused by the sudden rupture along a fault or weak zone in the earth's crust or mantle.

Earthquake Swarm   A series of minor earthquakes, none of which may be identified as the main shock, occurring in a limited area and time.

En échelon   A geologic feature that has a staggered or overlapping arrangement. An example would be surface fault rupture, where the rupture is in a linear form but there are individual features that are oblique to the main trace.

Epicenter   The location on the ground surface that is directly above the point where the initial earthquake motion originated.

Fault   A fracture or weak zone in the earth's crust or upper mantle along which movement has occurred. Faults are caused by earthquakes, and earthquakes are likely to recur on preexisting faults. Although definitions vary, a fault is often considered to be active if movement has occurred within the last 11,000 years (Holocene geologic time). Typical terms used to describe different types of faults are as follows:

    Blind Fault   A fault that has never extended upward to the ground surface. Blind faults often terminate in the upward region of an anticline.

    Blind Thrust Fault   A blind reverse fault where the dip is less than or equal to 45°.

Dip-Slip Fault   A fault which experiences slip only in the direction of its dip. In other words, the movement is perpendicular to the strike. Thus a fault could be described as a *dip-slip normal fault,* which would indicate that it is a normal fault (see Fig. 2.12) with the slip only in the direction of its dip.

Longitudinal Step Fault   A series of parallel faults. These parallel faults develop when the main fault branches upward into several subsidiary faults.

Normal Fault   A fault where the hangingwall block has moved downward with respect to the footwall block. Figure 2.12 illustrates a normal fault. The *hangingwall* is defined as the overlying side of a nonvertical fault.

Oblique-Slip Fault   A fault which experiences components of slip in both its strike and dip directions. A fault could be described as an *oblique-slip normal fault,* which would indicate that it is a normal fault (see Fig. 2.12) with components of slip in both the strike and dip directions.

Reverse Fault   A fault where the hangingwall block has moved upward with respect to the footwall block. Figure 2.13 illustrates a reverse fault.

Strike-Slip Fault   A fault on which the movement is parallel to the strike of the fault. A strike-slip fault is illustrated in Fig. 2.11.

Thrust Fault   A reverse fault where the dip is less than or equal to 45°.

Transform Fault   A fault that is located at a transform boundary (see Sec. 2.1). Yeats et al. (1997) define a transform fault as a strike-slip fault of plate-boundary dimensions that transforms into another plate-boundary structure at its terminus.

Fault Scarp   Generally a portion of the fault that has been exposed at ground surface due to ground surface fault rupture. The exposed portion of the fault often consists of a thin layer of *fault gouge,* which is a clayey seam that has formed during the slipping or shearing of the fault and often contains numerous slickensides.

First Arrival   The first recorded data attributed to seismic waves generated by the fault rupture.

Flow Slide   Phenomenon in which, if liquefaction occurs in or under a sloping soil mass, the entire mass could flow or translate laterally to the unsupported side. Such slides tend to develop in loose, saturated, cohesionless materials that liquefy during the earthquake.

Focal Depth   The distance between the focus and epicenter of the earthquake.

Focus   Also known as the *hypocenter* of an earthquake; the location within the earth that coincides with the initial slip of the fault. In essence, the focus is the location where the earthquake was initiated.

Foreshock   A small tremor that commonly precedes a larger earthquake or main shock by seconds to weeks and that originates in or near the rupture zone of the larger earthquake.

Gouge   The exposed portion of the fault often consisting of a thin layer of fault gouge, which is a clayey seam that has formed during the slipping or shearing of the fault and often contains numerous slickensides.

Graben   The dropping of a crustal block along faults. The crustal block usually has a length that is much greater than its width, resulting in the formation of a long, narrow valley. A graben can also be used to describe the down-dropping of the ground surface, such as a graben area associated with a landslide.

Hazard   A risk. An object or situation that has the possibility of injury or damage.

Hypocenter   See *Focus.*

Inactive Fault   Definitions vary, but in general an inactive fault that has had no displacement over a sufficiently long time in the geologic past that displacements in the foreseeable future are considered unlikely.

Intensity (of an Earthquake)   A measure based on the observations of damaged structures and the presence of secondary effects, such as earthquake-induced landslides, liquefaction, and ground cracking. The intensity of an earthquake is also based on the degree to which the earthquake was felt by individuals, which is determined through interviews. The most commonly used scale for the determination of the intensity of an earthquake is the modified Mercalli intensity scale (see Table 2.3).

Isolator Unit   A horizontally flexible and vertically stiff structural element that allows for large lateral deformation under the seismic load.

Isoseismal Line   A line connecting points on the earth's surface at which earthquake intensity is the same. It is usually a closed curve around the epicenter.

Leaking Mode   A surface seismic wave which is imperfectly trapped so that its energy leaks or escapes across a layer boundary, causing some attenuation or loss of energy.

Liquefaction   The sudden and large decrease of shear strength of a submerged cohesionless soil caused by contraction of the soil structure, produced by shock or earthquake-induced shear strains, associated with a sudden but temporary increase of pore water pressures. Liquefaction occurs when the increase in pore water pressures causes the effective stress to become equal to zero and the soil behaves as a liquid.

Lithosphere   The outermost layer of the earth. It commonly includes the crust and the more rigid part of the upper mantle.

Love Wave   Surface waves that are analogous to S waves in that they are transverse shear waves that travel close to the ground surface. It is named after A. E. H. Love, the English mathematician, who discovered it.

Low-Velocity Zone   Any layer in the earth in which seismic wave velocities are lower than in the layers above and below.

Magnitude (of the Earthquake)   A measure of the size of the earthquake at its source. Many different methods are used to determine the magnitude of an earthquake, such as the local magnitude scale, surface wave magnitude scale, the body wave magnitude scales, and the moment magnitude scale.

Major Earthquake   An earthquake having a magnitude of 7.0 or larger on the Richter scale.

Mantle   The layer of material that lies between the crust and the outer core of the earth. It is approximately 2900 km thick and is the largest of the earth's major layers.

Maximum Capable Earthquake   According to the *Uniform Building Code* (1997), in seismic zones 3 and 4, the level of earthquake ground motion that has a 10 percent probability of being exceeded in a 100-year period.

Maximum Credible Earthquake (MCE)   Often considered to be the largest earthquake that can reasonably be expected to occur based on known geologic and seismologic data. In essence, the maximum credible earthquake is the maximum earthquake that an active fault can produce, considering the geologic evidence of past movement and recorded seismic history of the area. According to Kramer (1996), other terms that have been used to describe similar worst-case levels of shaking include safe shutdown earthquake (used in the design of nuclear power plants), maximum capable earthquake, maximum design earthquake, contingency level earthquake, safe level earthquake, credible design earthquake, and contingency design earthquake. In general, these terms are used to describe the uppermost level of earthquake forces in the design of essential facilities.

The maximum credible earthquake is determined for particular earthquakes or levels of ground shaking. As such, the analysis used to determine the maximum credible earthquake is typically referred to as a *deterministic method.*

Maximum Probable Earthquake   Commonly the largest earthquake that a fault is predicted capable of generating within a specified time period of concern, say, 50 or 100 years. There are many different definitions of the maximum probable earthquake. The maximum probable earthquake is based on a study of nearby active faults. By using attenuation relationships, the maximum probable earthquake magnitude and maximum probable peak ground acceleration can be determined. Maximum probable earthquakes are most likely to occur within the time span of most developments and, therefore, are commonly used in assessing seismic risk.

Another commonly used definition of a maximum probable earthquake is an earthquake that will produce a peak ground acceleration a_{max} with a 50 percent probability of exceedence in 50 years.

According to Kramer (1996), other terms that have been used to describe earthquakes of similar size are operating basis earthquake, operating level earthquake, probable design earthquake, and strength level earthquake.

Microearthquake   An earthquake having a magnitude of 2 or less on the Richter scale.

Modified Mercalli Intensity Scale   See *Intensity (of an Earthquake)*.

Mohorovicic Discontinuity (or Moho Discontinuity)   The boundary surface or sharp seismic-velocity discontinuity that separates the earth's crust from the underlying mantle. Named for Andrija Mohorovicic, the Croatian seismologist who first suggested its existence.

Normal Fault   See *Fault*.

Paleomagnetism   The natural magnetic traces that reveal the intensity and direction of the earth's magnetic field in the geologic past. Also defined as the study of these magnetic traces.

Paleoseismology   The study of ancient (i.e., prehistoric) earthquakes.

Peak Ground Acceleration (PGA)   Also known as the maximum horizontal ground acceleration. The peak ground acceleration can be based on an analysis of historical earthquakes or based on probability (see Sec. 5.6). An attenuation relationship is used to relate the peak ground acceleration to the earthquake magnitude and closest distance between the site and the focus of the earthquake.

Period   The time interval between successive crests in a wave train. The period is the inverse of the frequency.

Plate Boundary   The location where two or more plates in the earth's crust meet.

Plate Tectonics   According to the plate tectonic theory, the earth's surface contains tectonic plates, also known as lithosphere plates, with each plate consisting of the crust and the more rigid part of the upper mantle. Depending on the direction of movement of the plates, there are three types of plate boundaries: divergent boundary, convergent boundary, and transform boundary (see Sec. 2.1).

Pseudostatic Analysis   A method that ignores the cyclic nature of the earthquake and treats it as if it applied an additional static force upon the slope or retaining wall.

P Wave   A body wave that is also known as the primary wave, compressional wave, or longitudinal wave. It is a seismic wave that causes a series of compressions and dilations of the materials through which it travels. The P wave is the fastest wave and is the first to arrive at a site. Being a compression-dilation type of wave, P waves can travel through both solids and liquids. Because soil and rock are relatively resistant to compression-dilation effects, the P wave usually has the least impact on ground surface movements.

Rayleigh Wave   Surface wave similar to the surface ripple produced by a rock thrown into a pond. These seismic waves produce both vertical and horizontal displacement of the ground as the surface waves propagate outward. They are usually felt as a rolling or rocking motion and, in the case of major earthquakes, can be seen as they approach. They are named after Lord Rayleigh, the English physicist who predicted their existence.

Recurrence Interval   The approximate length of time between earthquakes in a specific seismically active area.

Resonance   A condition where the frequency of the structure is equal to the natural frequency of the vibrating ground. At resonance, the structure will experience the maximum horizontal displacement.

Response Spectrum   See *Design Response Spectrum*.

Richter Magnitude Scale   Also known as the local magnitude scale; a system used to measure the strength of an earthquake. Professor Charles Richter developed this earthquake magnitude scale in 1935 as a means of categorizing local earthquakes.

Rift Valley   A long and linear valley formed by tectonic depression accompanied by extension. A divergent boundary between tectonic plates can create a rift valley. Earthquakes at a rift valley are often due to movement on normal faults. Examples of rift valleys are the East African rift and the Rhine Graben.

Risk   See *Seismic Risk.*

Rupture Zone   The area of the earth through which faulting occurred during an earthquake. For great earthquakes, the rupture zone may extend several hundred kilometers in length and tens of kilometers in width.

Sand Boil   Also known as sand blow, sand volcano, or silt volcano. The ejection of sand at ground surface, usually forming a cone shape, is caused by liquefaction of underlying soil during an earthquake. Sand boils can also be caused by piping (see Glossary 4).

Seiche   Identical to a tsunami, except that it occurs in an inland body of water, such as a lake. It can be caused by lake-bottom earthquake movements or by volcanic eruptions and landslides within the lake. A seiche has been described as being similar to the sloshing of water in a bathtub.

Seismic or Seismicity   Dealing with earthquake activity.

Seismic Belt   An elongated earthquake zone. Examples include the circum-Pacific, Mediterranean, and Rocky Mountain seismic belts.

Seismic Risk   The probability of human life and property loss due to an earthquake.

Seismogram   A written record of an earthquake that is produced by a seismograph.

Seismograph   An instrument that records the ground surface movement as a function of time caused by the seismic energy of an earthquake.

Seismology   The study of earthquakes.

Shear Wall   Sometimes referred to as a vertical diaphragm or structural wall, a shear wall is designed to resist lateral forces parallel to the plane of the wall. Shear walls are used to resist the lateral forces induced by the earthquake.

Spreading Center   An elongated region where two plates are being pulled away from each other. New crust is formed as molten rock is forced upward into the gap. An example is seafloor spreading, which has created the mid-Atlantic ridge. Another example is a rift valley, such as the East African rift.

Strike and Dip (of a Fault Plane)   A description of the orientation of the fault plane in space. Strike is the azimuth of a horizontal line drawn on the fault plane. The dip is measured in a direction perpendicular to the strike and is the angle between the inclined fault plane and a horizontal plane.

Strike-Slip Fault   See *Fault.*

Subduction Zone   An elongated region along which a plate descends relative to another plate. An example is the descent of the Nazca plate beneath the South American plate along the Peru-Chile trench.

S Wave   A body wave that is also known as the secondary wave, shear wave, or transverse wave. The S wave causes shearing deformations of the materials through which it travels. Because liquids have no shear resistance, S waves can only travel through solids. The shear resistance of soil and rock is usually less than the compression-dilation resistance, and thus an S wave travels more slowly through the ground than a P wave. Soil is weak in terms of its shear resistance, and S waves typically have the greatest impact on ground surface movements.

Syncline   Layers of rock that have been folded in a generally concave upward direction. The core of a syncline contains the younger rocks.

Travel Time   The time required for a seismic wave train to travel from its source to a point of observation.

Tsunami   A Japanese term that means *harbor wave.* It is a long-period ocean wave that can be created by seafloor earthquake movements or by submarine volcanic eruptions and landslides.

GLOSSARY REFERENCES

AASHTO (1996). *Standard Specifications for Highway Bridges,* 16th ed. Prepared by the American Association of State Highway and Transportation Officials (AASHTO), Washington.

ASTM (2000). *Annual Book of ASTM Standards,* vol. 04.08, *Soil and Rock (I).* Standard No. D 653-97, "Standard Terminology Relating to Soil, Rock, and Contained Fluids." Terms prepared jointly by the American Society of Civil Engineers and ASTM, West Conshohocken, PA, pp. 43–77.

ASTM (2000). *Annual Book of ASTM Standards,* vol. 04.09, *Soil and Rock (II), Geosynthetics.* Standard No. D 4439-99, "Standard Terminology for Geosynthetics," West Conshohocken, PA, pp. 852–855.

Asphalt Institute (1984). *Thickness Design—Asphalt Pavements for Highways and Streets.* Published by Asphalt Institute, College Park, MD.

Caterpillar Performance Handbook (1997), 28th ed. Prepared by Caterpillar, Inc., Peoria, IL.

Coduto, D. P. (1994). *Foundation Design, Principles and Practices.* Prentice-Hall, Englewood Cliffs, NJ.

Day, R. W. (1999). *Geotechnical and Foundation Engineering: Design and Construction.* McGraw-Hill, New York.

Holtz, R. D., and Kovacs, W. D. (1981). *An Introduction to Geotechnical Engineering.* Prentice-Hall, Englewood Cliffs, NJ.

Kramer, S. L. (1996). *Geotechnical Earthquake Engineering.* Prentice-Hall, Englewood Cliffs, NJ.

Krinitzsky, E. L., Gould, J. P., and Edinger, P. H. (1993). *Fundamentals of Earthquake-Resistant Construction.* John Wiley & Sons, New York.

Lambe, T. W., and Whitman, R. V. (1969). *Soil Mechanics.* John Wiley & Sons, New York.

McCarthy, D. F. (1977). *Essentials of Soil Mechanics and Foundations.* Reston Publishing Company, Reston, VA.

NAVFAC DM-7.1 (1982). *Soil Mechanics, Design Manual 7.1.* Department of the Navy, Naval Facilities Engineering Command, Alexandria, VA.

NAVFAC DM-7.2 (1982). *Foundations and Earth Structures, Design Manual 7.2.* Department of the Navy, Naval Facilities Engineering Command, Alexandria, VA.

NAVFAC DM-7.3 (1983). *Soil Dynamics, Deep Stabilization, and Special Geotechnical Construction, Design Manual 7.3.* Department of the Navy, Naval Facilities Engineering Command, Alexandria, VA.

Orange County Grading Manual (1993). *Orange County Grading Manual,* part of the *Orange County Grading and Excavation Code,* prepared by Orange County, CA.

Stokes, W. L., and Varnes, D. J. (1955). "Glossary of Selected Geologic Terms with Special Reference to Their Use in Engineering." *Colorado Scientific Society Proceedings,* vol. 16, Denver.

Terzaghi, K., and Peck, R. B. (1967). *Soil Mechanics in Engineering Practice,* 2d ed., John Wiley & Sons, New York.

Uniform Building Code (1997). International Conference of Building Officials, three volumes, Whittier, CA.

United States Geological Survey (2000). *Glossary of Some Common Terms in Seismology.* USGS Earthquake Hazards Program, National Earthquake Information Center, World Data Center for Seismology, Denver, CO. Glossary obtained from the Internet.

Yeats, R. S., Sieh, K., and Allen, C. R. (1997). *The Geology of Earthquakes.* Oxford University Press, New York.

EQSEARCH, EQFAULT, AND FRISKSP COMPUTER PROGRAMS

Appendix B presents data from the EQSEARCH, EQFAULT, and FRISKSP (Blake 2000 a, b, c) computer programs. These computer programs can be used to determine the peak ground acceleration (a_{max}) at the designated site. Each computer program is discussed below.

1. *EQSEARCH Computer Program (Figs. B.1 to B.11).* The purpose of this computer program is to perform a historical search of earthquakes. For this computer program, the input data are shown in Figure B.1 and include the job number, job name, site coordinates in terms of latitude and longitude, search parameters, attenuation relationship, and other earthquake parameters. The output data are shown in Figs. B.2 to B.11. As indicated in Figure B.4, the largest earthquake site acceleration from 1800 to 1999 is $a_{max} = 0.189g$.

The EQSEARCH computer program also indicates the number of earthquakes of a certain magnitude that have affected the site. For example, from 1800 to 1999, there were two earthquakes of magnitude 6.5 or larger that impacted the site (see Figure B.5).

2. *EQFAULT Computer Program (Figs. B.12 to B.19).* The EQFAULT computer program (Blake 2000a) was developed to determine the largest maximum earthquake site acceleration. For this computer program, the input data are shown in Fig. B.12 and include the job number, job name, site coordinates in terms of latitude and longitude, search radius, attenuation relationship, and other earthquake parameters. The output data are shown in Figs. B.13 to B.19. As indicated in Fig. B.13, the largest maximum earthquake site acceleration a_{max} is $0.4203g$.

3. *FRISKSP Computer Program (Figs. B.20 to B.25).* Figures B.20 to B.25 present a probabilistic analysis for the determination of the peak ground acceleration at the site using the FRISKSP computer program (Blake 2000c). Two probabilistic analyses were performed using different attenuation relationships. As shown in Figs. B.21 and B.23, the data are plotted in terms of the peak ground acceleration versus probability of exceedance for a specific design life of the structure.

```
**************************
*                        *
*    E Q S E A R C H      *
*                        *
*      Version 3.00       *
*                        *
**************************
```

ESTIMATION OF
PEAK ACCELERATION FROM
CALIFORNIA EARTHQUAKE CATALOGS

JOB NUMBER: 22132

                         DATE: 11-08-2000

JOB NAME: W.C.H. Medical Library

EARTHQUAKE-CATALOG-FILE NAME: ALLQUAKE.DAT

MAGNITUDE RANGE:
   MINIMUM MAGNITUDE: 4.00
   MAXIMUM MAGNITUDE: 9.00

SITE COORDINATES:
   SITE LATITUDE: 32.7812
   SITE LONGITUDE: 117.0111

SEARCH DATES:
        START DATE: 1800
        END DATE: 1999

SEARCH RADIUS:
        50.0 mi
        80.5 km

ATTENUATION RELATION: 10) Bozorgnia Campbell Niazi (1999) Hor.-Holocene Soil-Cor.
   UNCERTAINTY (M=Median, S=Sigma): S   Number of Sigmas: 1.0
   ASSUMED SOURCE TYPE: SS [SS=Strike-slip, DS=Reverse-slip, BT=Blind-thrust]
   SCOND: 0 Depth Source: A
   Basement Depth: 5.00 km   Campbell SSR: 0   Campbell SHR: 0
   COMPUTE PEAK HORIZONTAL ACCELERATION

MINIMUM DEPTH VALUE (km): 3.0

FIGURE B.1   EQSEARCH computer program. (*From Blake 2000b.*)

```
------------------------
EARTHQUAKE SEARCH RESULTS
------------------------
```

FILE CODE	LAT. NORTH	LONG. WEST	DATE	TIME (UTC) H M Sec	DEPTH (km)	QUAKE MAG.	SITE ACC. g	SITE INT.	APPROX. DISTANCE mi [km]
MGI	32.8000	117.1000	05/25/1803	0 0 0.0	0.0	5.00	0.189	VIII	5.3 (8.6)
PAS	32.6790	117.1510	06/18/1985	32228.7	5.7	4.00	0.067	VI	10.8 (17.3)
T-A	32.6700	117.1700	05/24/1865	0 0 0.0	0.0	5.00	0.101	VII	12.0 (19.3)
T-A	32.6700	117.1700	01/25/1863	1020 0.0	0.0	4.30	0.070	VI	12.0 (19.3)
T-A	32.6700	117.1700	12/00/1856	0 0 0.0	0.0	5.00	0.101	VII	12.0 (19.3)
T-A	32.6700	117.1700	10/21/1862	0 0 0.0	0.0	5.00	0.101	VII	12.0 (19.3)
T-A	32.6700	117.1700	04/15/1865	840 0.0	0.0	4.30	0.070	VI	12.0 (19.3)
MGI	32.7000	117.2000	05/20/1920	1330 0.0	0.0	4.00	0.059	VI	12.3 (19.8)
DMG	32.7000	117.2000	05/27/1862	20 0 0.0	0.0	5.90	0.173	VIII	12.3 (19.8)
MGI	32.7000	117.2000	04/19/1906	028 0.0	0.0	4.30	0.068	VI	12.3 (19.8)
MGI	32.7000	117.2000	09/08/1915	742 0.0	0.0	4.00	0.059	VI	12.3 (19.8)
DMG	32.8000	116.8000	10/23/1894	23 3 0.0	0.0	5.70	0.152	VIII	12.3 (19.8)
MGI	32.8000	116.8000	08/14/1927	1448 0.0	0.0	4.60	0.080	VII	12.3 (19.8)
PAS	32.6150	117.1520	10/29/1986	23815.3	14.6	4.10	0.055	VI	14.1 (22.7)
DMG	33.0000	117.0000	03/03/1906	2025 0.0	0.0	4.50	0.062	VI	15.1 (24.3)
MGI	33.0000	117.0000	09/21/1856	730 0.0	0.0	5.00	0.082	VII	15.1 (24.3)
MGI	33.0000	117.0000	12/29/1914	10 0 0.0	0.0	4.00	0.049	VI	15.1 (24.3)
MGI	32.7000	116.7000	03/21/1918	2325 0.0	0.0	4.00	0.039	V	18.9 (30.4)
DMG	33.0000	117.3000	11/22/1800	2130 0.0	0.0	6.50	0.138	VIII	22.6 (36.3)
GSP	33.0700	116.8000	12/04/1991	071057.5	15.0	4.20	0.035	V	23.4 (37.6)
PAS	32.6270	117.3770	06/29/1983	8 836.4	5.0	4.60	0.042	VI	23.8 (38.3)
MGI	33.1000	116.8000	06/22/1918	557 0.0	0.0	4.00	0.029	V	25.2 (40.5)
DMG	32.8500	117.4830	02/23/1943	92112.0	0.0	4.00	0.026	V	27.8 (44.7)
MGI	33.0000	116.6000	06/11/1917	354 0.0	0.0	4.00	0.026	V	28.2 (45.4)
MGI	33.2000	117.0000	07/20/1923	7 0 0.0	0.0	4.00	0.025	V	28.9 (46.5)
DMG	33.1000	116.6330	02/08/1952	174028.0	0.0	4.00	0.023	IV	31.1 (50.0)
MGI	32.6000	116.5000	05/03/1918	425 0.0	0.0	4.00	0.023	IV	32.2 (51.9)
MGI	33.1000	116.6000	03/04/1915	1250 0.0	0.0	4.00	0.022	IV	32.4 (52.2)
MGI	33.1000	116.6000	05/11/1915	1145 0.0	0.0	4.00	0.022	IV	32.4 (52.2)
MGI	33.1000	116.6000	08/19/1917	710 0.0	0.0	4.00	0.022	IV	32.4 (52.2)
MGI	33.1000	116.6000	08/10/1921	19 6 0.0	0.0	4.00	0.022	IV	32.4 (52.2)
MGI	33.1000	116.6000	08/10/1921	2151 0.0	0.0	4.00	0.022	IV	32.4 (52.2)
MGI	33.1000	116.6000	02/05/1922	1915 0.0	0.0	4.00	0.022	IV	32.4 (52.2)
MGI	33.1000	116.6000	05/28/1917	1017 0.0	0.0	4.00	0.022	IV	32.4 (52.2)
MGI	33.1000	116.6000	02/09/1920	220 0.0	0.0	4.00	0.022	IV	32.4 (52.2)
MGI	33.1000	116.6000	02/16/1915	1330 0.0	0.0	4.00	0.022	IV	32.4 (52.2)
DMG	33.2000	116.7200	05/12/1930	172548.5	0.0	4.20	0.024	V	33.5 (53.9)
DMG	33.2670	117.0170	06/07/1935	1633 0.0	0.0	4.00	0.022	IV	33.5 (54.0)
PAS	32.3020	116.8810	08/19/1978	931 5.7	19.8	4.10	0.022	IV	33.9 (54.6)
DMG	33.2000	116.7000	01/01/1920	235 0.0	0.0	5.00	0.036	V	34.1 (54.8)

FIGURE B.2   EQSEARCH computer program. (*From Blake 2000b.*)

```
DMG  |33.1500|116.5830|12/02/1935| 319 0.0|  0.0| 4.00| 0.020 | IV | 35.5( 57.2)
DMG  |33.1100|116.5230|01/24/1957|205449.9|  3.9| 4.60| 0.027 |  V | 36.3( 58.4)
DMG  |33.0020|116.4360|07/02/1957| 65638.5| 12.8| 4.10| 0.021 | IV | 36.7( 59.0)
DMG  |33.0000|116.4330|06/04/1940|1035 8.3|  0.0| 5.10| 0.035 |  V | 36.8( 59.2)
MGI  |33.2000|116.6000|10/12/1920|1748 0.0|  0.0| 5.30| 0.039 |  V | 37.5( 60.3)
PAS  |33.1380|116.5010|10/10/1984|212258.9| 11.6| 4.50| 0.024 |  V | 38.5( 61.9)
DMG  |32.9670|116.3830|10/31/1942|15 758.0|  0.0| 4.00| 0.019 | IV | 38.6( 62.1)
DMG  |32.6800|116.3540|01/21/1970|1124 0.4|  8.0| 4.10| 0.020 | IV | 38.8( 62.4)
DMG  |33.1000|116.4500|11/23/1953|1339 7.0|  0.0| 4.30| 0.021 | IV | 39.3( 63.2)
DMG  |33.0970|116.4440|08/18/1959|215221.3| 17.3| 4.30| 0.021 | IV | 39.4( 63.5)
DMG  |33.1670|116.5000|06/23/1932| 23037.1|  0.0| 4.00| 0.018 | IV | 39.8( 64.1)
DMG  |33.1670|116.5000|06/23/1932| 22552.7|  0.0| 4.00| 0.018 | IV | 39.8( 64.1)
DMG  |33.1670|116.4670|08/01/1960|193930.0|  0.0| 4.20| 0.019 | IV | 41.3( 66.4)
```

```
                      -------------------------
                       EARTHQUAKE SEARCH RESULTS
                      -------------------------
```

FILE CODE	LAT. NORTH	LONG. WEST	DATE	TIME (UTC) H M Sec	DEPTH (km)	QUAKE MAG.	SITE ACC. g	SITE MM INT.	APPROX. DISTANCE mi [km]
DMG	33.1170	116.4170	10/21/1940	64933.0	0.0	4.50	0.022	IV	41.5(66.8)
DMG	33.1170	116.4170	06/04/1940	103656.0	0.0	4.00	0.017	IV	41.5(66.8)
DMG	33.0380	116.3610	02/26/1957	211652.2	0.0	4.10	0.018	IV	41.6(67.0)
DMG	32.7000	116.3000	02/24/1892	720 0.0	0.0	6.70	0.084	VII	41.7(67.1)
GSP	33.1100	116.4000	04/01/1984	071702.3	11.0	4.00	0.017	IV	42.1(67.7)
DMG	32.6000	116.3170	06/15/1946	194653.0	0.0	4.80	0.026	V	42.2(68.0)
DMG	33.1670	116.4170	07/10/1938	18 6 0.0	0.0	4.00	0.016	IV	43.5(70.0)
DMG	33.1670	116.4170	10/14/1935	1550 0.0	0.0	4.00	0.016	IV	43.5(70.0)
DMG	33.1670	116.4170	12/05/1939	173352.0	0.0	4.00	0.016	IV	43.5(70.0)
PAS	32.9470	117.7360	01/15/1989	153955.2	6.0	4.20	0.018	IV	43.6(70.1)
DMG	32.9610	116.2900	08/25/1971	23 033.0	8.0	4.00	0.016	IV	43.6(70.2)
DMG	32.9230	116.2720	10/14/1969	131842.7	10.0	4.50	0.021	IV	44.0(70.8)
DMG	32.9520	116.2790	09/13/1973	173039.8	8.0	4.80	0.025	V	44.1(70.9)
DMG	32.3330	116.4670	01/13/1935	224 0.0	0.0	4.00	0.016	IV	44.3(71.2)
PAS	32.9050	116.2610	12/25/1975	71852.3	3.6	4.00	0.016	IV	44.3(71.4)
DMG	33.1210	116.3490	05/25/1971	10 252.9	8.0	4.10	0.017	IV	45.0(72.4)
DMG	33.0530	116.3060	04/02/1967	201538.6	1.0	4.30	0.018	IV	45.0(72.4)
DMG	32.9900	116.2680	11/08/1958	132044.1	2.4	4.10	0.017	IV	45.4(73.1)
DMG	32.9500	116.2500	11/14/1951	2355 3.0	0.0	4.10	0.016	IV	45.6(73.5)
DMG	33.1830	116.3830	10/14/1949	02925.0	0.0	4.10	0.016	IV	45.7(73.6)
T-A	32.2500	117.5000	01/13/1877	20 0 0.0	0.0	5.00	0.026	V	46.4(74.7)
DMG	33.4540	116.8980	07/29/1936	142252.8	10.0	4.00	0.015	IV	46.9(75.5)
DMG	33.4560	116.8960	06/16/1938	55916.9	10.0	4.00	0.015	IV	47.1(75.7)
MGI	32.8000	116.2000	07/23/1929	1155 0.0	0.0	4.30	0.018	IV	47.1(75.8)
DMG	32.8170	116.2000	11/22/1953	81138.0	0.0	4.10	0.016	IV	47.1(75.9)
DMG	33.0430	116.2600	08/22/1961	231933.6	12.1	4.40	0.019	IV	47.1(75.9)
PAS	32.9450	117.8060	09/07/1984	11 313.4	6.0	4.30	0.017	IV	47.5(76.4)
PAS	33.4200	116.6980	06/05/1978	16 3 3.9	11.9	4.40	0.018	IV	47.7(76.7)
DMG	32.8000	117.8330	01/24/1942	214148.0	0.0	4.00	0.015	IV	47.7(76.8)
PAS	32.9700	117.8030	07/14/1986	03246.2	10.0	4.00	0.015	IV	47.7(76.8)
DMG	32.5830	117.8000	04/19/1939	741 0.0	0.0	4.50	0.019	IV	47.8(77.0)
DMG	32.7170	117.8330	11/06/1950	205546.0	0.0	4.40	0.018	IV	47.9(77.1)
DMG	32.1130	116.7850	04/23/1968	131825.4	10.0	4.20	0.016	IV	48.0(77.2)
GSP	32.9700	117.8100	04/04/1990	085439.3	6.0	4.00	0.015	IV	48.1(77.4)

FIGURE B.3   EQSEARCH computer program. (*From Blake 2000b.*)

```
DMG   32.0830 117.0000 05/10/1948  34925.0    0.0  4.00   0.015    IV    48.2( 77.6)
DMG   32.2000 116.5500 11/04/1949 204238.0    0.0  5.70   0.038     V    48.3( 77.7)
DMG   32.2000 116.5500 11/05/1949 20 2 7.0    0.0  4.00   0.015    IV    48.3( 77.7)
DMG   32.2000 116.5500 11/11/1949 1354 0.0    0.0  4.20   0.016    IV    48.3( 77.7)
DMG   32.2000 116.5500 11/06/1949 23 510.0    0.0  4.00   0.015    IV    48.3( 77.7)
DMG   32.2000 116.5500 11/05/1949  43524.0    0.0  5.10   0.026     V    48.3( 77.7)
DMG   33.0330 116.2330 09/20/1961  5 410.0    0.0  4.00   0.015    IV    48.3( 77.8)
DMG   33.0190 116.2250 08/20/1969 152957.2    0.6  4.00   0.015    IV    48.4( 77.9)
DMG   33.0500 116.2380 08/23/1961  1 047.8   11.9  4.70   0.021    IV    48.5( 78.0)
DMG   33.0210 116.2230 01/13/1963  23938.9   13.0  4.20   0.016    IV    48.6( 78.2)
GSP   32.8220 116.1750 05/24/1992 122225.8   12.0  4.10   0.015    IV    48.6( 78.2)
DMG   33.2670 116.4000 06/06/1940 2321 4.0    0.0  4.00   0.015    IV    48.7( 78.4)
GSP   32.9850 117.8180 06/21/1995 211736.2    6.0  4.30   0.017    IV    48.9( 78.6)
PAS   32.9450 117.8310 07/29/1986  81741.8   10.0  4.10   0.015    IV    48.9( 78.6)
PAS   32.9330 117.8410 07/29/1986  81741.6   10.0  4.30   0.017    IV    49.3( 79.3)
USG   33.0170 117.8170 07/16/1986 1247 3.7   10.0  4.11   0.015    IV    49.5( 79.6)
USG   33.0170 117.8170 07/14/1986  11112.6   10.0  4.12   0.015    IV    49.5( 79.6)
DMG   33.5000 117.0000 08/08/1925 1013 0.0    0.0  4.50   0.018    IV    49.6( 79.9)
T-A   33.5000 117.0700 12/29/1880  7 0 0.0    0.0  4.30   0.017    IV    49.7( 80.0)
```

```
                     -----------------------
                     EARTHQUAKE SEARCH RESULTS
                     -----------------------
```

```
-----------------------------------------------------------------------------
                                     TIME            SITE SITE   APPROX.
FILE   LAT.     LONG.      DATE      (UTC)  DEPTH QUAKE ACC.  MM    DISTANCE
CODE   NORTH    WEST                 H M Sec (km)  MAG.   g   INT.  mi  [km]
----+-------+--------+----------+-------+-----+-----+-------+----+-----------
DMG   33.4000 116.5670 02/04/1953  43616.0   0.0  4.30   0.017    IV    49.8( 80.2)
DMG   33.4500 116.6830 04/25/1955  25515.0   0.0  4.00   0.014    IV    49.9( 80.3)
DMG   33.5000 116.9170 11/04/1935  355 0.0   0.0  4.50   0.018    IV    49.9( 80.3)
```

****
-END OF SEARCH-   109 EARTHQUAKES FOUND WITHIN THE SPECIFIED SEARCH AREA.

TIME PERIOD OF SEARCH: 1800   TO   1999

LENGTH OF SEARCH TIME: 200 years

THE EARTHQUAKE CLOSEST TO THE SITE IS ABOUT 5.3 MILES (8.6 km) AWAY.

LARGEST EARTHQUAKE MAGNITUDE FOUND IN THE SEARCH RADIUS: 6.7

LARGEST EARTHQUAKE SITE ACCELERATION FROM THIS SEARCH: 0.189 g

COEFFICIENTS FOR GUTENBERG & RICHTER RECURRENCE RELATION:
   a-value= 3.052
   b-value= 0.844
   beta-value= 1.944

```
--------------------------------------
TABLE OF MAGNITUDES AND EXCEEDANCES:
--------------------------------------
```

   Earthquake | Number of Times | Cumulative

FIGURE B.4   EQSEARCH Computer Program. (*From Blake 2000b.*)

Magnitude	Exceeded	No. / Year
4.0	109	0.54500
4.5	28	0.14000
5.0	15	0.07500
5.5	5	0.02500
6.0	2	0.01000
6.5	2	0.01000

FIGURE B.5   EQSEARCH computer program. (*From Blake 2000b.*)

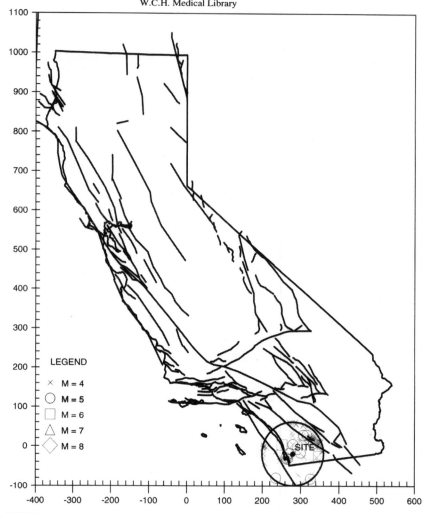

EARTHQUAKE EPICENTER MAP
W.C.H. Medical Library

LEGEND

× M = 4
○ M = 5
□ M = 6
△ M = 7
◇ M = 8

FIGURE B.6   EQSEARCH computer program. (*From Blake 2000b.*)

STRIKE-SLIP FAULTS
10) Bozorgnia Campbell Niazi (1999) Hor.-Holocene Soil-Cor.

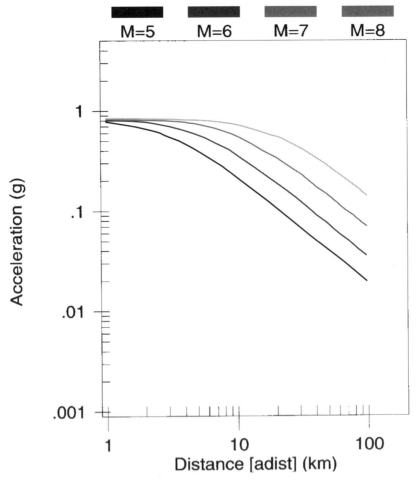

FIGURE B.7 EQSEARCH Computer Program. (*From Blake 2000b.*)

DIP-SLIP FAULTS
10) Bozorgnia Campbell Niazi (1999) Hor.-Holocene Soil-Cor.

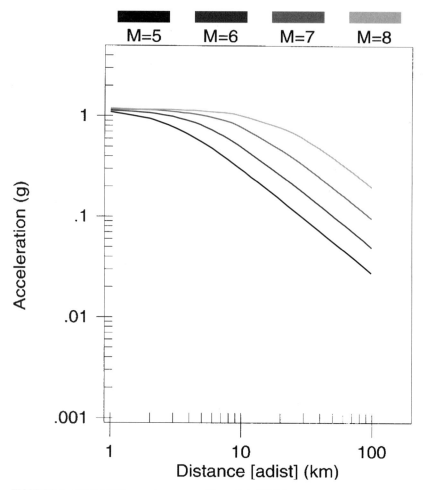

FIGURE B.8   EQSEARCH computer program. (*From Blake 2000b.*)

BLIND-THRUST FAULTS

10) Bozorgnia Campbell Niazi (1999) Hor.-Holocene Soil-Coi

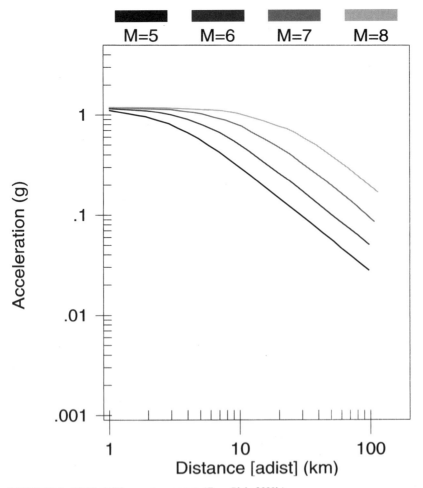

FIGURE B.9   EQSEARCH computer program. (*From Blake 2000b.*)

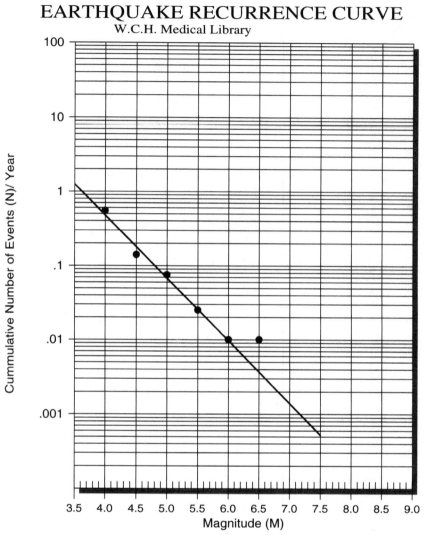

FIGURE B.10 EQSEARCH computer program. (*From Blake 2000b.*)

Number of Earthquakes (N) Above Magnitude (M)
W.C.H. Medical Library

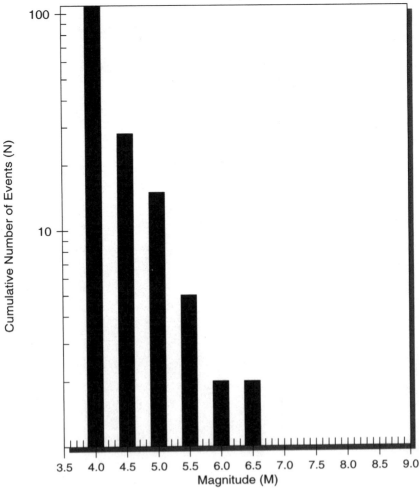

FIGURE B.11   EQSEARCH computer program. (*From Blake 2000b.*)

```
************************
*                      *
*    E Q F A U L T     *
*                      *
*    Version 3.00      *
*                      *
************************
```

DETERMINISTIC ESTIMATION OF
PEAK ACCELERATION FROM DIGITIZED FAULTS

JOB NUMBER: 22132

DATE: 11-08-2000

JOB NAME: W.C.H. Medical Library

CALCULATION NAME: W.C.H. Medical Library

FAULT-DATA-FILE NAME: CDMGFLTE.DAT

SITE COORDINATES:
    SITE LATITUDE: 32.7812
    SITE LONGITUDE: 117.0111

SEARCH RADIUS: 50 mi

ATTENUATION RELATION: 10) Bozorgnia Campbell Niazi (1999) Hor.-Holocene Soil-Cor.
    UNCERTAINTY (M=Median, S=Sigma): S   Number of Sigmas: 1.0
    DISTANCE MEASURE: cdist
    SCOND: 0
    Basement Depth: 5.00 km   Campbell SSR: 0   Campbell SHR: 0
    COMPUTE PEAK HORIZONTAL ACCELERATION

FAULT-DATA FILE USED: CDMGFLTE.DAT

MINIMUM DEPTH VALUE (km): 3.0

FIGURE B.12   EQFAULT computer program. (*From Blake 2000a.*)

```
                         ---------------
                         EQFAULT SUMMARY
                         ---------------



                 ------------------------------
                 DETERMINISTIC SITE PARAMETERS
                 ------------------------------

Page  1
--------------------------------------------------------------------------
                                    |ESTIMATED MAX. EARTHQUAKE EVENT
                             APPROXIMATE  ------------------------------
       ABBREVIATED          DISTANCE  | MAXIMUM  |  PEAK    |EST. SITE
       FAULT  NAME          mi   (km) |EARTHQUAKE|  SITE    |INTENSITY
                                    | MAG.(Mw) |ACCEL. g  |MOD.MERC.
=================================== |==============|==========|==========|========
ROSE CANYON                    8.9(  14.3)|   6.9    |  0.420   |    X
CORONADO BANK                 22.8(  36.7)|   7.4    |  0.248   |    IX
ELSINORE-JULIAN               32.9(  52.9)|   7.1    |  0.141   |   VIII
NEWPORT-INGLEWOOD (Offshore)  35.5(  57.2)|   6.9    |  0.113   |   VII
EARTHQUAKE VALLEY             37.3(  60.0)|   6.5    |  0.082   |   VII
ELSINORE-COYOTE MOUNTAIN      39.8(  64.1)|   6.8    |  0.094   |   VII
ELSINORE-TEMECULA             41.2(  66.3)|   6.8    |  0.091   |   VII
***************************************************************************
-END OF SEARCH-    7 FAULTS FOUND WITHIN THE SPECIFIED SEARCH RADIUS.

THE ROSE CANYON                     FAULT IS CLOSEST TO THE SITE.
IT IS ABOUT 8.9 MILES (14.3 km) AWAY.

LARGEST MAXIMUM-EARTHQUAKE SITE ACCELERATION: 0.4203 g
```

FIGURE B.13   EQFAULT computer program. (*From Blake 2000a.*)

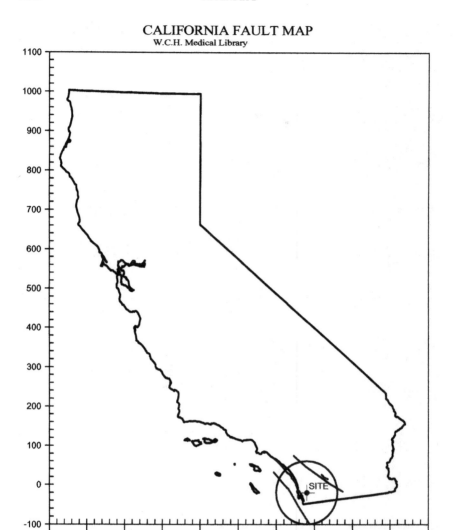

FIGURE B.14 EQFAULT computer program. (*From Blake 2000a.*)

STRIKE-SLIP FAULTS
10) Bozorgnia Campbell Niazi (1999) Hor.-Holocene Soil-Cor.

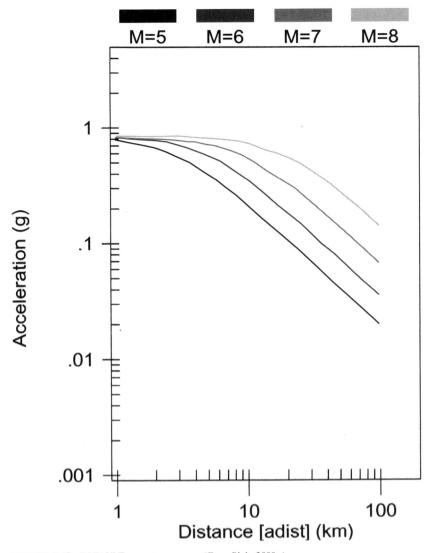

FIGURE B.15 EQFAULT computer program. (*From Blake 2000a.*)

DIP-SLIP FAULTS
10) Bozorgnia Campbell Niazi (1999) Hor.-Holocene Soil-Cor.

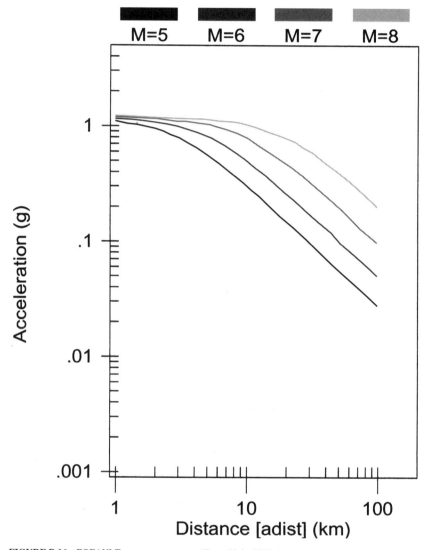

FIGURE B.16   EQFAULT computer program. (*From Blake 2000a.*)

BLIND-THRUST FAULTS
10) Bozorgnia Campbell Niazi (1999) Hor.-Holocene Soil-Cor.

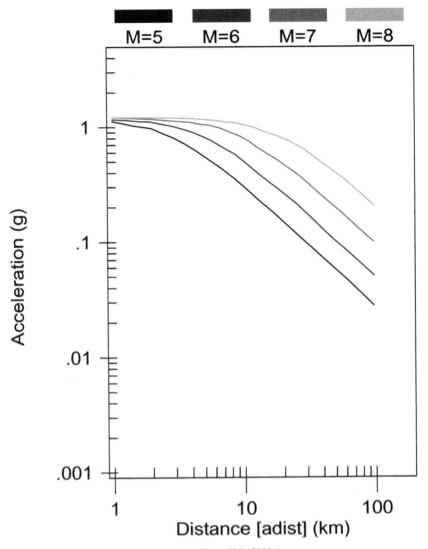

FIGURE B.17   EQFAULT computer program. (*From Blake 2000a.*)

MAXIMUM EARTHQUAKES
W.C.H. Medical Library

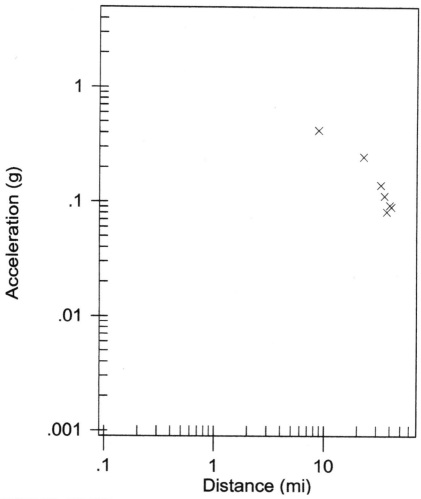

FIGURE B.18 EQFAULT computer program. (*From Blake 2000a.*)

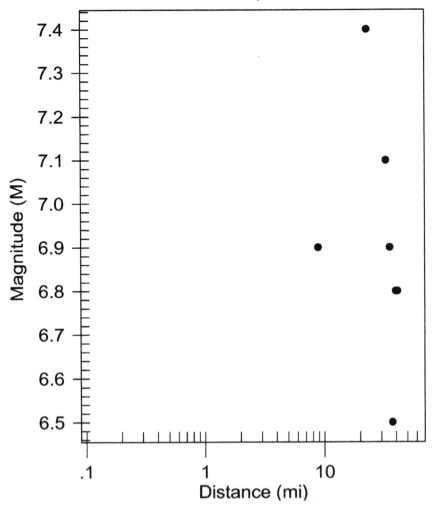

EARTHQUAKE MAGNITUDES & DISTANCES
W.C.H. Medical Library

FIGURE B.19   EQFAULT computer program. (*From Blake 2000b.*)

```
***********************************************
*                                             *
*         FRISKSP - IBM-PC VERSION            *
*                                             *
*   Modified from *FRISK* (McGuire 1978)      *
*     To Perform Probabilistic Earthquake     *
*     Hazard Analyses Using Multiple Forms    *
*   of Ground-Motion-Attenuation Relations    *
*                                             *
*     Modifications by:  Thomas F. Blake      *
*              - 1988-2000 -                   *
*                                             *
*              VERSION 4.00                    *
*             (Visual Fortran)                 *
***********************************************
```

FIGURE B.20   FRISKSP computer program (*From Blake 2000c.*)

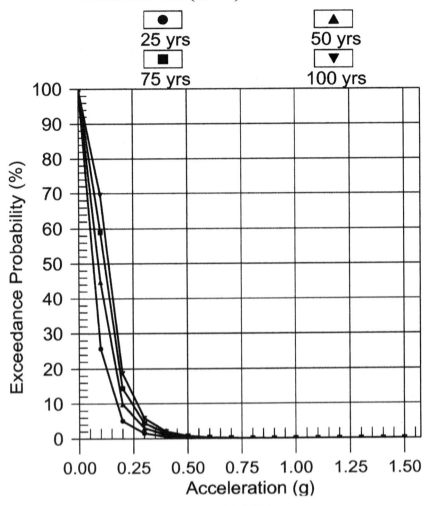

FIGURE B.21   FRISKSP computer program. (*From Blake 2000c.*)

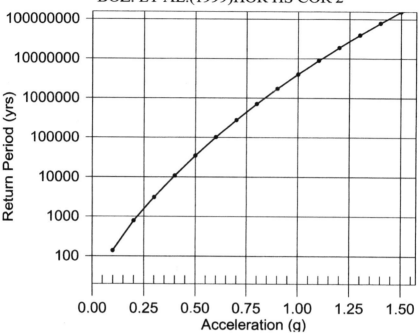

RETURN PERIOD vs. ACCELERATION
BOZ. ET AL.(1999)HOR HS COR 2

FIGURE B.22   FRISKSP computer program. (*From Blake 2000c.*)

PROBABILITY OF EXCEEDANCE
BOZ. ET AL.(1999)HOR HS COR 2

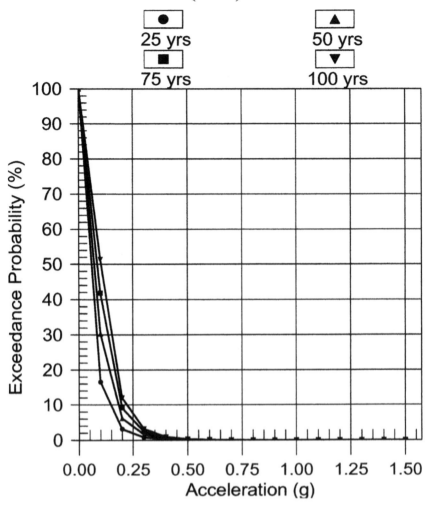

FIGURE B.23 FRISKSP computer program. (*From Blake 2000c.*)

RETURN PERIOD vs. ACCELERATION
BOZ. ET AL.(1999)HOR HS COR 1

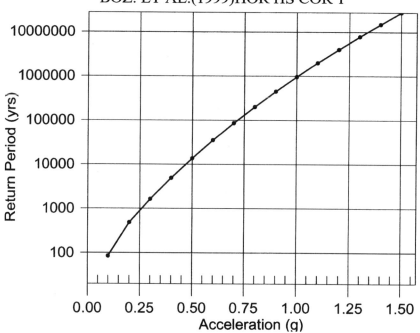

FIGURE B.24   FRISKSP Computer Program. (*From Blake 2000c.*)

CALIFORNIA FAULT MAP
W.C.H. Medical Library

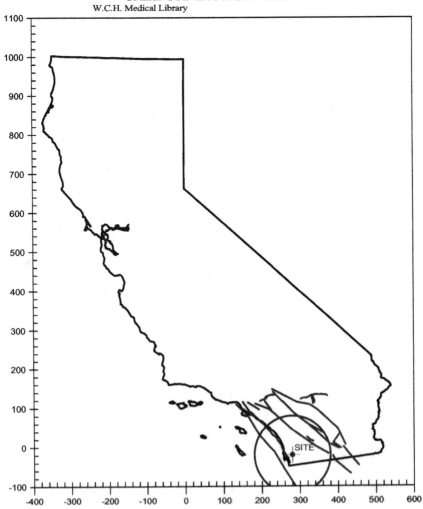

FIGURE B.25   FRISKSP computer program. (*From Blake 2000c.*)

APPENDIX C
CONVERSION FACTORS

From	Multiply by*	Converts to†
Area, acres	4046.9	square meters
Area, square yards	0.8361	square meters
Area, square feet	0.0929	square meters
Area, square inches	0.0006451	square meters
Bending moment, lb force-foot	1.3558	newton-meter
Density, pounds/cubic yard	0.5932	kilograms/cubic meter
Density, pounds/cubic foot	16.0185	kilograms/cubic meter
Force, kips	4.4482	kilonewtons
Force, pounds	4.4482	newtons
Length, miles	1609.344	meter
Length, yards	0.9144	meter
Length, feet	0.3048	meter
Length, inches	0.0254	meter
Force/length, pounds/foot	14.5939	newtons/meter
Force/length, pounds/inch	175.127	newtons/meter
Mass, tons	907.184	kilogram
Mass, pounds	0.4536	kilogram
Mass, ounces	28.35	gram
Pressure or stress, pounds/square foot	47.8803	pascal
Pressure or stress, pounds/square inch	6.8947	kilopascal
Temperature, °F	$(t_F - 32)/1.8 = t_C$	°C
Volume, cubic yards	0.7646	cubic meters
Volume, cubic feet	0.02831	cubic meters
Volume, cubic inches	1.6387×10^{-5}	cubic meters

*The precision of a measurement converted to other units can never be greater than that of the original. To go from SI units to U.S. customary system units, divide by the given constant. ASTM E 380 provides guidance on use of SI.

†The common SI prefixes are

mega	M	1,000,000
kilo	k	1,000
centi	c	0.01
milli	m	0.001
micro	μ	0.000001

APPENDIX D
EXAMPLE OF A GEOTECHNICAL REPORT DEALING WITH EARTHQUAKE ENGINEERING

INTRODUCTION

The following report has been prepared to present the findings and recommendations resulting from our geotechnical earthquake investigation of the subject site. The purpose of the investigation was to assess the feasibility of the project and to provide geotechnical earthquake engineering parameters and recommendations for the design of the foundation for the medical library.

Scope of Services

The scope of services for the project included the following:

- Screening investigation consisting of a review of published and unpublished geologic, seismicity, and soil engineering maps and reports pertinent to the earthquake engineering aspects of the project

- Quantitative analysis, including subsurface exploration consisting of eight auger borings excavated to a maximum depth of 40 ft

- Logging and sampling of three exploratory trenches to evaluate the soil conditions and to obtain samples for laboratory testing

- Laboratory testing of soil samples obtained during the field investigation

- Geologic and soil engineering evaluations of field and laboratory data which provide the basis for the geotechnical earthquake engineering conclusions and recommendations

- Preparation of this report and other graphics presenting the findings, conclusions, and recommendations

Site Description

The subject site is located at the southeast corner of the intersection of Wake Street and Highway 1. The site and surrounding areas are level (i.e., level-ground site). It is our understanding that the proposed development is to consist of a medical library. It is also our understanding that there will be three buildings, containing a library, administration offices, and a conference center. In addition, there will be a plaza to be constructed adjacent to the building. The proposed site development will also include the construction of an entrance roadway and parking facilities. The final size and the location of the building, plaza, and parking facilities are in the planning stages. No details on the foundation loads were available at the time of preparation of this report.

SITE GEOLOGY AND SEISMICITY

Geology

The site is located in the peninsular ranges geomorphic province of California near the western limits of the southern California batholith. The topography at the edge of the batholith changes from the typically rugged landforms developed over the granitic rocks to the more subdued landforms characteristic of the sedimentary bedrock of the local embayment.

The subject site contains a 10-ft-thick upper layer of cohesive soil that is underlain by a 25-ft-thick layer of soft clay and submerged loose sand. It is our understanding that the site was used as a reservoir in the past and that the 25-ft-thick layer is an old lake deposit. The top of the groundwater table is approximately at the top of the lake deposit layer.

Underlying the lake deposit (at 35-ft depth), there is Eocene aged stadium conglomerate bedrock. The stadium conglomerate bedrock is a very dense sedimentary rock composed of hard and rounded cobble-size particles embedded within an orange to yellow, cemented sandstone matrix. The cobbles typically comprise approximately 30 to 50 percent of the stadium conglomerate bedrock. The nature of the material (i.e., cemented stadium conglomerate) makes it an adequate bearing material.

Seismicity

The site can be considered a seismically active area, as can all of southern California. There are, however, no active faults on or adjacent to the site. Seismic risk is considered moderate compared to other areas of southern California. Seismic hazards within the site can be attributed to ground shaking resulting from events on distant active faults. There are several active and potentially active faults which can significantly affect the site. The EQSEARCH Version 2.01 (Thomas Blake) computer program was used to estimate the peak ground acceleration at the site due to earthquake shaking on known active faults. Based on an analysis of possible earthquake accelerations at the site, the most significant event is a 6.5 magnitude event on the La Nacion fault, which lies approximately 4 mi to the southwest of the site. The ground surface accelerations produced at the site by such an event would exceed those events on any other known fault. The Rose Canyon fault zone is the closest active fault, which lies approximately 10 mi to the west of the site.

Based on an analysis of the earthquake data, the peak ground acceleration used for the geotechnical engineering analyses is $0.20g$. As discussed in the next section, it is anticipated that the sandy soil layers located from a depth of 10 to 35 ft will liquefy during the design earthquake.

GEOTECHNICAL EARTHQUAKE ENGINEERING

The results of the subsurface exploration and laboratory testing indicate that there are three predominant materials at the site:

1. *Fill:*   The upper 10 ft of the site contains fill. The fill is classified as a clay of low plasticity (CL) and is considered to be a competent bearing material. It will provide an adequate thickness of surface material to prevent surface fissuring and sand boils (due to liquefaction of the underlying lake deposits).

2. *Lake deposits:*   Underneath the fill, there are layers of soft clay and loose sand. The groundwater table corresponds to the top of this soil layer. It is our understanding that the site was originally used as a reservoir and that the soil can be considered to be an old lake deposit. This material is not suitable for supporting structural loads, and as such all foundation elements (i.e., piles) would have to penetrate this material and be embedded in the underlying stadium conglomerate bedrock. In addition, the soil has a high concentration of sulfate (0.525 percent), indicating a "severe" classification of sulfate exposure. All piles that penetrate this material will need to be sulfate-resistant (i.e., per the *Uniform Building Code,* type V cement and a maximum water cement ratio = 0.45 are required).

3. *Stadium conglomerate:*   At a depth of about 35 ft, the stadium conglomerate was encountered. As previously mentioned, the stadium conglomerate bedrock is a very dense sedimentary rock composed of hard and rounded cobble-size particles embedded within an orange to yellow, cemented sandstone matrix. The upper few feet of the bedrock are typically weathered.

CONCLUSIONS

The proposed construction is feasible from a geotechnical aspect. However, during design and construction, there is one constraint on the proposed construction that must be considered. This is the presence of lake deposits which are soft and potentially liquefiable during the design earthquake. The site is a level-ground site, and thus lateral spreading or flow failures will not occur.

Because of the lake deposits, the buildings should be supported on driven prestressed concrete piles embedded into the stadium conglomerate bedrock. The piles could support grade beams which in turn support a structural floor slab that is capable of transferring dead and live loads to the piles. Given the presence of groundwater, removal and recompaction of these lake deposits are not considered an economical option.

Foundation Design Parameters

For driven piles founded in intact stadium conglomerate bedrock, the allowable pile load is 100 kips, assuming 12-in-square prestressed concrete piles. Because the weight of the concrete is approximately equal to the weight of the displaced soil, the above allowable pile capacity should be considered to be the *net allowable load* (i.e., neglect the weight of that portion of the pile below the ground surface). Note that a portion of the net allowable load should be used to resist down-drag loads of 20 kips per pile due to possible liquefaction during the design earthquake.

In designing to resist lateral loads, an allowable passive resistance of 150 lb/ft^2 per foot of depth to a maximum value of 1500 lb/ft^2 and a coefficient of friction equal to 0.25 may be utilized for embedment within the upper clayey fill layer. The passive resistance should be neglected for the lake deposits. The above allowable passive resistance values in the fill may be increased by one-third for transient loads such as wind and seismic activity (earthquake loads).

The structural engineer or architect should determine the steel reinforcement required for the foundation based on structural loadings, shrinkage, temperature stresses, and dynamic loads from the design earthquake.

Foundation Settlement

As previously mentioned, it is recommended that the buildings be supported by foundations embedded in the stadium conglomerate bedrock. By using the deep foundation system, the maximum differential settlement is expected to be less than 0.25 in. Settlement should consist of deformation of the stadium conglomerate bedrock and should occur during construction. There may be differential settlement between the buildings which are supported by the underlying bedrock and the adjacent common areas. Thus flexible utility connections should be utilized at the location where they enter the buildings.

Floor Slabs

A structural floor slab will be required that transfers dead and live loads to the piles. This means that the slab will be self-supporting and will be able to transfer all loads to the grade beams and piles.

A moisture barrier should also be placed below the floor slabs. The first step in the construction of the moisture barrier should be the placement of sand to act as a leveling surface. Then the sand should be overlain by a 10-mil visqueen moisture barrier. The visqueen moisture barrier should be properly lapped and sealed at all splices. It should also be sealed at all plumbing or other penetrations. In addition, the visqueen should be draped into the footing excavations and should extend to near the bottom of both interior and exterior footings. The visqueen in slab areas should then be covered with a 4-in-thick layer of $^1/_2$- to $^3/_4$-in open-graded gravel (preferably rounded gravel). Care should be taken so that the open-graded gravel does not puncture the visqueen. The reinforced concrete floor slabs can then be constructed on top of the open-graded gravel.

Plaza Area and Pavement Areas

To mitigate the potential damage to appurtenant structures due to possible settlement from the lake deposits, it is recommended that a flexible joint be provided between the building and all appurtenant structures that abut the building. In addition, the appurtenant structures should be as flexible as possible. For example, the paving areas should be constructed of asphalt concrete rather than Portland cement concrete. Asphalt concrete is more flexible, and any cracks that may develop because of underlying lake deposit settlement can be patched. It is important that positive drainage be provided for the pavement areas so that if settlement depressions do develop, water will not pond as easily on the pavement surface.

The actual pavement recommendations should be developed when the subgrade is exposed (the pavement thickness will depend on the underlying bearing material). For concrete in the plaza area, it is best to provide the concrete with numerous joints to provide locations for crack control. In addition, the concrete can be reinforced with steel reinforcement. To reduce the possibility of differential movement of concrete sections, steel dowels can be placed within the concrete and across the joints. Any structure in the plaza area that is settlement-sensitive (such as a water fountain or statue) can be placed on piles embedded in the stadium conglomerate bedrock.

Site Drainage

Proper surface drainage is required to help reduce water migration adjacent to the foundations. As a minimum, the following standard drainage guidelines should be considered during final plan preparation and/or construction:

1. Roof drains should be installed on the building and tied via a *tight line* to a drain system that empties to the street, a storm drain, or terrace drain.

2. Surface water should flow away from structures and be directed to suitable (maintained) disposal systems such as yard drains, drainage swales, and street gutters. Five percent drainage directed away from the building is recommended, and 2 percent minimum is recommended over soil areas. Planter areas adjacent to the foundation should be minimized. Preferably within 5 ft of the building, the planters should be self-contained with appropriate drainage outlets (i.e., drainage outlets tied via a tight line to a yard drain system).

3. No drains should be allowed to empty adjacent to the building.

4. PVC Schedule 40, ABS, or equivalent is preferred for yard drains. A corrugated plastic yard drain should not be used.

Seismic Design Parameters

A risk common to all southern California areas which should not be overlooked is the potential for damage resulting from seismic events (earthquakes). Even if the structural engineer or architect designs in accordance with applicable codes for seismic design, the possibility of damage occurring cannot be ruled out if moderate shaking occurs as a result of a large earthquake. This is the case for essentially all buildings in southern California. The building should be designed in accordance with the latest *Uniform Building Code* (1997) criteria for seismic design. The site area should be categorized as seismic zone 4.

The following parameters may be used by the structural engineer or architect for seismic design. These parameters are based upon the 1997 *Uniform Building Code* (chapter 16) and recent earthquake and fault studies for the general area of the site. In determining these values, it was assumed that the nearest active fault zone is Rose Canyon, which is assumed to be located at a distance greater than 15 km from the site. In addition, a slip rate of 2 mm/yr was assumed. The following parameters are considered minimums for design of the building, and they were developed assuming that all the building foundation elements are supported on piles embedded in stadium conglomerate bedrock:

- Soil profile type S_A
- Seismic zone factor 0.40
- Seismic source type B
- Near-source factor of 1.0

General Recommendations

This office should be contacted for review of plans for improvements and should be involved during construction to monitor the geotechnical aspects of the development (i.e., pile installation, foundation excavations, etc.). It is recommended that a pile load test be performed in order to verify the 100-kip allowable design value for the piles.

During construction, it is recommended that this office verify site geotechnical conditions and conformance with the intentions of the recommendations for construction. Although not all possible geotechnical observation and testing services are required by the governing agencies, the more site reviews performed, the lower the risk of future problems.

The contractor is the party responsible for providing a safe site. We will not direct the contractor's operations and cannot be responsible for the safety of personnel other than our own representatives on site. The contractor should notify the owner if he or she is aware of, and/or anticipates, unsafe conditions. At the time of construction, if the geotechnical consultant considers conditions unsafe, the contractor, as well as the owner's representative, will be notified. Within this report the term *safe* or *safety* has been used to imply low risk. Some risk will remain, however, as is always the case.

CLOSURE

The geotechnical investigation was performed using the degree of care and skill ordinarily exercised, under similar circumstances, by geotechnical engineers and geologists practicing in this or similar localities. No warranty, expressed or implied, is made as to the conclusions and professional advice included in this report.

The samples taken and used for testing and the observations are believed to be representative of the entire area. However, soil and geologic conditions can vary significantly between borings and surface outcrops. As in many developments, conditions revealed by excavations may be at variance with preliminary findings. If this occurs, the changed conditions must be evaluated by the geotechnical engineer and designs adjusted or alternate designs recommended.

APPENDIX E
SOLUTION TO PROBLEMS

CHAPTER 2

2.1 Since the fault has moved solely in the dip direction, it is termed a *dip-slip fault*. Because the hangingwall block has moved upward with respect to the footwall block, the fault is either a reverse fault or a thrust fault. Based on the orientation of the fault plane (5 NW, 34 W), the dip of the fault plane is 34°. Thrust faults are defined as having a dip that is less than or equal to 45°, and therefore the fault is termed a *dip-slip thrust fault*.

2.2 In Fig. 2.21, the hangingwall block has moved downward with respect to the footwall block. The fault is thus termed a *normal fault*.

2.3 From Fig. 2.14, the maximum displacement recorded by the seismograph is 14.9 cm. Assuming this displacement represents the trace data from a standard Wood-Anderson seismograph and that the instrument is exactly 100 km from the epicenter, then $A = 14.9$ cm $= 149$ mm. Using Eq. (2.1) with $A_0 = 0.001$ mm gives

$$M_L = \log \frac{A}{A_0} = \log \frac{149}{0.001} = 5.2$$

2.4

$$\text{Circumference of the earth} = 4.0 \times 10^7 \text{ m } (360°)$$

$$\text{Distance to seismograph} = 1200 \text{ km} = 1.2 \times 10^6 \text{ m}$$

$$\Delta = \frac{1.2 \times 10^6}{4.0 \times 10^7} (360°) = 10.8°$$

Using Eq. (2.2) and $A' = 15.6$ mm $= 15,600$ μm gives

$$M_s = \log A' + 1.66 \log \Delta + 2.0$$

$$= \log 15,600 + 1.66 \log 10.8 + 2.0 = 7.9$$

2.5 Use Eq. (2.3)

$$M_0 = \mu A_f D$$

where M_0 = seismic moment, N · m

μ = shear modulus of material along fault plane = 3×10^{10} N/m²

A_f = area of fault plane undergoing slip = 15 (600) = 9000 km² = 9×10^9 m²

D = average displacement of ruptured segment of fault = 2.5 m

Therefore,

$$M_0 = \mu A_f D = (3 \times 10^{10} \text{ N/m}^2) (9 \times 10^9 \text{ m}^2) (2.5 \text{ m}) = 6.75 \times 10^{20} \text{ N} \cdot \text{m}$$

Using Eq. (2.4) gives

$$M_w = -6.0 + 0.67 \log M_0 = -6.0 + 0.67 \log (6.75 \times 10^{20}) = 8.0$$

2.6 Half Moon Bay is located on the California coast, about 40 km due south of San Francisco. Based on Fig. 2.20, it should expect a modified Mercalli level of damage of IX, which corresponds to heavy damage. Per Table 2.3, at a level of IX, well-designed frame structures are thrown out of plumb.

CHAPTER 5

5.1 N value = 8 + 9 = 17 per Table 5.2, the sand has a medium density.

5.2

$$\sigma'_{v0} = \sigma_v - u = \gamma_t (1.5) + \gamma_b (1.5) = 18.9 (1.5) + 9.84 (1.5) = 43 \text{ kPa}$$

$$N_{60} = 1.67 \, E_m C_b C_r N \qquad \text{Eq. (5.1)}$$

where $E_m = 0.45$ (doughnut hammer)
$C_b = 1.0$ (100-mm-diameter hole)
$C_r = 0.75$ (3-m length of drill rods)
$N = 4 + 5 = 9$ (N value)

Substituting these values into the equation gives

$$N_{60} = 1.67 \, (0.45) \, (1.0) \, (0.75) \, (9) = 5.1$$

5.3 From Prob. 5.2, the N value corrected for field testing procedures is $N_{60} = 5.1$. Using Eq. (5.2) with $\sigma'_{v0} = 43$ kPa and $N_{60} = 5.1$, then $(N_1)_{60} = 7.8$. Per Table 5.3, for $(N_1)_{60} = 7.8$, the sand has a medium density

5.4 Use $q_c = 40$ kg/cm² and $\sigma'_{v0} = 43$ kPa (Prob. 5.2). Therefore,

$$q_{c1} = \frac{1.8 q_c}{0.8 + \sigma'_{v0}/100} = \frac{1.8 \, (40)}{0.8 + 43/100} = 59 \text{ kg/cm}^2$$

5.5 In Fig. 5.12, using $N_{60} = 5$ and $\sigma'_{v0} = 43$ kPa, we get $\phi = 30°$.

5.6 In Fig. 5.14, using $q_c = 40$ kg/cm² and $\sigma'_{v0} = 43$ kPa, we get $\phi = 40°$.

CHAPTER 6

6.1 *Soil type no. 1:* This soil is described as a crushed limestone from Tennessee. The soil has a liquid limit of 18 and a plasticity index of 7. Because of the plastic nature of the soil, it would typically be termed a cohesive soil. Per Sec. 6.3, in order for a cohesive soil to liquefy, it must meet all the following three criteria:

- Percent finer at 0.005 mm < 15 percent: Since the percent passing the No. 200 sieve is 11 percent, this criterion is met.
- Liquid limit < 35: Since the liquid limit is equal to 18, this criterion is also met.
- Water content > 0.9 times the liquid limit: To be susceptible to liquefaction, the soil must have an in situ water content that is greater than 0.9 times the liquid limit, or the water content must be greater than 16 percent. No data on the water content of the in situ soil are provided.

Since the soil meets the first and second criteria listed above, the soil is susceptible to liquefaction provided that the in situ water content is greater than 16 percent.

Soil type no. 2: This soil is described as silty gravel derived from weathered gabbro from Oman. The soil has a liquid limit of 23 and a plasticity index of 3. Because of the plastic nature of the soil, it would typically be termed a cohesive soil. Per Sec. 6.3, in order for a cohesive soil to liquefy, it must meet all the following three criteria:

- Percent finer at 0.005 mm < 15 percent: Since the percent passing the No. 200 sieve is 10 percent, this criterion is met.
- Liquid limit < 35: Since the liquid limit is equal to 23, this criterion is also met.
- Water content > 0.9 times the liquid limit: To be susceptible to liquefaction, the soil must have an in situ water content that is greater than 0.9 times the liquid limit, or the water content must be greater than 21 percent. No data on the water content of the in situ soil are provided.

Since the soil meets the first and second criteria listed above, the soil is susceptible to liquefaction provided that the in situ water content is greater than 21 percent.

Soil type no. 3: This soil is described as alluvial gravelly sand from Mississippi. The soil is nonplastic. Because of the nonplastic nature of the soil, it would typically be termed a cohesionless soil. Such a soil is susceptible to liquefaction.

Soil type no. 4: This soil is described as eolian sand from Oman. The soil is nonplastic. Because of the nonplastic nature of the soil, it would typically be termed a cohesionless soil. Such a soil is susceptible to liquefaction.

Soil type no. 5: This soil is described as glacial till from Illinois. The soil has a liquid limit of 25 and a plasticity index of 10. Because of the plastic nature of the soil, it would typically be termed a cohesive soil. Per Sec. 6.3, in order for a cohesive soil to liquefy, it must meet all the following three criteria:

- Percent finer at 0.005 mm <15 percent: Since the percent finer at a grain size of 0.005 mm is equal to 26 percent, this criterion is not met.
- Liquid limit < 35: Since the liquid limit is equal to 25, this criterion is met.
- Water content > 0.9 times the liquid limit: No data on the water content of the in situ soil are provided.

Since the soil does not meet the first criterion listed above, the soil is not susceptible to liquefaction.

Soil type no. 6: This soil is described as Wewahitchka sandy clay from Florida. The soil has a liquid limit of 65 and a plasticity index of 41. Because of the plastic nature of the soil, it would typically be termed a cohesive soil. Per Sec. 6.3, in order for a cohesive soil to liquefy, it must meet all the following three criteria:

- Percent finer at 0.005 mm < 15 percent: Since the percent finer at a grain size of 0.005 mm is equal to 50 percent, this criterion is not met.
- Liquid limit < 35: Since the liquid limit is equal to 41, this criterion is also not met.
- Water content > 0.9 times the liquid limit: No data on the water content of the in situ soil are provided.

Since the soil does not meet the first and second criteria listed above, the soil is not susceptible to liquefaction.

Soil type no. 7: This soil is described as loess from Mississippi. The soil has a liquid limit of 29 and a plasticity index of 5. Because of the plastic nature of the soil, it would typically be termed a cohesive soil. Per Sec. 6.3, in order for a cohesive soil to liquefy, it must meet all the following three criteria:

- Percent finer at 0.005 mm < 15 percent: Since the percent finer at a grain size of 0.005 mm is equal to 14 percent, this criterion is met.
- Liquid limit < 35: Since the liquid limit is equal to 29, this criterion is also met.
- Water content > 0.9 times the liquid limit: No data on the water content of the in situ soil are provided.

Since the soil meets the first and second criteria listed above, the soil is susceptible to liquefaction provided that the in situ water content is greater than 26 percent.

Soil type no. 8: This soil is described as backswamp deposit from the Mississippi River. The soil has a liquid limit of 59 and a plasticity index of 41. Because of the plastic nature of the soil, it would typically be termed a cohesive soil. Per Sec. 6.3, in order for a cohesive soil to liquefy, it must meet all the following three criteria:

- Percent finer at 0.005 mm < 15 percent: Since the percent finer at a grain size of 0.005 mm is equal to 53 percent, this criterion is not met.
- Liquid limit < 35: Since the liquid limit is equal to 59, this criterion is also not met.
- Water content > 0.9 times the liquid limit: No data on the water content of the in situ soil are provided.

Since the soil does not meet the first and second criteria listed above, the soil is not susceptible to liquefaction.

6.2

$$\sigma_v = 58 \text{ kPa} + 20 \text{ kPa} = 78 \text{ kPa}$$

$$\sigma'_v = 43 \text{ kPa} + 20 \text{ kPa} = 63 \text{ kPa}$$

$$\text{CSR} = 0.65 r_d \left(\frac{\sigma_{v0}}{\sigma'_{v0}} \right) \left(\frac{a_{max}}{g} \right) \qquad \text{Eq. (6.6)}$$

$$= 0.65 \,(0.96) \left(\frac{78}{63} \right) (0.40) = 0.31$$

6.3

$$CSR = 0.65r_d \frac{\sigma_{v0}}{\sigma'_{v0}} \frac{a_{max}}{g} \qquad \text{Eq. (6.6)}$$

$$= 0.65\,(0.96)\,(1.35)\,(0.10) = 0.084$$

Per Fig. 6.6, for $(N_1)_{60} = 7.7$, intersecting 15 percent fines curve, CRR $= 0.14$

$$FS = \frac{CRR}{CSR} = \frac{0.14}{0.084} = 1.67$$

6.4

$$CSR = 0.65r_d \frac{\sigma_{v0}}{\sigma'_{v0}} \frac{a_{max}}{g} \qquad \text{Eq. (6.6)}$$

$$= 0.65\,(0.96)\,(1.35)\,(0.20) = 0.17$$

Per Fig. 6.6, for $(N_1)_{60} = 7.7$, intersecting clean sand curve, and CRR $= 0.09$.
Using magnitude scaling factor (MSF) $= 1.50$ (Table 6.2, $5^1/4$ magnitude earthquake) gives

$$\text{Adjusted CRR} = (1.50)\,(0.09) = 0.14$$

$$FS = \frac{CRR}{CSR} = \frac{0.14}{0.17} = 0.82$$

6.5

$$CSR = 0.65r_d \frac{\sigma_{v0}}{\sigma'_{v0}} \frac{a_{max}}{g} \qquad \text{Eq. (6.6)}$$

$$= 0.65\,(0.96)\,(1.35)\,(0.40) = 0.34$$

Using Eq. (5.3), for $q_c = 3.9$ MPa, $q_{c1} = 5.8$ MPa. Per Fig. 6.8, for $q_{c1} = 5.8$ MPa, intersecting clean sand curve, CRR $= 0.09$, and

$$FS = \frac{CRR}{CSR} = \frac{0.09}{0.34} = 0.26$$

6.6

$$CSR = 0.65r_d \frac{\sigma_{v0}}{\sigma'_{v0}} \frac{a_{max}}{g} \qquad \text{Eq. (6.6)}$$

$$= 0.65\,(0.96)\,(1.35)\,(0.40) = 0.34$$

Using Eq. (6.9) with $V_s = 150$ m/s and $\sigma'_{v0} = 43$ kPa, then $V_{s1} = 185$ m/s. Per Fig. 6.8, for $V_{s1} = 185$ m/s, intersecting clean sand curve, and CRR $= 0.16$

$$FS = \frac{CRR}{CSR} = \frac{0.16}{0.34} = 0.47$$

6.7

$$CSR = 0.65 r_d \frac{\sigma_{v0}}{\sigma'_{v0}} \frac{a_{max}}{g} \qquad \text{Eq. (6.6)}$$

$$= 0.65\,(0.96)\,(1.35)\,(0.40) = 0.34$$

Crushed limestone has more than 50 percent gravel-size particles with about 11 percent fines. Given plasticity characteristics (liquid limit = 18 and plasticity index = 7), use silty gravel curve in Fig. 6.9. For $q_{c1} = 5.0$ MPa and intersecting silty gravel curve, CRR = 0.18.

$$FS = \frac{CRR}{CSR} = \frac{0.18}{0.34} = 0.53$$

6.8

$$CSR = 0.65 r_d \frac{\sigma_{v0}}{\sigma'_{v0}} \frac{a_{max}}{g} \qquad \text{Eq. (6.6)}$$

$$= 0.65\,(0.96)\,(1.35)\,(0.40) = 0.34$$

Silty gravel has more than 50 percent gravel-size particles with about 10 percent fines. Given plasticity characteristics (liquid limit = 23 and plasticity index = 3), use silty gravel curve in Fig. 6.9. For $q_{c1} = 7.5$ MPa and intersecting silty gravel curve, CRR = 0.27.

$$FS = \frac{CRR}{CSR} = \frac{0.27}{0.34} = 0.79$$

6.9

$$CSR = 0.65 r_d \frac{\sigma_{v0}}{\sigma'_{v0}} \frac{a_{max}}{g} \qquad \text{Eq. (6.6)}$$

$$= 0.65\,(0.96)\,(1.35)\,(0.40) = 0.34$$

The gravelly sand does not have any fines (hence clean soil). The soil type is approximately midway between clean sand and clean gravel. In Fig. 6.9, using $q_{c1} = 14$ MPa and at a midpoint between the clean sand and clean gravel curves, CRR = 0.44.

$$FS = \frac{CRR}{CSR} = \frac{0.44}{0.34} = 1.29$$

6.10

$$CSR = 0.65 r_d \frac{\sigma_{v0}}{\sigma'_{v0}} \frac{a_{max}}{g} \qquad \text{Eq. (6.6)}$$

$$= 0.65\,(0.96)\,(1.35)\,(0.40) = 0.34$$

The eolian sand has about 6 percent fines (i.e., 6 percent passing No. 200 sieve). This is very close to the clean sand condition. Per Fig. 6.6, for $(N_1)_{60} = 7.7$ and intersecting the clean sand curve, CRR = 0.09.

$$FS = \frac{CRR}{CSR} = \frac{0.09}{0.34} = 0.26$$

(Note that since the soil is essentially a clean sand, the answer is identical to the example problem in Sec. 6.4.5.)

6.11

$$CSR = 0.65 r_d \frac{\sigma_{v0}}{\sigma'_{v0}} \frac{a_{max}}{g} \qquad \text{Eq. (6.6)}$$

$$= 0.65 \, (0.96) \, (1.35) \, (0.40) = 0.34$$

The noncemented loess has about 95 percent fines (i.e., 95 percent passing No. 200 sieve). In Fig. 6.6, there is no curve available for such a high fines content. Taking a conservative approach by using the 35 percent fines curve in Fig. 6.6, for $(N_1)_{60} = 7.7$ and intersecting the 35 percent fines curve, CRR = 0.18. Thus

$$FS = \frac{CRR}{CSR} = \frac{0.18}{0.34} = 0.53$$

6.12   See next page.

6.13

Depth below ground surface, m	Cyclic resistance ratio (CRR) from laboratory tests	Cyclic stress ratio (CSR) from Prob. 6.12	FS = CRR/CSR
2.3	0.18	0.13	1.39
5.5	0.16	0.15	1.07
7.0	0.15	0.16	0.94
8.5	0.17	0.16	1.06
9.5	0.17	0.16	1.06
11.5	0.16	0.16	1.00
13.0	0.16	0.16	1.00
14.5	0.24	0.16	1.50

6.14   Per Fig. 6.14, the standard penetration test data indicate that there are three zones of liquefaction from about 2 to 11 m, 12 to 15 m, and 17 to 20 m below ground surface. Per Fig. 6.14, the laboratory cyclic strength tests indicate that there are two zones of liquefaction from about 6 to 8 m and from 10 to 14 m below ground surface.

6.15   See page E.9.

6.16

Depth below ground surface, m	Cyclic resistance ratio from laboratory tests	Cyclic stress ratio from Prob. 6.15	FS = CRR/CSR
2.0	0.20	0.17	1.18
3.5	0.20	0.18	1.11
5.0	0.21	0.18	1.17
8.0	0.28	0.18	1.56
11.0	0.29	0.18	1.61

6.12

Depth, m	σ_v, kPa	σ_v', kPa	σ_v/σ_v'	r_d	CSR	N value	C_r	N_{60}	C_N	$(N_1)_{60}$	CRR	FS = CRR/CSR
		Cyclic stress ratio (CSR)					N value corrections					
1.5	27.5	27.5	1.00	0.98	0.10	8	0.75	6.0	1.91	11	0.12	1.18
2.5	47.0	37.2	1.26	0.97	0.13	5	0.75	3.8	1.64	6.2	0.07	0.55
3.5	66.5	46.9	1.42	0.96	0.14	4	0.75	3.0	1.46	4.4	0.05	0.35
4.5	86.0	56.6	1.52	0.95	0.15	5	0.85	4.3	1.33	5.7	0.06	0.40
5.5	105	66.3	1.58	0.93	0.15	9	0.85	7.7	1.23	9.5	0.11	0.72
6.5	125	76.0	1.64	0.92	0.16	10	0.95	9.5	1.15	11	0.12	0.76
7.5	144	85.7	1.68	0.91	0.16	12	0.95	11	1.08	12	0.13	0.82
8.5	164	95.4	1.72	0.90	0.16	12	0.95	11	1.02	11	0.12	0.75
9.5	183	105	1.74	0.89	0.16	15	0.95	14	0.98	14	0.16	1.00
10.5	203	115	1.77	0.87	0.16	11	1.00	11	0.93	10	0.11	0.69
11.5	222	124	1.79	0.86	0.16	23	1.00	23	0.90	21	0.23	1.44
12.5	242	134	1.81	0.85	0.16	11	1.00	11	0.86	9.5	0.11	0.69
13.5	261	144	1.81	0.84	0.16	10	1.00	10	0.83	8.3	0.09	0.57
14.5	281	154	1.82	0.83	0.16	10	1.00	10	0.81	8.1	0.09	0.57
15.5	300	163	1.84	0.81	0.16	25	1.00	25	0.78	20	0.23	1.48
16.5	320	173	1.85	0.80	0.15	27	1.00	27	0.76	21	0.24	1.56
17.5	339	183	1.85	0.79	0.15	4	1.00	4	0.74	3.0	0.03	0.20
18.5	359	192	1.87	0.78	0.15	5	1.00	5	0.72	3.6	0.04	0.26
19.5	378	202	1.87	0.77	0.15	3	1.00	3	0.70	2.1	0.02	0.13
20.5	398	212	1.88	0.75	0.15	38	1.00	38	0.69	26	0.30	2.05

Notes: Cyclic stress ratio: $a_{max} = 0.16g$, r_d from Eq. (6.7). N value corrections: $E_m = 0.6$, $C_b = 1.0$, C_N from Eq. (5.2). CRR from Fig. 6.6.

6.15

	Cyclic stress ratio					N value corrections						FS =
Depth, m	σ_v, kPa	σ'_v, kPa	σ_v/σ'_v	r_d	CSR	N value	C_r	N_{60}	C_N	$(N_1)_{60}$	CRR	CRR/CSR
1.2	22.9	15.1	1.52	0.99	0.16	4	0.75	3.0	2.57	7.7	0.09	0.56
2.2	42.4	24.8	1.71	0.97	0.17	6	0.75	4.5	2.01	9.0	0.10	0.59
3.2	61.9	34.5	1.79	0.96	0.18	5	0.75	3.8	1.70	6.5	0.07	0.39
4.2	81.5	44.2	1.84	0.95	0.18	8	0.85	6.8	1.50	10	0.11	0.61
5.2	101	53.9	1.87	0.94	0.18	7	0.85	6.0	1.36	8.2	0.09	0.50
6.2	120	63.6	1.89	0.93	0.18	13	0.95	12	1.25	15	0.16	0.89
7.2	140	73.3	1.91	0.91	0.18	36	0.95	34	1.17	40	>0.5	>2.8
8.2	159	83.0	1.92	0.90	0.18	24	0.95	23	1.09	25	0.29	1.61
9.2	179	92.7	1.93	0.89	0.18	35	0.95	33	1.04	34	>0.5	>2.8
10.2	199	102	1.95	0.88	0.18	30	1.00	30	0.99	30	0.50	2.78
11.2	218	112	1.95	0.87	0.18	28	1.00	28	0.94	26	0.30	1.67
12.2	238	122	1.95	0.85	0.17	32	1.00	32	0.91	29	0.45	2.65
13.2	257	131	1.96	0.84	0.17	16	1.00	16	0.87	14	0.16	0.94
14.2	277	141	1.96	0.83	0.17	28	1.00	28	0.84	24	0.28	1.65
15.2	296	151	1.96	0.82	0.17	27	1.00	27	0.81	22	0.25	1.47
16.2	316	161	1.96	0.81	0.17	23	1.00	23	0.79	18	0.20	1.18
17.2	335	170	1.97	0.79	0.16	38	1.00	38	0.77	29	0.45	2.81
18.2	355	180	1.97	0.78	0.16	32	1.00	32	0.75	24	0.28	1.75
19.2	374	190	1.97	0.77	0.16	47	1.00	47	0.73	34	>0.5	>3.1

Notes: Cyclic stress ratio: $a_{max} = 0.16g$, r_d from Eq. (6.7). N value corrections: $E_m = 0.6$, $C_b = 1.0$, C_N from Eq. (5.2). CRR from Fig. 6.6.

6.17   Per Fig. 6.16, the standard penetration test data indicate that there are two zones of liquefaction from about 1.2 to 6.7 m and from 12.7 to 13.7 m below ground surface. Per Fig. 6.16, the laboratory cyclic strength tests indicate that the soil has a factor of safety against liquefaction in excess of 1.0.

6.18   See pages E.11 and E.12.

CHAPTER 7

7.1   *Solution using Fig. 7.1:*   Since the soil has 15 percent fines, use the correction in Sec. 7.2.2. Extrapolating for 15 percent fines, $N_{corr} = 1.3$. Therefore, for Fig. 7.1, $(N_1)_{60} = (N_1)_{60} + N_{corr} = 7.7 + 1.3 = 9$. Assume that the Japanese N_1 value is approximately equal to this $(N_1)_{60}$ value, or use a Japanese N_1 value of 9. For FS = 1.67 and a Japanese N_1 value of 9, the volumetric strain ε_v is equal to 0.15 percent. Since the in situ soil layer is 1.0 m thick, the ground surface settlement of this soil layer is equal to

$$\text{Settlement} = \left(\frac{\varepsilon_v}{100} \right)(H) = \left(\frac{0.15}{100} \right)(1.0 \text{ m}) = 0.0015 \text{ m} = 0.15 \text{ cm}$$

Solution using Fig. 7.2:   Since the soil has 15 percent fines, use the correction in Sec. 7.2.2. Extrapolating for 15 percent fines, $N_{corr} = 1.3$. Therefore, for Fig. 7.2, $(N_1)_{60} = (N_1)_{60} + N_{corr} = 7.7 + 1.3 = 9$. The cyclic stress ratio from Eq. (6.6) is equal to 0.084. Entering Fig. 7.2 with CSR = 0.084 and $(N_1)_{60} = 9$, the volumetric strain is equal to 0.15 percent. Since the in situ soil layer is 1.0 m thick, the ground surface settlement of this soil layer is equal to

$$\text{Settlement} = \left(\frac{\varepsilon_v}{100} \right)(H) = \left(\frac{0.15}{100} \right)(1.0 \text{ m}) = 0.0015 \text{ m} = 0.15 \text{ cm}$$

7.2   *Solution using Fig. 7.1:*   For Fig. 7.1, assume that the Japanese N_1 value is approximately equal to the $(N_1)_{60}$ value from Eq. (5.2), or use Japanese $N_1 = 7.7$. Using FS = 0.82 and then extrapolating between the curves for an N_1 value of 7.7, the volumetric strain ε_v is equal to 4.1 percent. Since the in situ liquefield soil layer is 1.0 m thick, the ground surface settlement of the liquefied soil is equal to

$$\text{Settlement} = \left(\frac{\varepsilon_v}{100} \right)(H) = \left(\frac{4.1}{100} \right)(1.0 \text{ m}) = 0.041 \text{ m} = 4.1 \text{ cm}$$

Solution using Fig. 7.2:   The cyclic stress ratio from Eq. (6.6) is equal to 0.17, and the calculated value of $(N_1)_{60}$ determined at a depth of 3 m below ground surface is equal to 7.7. Entering Fig. 7.2 with CSR = 0.17 and $(N_1)_{60} = 7.7$, the volumetric strain is equal to 3.0 percent. Since the in situ liquefied soil layer is 1.0 m thick, the ground surface settlement of the liquefied soil is equal to

$$\text{Settlement} = \left(\frac{\varepsilon_v}{100} \right)(H) = \left(\frac{3.0}{100} \right)(1.0 \text{ m}) = 0.030 \text{ m} = 3.0 \text{ cm}$$

For the given earthquake magnitude of $5^{1}/_4$, the magnitude scaling factor = 1.5 (Table 6.2). Thus the corrected CSR is equal to 0.17 divided by 1.5, or 0.11. Entering Fig. 7.2 with

6.18 Before Improvement

Depth, m	Cyclic stress ratio					N value corrections						CRR (see Notes)	FS = CRR/CSR
	σ_v, kPa	σ_v', kPa	σ_v/σ_v'	r_d	CSR	N value	C_r	N_{60}	C_N	$(N_1)_{60}$			
0.5	9.45	9.45	1.00	0.99	0.26	16	0.75	12	3.24	39	>0.5	>2
1.5	29.1	19.3	1.51	0.98	0.38	11	0.75	8.3	2.28	19	0.25	0.66
2.5	48.7	29.1	1.67	0.97	0.42	9	0.75	6.8	1.85	13	0.18	0.43
3.5	68.3	38.9	1.76	0.96	0.44	12	0.75	9.0	1.60	14	0.19	0.42
4.5	87.9	48.7	1.80	0.95	0.44	19	0.85	16	1.43	23	0.33	0.76
5.5	108	58.5	1.85	0.93	0.45	11	0.85	9.4	1.31	12	0.16	0.36
6.5	127	68.3	1.86	0.92	0.44	9	0.95	8.6	1.21	10	0.14	0.32
7.5	147	78.1	1.88	0.91	0.44	19	0.95	18	1.13	20	0.26	0.59
8.5	166	87.9	1.89	0.90	0.44	10	0.95	9.5	1.07	10	0.14	0.32
9.5	186	97.7	1.90	0.89	0.44	10	0.95	9.5	1.01	10	0.14	0.32
10.5	206	107	1.93	0.88	0.44	11	1.00	11	0.97	11	0.15	0.34
11.5	225	117	1.93	0.86	0.43	4	1.00	4.0	0.92	3.7	0.07	0.16
12.5	245	127	1.93	0.85	0.43	10	1.00	10	0.89	8.9	0.13	0.30
13.5	264	137	1.93	0.84	0.42	11	1.00	11	0.85	9.4	0.13	0.31
14.5	284	147	1.93	0.83	0.42	12	1.00	12	0.82	10	0.14	0.33
15.5	304	156	1.95	0.81	0.41	13	1.00	13	0.80	10	0.14	0.34

Notes: Cyclic stress ratio: $a_{max} = 0.40g$, r_d from Eq. (6.7). N value corrections: $E_m = 0.6$, $C_b = 1.0$, C_N from Eq. (5.2). CRR from Fig. 6.6 (silty sand with 15 percent fines). The values of CRR from Fig. 6.6 were multiplied by a magnitude scaling factor = 0.89 (see Table 6.2).

6.18 After Improvement

Depth, m	Cyclic stress ratio					N value	N value corrections				CRR (see Notes)	FS = CRR/CSR
	σ_v, kPa	σ_v', kPa	σ_v/σ_v'	r_d	CSR	N value	C_r	N_{60}	C_N	$(N_1)_{60}$		
0.5	9.45	9.45	1.00	0.99	0.26	19	0.75	14	3.24	46	>0.5	>2
1.5	29.1	19.3	1.51	0.98	0.38	21	0.75	16	2.28	36	>0.5	>1.3
2.5	48.7	29.1	1.67	0.97	0.42	19	0.75	14	1.85	26	>0.5	>1.2
3.5	68.3	38.9	1.76	0.96	0.44	21	0.75	16	1.60	25	>0.5	>1.1
4.5	87.9	48.7	1.80	0.95	0.44	25	0.85	21	1.43	30	>0.5	>1.1
5.5	108	58.5	1.85	0.93	0.45	31	0.85	26	1.31	35	>0.5	>1.1
6.5	127	68.3	1.86	0.92	0.44	33	0.95	31	1.21	38	>0.5	>1.1
7.5	147	78.1	1.88	0.91	0.44	31	0.95	29	1.13	33	>0.5	>1.1
8.5	166	87.9	1.89	0.90	0.44	37	0.95	35	1.07	38	>0.5	>1.1
9.5	186	97.7	1.90	0.89	0.44	41	0.95	39	1.01	39	>0.5	>1.1
10.5	206	107	1.93	0.88	0.44	35	1.00	35	0.97	34	>0.5	>1.1
11.5	225	117	1.93	0.86	0.43	36	1.00	36	0.92	33	>0.5	>1.2
12.5	245	127	1.93	0.85	0.43	39	1.00	39	0.89	35	>0.5	>1.2
13.5	264	137	1.93	0.84	0.42	45	1.00	45	0.85	38	>0.5	>1.2
14.5	284	147	1.93	0.83	0.42	40	1.00	40	0.82	33	>0.5	>1.2

Notes: Cyclic stress ratio: $a_{max} = 0.40g$, r_d from Eq. (6.7). N value corrections: $E_m = 0.6$, $C_b = 1.0$, C_N from Eq. (5.2). CRR from Fig. 6.6 (silty sand with 15 percent fines). The values of CRR from Fig. 6.6 were multiplied by a magnitude scaling factor = 0.89 (see Table 6.2).

the modified CSR = 0.11 and $(N_1)_{60} = 7.7$, the volumetric strain is equal to 2.9 percent. Since the in situ liquefied soil layer is 1.0 m thick, the ground surface settlement of the liquefied soil is equal to

$$\text{Settlement} = \left(\frac{\varepsilon_v}{100}\right)(H) = \left(\frac{2.9}{100}\right)(1.0 \text{ m}) = 0.029 \text{ m} = 2.9 \text{ cm}$$

Summary: Note that in this case, the different magnitude earthquakes do make a small difference in the settlement based on Fig. 7.2.

7.3 *Solution using Fig. 7.1:* Using Eq. (5.3), for $q_c = 3.9$ MPa, the value of $q_{c1} = 5.8$ MPa = 58 kg/cm². Using FS = 0.26 and then extrapolating between the curves for a q_{c1} value of 58 kg/cm², the volumetric strain ε_v is equal to 3.6 percent. Since the in situ liquefied soil layer is 1.0 m thick, the ground surface settlement of the liquefied soil is equal to

$$\text{Settlement} = \left(\frac{\varepsilon_v}{100}\right)(H) = \left(\frac{3.6}{100}\right)(1.0 \text{ m}) = 0.036 \text{ m} = 3.6 \text{ cm}$$

Solution using Fig. 7.2: The cyclic stress ratio from Eq. (6.6) is equal to 0.34. Using Fig. 6.6, with CRR = 0.09 and intersecting the clean sand curve, $(N_1)_{60} = 7.7$. Entering Fig. 7.2 with CSR = 0.34 and $(N_1)_{60} = 7.7$, the volumetric strain is equal to 3.0 percent. Since the in situ liquefied soil layer is 1.0 m thick, the ground surface settlement of the liquefied soil is equal to

$$\text{Settlement} = \left(\frac{\varepsilon_v}{100}\right)(H) = \left(\frac{3.0}{100}\right)(1.0 \text{ m}) = 0.030 \text{ m} = 3.0 \text{ cm}$$

7.4 *Solution using Fig. 7.1:* Using Eq. (6.9), for $V_s = 150$ m/s and $\sigma'_{v0} = 43$ kPa, then $V_{s1} = 185$ m/s. Using Fig. 6.6, with CRR = 0.16 and intersecting the clean sand curve, $(N_1)_{60} = 14$. Assume that the Japanese N_1 value is approximately equal to the $(N_1)_{60}$ value, or use Japanese N_1 value of 14. Using FS = 0.47 and then extrapolating between the curves for an N_1 value of 14, the volumetric strain ε_v is equal to 2.8 percent. Since the in situ liquefied soil layer is 1.0 m thick, the ground surface settlement of the liquefied soil is equal to

$$\text{Settlement} = \left(\frac{\varepsilon_v}{100}\right)(H) = \left(\frac{2.8}{100}\right)(1.0 \text{ m}) = 0.028 \text{ m} = 2.8 \text{ cm}$$

Solution using Fig. 7.2: The cyclic stress ratio from Eq. (6.6) is equal to 0.34. Entering Fig. 7.2 with CSR = 0.34 and $(N_1)_{60} = 14$, the volumetric strain is equal to 2.1 percent. Since the in situ liquefied soil layer is 1.0 m thick, the ground surface settlement of the liquefied soil is equal to

$$\text{Settlement} = \left(\frac{\varepsilon_v}{100}\right)(H) = \left(\frac{2.1}{100}\right)(1.0 \text{ m}) = 0.021 \text{ m} = 2.1 \text{ cm}$$

7.5 Figures 7.1 and 7.2 were developed for clean sand. However, using the figures for the crushed limestone, we get the following.
Solution using Fig. 7.1: The value of $q_{c1} = 5.0$ MPa = 50 kg/cm². Using FS = 0.53 and then extrapolating between the curves for q_{c1} value of 50 kg/cm², the volumetric strain ε_v is equal to 4.2 percent. Since the in situ liquefied soil layer is 1.0 m thick, the ground surface settlement of the liquefied soil is equal to

$$\text{Settlement} = \left(\frac{\varepsilon_v}{100}\right)(H) = \left(\frac{4.2}{100}\right)(1.0 \text{ m}) = 0.042 \text{ m} = 4.2 \text{ cm}$$

Solution using Fig. 7.2: In Fig. 7.1, each curve was developed for a Japanese N_1 value and corresponding q_{c1} value. In Fig. 7.1, a q_{c1} value of 50 kg/cm^2 corresponds to a Japanese N_1 value equal to 7.3. Assuming that the Japanese N_1 value is approximately equal to the $(N_1)_{60}$ value, then use $(N_1)_{60} = 7.3$. Entering Fig. 7.2 with CSR = 0.34 and $(N_1)_{60} = 7.3$, the volumetric strain is equal to 3.1 percent. Since the in situ liquefied soil layer is 1.0 m thick, the ground surface settlement of the liquefied soil is equal to

$$\text{Settlement} = \left(\frac{\varepsilon_v}{100} \right) (H) = \left(\frac{3.1}{100} \right) (1.0 \text{ m}) = 0.031 \text{ m} = 3.1 \text{ cm}$$

Summary: The crushed limestone has 11 percent fines and 55 percent gravel-size particles (see Fig. 6.12). Both the fines and gravel size particles would tend to lower the volumetric strain, and therefore the above settlement values are probably too high.

7.6 Figures 7.1 and 7.2 were developed for clean sand. However, using the figures for the silty gravel gives the following.

Solution using Fig. 7.1: The value of $q_{c1} = 7.5$ MPa = 75 kg/cm^2. Using FS = 0.79 and then extrapolating between the curves for a q_{c1} value of 75 kg/cm^2, the volumetric strain ε_v is equal to 3.0 percent. Since the in situ liquefied soil layer is 1.0 m thick, the ground surface settlement of the liquefied soil is equal to

$$\text{Settlement} = \left(\frac{\varepsilon_v}{100} \right) (H) = \left(\frac{3.0}{100} \right) (1.0 \text{ m}) = 0.030 \text{ m} = 3.0 \text{ cm}$$

Solution using Fig. 7.2: In Fig. 7.1, each curve was developed for a Japanese N_1 value and corresponding q_{c1} value. In Fig. 7.1, a q_{c1} value of 75 kg/cm^2 corresponds to a Japanese N_1 value equal to 13. Assuming that the Japanese N_1 value is approximately equal to the $(N_1)_{60}$ value, then use $(N_1)_{60} = 13$. Entering Fig. 7.2 with CSR = 0.34 and $(N_1)_{60} = 13$, the volumetric strain is equal to 2.2 percent. Since the in situ liquefied soil layer is 1.0 m thick, the ground surface settlement of the liquefied soil is equal to

$$\text{Settlement} = \left(\frac{\varepsilon_v}{100} \right) (H) = \left(\frac{2.2}{100} \right) (1.0 \text{ m}) = 0.022 \text{ m} = 2.2 \text{ cm}$$

Summary: The silty gravel has 10 percent fines and 52 percent gravel-size particles (see Fig. 6.12). Both the fines and gravel-size particles would tend to lower the volumetric strain, and therefore the above settlement values are probably too high.

7.7 Note that Figs. 7.1 and 7.2 were developed for clean sand. However, using the figures for the gravelly sand gives the following:

Solution using Fig. 7.1: The value of $q_{c1} = 14$ MPa = 140 kg/cm^2. Using FS = 1.29 and a q_{c1} value of 140 kg/cm^2, the volumetric strain ε_v is equal to 0.3 percent. Since the in situ soil layer is 1.0 m thick, the ground surface settlement of the soil layer is equal to

$$\text{Settlement} = \left(\frac{\varepsilon_v}{100} \right) (H) = \left(\frac{0.3}{100} \right) (1.0 \text{ m}) = 0.003 \text{ m} = 0.3 \text{ cm}$$

Solution using Fig. 7.2: In Fig. 7.1, each curve was developed for a Japanese N_1 value and corresponding q_{c1} value. In Fig. 7.1, a q_{c1} value of 140 kg/cm^2 corresponds to a Japanese N_1 value equal to 24. Assuming that the Japanese N_1 value is approximately equal to the $(N_1)_{60}$ value, then use $(N_1)_{60} = 24$. Entering Fig. 7.2 with CSR = 0.34 and $(N_1)_{60} = 24$, the volumetric strain is equal to 1.2 percent. Since the in situ soil layer is 1.0 m thick, the ground surface settlement of the soil layer is equal to

$$\text{Settlement} = \left(\frac{\varepsilon_v}{100}\right)(H) = \left(\frac{1.2}{100}\right)(1.0 \text{ m}) = 0.012 \text{ m} = 1.2 \text{ cm}$$

Summary: In Fig. 7.2, the data plot within the portion of the graph that corresponds to a factor of safety against liquefaction that is less than 1.0. Since the actual value of the factor of safety against liquefaction is greater than 1.0 (that is, FS = 1.29), the settlement value of 1.2 cm is too high. The actual settlement would probably be closer to 0.3 cm.

7.8 The eolian sand has about 6 percent fines (that is, 6 percent passing the No. 200 sieve). This is very close to the clean sand condition.

Solution using Fig. 7.1: For Fig. 7.1, assume that the Japanese N_1 value is approximately equal to the $(N_1)_{60}$ value from Eq. (5.2), or use Japanese $N_1 = 7.7$. The Japanese N_1 curves labeled 6 and 10 are extended straight downward to FS = 0.26, and then after extrapolating between the curves for an N_1 value of 7.7, the volumetric strain is equal to 4.1 percent. Since the in situ liquefied soil layer is 1.0 m thick, the ground surface settlement of the liquefied soil is equal to

$$\text{Settlement} = \left(\frac{\varepsilon_v}{100}\right)(H) = \left(\frac{4.1}{100}\right)(1.0 \text{ m}) = 0.041 \text{ m} = 4.1 \text{ cm}$$

Solution using Fig. 7.2: The cyclic stress ratio from Eq. (6.6) is equal to 0.34, and the calculated value of $(N_1)_{60}$ determined at a depth of 3 m below ground surface is equal to 7.7. Entering Fig. 7.2 with CSR = 0.34 and $(N_1)_{60} = 7.7$, the volumetric strain is equal to 3.0 percent. Since the in situ liquefied soil layer is 1.0 m thick, the ground surface settlement of the liquefied soil is equal to

$$\text{Settlement} = \left(\frac{\varepsilon_v}{100}\right)(H) = \left(\frac{3.0}{100}\right)(1.0 \text{ m}) = 0.030 \text{ m} = 3.0 \text{ cm}$$

Summary: Since the soil is essentially a clean sand, the answers are identical to the example problem in Sec. 7.2.2.

7.9 *Solution using Fig. 7.1:* Since the soil has 96 percent fines, use the correction in Sec. 7.2.2. Using the value for 75 percent fines, $N_{corr} = 5$. Therefore, $(N_1)_{60}$ for Fig. 7.1 is $(N_1)_{60} + N_{corr} = 7.7 + 5 = 13$. Assume that the Japanese N_1 value is approximately equal to this $(N_1)_{60}$ value, or use a Japanese N_1 value of 13. For FS = 0.53 and a Japanese N_1 value of 13, the volumetric strain ε_v is equal to 3.0 percent. Since the in situ soil layer is 1.0 m thick, the ground surface settlement of this soil layer is equal to

$$\text{Settlement} = \left(\frac{\varepsilon_v}{100}\right)(H) = \left(\frac{3.0}{100}\right)(1.0 \text{ m}) = 0.030 \text{ m} = 3.0 \text{ cm}$$

Solution using Fig. 7.2: Since the soil has 96 percent fines, use the correction in Sec. 7.2.2. Using the value for 75 percent fines, $N_{corr} = 5$. Therefore, $(N_1)_{60}$ for Fig. 7.2 is $(N_1)_{60} + N_{corr} = 7.7 + 5 = 13$. The cyclic stress ratio from Eq. (6.6) is equal to 0.34. Entering Fig. 7.2 with CSR = 0.34 and $(N_1)_{60} = 13$, the volumetric strain is equal to 2.3 percent. Since the in situ soil layer is 1.0 m thick, the ground surface settlement of this soil layer is equal to

$$\text{Settlement} = \left(\frac{\varepsilon_v}{100}\right)(H) = \left(\frac{2.3}{100}\right)(1.0 \text{ m}) = 0.023 \text{ m} = 2.3 \text{ cm}$$

7.10 *Solution using Fig. 7.1:* First determine the factor of safety against liquefaction using Fig. 6.6, or

Layer depth, m	CSR	CRR (Fig. 6.6)	FS = CRR/CSR
2–3	0.18	0.11	0.61
3–5	0.20	0.06	0.30
5–7	0.22	0.08	0.36

Assume that the Japanese N_1 value is approximately equal to the $(N_1)_{60}$ value. Using the above factors of safety against liquefaction and the given $(N_1)_{60}$ value gives

For the 2- to 3-m layer:

$$\varepsilon_v = 3.5\% \quad \text{or} \quad \text{settlement} = 0.035\,(1.0) = 0.035 \text{ m}$$

For the 3- to 5-m layer:

$$\varepsilon_v = 4.8\% \quad \text{or} \quad \text{settlement} = 0.048\,(2.0) = 0.096 \text{ m}$$

For the 5- to 7-m layer:

$$\varepsilon_v = 4.3\% \quad \text{or} \quad \text{settlement} = 0.043\,(2.0) = 0.086 \text{ m}$$

$$\text{Total settlement} = 0.035 + 0.096 + 0.086 = 0.22 \text{ m} = 22 \text{ cm}$$

Solution using Fig. 7.2: Enter the curve with the given $(N_1)_{60}$ and CSR values:

For the 2- to 3-m layer:

$$\varepsilon_v = 2.6\% \quad \text{or} \quad \text{settlement} = 0.026\,(1.0) = 0.026 \text{ m}$$

For the 3- to 5-m layer:

$$\varepsilon_v = 4.2\% \quad \text{or} \quad \text{settlement} = 0.042\,(2.0) = 0.084 \text{ m}$$

For the 5- to 7-m layer:

$$\varepsilon_v = 3.2\% \quad \text{or} \quad \text{settlement} = 0.032\,(2.0) = 0.064 \text{ m}$$

$$\text{Total settlement} = 0.026 + 0.084 + 0.064 = 0.174 \text{ m} = 17 \text{ cm}$$

7.11 See next page.

7.12 See page E.18.

7.13 See pages E.19 and E.20.

7.14 See pages E.21 and E.22.

7.15 See pages E.23 and E.24.

7.11

Depth, m	Figure 7.1					Figure 7.2				
	$(N_1)_{60}$	FS	ε_v, percent	H, m	Settlement, cm	$(N_1)_{60}$	CSR	ε_v, percent	H, m	Settlement, cm
1.5	11	1.18	0.6	0.5	0.3	11	0.10	0.1	0.5	0.05
2.5	6.2	0.55	4.5	1.0	4.5	6.2	0.13	3.7	1.0	3.7
3.5	4.4	0.35	5.1	1.0	5.1	4.4	0.14	4.5	1.0	4.5
4.5	5.7	0.40	4.6	1.0	4.6	5.7	0.15	4.0	1.0	4.0
5.5	9.5	0.72	3.6	1.0	3.6	9.5	0.15	2.7	1.0	2.7
6.5	11	0.76	3.3	1.0	3.3	11	0.16	2.4	1.0	2.4
7.5	12	0.82	3.1	1.0	3.1	12	0.16	2.2	1.0	2.2
8.5	11	0.75	3.3	1.0	3.3	11	0.16	2.4	1.0	2.4
9.5	14	1.00	1.1	1.0	1.1	14	0.16	1.2	1.0	1.2
10.5	10	0.69	3.5	1.0	3.5	10	0.16	2.6	1.0	2.6
11.5	21	1.44	0.2	1.0	0.2	21	0.16	0	1.0	0
12.5	9.5	0.69	3.6	1.0	3.6	9.5	0.16	2.7	1.0	2.7
13.5	8.3	0.57	3.9	1.0	3.9	8.3	0.16	2.9	1.0	2.9
14.5	8.1	0.57	4.0	1.0	4.0	8.1	0.16	2.9	1.0	2.9
15.5	20	1.48	0.2	1.0	0.2	20	0.16	0	1.0	0
16.5	21	1.56	0.2	1.0	0.2	21	0.15	0	1.0	0
17.5	3.0	0.20	5.5	1.0	5.5	3.0	0.15	6.0	1.0	6.0
18.5	3.6	0.26	5.3	1.0	5.3	3.6	0.15	5.0	1.0	5.0
19.5	2.1	0.13	6.0	1.0	6.0	2.1	0.15	8.0	1.0	8.0
20.5	26	2.05	0	1.0	0	26	0.15	0	1.0	0
					Total = 61 cm					Total = 53 cm

Notes: $(N_1)_{60}$, FS, and CSR obtained from Prob. 6.12. For Fig. 7.1, assume Japanese $N_1 = (N_1)_{60}$. H = thickness of soil layer.

7.12

| | Figure 7.1 | | | | | Figure 7.2 | | | | |
Depth, m	$(N_1)_{60}$	FS	ε_v, percent	H, m	Settlement, cm	$(N_1)_{60}$	CSR	ε_v, percent	H m	Settlement, cm
1.2	7.7	0.56	4.1	0.5	2.0	7.7	0.16	3.2	0.5	1.6
2.2	9.0	0.59	3.7	1.0	3.7	9.0	0.17	2.8	1.0	2.8
3.2	6.5	0.39	4.4	1.0	4.4	6.5	0.18	3.6	1.0	3.6
4.2	10	0.61	3.5	1.0	3.5	10	0.18	2.6	1.0	2.6
5.2	8.2	0.50	4.0	1.0	4.0	8.2	0.18	3.0	1.0	3.0
6.2	15	0.89	1.8	1.0	1.8	15	0.18	1.7	1.0	1.7
7.2	40	>2.8	0	1.0	0	40	0.18	0	1.0	0
8.2	25	1.61	0.2	1.0	0.2	25	0.18	0	1.0	0
9.2	34	>2.8	0	1.0	0	34	0.18	0	1.0	0
10.2	30	2.78	0	1.0	0	30	0.18	0	1.0	0
11.2	26	1.67	0.1	1.0	0.1	26	0.18	0	1.0	0
12.2	29	2.65	0	1.0	0	29	0.17	0	1.0	0
13.2	14	0.94	1.6	1.0	1.6	14	0.17	1.6	1.0	1.6
14.2	24	1.65	0.1	1.0	0.1	24	0.17	0	1.0	0
15.2	22	1.47	0.2	1.0	0.2	22	0.17	0	1.0	0
16.2	18	1.18	0.5	1.0	0.5	18	0.17	0.1	1.0	0.1
17.2	29	2.81	0	1.0	0	29	0.16	0	1.0	0
18.2	24	1.75	0.1	1.0	0.1	24	0.16	0	1.0	0
19.2	34	>3.1	0	1.3	0	34	0.16	0	1.3	0
					Total = 22 cm					Total = 17 cm

Notes: $(N_1)_{60}$, FS, and CSR obtained from Prob. 6.15. For Fig. 7.1, assume Japanese $N_1 = (N_1)_{60}$. H = thickness of soil layer in meters.

E.18

7.13

Depth, m	Cyclic stress ratio					N value	N value corrections				CRR	FS = CRR/CSR
	σ_v, kPa	σ_v', kPa	σ_v/σ_v'	r_d	CSR		C_r	N_{60}	C_N	$(N_1)_{60}$		
1.0	18.7	17.2	1.09	0.988	0.14	3	0.75	2.3	2.41	5.5	0.06	0.43
2.0	38.3	27.0	1.42	0.976	0.18	3	0.75	2.3	1.92	4.4	0.05	0.28
3.5	67.7	41.7	1.62	0.958	0.20	6	0.75	4.5	1.55	7.0	0.08	0.40
4.5	87.3	51.5	1.70	0.946	0.21	6	0.85	5.1	1.39	7.1	0.08	0.38
5.5	107	61.3	1.75	0.934	0.21	7	0.85	6.0	1.28	7.7	0.14*	0.67
6.5	127	71.1	1.79	0.922	0.21	6	0.95	5.7	1.19	6.8	0.13*	0.62
7.5	146	80.9	1.80	0.910	0.21	10	0.95	9.5	1.11	11	0.16*	0.76
8.5	166	90.7	1.83	0.898	0.21	21	0.95	20	1.05	21	0.23	1.10
9.5	185	100	1.85	0.886	0.21	35	0.95	33	1.00	33	>0.5	>2.4
10.5	205	110	1.86	0.874	0.21	35	1.00	35	0.95	33	>0.5	>2.4
11.5	225	120	1.88	0.862	0.21	31	1.00	31	0.91	28	0.38	1.81
12.5	244	130	1.88	0.850	0.21	46	1.00	46	0.88	40	>0.5	>2.4
13.5	264	140	1.89	0.838	0.21	31	1.00	31	0.85	26	0.30	1.43
14.5	283	149	1.90	0.826	0.20	36	1.00	36	0.82	30	>0.5*	>2.5
15.5	303	159	1.91	0.814	0.20	39	1.00	39	0.79	31	>0.5*	>2.5

Notes: Cyclic stress ratio: $a_{max} = 0.20g$, r_d from Eq. (6.7). N value corrections: $E_m = 0.6$, $C_b = 1.0$, C_N from Eq. (5.2). CRR from Fig. 6.6, asterisk means 15 percent fines curve was used.

7.13 (*Continued*)

Depth, m	Figure 7.1					Figure 7.2				
	$(N_1)_{60}$	FS	ε_v, percent	H, m	Settlement, cm	$(N_1)_{60}$	CSR	ε_v, percent	H, m	Settlement, cm
1.0	5.5	0.43	4.7	0.65	3.1	5.5	0.14	4.0	0.65	2.6
2.0	4.4	0.28	5.0	1.5	7.5	4.4	0.18	4.5	1.5	6.8
3.5	7.0	0.40	4.2	1.0	4.2	7.0	0.20	3.4	1.0	3.4
4.5	7.1	0.38	4.2	0.7	2.9	7.1	0.21	3.4	0.7	2.4
5.5	9.0*	0.67	3.7	1.3	4.8	9.0*	0.21	2.8	1.3	3.6
6.5	8.1*	0.62	4.0	1.0	4.0	8.1*	0.21	3.0	1.0	3.0
7.5	12*	0.76	3.1	0.8	2.5	12*	0.21	2.2	0.8	1.8
8.5	21	1.10	0.5	1.2	0.6	21	0.21	0.2	1.2	0.2
9.5	33	>2.4	0	1.0	0	33	0.21	0	1.0	0
10.5	33	>2.4	0	1.0	0	33	0.21	0	1.0	0
11.5	28	1.81	0.1	1.0	0.1	28	0.21	0	1.0	0
12.5	40	>2.4	0	1.0	0	40	0.21	0	1.0	0
13.5	26	1.43	0.2	0.8	0.2	26	0.21	0	0.8	0
14.5	31*	>2.5	0	1.2	0	31*	0.20	0	1.2	0
15.5	32*	>2.5	0	0.5	0	32*	0.20	0	0.5	0
					Total = 30 cm					Total = 24 cm

*Per Sec. 7.2.2, N_{corr} = 1.3 for 15 percent fines.

Notes: $(N_1)_{60}$, FS, and CSR obtained from prior table. For Fig. 7.1, assume Japanese $N_1 = (N_1)_{60}$; H = thickness of soil layer in meters (values vary because of soil layer thickness, see Fig. 7.12).

7.14

	Cyclic stress ratio					N value corrections						
Depth, m	σ_v, kPa	σ_v', kPa	σ_v/σ_v'	r_d	CSR	N value	C_r	N_{60}	C_N	$(N_1)_{60}$	CRR	FS = CRR/CSR
2.5†	46.3	46.3	1.00	0.97	0.13	2	0.75	1.5	1.47	2.2	0.08*	0.62
3.0	56.1	51.2	1.10	0.96	0.14	3	0.75	2.3	1.40	3.2	0.10*	0.71
4.0	75.7	61.0	1.24	0.95	0.15	14	0.75	11	1.28	14	0.16	1.07
5.2	99.2	72.7	1.36	0.94	0.17	9	0.85	7.7	1.17	9.0	0.10	0.59
6.0	115	80.6	1.43	0.93	0.17	11	0.85	9.4	1.11	10	0.19*	1.12
7.0	134	90.4	1.48	0.92	0.18	18	0.95	17	1.05	18	0.20	1.11
8.0	154	100	1.54	0.90	0.18	13	0.95	12	1.00	12	0.14	0.78
9.0	174	110	1.58	0.89	0.18	9	0.95	8.6	0.95	8.2	0.10	0.56
10.2	197	122	1.61	0.88	0.18	3	1.00	3.0	0.91	2.7	0.08*	0.44
11.0	213	130	1.64	0.87	0.19	11	1.00	11	0.88	9.7	0.18*	0.95
12.2	236	141	1.67	0.85	0.18	8	1.00	8.0	0.84	6.7	0.07	0.39
13.0	252	149	1.69	0.84	0.18	17	1.00	17	0.82	14	0.16	0.89
14.2	276	161	1.71	0.83	0.18	20	1.00	20	0.79	16	0.17	0.94

*Curve for 35 percent fines used in CRR from Fig. 6.6.

Notes: †SPT at 2 m used at groundwater table.

Cyclic stress ratio: $a_{max} = 0.20g$. r_d from Eq. (6.7). N value corrections: $E_m = 0.6$, $C_b = 1.0$, C_N from Eq. (5.2).

7.14 *(Continued)*

Depth, m	Figure 7.1					Figure 7.2				
	$(N_1)_{60}$	FS	ε_v, percent	H, m	Settlement, cm	$(N_1)_{60}$	CSR	ε_v, percent	H, m	Settlement, cm
2.5†	7.2*	0.62	4.2	0.2	0.8	7.2*	0.13	3.0	0.2	0.6
3.0	7.2*	0.71	4.2	1.3	5.5	7.2*	0.14	3.0	1.3	3.9
4.0	14	1.07	0.7	0.5	0.4	14	0.15	1.0	0.5	0.5
5.2	9.0	0.59	3.7	1.3	4.8	9.0	0.17	2.8	1.3	3.6
6.0	15*	1.12	0.6	0.4	0.2	15*	0.17	1.4	0.4	0.6
7.0	18	1.11	0.6	1.3	0.8	18	0.18	0.2	1.3	0.3
8.0	12	0.78	3.1	1.0	3.1	12	0.18	2.2	1.0	2.2
9.0	8.2	0.56	3.9	1.6	6.2	8.2	0.18	2.9	1.6	4.6
10.2	7.7*	0.44	4.1	0.4	1.6	7.7*	0.18	3.1	0.4	1.2
11.0	15*	0.95	1.3	0.8	1.0	15*	0.19	1.8	0.8	1.4
12.2	6.7	0.39	4.3	1.2	5.2	6.7	0.18	3.5	1.2	4.2
13.0	14	0.89	2.1	1.0	2.1	14	0.18	1.9	1.0	1.9
14.2	16	0.94	1.3	1.5	2.0	16	0.18	1.5	1.5	2.3
					Total = 34 cm					Total = 27 cm

*Per Sec. 7.2.2, N_{corr} used to account for fines.

†SPT at 2 m used at groundwater table.

Notes: $(N_1)_{60}$, FS, and CSR obtained from prior table. For Fig. 7.1, assume Japanese $N_1 = (N_1)_{60}$. H = thickness of soil layer in meters (values vary because of soil layer thickness, see Fig. 7.13).

7.15 Before Improvement

Depth, m	(N₁)₆₀ (see Notes)	Figure 7.1 FS	ε_v, percent	H, m	Settlement, cm	$(N_1)_{60}$	Figure 7.2 CSR	ε_v, percent	H, m	Settlement, cm
0.5	40	>2	0	0.5	0	40	0.26	0	0.5	0
1.5	20	0.66	2.2	1.0	2.2	20	0.38	1.6	1.0	1.6
2.5	14	0.43	2.8	1.0	2.8	14	0.42	2.1	1.0	2.1
3.5	15	0.42	2.7	1.0	2.7	15	0.44	2.0	1.0	2.0
4.5	24	0.76	1.4	1.0	1.4	24	0.44	1.3	1.0	1.3
5.5	13	0.36	2.9	1.0	2.9	13	0.45	2.2	1.0	2.2
6.5	11	0.32	3.3	1.0	3.3	11	0.44	2.5	1.0	2.5
7.5	21	0.59	2.1	1.0	2.1	21	0.44	1.5	1.0	1.5
8.5	11	0.32	3.3	1.0	3.3	11	0.44	2.5	1.0	2.5
9.5	11	0.32	3.3	1.0	3.3	11	0.44	2.5	1.0	2.5
10.5	12	0.34	3.1	1.0	3.1	12	0.44	2.3	1.0	2.3
11.5	5	0.16	4.8	1.0	4.8	5	0.43	4.2	1.0	4.2
12.5	10	0.30	3.5	1.0	3.5	10	0.43	2.6	1.0	2.6
13.5	11	0.31	3.3	1.0	3.3	11	0.42	2.5	1.0	2.5
14.5	11	0.33	3.3	1.0	3.3	11	0.42	2.5	1.0	2.5
15.5	11	0.34	3.3	1.0	3.3	11	0.41	2.5	1.0	2.5
					Total = 45 cm					Total = 35 cm

7.15 After Improvement

Depth, m	$(N_1)_{60}$	FS (estimated)	Figure 7.1 ε_v, percent	H, m	Settlement, cm	$(N_1)_{60}$	CSR	Figure 7.2 ε_v, percent	H, m	Settlement, cm
0.5	47	>2.0	0	0.5	0	47	0.26	0	0.5	0
1.5	37	>2.0	0	1.0	0	37	0.38	0	1.0	0
2.5	27	1.7	0.1	1.0	0.1	27	0.42	1.0*	1.0	1.0
3.5	26	1.4	0.2	1.0	0.2	26	0.44	1.2*	1.0	1.2
4.5	31	>2.0	0	1.0	0	31	0.44	0.5	1.0	0.5
5.5	36	>2.0	0	1.0	0	36	0.45	0	1.0	0
6.5	39	>2.0	0	1.0	0	39	0.44	0	1.0	0
7.5	34	>2.0	0	1.0	0	34	0.44	0	1.0	0
8.5	39	>2.0	0	1.0	0	39	0.44	0	1.0	0
9.5	40	>2.0	0	1.0	0	40	0.44	0	1.0	0
10.5	35	>2.0	0	1.0	0	35	0.44	0	1.0	0
11.5	34	>2.0	0	1.0	0	34	0.43	0	1.0	0
12.5	36	>2.0	0	1.0	0	36	0.43	0	1.0	0
13.5	39	>2.0	0	1.0	0	39	0.42	0	1.0	0
14.5	34	>2.0	0	2.0	0	34	0.42	0	2.0	0
					Total = 0.3 cm					Total = 2.7 cm

*In Fig. 7.2, these data points plot within the portion of the graph that corresponds to a factor of safety against liquefaction that is less than 1.0. Since the actual value of the factor of safety against liquefaction is greater than 1.0, the settlement value of 2.7 cm is too high. The actual settlement would probably be closer to 0.3 cm.

Notes: FS and CSR obtained from Prob. 6.18. $(N_1)_{60}$ values increased slightly to account for 15 percent fines (that is, N_{corr}). For Fig. 7.1, assume Japanese $N_1 = (N_1)_{60}$. CSR values not adjusted for $M = 8.5$ earthquake. H = thickness of soil layer in meters.

E.24

7.16 Assume that with soil improvement there will be no settlement in the upper 15 m of the soil deposit. The only remaining liquefiable soil layer is at a depth of 17 to 20 m. The stress increase due to the building load of 50 kPa at a depth of 17 m can be estimated from the 2:1 approximation, or

$$\Delta\sigma_v = \frac{qBL}{(B + z)(L + z)}$$

where z = depth from bottom of footing to top of liquefied soil layer
$$= 17 - 1 = 16 \text{ m}$$
$$L = 20 \text{ m} \qquad B = 10 \text{ m} \qquad q = 50 \text{ kPa}$$

Therefore,

$$\Delta\sigma_v = \frac{(50 \text{ kPa})(20)(10)}{(10 + 16)(20 + 16)} = 10.7 \text{ kPa}$$

Or, in terms of a percentage increase in σ'_{v0},

$$\text{Percent increase in } \sigma'_{v0} = \frac{10.7}{178} = 6\%$$

This is a very low percentage increase in vertical stress due to the foundation load. Thus the shear stress caused by the building load should not induce any significant additional settlement of the liquefied soil. Using the data from Prob. 7.11 at a depth of 15 to 20 m, we find the results shown on the next page.

7.17 Since the tank is in the middle of a liquefied soil layer, it is expected that the empty tank will not settle, but rather will float to the ground surface.

7.18 The thickness of the liquefiable sand layer H_2 is equal to 4 m. Entering Fig. 7.6 with $H_2 = 4$ m and intersecting the $a_{max} = 0.4g$ curve, the minimum thickness of the surface layer H_1 needed to prevent surface damage is 8.3 m. Since the surface layer of unliquefiable soil is only 6 m thick, there will be liquefaction-induced ground damage.

7.19 Problem 6.12 was solved based on a peak ground acceleration $a_{max} = 0.16g$. For a peak ground acceleration $a_{max} = 0.2g$, the factor of safety against liquefaction at a depth of 1.5 m is as follows:

$$\text{CSR} = \left(\frac{0.20}{0.16}\right)(0.10) = 0.13$$

Therefore,

$$\text{FS} = \frac{\text{CRR}}{\text{CSR}} = \frac{0.12}{0.13} = 0.92$$

For a peak ground acceleration of 0.2g, the zone of liquefaction will extend from a depth of 1.5 to 11 m. The thickness of the liquefiable sand layer H_2 is equal to 9.5 m. Entering Fig. 7.6 with $H_2 = 9.5$ m and extending the $a_{max} = 0.2g$ curve, the minimum thickness of

7.16

Depth, m	Figure 7.1					Figure 7.2				
	$(N_1)_{60}$	FS	ε_v, percent	H, m	Settlement, cm	$(N_1)_{60}$	CSR	ε_v, percent	H, m	Settlement, cm
15.5	20	1.48	0.2	1.0	0.2	20	0.16	0	1.0	0
16.5	21	1.56	0.2	1.0	0.2	21	0.15	0	1.0	0
17.5	3.0	0.20	5.5	1.0	5.5	3.0	0.15	6.0	1.0	6.0
18.5	3.6	0.26	5.3	1.0	5.3	3.6	0.15	5.0	1.0	5.0
19.5	2.1	0.13	6.0	1.0	6.0	2.1	0.15	8.0	1.0	8.0
20.5	26	2.05	0	1.0	0	26	0.15	0	1.0	0
					Total = 17 cm					Total = 19 cm

Note: All data obtained from Prob. 7.11.

E.26

the surface layer H_1 needed to prevent surface damage is 3 m. Since the surface layer of unliquefiable soil is only 1.5 m thick, there will be liquefaction-induced ground damage.

7.20 Since the zone of liquefaction extends from a depth of 0.85 to 8 m, the thickness of the liquefiable sand layer H_2 is equal to 7.2 m. Entering Fig. 7.6 with $H_2 = 7.2$ m and intersecting the $a_{max} = 0.2g$ curve, the minimum thickness of the surface layer H_1 needed to prevent surface damage is 3 m. Since the surface layer of unliquefiable soil is only 0.85 m thick, the minimum thickness of the surface fill layer is equal to $3 - 0.85 = 2.2$ m.

7.21 For a peak ground acceleration of $0.2g$, the zone of liquefaction will extend from a depth of 2.5 to 3.5 m. The thickness of the liquefiable sand layer H_2 is equal to 1 m. Entering Fig. 7.6 with $H_2 = 1$ m and intersecting the $a_{max} = 0.2g$ curve, the minimum thickness of the surface layer H_1 needed to prevent surface damage is 0.9 m. Since the surface layer of unliquefiable soil is 2.5 m thick, there will not be liquefaction-induced ground damage.

Note that the soils at depths of 4, 6, and 7 m have factors of safety against liquefaction that are only slightly in excess of 1.0. If these soils were to liquefy during the earthquake (such as due to the upward flow of groundwater), then the thickness of the liquefiable zone would extend from a depth of 2.5 to 14.5 m (or $H_2 = 12$ m). In this case, there would be liquefaction-induced ground damage.

7.22   See next page.

7.23   See page E.29.

7.24   See page E.30.

7.25   See page E.31.

CHAPTER 8

8.1   Use the following values:

$\text{FS} = 5.0$ $T = 2.5$ m $\tau_f = 50$ kPa $= 50$ kN/m^2 $P = 500$ kN $B = L$

Inserting the above values into Eq. (8.1b) (spread footing) yields

$$\text{FS} = \frac{R}{P} = \frac{2\,(B + L)\,T\tau_f}{P}$$

$$5.0 = \frac{2\,(2B)\,(2.5\text{ m})\,(50\text{ kN/m}^2)}{500\text{ kN}}$$

$$B = L = 5\text{ m}$$

8.2   Use the following values:

$\text{FS} = 5.0$ $T = 2.5$ m $\tau_f = 50$ kPa $= 50$ kN/m^2 $B = L = 2$ m

Inserting the above values into Eq. (8.1b) (spread footing) gives

$$\text{FS} = \frac{2\,(B + L)\,T\tau_f}{P}$$

7.22 Settlement Calculations Using the Tokimatsu and Seed (1987) Method

Layer number (1)	Layer thickness, ft (2)	$\sigma'_{v0} = \sigma_v$, lb/ft² (3)	$(N_1)_{60}$ (4)	G_{max} [Eq. (7.5)], kip/ft² (5)	$\gamma_{eff}(G_{eff}/G_{max})$ [Eq. (7.4)] (6)	γ_{eff} (Fig. 7.8) (7)	$\%\gamma_{cyc} = 100\gamma_{eff}$ (8)	ε_v (Fig. 7.9), percent (9)	Multi-directional shear = $2\varepsilon_v$ (10)	Multiply by VSR (11)	Settlement, in (12)
1	5	238	5	425	1.6×10^{-4}	8×10^{-4}	8×10^{-2}	0.38	0.76	0.61	0.4
2	5	713	5	736	2.8×10^{-4}	2.2×10^{-3}	2.2×10^{-1}	1.1	2.2	1.8	1.1
3	10	1425	5	1040	3.8×10^{-4}	2.3×10^{-3}	2.3×10^{-1}	1.2	2.4	1.9	2.3
4	10	2375	5	1340	4.7×10^{-4}	2.6×10^{-3}	2.6×10^{-1}	1.4	2.8	2.2	2.6
5	10	3325	5	1590	5.3×10^{-4}	2.3×10^{-3}	2.3×10^{-1}	1.2	2.4	1.9	2.3
6	10	4275	5	1800	5.8×10^{-4}	2.2×10^{-3}	2.2×10^{-1}	1.1	2.2	1.8	2.2

Total = 11 in

7.23 Settlement Calculations Using the Tokimatsu and Seed (1987) Method and the Chart Shown in Fig. 7.7

Layer number (1)	Layer thickness, ft (2)	$\sigma'_{v0} = \sigma_{v'}$, lb/ft² (3)	$(N_1)_{60}$ (4)	G_{max} [Eq. (7.5)], kip/ft² (5)	$\gamma_{eff}(G_{eff}/G_{max})$ [Eq. (7.4)] (6)	γ_{eff} (Fig. 7.8) (7)	$\%\gamma_{cyc} = 100\gamma_{eff}$ (8)	ε_v (Fig. 7.9), percent (9)	Multi-directional shear = $2\varepsilon_v$ (10)	Multiply by VSR (11)	Settlement, in (12)
1	5	238	5	613	1.1×10^{-4}	3.0×10^{-4}	3.0×10^{-2}	0.045	0.09	0.07	0.04
2	5	713	5	1060	1.9×10^{-4}	5.5×10^{-4}	5.5×10^{-2}	0.08	0.16	0.13	0.08
3	10	1425	5	1500	2.6×10^{-4}	8.5×10^{-4}	8.5×10^{-2}	0.13	0.26	0.21	0.25
4	10	2375	5	1940	3.3×10^{-4}	1.0×10^{-3}	1.0×10^{-1}	0.16	0.32	0.26	0.31
5	10	3325	5	2290	3.7×10^{-4}	9.5×10^{-4}	9.5×10^{-2}	0.15	0.30	0.24	0.29
6	10	4275	5	2600	4.0×10^{-4}	9.5×10^{-4}	9.5×10^{-2}	0.15	0.30	0.24	0.29
											Total = 1.3 in

Using Fig. 7.7, with $a_p/g = 0.45$ and $N = 15$, $\varepsilon_v = 0.15$ percent. Therefore, settlement = $(0.15/100)(50 \text{ ft})(12 \text{ in}) = 0.9$ in (2 cm)

7.24 Settlement Calculations Using the Tokimatsu and Seed (1987) Method and the Chart Shown in Fig. 7.7

Layer number (1)	Layer thickness, ft (2)	$\sigma'_{v0} = \sigma_v$, lb/ft² (3)	$(N_1)_{60}$ (4)	G_{max} [Eq. (7.5)], kip/ft² (5)	$\gamma_{eff}(G_{eff}/G_{max})$ [Eq. (7.4)] (6)	γ_{eff} (Fig. 7.8) (7)	$\%\gamma_{cyc} = 100\gamma_{eff}$ (8)	ε_v (Fig. 7.9), percent (9)	Multi-directional shear = $2\varepsilon_v$ (10)	Multiply by VSR (11)	Settlement, in (12)
1	5	238	9	517	5.9×10^{-5}	1.0×10^{-4}	1.0×10^{-2}	0.03	0.06	0.06	0.04
2	5	713	9	896	1.0×10^{-4}	1.7×10^{-4}	1.7×10^{-2}	0.05	0.10	0.10	0.06
3	10	1425	9	1270	1.4×10^{-4}	2.3×10^{-4}	2.3×10^{-2}	0.07	0.14	0.14	0.17
4	10	2375	9	1630	1.7×10^{-4}	2.6×10^{-4}	2.6×10^{-2}	0.08	0.16	0.16	0.19
5	10	3325	9	1930	2.0×10^{-4}	2.9×10^{-4}	2.9×10^{-2}	0.09	0.18	0.18	0.22
6	10	4275	9	2190	2.1×10^{-4}	3.0×10^{-4}	3.0×10^{-2}	0.095	0.19	0.19	0.23
											Total = 0.9 in

Using Fig. 7.7, with $a_p/g = 0.20$ and $N = 9$, $\varepsilon_v = 0.1$ percent. Therefore, settlement = $(0.1/100)(50 \text{ ft})(12 \text{ in}) = 0.6$ in (1.5 cm)

7.25 Settlement Calculations Using the Tokimatsu and Seed (1987) Method

Layer number (1)	Layer thickness, ft (2)	$\sigma'_{v0} = \sigma_v$, lb/ft² (3)	$(N_1)_{60}$ (4)	G_{max} [Eq. (7.5)], kip/ft² (5)	$\gamma_{eff}(G_{eff}/G_{max})$ [Eq. (7.4)] (6)	γ_{eff} (Fig. 7.8) (7)	$\%\gamma_{cyc} = 100\gamma_{eff}$ (8)	ε_v (Fig. 7.9), percent (9)	Multi-directional shear = $2\varepsilon_v$ (10)	Multiply by VSR (11)	Settlement, in (12)
1	5	238	5	425	7.2×10^{-5}	1.5×10^{-4}	1.5×10^{-2}	0.08	0.16	0.16	0.10
2	5	713	5	736	1.2×10^{-4}	2.2×10^{-4}	2.2×10^{-2}	0.11	0.22	0.22	0.13
3	10	1425	5	1040	1.7×10^{-4}	3.2×10^{-4}	3.2×10^{-2}	0.17	0.34	0.34	0.41
4	10	2375	5	1340	2.1×10^{-4}	3.6×10^{-4}	3.6×10^{-2}	0.18	0.36	0.36	0.43
5	10	3325	5	1590	2.4×10^{-4}	4.4×10^{-4}	4.4×10^{-2}	0.21	0.42	0.42	0.50
6	10	4275	5	1800	2.6×10^{-4}	4.2×10^{-4}	4.2×10^{-2}	0.20	0.40	0.40	0.48

Total = 2 in

$$5.0 = \frac{2\,(2+2)\,(2.5\text{ m})\,(50\text{ kN/m}^2)}{P}$$

$$P = 200\text{ kN}$$

8.3   Use the following values:

$$FS = 5.0 \qquad T = 2.5\text{ m} \qquad \tau_f = 10\text{ kPa} = 10\text{ kN/m}^2$$

Inserting the above values into Eq. (8.1a) (strip footing) yields

$$FS = \frac{2T\tau_f}{P}$$

$$5.0 = \frac{2\,(2.5\text{ m})\,(10\text{ kN/m}^2)}{P}$$

$$P = 10\text{ kN/m}$$

8.4   Use the following values:

$$FS = 5.0 \qquad T = 2.5\text{ m} \qquad \tau_f = 10\text{ kPa} = 10\text{ kN/m}^2 \qquad B = L = 2\text{ m}$$

Inserting the above values into Eq. (8.1b) (spread footing) gives

$$FS = \frac{2\,(B+L)\,T\tau_f}{P}$$

$$5.0 = \frac{2\,(2+2)\,(2.5\text{ m})\,(10\text{ kN/m}^2)}{P}$$

$$P = 40\text{ kN}$$

8.5   The first step is to check the two requirements in Table 8.3. Since the footings will be located within the upper unliquefiable cohesionless soil, the first requirement is met. As indicated in Fig. 7.6, the surface unliquefiable soil layer must be at least 3 m thick to prevent liquefaction-induced ground damage. Since a fill layer equal to 1.5 m is proposed for the site, the final thickness of the unliquefiable soil will be 3 m. Thus the second requirement is met.

To calculate the allowable bearing pressure for the strip and spread footings, the following values are used:

$$FS = 5.0$$

$T = 2.7$ m   (total thickness of unliquefiable soil layer minus footing embedment depth $= 3 - 0.3 = 2.7$ m)

$$\sigma'_{v0} = \sigma_v - u$$

Since the soil is above the groundwater table, assume $u = 0$. Use a total unit weight of 18.3 kN/m^3 (Prob. 6.12) and an average depth of 1.65 m [that is, $(0.3 + 3.0)/2 = 1.65$ m] or

$\sigma'_{v0} = 18.3 \times 1.65 = 30$ kPa

$\tau_f = k_0 \sigma'_{v0} \tan \phi' = 0.5$ (30 kPa) (tan 33°) $= 9.8$ kPa $= 9.8$ kN/m²   [Eq. (8.2c)]

$B = L = 1$ m

Using Eqs. (8.1a) and (8.1b) and the above values, we have the following:

For the strip footings:

$$P = q_{all} B = \frac{2T\tau_f}{FS}$$

$$q_{all} = \frac{2T\tau_f}{(FS)(B)} = \frac{2 \, (2.7 \text{ m}) \, (9.8 \text{ kPa})}{5 \, (1 \text{ m})} = 10 \text{ kPa}$$

For the spread footings:

$$P = q_{all} B^2 = \frac{2 \, (B + L) \, T\tau_f}{FS}$$

$$q_{all} = \frac{2 \, (B + L) \, T\tau_f}{FS \, (B^2)} = \frac{2 \, (1 + 1) \, (2.7 \text{ m}) \, (9.8 \text{ kPa})}{5 \, (1 \text{ m})^2} = 21 \text{ kPa}$$

Use $q_{all} = 20$ kPa.

8.6 The first step is to check the two requirements in Table 8.3. Since the footings will be located within the upper unliquefiable cohesive soil, the first requirement is met. As indicated in Fig. 7.6, the surface unliquefiable soil layer must be at least 3 m thick to prevent liquefaction-induced ground damage. Since a fill layer equal to 1.5 m is proposed for the site, the final thickness of the unliquefiable soil will be equal to 3 m. Thus the second requirement is met.

To calculate the allowable bearing capacity for the strip and spread footings, the following values are used:

FS = 5.0

$T = 2.7$ m   (total thickness of unliquefiable soil layer minus footing embedment depth $= 3 - 0.3 = 2.7$ m)

$\tau_f = s_u = 20$ kPa

$B = L = 1$ m

Using Eqs. (8.1a) and (8.1b) and the above values gives the following:

For the strip footings:

$$P = q_{all} \, B = \frac{2T\tau_f}{FS}$$

$$q_{all} = \frac{2T\tau_f}{FS \, (B)} = \frac{2 \, (2.7 \text{ m}) \, (20 \text{ kPa})}{5 \, (1 \text{ m})} = 21.6 \text{ kPa}$$

Use q_{all} = 20 kPa.

For the spread footings:

$$P = q_{all}B^2 = \frac{2\,(B + L)\,T\tau_f}{FS}$$

$$q_{all} = \frac{2\,(B + L)\,T\tau_f}{FS\,(B^2)} = \frac{2\,(1 + 1)\,(2.7\text{ m})\,(20\text{ kPa})}{5\,(1\text{ m})^2} = 43.5\text{ kPa}$$

Use q_{all} = 40 kPa.

8.7 For the sand, $c = 0$ and we neglect the third term in Eq. (8.3). Therefore,

$$q_{ult} = {}^1\!/_2\gamma_t BN_\gamma$$

Using Fig. 8.11, for $\phi = 33°$ and $N_\gamma = 26$, $T = 2.7$ m (i.e., total thickness of the unliquefiable soil layer minus footing embedment depth = $3 - 0.3 = 2.7$ m). Since $T/B = 2.7/1.0 = 2.7$, a reduction in N_γ would tend to be small for such a high ratio of T/B. Therefore,

For the strip footings:

$$q_{ult} = {}^1\!/_2\gamma_t BN_\gamma = {}^1\!/_2\,(18.3\text{ kN/m}^3)\,(1\text{ m})\,(26) = 238\text{ kPa}$$

$$q_{all} = \frac{q_{ult}}{FS} = \frac{238}{5} = 48\text{ kPa}$$

For the spread footings [using Eq. (8.4)]:

$$q_{ult} = 0.4\gamma_t BN_\gamma = 0.4\,(18.3\text{ kN/m}^3)\,(1\text{ m})\,(26) = 190\text{ kPa}$$

$$q_{all} = \frac{q_{ult}}{FS} = \frac{190}{5} = 38\text{ kPa}$$

Summary: $q_{all} = 48$ kPa for the 1-m-wide strip footings and $q_{all} = 38$ kPa for the 1-m by 1-m spread footings. For the design of the footings, use the lower values calculated in Prob. 8.5.

8.8 Total thickness of the unliquefiable soil layer minus footing embedment depth $T = 3 - 0.3 = 2.7$ m. Since $T/B = 2.7/1.0 = 2.7$, $N_c = 5.5$ per Fig. 8.8.

For the strip footings [using Eq. (8.6a)]:

$$q_{ult} = s_u N_c = (20\text{ kPa})\,(5.5) = 110\text{ kPa}$$

$$q_{all} = \frac{q_{ult}}{FS} = \frac{110}{5} = 22\text{ kPa}$$

For the spread footings [using Eq. (8.6b)]:

$$q_{ult} = s_u N_c \left(1 + 0.3\,\frac{B}{L}\right) = 1.3\,(20\text{ kPa})\,(5.5) = 143\text{ kPa}$$

$$q_{all} = \frac{q_{ult}}{FS} = \frac{143}{5} = 29 \text{ kPa}$$

Use $q_{all} = 30$ kPa.

Summary: $q_{all} = 22$ kPa for the 1-m-wide strip footings, and $q_{all} = 30$ kPa for the 1-m by 1-m spread footings. For the design of the strip footings, use the value from Prob. 8.6 ($q_{all} = 20$ kPa). For the design of the spread footings, use the lower value calculated in this problem ($q_{all} = 30$ kPa).

8.9 *Method from Sec. 8.2.2:* To calculate the allowable bearing pressure for the spread footings, the following values are used:

$$FS = 5.0$$

$T = 2.7$ m   (total thickness of unliquefiable soil layer minus footing embedment depth $= 3 - 0.3 = 2.7$ m)

$\tau_f = s_u = 20$ kPa

$B = L = 3$ m

Using Eq. (8.1*b*) and the above values gives

$$P = q_{all}B^2 = \frac{2(B + L)T\tau_f}{FS}$$

$$q_{all} = \frac{2(B + L)T\tau_f}{FS(B^2)} = \frac{2(3 + 3)(2.7 \text{ m})(20 \text{ kPa})}{5(3 \text{ m})^2} = 14 \text{ kPa}$$

Method from Sec. 8.2.3: To calculate the allowable bearing pressure for the spread footings, the following values are used:

$T = 2.7$ m   (total thickness of unliquefiable soil layer minus footing embedment depth $= 3 - 0.3 = 2.7$ m)

Since $T/B = 2.7/3.0 = 0.9$, $N_c = 2.3$ per Fig. 8.8. Using Eq. (8.6*b*) gives

$$q_{ult} = s_u N_c \left(1 + 0.3 \frac{B}{L}\right) = 1.3(20 \text{ kPa})(2.3) = 60 \text{ kPa}$$

$$q_{all} = \frac{q_{ult}}{FS} = \frac{60}{5} = 12 \text{ kPa}$$

Summary: From Eq. (8.1*b*), $q_{all} = 14$ kPa. From method in Sec. 8.2.3, $q_{all} = 12$ kPa. Use the lower value of 12 kPa for the design of the 3-m by 3-m spread footings.

8.10

$$FS = \frac{2(B + L)T\tau_f}{P}$$

where $B = L = 20$ m
$T = 2$ m   (distance from pile tips to top of liquefied soil layer)
$P = 50$ MN $= 5 \times 10^4$ kN
$\tau_f = k_0 \sigma'_{v0} \tan \phi'$

Average depth = 16 m, and use γ_t = 18.3 kN/m³ and γ_b = 9.7 kN/m³. At 18 m, σ'_{v0} = 1.5(18.3) + (16 − 1.5)(9.7) = 168 kPa = 168 kN/m².

$$\tau_f = 0.6 \,(168 \text{ kN/m}^2) \,(\tan 34°) = 68 \text{ kN/m}^2$$

$$FS = \frac{2\,(20 \text{ m} + 20 \text{ m})\,(2 \text{ m})\,(68 \text{ kN/m}^2)}{5 \times 10^4 \text{ kN}} = 0.20$$

Therefore, the pile foundation will punch down into the liquefied soil layer located at a depth of 17 to 20 m below ground surface.

8.11

$$FS = \frac{2\,(B' + L')\,T\tau_f}{P}$$

$z = \frac{1}{3}L = \frac{1}{3}\,(15 \text{ m}) = 5 \text{ m}$

$L' = L + z = 20 + 5 = 25 \text{ m}$ $B' = B + z = 20 + 5 = 25 \text{ m}$

$T = 2 \text{ m}$ (distance from pile tips to top of liquefied soil layer)

$P = 50 \text{ MN} = 5 \times 10^4 \text{ kN}$

$\tau_f = k_0 \sigma'_{v0} \tan \phi'$

Average depth = 16 m, and use γ_t = 18.3 kN/m³ and γ_b = 9.7 kN/m³. At 18 m, σ'_{v0} = 1.5(18.3) + (16 − 1.5)(9.7) = 168 kPa = 168 kN/m².

$$\tau_f = 0.6 \,(168 \text{ kN/m}^2) \,(\tan 34°) = 68 \text{ kN/m}^2$$

$$FS = \frac{2\,(25 \text{ m} + 25 \text{ m})\,(2 \text{ m})\,(68 \text{ kN/m}^2)}{5 \times 10^4 \text{ kN}} = 0.27$$

Therefore, the pile foundation will punch down into the liquefied soil layer located at a depth of 17 to 20 m below ground surface.

8.12   *Strip footing using Terzaghi bearing capacity equation:*

$e = 0.10 \text{ m}$ (for middle third of footing, e cannot exceed 0.17 m, so e is within middle third of footing)

$T = 2.5 \text{ m}$ (total thickness of unliquefiable soil layer minus footing embedment depth = 3 − 0.5 = 2.5 m)

$c_1 = s_u = 50 \text{ kPa} = 50 \text{ kN/m}^2$ (upper cohesive soil layer)

$c_2 = 0 \text{ kPa} = 0 \text{ kN/m}^2$ (liquefied soil layer)

$B = 1 \text{ m}$

$N_c = 5.5$ (using Fig. 8.8 with $T/B = 2.5/1.0 = 2.5$ and $c_2/c_1 = 0$)

Using the Terzaghi bearing capacity equation to calculate q_{ult} gives

$$q_{ult} = cN_c = s_u N_c = (50 \text{ kN/m}^2)\,(5.5) = 275 \text{ kN/m}^2$$

$$FS = q_{ult}/q' \quad \text{so} \quad q' = \frac{275 \text{ kN/m}^2}{5} = 55 \text{ kN/m}^2$$

$$q' = \frac{Q(B + 6e)}{B^2}$$

$$55 \text{ kN/m}^2 = \frac{Q[1 + 6(0.1)]}{1^2}$$

$$Q = 34 \text{ kN/m}$$

$$e = \frac{M}{Q}$$

$$M = eQ = 0.1 (34) = 3.4 \text{ kN} \cdot \text{m/m}$$

Strip footing using Fig. 8.9:

$$B' = B - 2e = 1 - 2(0.10) = 0.8 \text{ m}$$

$$T = 2.5 \text{ m}$$

$$c_1 = s_u = 50 \text{ kPa} = 50 \text{ kN/m}^2 \quad \text{(upper cohesive soil layer)}$$

$$c_2 = 0 \text{ kPa} = 0 \text{ kN/m}^2 \quad \text{(liquefied soil layer)}$$

$$N_c = 5.5 \quad \text{(using Fig. 8.8 with } T/B = 2.5/1.0 = 2.5 \text{ and } c_2/c_1 = 0)$$

Using the Terzaghi bearing capacity equation to calculate q_{ult} gives

$$q_{ult} = cN_c = s_u N_c = (50 \text{ kN/m}^2)(5.5) = 275 \text{ kN/m}^2$$

$$Q_{ult} = q_{ult}B' = (275 \text{ kN/m}^2)(0.8 \text{ m}) = 220 \text{ kN/m}$$

$$FS = \frac{Q_{ult}}{Q} \quad \text{or} \quad Q = \frac{220 \text{ kN/m}}{5} = 44 \text{ kN/m}$$

$$e = \frac{M}{Q}$$

$$M = eQ = 0.1 (44) = 4.4 \text{ kN} \cdot \text{m/m}$$

Use the lower values of $Q = 34$ kN/m and $M = 3.4$ kN·m/m calculated from the strip footing using the Terzaghi bearing capacity equation.

Spread footing using Terzaghi bearing capacity equation:

$$e = \frac{M}{Q} = 0.30 \text{ m} \qquad \text{(for middle third of footing, } e \text{ cannot exceed 0.33 m,}$$
$$\text{so } e \text{ } is \text{ within middle third of footing)}$$

$$T = 2.5 \text{ m}$$

$$c_1 = s_u = 50 \text{ kPa} = 50 \text{ kN/m}^2 \quad \text{(upper cohesive soil layer)}$$

$$c_2 = 0 \text{ kPa} = 0 \text{ kN/m}^2 \quad \text{(liquefied soil layer)}$$

$B = 2$ m

$N_c = 3.2$ (for spread footing, using Fig. 8.8 with $T/B = 2.5/2 = 1.25$ and $c_2/c_1 = 0$)

Use the Terzaghi beairng capacity equation to calculate q_{ult}:

$$q_{ult} = s_u N_c \left(1 + 0.3\frac{B}{L}\right) = 1.3 s_u N_c = 1.3(50 \text{ kN/m}^2)(3.2) = 208 \text{ kN/m}^2$$

$$FS = q_{ult}/q' \quad \text{or} \quad q' = \frac{208 \text{ kN/m}^2}{5} = 41.6 \text{ kN/m}^2$$

$$q' = \frac{Q(B + 6e)}{B^2}$$

$$41.6 \text{ kN/m}^2 = \frac{Q[2 + (6)(0.3)]}{(2)^2}$$

$$Q = 43.8 \text{ kN/m}$$

Convert Q to a load per the entire length of footing:

$$Q = (43.8 \text{ kN})(2 \text{ m}) = 88 \text{ kN}$$

$$e = \frac{M}{Q} \quad \text{or} \quad M = eQ = 0.3(88) = 26 \text{ kN} \cdot \text{m}$$

Spread footing using Fig. 8.9:

$B' = B - 2e = 2 - 2(0.30) = 1.4$ m

$L' = L = 2$ m   (moment only in B direction of footing)

$T = 2.5$ m

$c_1 = s_u = 50$ kPa $= 50$ kN/m²   (upper cohesive soil layer)

$c_2 = 0$ kPa $= 0$ kN/m²   (liquefied soil layer)

$N_c = 3.2$ (for spread footing, using Fig. 8.8 with $T/B = 2.5/2 = 1.25$ and $c_2/c_1 = 0$)

Use the Terzaghi bearing capacity equation to calculate q_{ult}:

$$q_{ult} = s_u N_c \left(1 + 0.3\frac{B'}{L'}\right) = 1.2 s_u N_c = 1.2(50 \text{ kN/m}^2)(3.2) = 190 \text{ kN/m}^2$$

$$Q_{ult} = q_{ult} B' L' = (190 \text{ kN/m}^2)(1.4 \text{ m})(2 \text{ m}) = 530 \text{ kN}$$

$$FS = \frac{Q_{ult}}{Q} \quad \text{or} \quad Q = \frac{530 \text{ kN}}{5} = 106 \text{ kN}$$

$$e = \frac{M}{Q} \quad \text{or} \quad M = eQ = 0.3(106) = 32 \text{ kN} \cdot \text{m}$$

Use the lower values of $Q = 88$ kN and $M = 26$ kN·m calculated earlier.

8.13

$$B' = B - 2e = 2 - 2 (0.30) = 1.4 \text{ m}$$

$$L' = L - 2e = 2 - 2 (0.30) = 1.4 \text{ m}$$

$$T = 2.5 \text{ m}$$

$$c_1 = s_u = 50 \text{ kPa} = 50 \text{ kN/m}^2 \qquad \text{(upper cohesive soil layer)}$$

$$c_2 = 0 \text{ kPa} = 0 \text{ kN/m}^2 \qquad \text{(liquefied soil layer)}$$

$$N_c = 3.2 \qquad \text{(for spread footing, using Fig. 8.8 with } T/B = 2.5/2 = 1.25 \\ \text{and } c_2/c_1 = 0)$$

Using the Terzaghi bearing capacity equation to calculate q_{ult} gives

$$q_{ult} = s_u N_c \left(1 + 0.3 \frac{B'}{L'}\right) = 1.3 s_u N_c = 1.3 \, (50 \text{ kN/m}^2) \, (3.2) = 208 \text{ kN/m}^2$$

$$Q_{ult} = q_{ult} B'L' = (208 \text{ kN/m}^2) \, (1.4 \text{ m}) \, (1.4 \text{ m}) = 408 \text{ kN}$$

$$FS = \frac{Q_{ult}}{Q} = \frac{408 \text{ kN}}{500 \text{ kN}} = 0.82$$

8.14 *Solution using Eq. (8.7):*

$$Q = P = 15{,}000 \text{ kips} \qquad \text{for mat foundation}$$

$$e = 5 \text{ ft} \qquad \text{(for middle third of mat, } e \text{ cannot exceed 16.7 ft, so } e \text{ is} \\ \text{within middle third of mat)}$$

Converting Q to a load per unit length of the mat gives

$$Q = \frac{15{,}000 \text{ kips}}{100 \text{ ft}} = 150 \text{ kips/ft} = 150{,}000 \text{ lb/ft}$$

$$q' = \frac{Q (B + 6e)}{B^2} = \frac{150{,}000 \, [100 + 6 \, (5)]}{100^2} = 1950 \text{ lb/ft}^2$$

Since $1950 < 2500$ lb/ft², design is acceptable.

Solution using Fig. 8.9:

$$Q = P = 15{,}000 \text{ kips} = 15{,}000{,}000 \text{ lb} \qquad \text{for mat foundation}$$

$$e = 5 \text{ ft}$$

$$B' = B - 2e = 100 - 2 \, (5) = 90 \text{ ft}$$

$$L' = L = 100 \text{ ft} \qquad \text{(moment only in } B \text{ direction of mat)}$$

$$q' = \frac{Q}{B'L'} = \frac{15{,}000{,}000}{100 \, (90)} = 1670 \text{ lb/ft}^2$$

Since $1670 < 2500$ lb/ft^2, design is acceptable.

8.15 *Strip footing:*

$$q_{all} = 24 \text{ kPa} = 24 \text{ kN/m}^2$$

Using $B = 1.5$ m yields

$$Q_{all} = (24 \text{ kN/m}^2) (1.5 \text{ m}) = 36 \text{ kN/m}$$

Spread footing:

$$q_{all} = 32 \text{ kPa} = 32 \text{ kN/m}^2$$

Using $B = L = 2.5$ m gives

$$Q_{all} = q_{all}BL = (32 \text{ kN/m}^2) (2.5 \text{ m}) (2.5 \text{ m}) = 200 \text{ kN}$$

8.16 Use the following values:

$\gamma_b = 9.7$ kN/m^3

$N_\gamma = 21$ (entering Fig. 8.11 with $\phi' = 32°$ and intersecting N_γ curve, N_γ from vertical axis is 21)

$B = 1.5$ m   for strip footings
 2.5 m   for spread footings

$r_u = 0.3$ (entering Fig. 5.15 with factor of safety against liquefaction = 1.2, r_u for sand varies from 0.12 to 0.46. Using an average value, $r_u = 0.3$)

Insert the above values into Eq. (8.10).

For the strip footings:

$$q_{ult} = \tfrac{1}{2} (1 - r_u) \gamma_b B N_\gamma = \tfrac{1}{2} (1 - 0.3) (9.7 \text{ kN/m}^3) (1.5 \text{ m}) (21) = 107 \text{ kPa}$$

and using a factor of safety of 5.0 gives

$$q_{all} = \frac{q_{ult}}{FS} = \frac{107 \text{ kPa}}{5.0} = 21 \text{ kPa}$$

For the spread footings:

$$q_{ult} = 0.4 (1 - r_u) \gamma_b B N_\gamma = 0.4 (1 - 0.3) (9.7 \text{ kN/m}^3) (2.5 \text{ m}) (21) = 140 \text{ kPa}$$

and using a factor of safety of 5.0 gives

$$q_{all} = \frac{q_{ult}}{FS} = \frac{140 \text{ kPa}}{5.0} = 28 \text{ kPa}$$

Thus, provided the strip and spread footings are at least 1.5 and 2.5 m wide, respectively, the allowable bearing capacity is equal to 21 kPa for the strip footings and 28 kPa for the spread footings. These allowable bearing pressures would be used to determine the size of the footings based on the anticipated dead, live, and seismic loads.

8.17 From Fig. 8.12, the fully weakened shear strength = 25 lb/in² = 3600 lb/ft² and

$$q = \frac{Q}{BL} = \frac{1000\ (20,000\ \text{kips})}{75\ (50)} = 5300\ \text{lb/ft}^2$$

Using Eq. (8.11b) gives

$$q_{ult} = cN_c \left(1 + 0.3\ \frac{B}{L}\right) = 5.5s_u \left(1 + 0.3\ \frac{B}{L}\right)$$

$$= 5.5s_u \left[1 + 0.3\left(\frac{50}{75}\right)\right] = 6.6s_u = (6.6)\ (3600) = 23,800\ \text{lb/ft}^2$$

$$\text{FS} = \frac{q_{ult}}{q} = \frac{23,800}{5300} = 4.5$$

8.18 *Zone of liquefaction:* The following table shows the liquefaction calculations. The data indicate that the zone of liquefaction extends from a depth of 2.3 to 18 m.

Fill layer: Since the zone of liquefation extends from a depth of 2.3 to 18 m, the thickness of the liquefiable sand layer H_2 is equal to 15.7 m. Entering Fig. 7.6 with $H_2 = 15.7$ m and extending the $a_{max} = 0.2g$ curve, the minimum thickness of the surface layer H_1 needed to prevent surface damage is 3 m. Since the surface layer of unliquefiable soil is 2.3 m thick, there will be liquefaction-induced ground damage. The required fill layer to be added at ground surface is equal to 0.7 m (3 m − 2.3 m = 0.7 m).

Settlement: The following table shows the calculations for the liquefaction-induced settlement of the ground surface. The calculated settlement is 54 to 66 cm using Figs. 7.1 and 7.2. The settlement calculations should include the 0.7-m fill layer, but its effect is negligible. The settlement calculations should also include the weight of the oil in the tank, which could cause the oil tank to punch through or deform downward the upper clay layer, resulting in substantial additional settlement. As indicated in the next section, the factor of safety for a bearing capacity failure is only 1.06, and thus the expected liquefaction-induced settlement will be significantly greater than 66 cm.

Bearing capacity:

$$P = \pi \left(\frac{D^2}{4}\right) (H)\ (\gamma_{oil})$$

where D = diameter of the tank, H = height of oil in the tank, and γ_{oil} = unit weight of oil. Therefore,

$$P = \pi \left[\frac{(20\ \text{m})^2}{4}\right] (3\ \text{m})\ (9.4\ \text{kN/m}^3) = 8860\ \text{kN}$$

$$\tau_f = 50\ \text{kPa} = 50\ \text{kN/m}^2$$

For the circular tank:

$$\text{FS} = \frac{R}{P} = \frac{\pi DT\tau_f}{P} = \frac{\pi\ (20\ \text{m})\ (3\ \text{m})\ (50\ \text{kN/m}^2)}{8860\ \text{kN}} = 1.06$$

8.18 (Continued)

Liquefaction analysis

Depth, m	Cyclic stress ratio						N value corrections						
	σ_v, kPa	σ_v', kPa	σ_v/σ_v'	r_d	CSR	N value	C_r	N_{60}	C_N	$(N_1)_{60}$	CRR	FS = CRR/CSR
4	77.3	47.4	1.631	0.952	0.202	9	0.85	7.7	1.45	11	0.12	0.59
8	155	86.7	1.788	0.904	0.210	5	0.95	4.8	1.07	5.2	0.06	0.29
10.4	202	110	1.836	0.875	0.209	8	1.00	8	0.95	7.6	0.08	0.38
12.1	235	126	1.865	0.855	0.207	4	1.00	4	0.89	3.6	0.10	0.48
14.8	287	152	1.888	0.822	0.202	2	1.00	2	0.81	1.6	0.07	0.35
17.4	338	177	1.910	0.791	0.196	9	1.00	9	0.75	6.8	0.15	0.77

Liquefaction-induced settlement analysis

Depth, m	Figure 7.1					Figure 7.2				
	$(N_1)_{60}$	FS	ε_v, percent	H, m	Settlement, cm	$(N_1)_{60}$	CSR	ε_v, percent	H, m	Settlement, cm
4	11	0.59	3.3	3.7	12.2	11	0.202	2.5	3.7	9.3
8	5.2	0.29	4.8	3.2	15.4	5.2	0.210	4.2	3.2	13.4
10.4	7.6	0.38	4.1	2.0	8.2	7.6	0.209	3.2	2.0	6.4
12.1	6.4	0.48	4.4	2.3	10.1	6.4	0.207	3.6	2.3	8.3
14.8	4.4	0.35	5.1	2.6	13.3	4.4	0.202	4.5	2.6	11.7
17.4	9.6	0.77	3.6	1.9	6.8	9.6	0.196	2.6	1.9	4.9
					Total = 66 cm					Total = 54 cm

Notes: Cyclic stress ratio: a_{max} = 0.20g, r_d from Eq. (6.7). N value corrections: E_m = 0.6, C_b = 1.0, C_N from Eq. (5.2). CRR from Fig. 6.6. Settlement analysis: Assume Japanese $N_1 = (N_1)_{60}$ with N_{corr} = 2.8 for the silty sand layer. H = thickness of soil layer in meters.

CHAPTER 9

9.1 The area of the wedge is first determined from simple geometry and is equal to 41.4 m² (450 ft²). For a unit length of the slope, the total weight W of the wedge equals the area times total unit weight, or 750 kN per meter of slope length (52,000 lb per foot of slope length).

Static case: Use Eq. (9.2a) with $F_h = 0$ and the following values.

$$c = 14.5 \text{ kPa} = 14.5 \text{ kN/m}^2 \ (300 \text{ lb/ft}^2)$$

$$\phi = 0$$

Length of slip surface $L = 29$ m (95 ft)

Slope inclination $\alpha = 18°$

Total weight of wedge $W = 750$ kN/m (51,700 lb/ft)

$$FS = \frac{cL + (W \cos \alpha) (\tan \phi)}{W \sin \alpha} = \frac{(14.5 \text{ kN/m}^2) \ (29 \text{ m})}{(750 \text{ kN/m}) \ (\sin 18°)} = 1.8$$

Earthquake case:

$$FS = \frac{cL + (W \cos \alpha - F_h \sin \alpha) (\tan \phi)}{W \sin \alpha + F_h \cos \alpha}$$

$$\phi = 0$$

$$FS = \frac{(14.5 \text{ kN/m}^2) \ (29 \text{ m})}{750 \sin 18° + 0.3 \ (750) \ (\cos 18°)} = 0.94$$

9.2 The area of the wedge $= 41.4$ m² (450 ft²), and for a unit length of the slope, the total weight W of the wedge $= 750$ kN per meter of slope length (52,000 lb per foot of slope length).

Static case: Use Eq. (9.2a) with $F_h = 0$ and the following values:

$$c' = 3.4 \text{ kPa} = 3.4 \text{ kN/m}^2 \ (70 \text{ lb/ft}^2)$$

$$\phi' = 29°$$

Length of slip surface $L = 29$ m (95 ft)

Slope inclination $\alpha = 18°$

Total weight of wedge $W = 750$ kN/m (51,700 lb/ft)

Average pore water pressure acting on slip surface $u = 2.4$ kPa (50 lb/ft²)

$$FS = \frac{cL + (W \cos \alpha - uL) (\tan \phi)}{W \sin \alpha}$$

$$= \frac{(3.4 \text{ kN/m}^2) \ (29 \text{ m}) + [750 \cos 18° - (2.4) \ (29 \text{ m})] \ (\tan 29°)}{(750 \text{ kN/m}) \ (\sin 18°)} = 1.95$$

Earthquake case:

$$FS = \frac{c'L + (W\cos\alpha - F_h\sin\alpha - uL)(\tan\phi')}{W\sin\alpha + F_h\cos\alpha}$$

$$= \frac{(3.4\text{ kN/m}^2)(29\text{ m}) + [750\cos 18° - (0.2)(750)(\sin 18°) - (2.4)(29)](\tan 29°)}{750\sin 18° + (0.2)(750)(\cos 18°)}$$

$$= 1.15$$

9.3   Using Eq. (9.2b) gives

$$FS = \frac{c'L + N'\tan\phi'}{W\sin\alpha + F_h\cos\alpha} = \frac{c'L + (W\cos\alpha - F_h\sin\alpha - uL)(\tan\phi')}{W\sin\alpha + F_h\cos\alpha}$$

where $W = (20\text{ ft})(10\text{ ft})(140\text{ lb/ft}^3) = 28{,}000$ lb/ft

$$F_h = 0.50(28{,}000) = 14{,}000\text{ lb/ft}$$

Since $c' = 0$, $u = 0$, and $\alpha = 0$, Eq. (9.2b) reduces to

$$FS = \frac{W\tan\phi'}{F_h} = \frac{28{,}000\tan 40°}{14{,}000} = 1.68$$

9.4   Available data from secs. 9.2.7 and 9.3.2:

$$FS = 1.0 \qquad k_h = 0.22$$

$$FS = 0.734 \qquad k_h = 0.40$$

Using a linear interpolation gives

$$\frac{1 - 0.734}{0.22 - 0.40} = \frac{FS - 0.734}{0.3 - 0.4}$$

$$FS = 0.88 \qquad \text{for } k_h = 0.30$$

9.5

$$FS = \frac{cL + (W\cos\alpha - F_h\sin\alpha)(\tan\phi)}{W\sin\alpha + F_h\cos\alpha}$$

$$\phi = 0$$

therefore,

$$FS = 1.0 = \frac{(14.5\text{ kN/m}^2)(29\text{ m})}{750\sin 18° + k_h(750)(\cos 18°)}$$

Solving the above equation for k_h, therefore for FS = 1, gives $k_h = a_y/g = 0.26$. From Prob. 9.1, $a_{max} = 0.30g$. Therefore, $a_y/a_{max} = 0.26g/0.30g = 0.867$. Using Eq. (9.3) with $a_y/a_{max} = 0.867$ gives

$$\log d = 0.90 + \log \left[\left(1 - \frac{a_y}{a_{max}}\right)^{2.53} \left(\frac{a_y}{a_{max}}\right)^{-1.09}\right]$$

$$= 0.90 + \log \left[(1 - 0.867)^{2.53} \, (0.867)^{-1.09}\right]$$

$$= 0.90 - 2.15 = -1.25$$

Solving for d gives $d = 0.06$ cm.

9.6 Since the pseudostatic factor of safety is greater than 1.0, per the Newmark method, the displacement of the slope is equal to zero ($d = 0$).

9.7

$$FS = \frac{c'L + (W \cos \alpha - F_h \sin \alpha - uL)(\tan \phi')}{W \sin \alpha + F_h \cos \alpha}$$

$$= \frac{(3.4 \text{ kN/m}^2)\,(29 \text{ m}) + [750 \cos 18° - (0.5)\,(750)\,(\sin 18°) - (2.4)\,(29)]\,(\tan 29°)}{750 \sin 18° + (0.5)\,(750)\,(\cos 18°)}$$

$$= 0.665$$

For FS $= 1.0$:

$$FS = 1.0 = \frac{3.4\,(29 \text{ m}) + [750 \cos 18° - F_h\,(750)\,(\sin 18°) - (2.4)\,(29)]\,(\tan 29°)}{750 \sin 18° + F_h\,(750 \cos 18°)}$$

Solving the above equation for k_h, therefore for FS $= 1$, gives $k_h = a_y/g = 0.266$. Per the problem statement, $a_{max} = 0.50g$. Therefore, $a_y/a_{max} = 0.266g/0.50g = 0.532$. Using Eq. (9.3) with $a_y/a_{max} = 0.532$ gives

$$\log d = 0.90 + \log \left[\left(1 - \frac{a_y}{a_{max}}\right)^{2.53} \left(\frac{a_y}{a_{max}}\right)^{-1.09}\right]$$

$$= 0.90 + \log \left[(1 - 0.532)^{2.53} \, (0.532)^{-1.09}\right]$$

$$= 0.90 - 0.54 = 0.36$$

Solving for d gives $d = 2.3$ cm.

9.8

$$\text{Slope height} = 9 \text{ m} + 1.5 \text{ m} = 10.5 \text{ m}.$$

Consider possible liquefaction of the soil from ground surface to a depth of 10.5 m. Based on the table at the top of the next page, the portion of the riverbank that has a factor of safety against liquefaction ≤ 1.0 is from a depth of 2 m to the toe of the slope (11.5 m below ground surface). Thus most of the 3 : 1 sloping riverbank is expected to liquefy during the design earthquake. As indicated in Sec. 9.4.1, if most of the sloping mass is susceptible to failure, a mass liquefaction flow slide can be expected during the earthquake.

Depth, m	$(N_1)_{60}$	D_r, percent (see Note 1)	K_α (see Note 2)	FS (see Note 3)	FS (see Note 4)
1.5	11	47	1.0	1.18	1.18
2.5	6.2	37	0.4	0.55	0.22
3.5	4.4	31	0.2	0.35	0.07
4.5	5.7	36	0.3	0.40	0.12
5.5	9.5	44	0.9	0.72	0.65
6.5	11	47	1.0	0.76	0.76
7.5	12	49	1.0	0.82	0.82
8.5	11	47	1.0	0.75	0.75
9.5	14	53	1.0	1.00	1.00
10.5	10	45	1.0	0.69	0.69

Notes:
1. D_r is based on $(N_1)_{60}$ correlation in Table 5.3.
2. K_α based on suggested guidelines in Sec. 9.4.2 and Fig. 9.24. For a 3:1 (horizontal:vertical) slope, $\alpha = \frac{1}{3} = 0.33$.
3. Factor of safety against liquefaction for level-ground site (see Prob. 6.12).
4. Factor of safety against liquefaction for the 3:1 sloping riverbank, or FS for sloping ground = (FS for level ground)(K_α).

9.9

$$\text{Slope height} = 0.4 \text{ m} + 5 \text{ m} = 5.4 \text{ m.}$$

Consider possible liquefaction of the soil from ground surface to a depth of 6.2 m.

Depth, m	$(N_1)_{60}$	D_r, percent (see Note 1)	K_α (see Note 2)	FS (see Note 3)	FS (see Note 4)
1.2	7.7	40	0.8	0.56	0.45
2.2	9.0	43	0.9	0.59	0.53
3.2	6.5	38	0.7	0.39	0.27
4.2	10	45	1.0	0.61	0.61
5.2	8.2	41	0.8	0.50	0.40
6.2	15	55	1.0	0.89	0.89

Notes:
1. D_r is based on $(N_1)_{60}$ correlation in Table 5.3.
2. K_α based on suggested guidelines in Sec. 9.4.2 and Fig. 9.24. For a 4:1 (horizontal:vertical) slope, $\alpha = \frac{1}{4} = 0.25$.
3. Factor of safety against liquefaction for level-ground site (see Prob. 6.15).
4. Factor of safety against liquefaction for the 4:1 sloping riverbank, i.e., FS for sloping ground = (FS for level ground)(K_α).

Based on the above table, the portion of the riverbank that has a factor of safety against liquefaction ≤ 1.0 is from a depth of 1.2 m to the toe of the slope (5.4 m below ground surface). Thus most of the 4:1 sloping riverbank is expected to liquefy during the design earthquake. As indicated in Sec. 9.4.1, if most of the sloping mass is susceptible to failure, a mass liquefaction flow slide can be expected during the earthquake.

9.10 *Before improvement:*

$$\text{Slope height} = 0.5 \text{ m} + 5 \text{ m} = 5.5 \text{ m}$$

Consider possible liquefaction of the soil from ground surface to a depth of 5.5 m.

Depth, m	$(N_1)_{60}$	D_r, percent (see Note 1)	K_α (see Note 2)	FS (see Note 3)	FS (see Note 4)
0.5	39	>85	1.0	>2	>2
1.5	19	63	1.0	0.66	0.66
2.5	13	51	1.0	0.43	0.43
3.5	14	53	1.0	0.42	0.42
4.5	23	69	1.0	0.76	0.76
5.5	12	49	1.0	0.36	0.36

Notes:
1. D_r is based on $(N_1)_{60}$ correlation in Table 5.3.
2. K_α based on suggested guidelines in Sec. 9.4.2 and Fig. 9.24. For a 3:1 (horizontal:vertical) slope, $\alpha = \frac{1}{3} = 0.33$.
3. Factor of safety against liquefaction for level-ground site (see Prob. 6.18).
4. Factor of safety against liquefaction for the 3:1 sloping riverbank, i.e., FS for sloping ground = (FS for level ground)(K_α).

Based on the above table, the portion of the riverbank that has a factor of safety against liquefaction ≤1.0 is from a depth of 1 m to the toe of the slope (5.5 m below ground surface). Thus most of the 3:1 sloping riverbank is expected to liquefy during the design earthquake. As indicated in Sec. 9.4.1, if most of the sloping mass is susceptible to failure, a mass liquefaction flow slide can be expected during the earthquake.

After improvement:

$$\text{Slope height} = 0.5 \text{ m} + 5 \text{ m} = 5.5 \text{ m}$$

Consider possible liquefaction of the soil from ground surface to a depth of 5.5 m.

Depth, m	$(N_1)_{60}$	D_r, percent (see Note 1)	K_α (see Note 2)	FS (see Note 3)	FS (see Note 4)
0.5	46	>85	1.0	>2	>2
1.5	36	>85	1.0	>1.3	>1.3
2.5	26	73	1.0	>1.2	>1.2
3.5	25	72	1.0	>1.1	>1.1
4.5	30	78	1.0	>1.1	>1.1
5.5	35	85	1.0	>1.1	>1.1

Notes:
1. D_r is based on $(N_1)_{60}$ correlation in Table 5.3.
2. K_α based on suggested guidelines in Sec. 9.4.2 and Fig. 9.24. For a 3:1 (horizontal:vertical) slope, $\alpha = \frac{1}{3} = 0.33$.
3. Factor of safety against liquefaction for level-ground site (see Prob. 6.18).
4. Factor of safety against liquefaction for the 3:1 sloping riverbank, i.e., FS for sloping ground = (FS for level ground)(K_α).

After improvement, the entire slope has a factor of safety against liquefaction >1.0. Therefore, the slope will not liquefy, and the riverbank is not susceptible to a flow failure.

9.11 Use the following conditions:

- Factor of safety against a flow slide at site >1
- Earthquake magnitude $M = 7.5$
- Distance from expected fault rupture to site $R = 50$ km
- Slope gradient $S = 1$ percent

- Thickness of layer T: The submerged sand having $(N_1)_{60} < 15$ is at a depth of 1 m and extends to a depth of 6 m, and therefore $T = 5$ m. This layer is also expected to liquefy during the design earthquake.
- Soil properties: Assume the soil comprising layer T has the same grain-size curve as soil no. 4 in Fig. 6.12.

Using the grain-size curve for soil no. 4 in Fig. 6.12, the percent passing the No. 200 sieve $F = 6$ percent, and the grain size corresponding to 50 percent finer $D_{50} = 0.38$ mm. For the sloping ground conditions:

$$\log D_H = -15.787 + 1.178\,(7.5) - 0.927 \log 50 - 0.013\,(50) + 0.429 \log 1 \\ + 0.348 \log 5 + 4.527 \log\,(100 - 6) - 0.922\,(0.38)$$

or $\log D_H = -0.35$ or $D_H = 0.45$ m

Bartless and Youd (1995) recommended that the value of D_H from Eq. (9.6) be multiplied by 2 in order to obtain a conservative design estimate of the lateral spreading. Thus the final expected horizontal deformation at the site due to lateral spreading $D_H = (0.45 \text{ m})(2) = 0.9$ m.

9.12 Use the following conditions:

- Factor of safety against a flow slide at site > 1
- Earthquake magnitude $M = 7.5$
- Distance from expected fault rupture to site $R = 50$ km
- Slope gradient $S = 6$ percent
- Thickness of layer $T = 1$ m
- Soil properties: Assume the soil comprising layer T has the same grain-size curve as soil no. 4 in Fig. 6.12.

Using the grain-size curve for soil no. 4 in Fig. 6.12, the percent passing the No. 200 sieve $F = 6$ percent, and the grain size corresponding to 50 percent finer $D_{50} = 0.38$ mm. For the sloping ground conditions:

$$\log D_H = -15.787 + 1.178\,(7.5) - 0.927 \log 50 - 0.013\,(50) + 0.429 \log 5 \\ + 0.348 \log 1 + 4.527 \log\,(100 - 6) - 0.922\,(0.38)$$

or $\log D_H = -0.30$ or $D_H = 0.51$ m

Bartlett and Youd (1995) recommended that the value of D_H from Eq. (9.6) be multiplied by 2 in order to obtain a conservative design estimate of the lateral spreading. Thus the final expected horizontal deformation at the site due to lateral spreading $D_H = (0.51 \text{ m})(2) = 1.0$ m.

9.13 The table on next page.

Lateral spreading analysis: Use the following conditions:

- Factor of safety against a flow slide at site > 1
- Earthquake magnitude $M = 7.5$
- Distance from expected fault rupture to site $R = 37$ km
- Slope gradient $S = 6$ percent

9.13

Liquefaction analysis

| Depth, m | Cyclic stress ratio | | | | | N value | N value corrections | | | | CRR | FS = CRR/CSR |
	σ_v, kPa	σ_v', kPa	σ_v/σ_v'	r_d	CSR		C_r	N_{60}	C_N	$(N_1)_{60}$		
2	37.2	32.3	1.13	0.98	0.18	5	0.75	3.8	1.76	6.7	0.07	0.39
4	76.2	51.7	1.42	0.95	0.22	6	0.75	4.5	1.39	6.3	0.07	0.32

Notes: Cyclic stress ratio: $a_{max} = 0.25g$, r_d from Eq. (6.7). N value corrections: $E_m = 0.6$, $C_b = 1.0$, C_N from Eq. (5.2). CRR from Fig. 6.6.

- Thickness of layer T: The submerged sand having $(N_1)_{60} < 15$ is at a depth of 1.5 m and extends to a depth of 6 m, and therefore $T = 4.5$ m. Per the liquefaction analysis, this layer is also expected to liquefy during the design earthquake.
- Soil properties: The percent passing the No. 200 sieve $F = 2$ percent, and the grain size corresponding to 50 percent finer $D_{50} = 0.6$ mm.

For the sloping ground conditions:

$$\log D_H = -15.787 + 1.178\ (7.5) - 0.927 \log 37 - 0.013\ (37) + 0.429 \log 6$$
$$+ 0.348 \log 4.5 + 4.527 \log\ (100 - 2) - 0.922\ (0.6)$$

or $\qquad\qquad \log D_H = 0.136 \qquad$ or $\qquad D_H = 1.37$ m

Bartlett and Youd (1995) recommended that the value of D_H from Eq. (9.6) be multiplied by 2 in order to obtain a conservative design estimate of the lateral spreading. Thus the final expected horizontal deformation at the site due to lateral spreading $D_H = (1.37$ m$)(2) = 2.7$ m.

9.14 *Static analysis:* Using the peak point on the stress-strain curve gives

$P_f = 105$ lb $\qquad H_0 = 6.98$ in $\qquad \Delta H = 0.14$ in $\qquad D_0 = 2.5$ in, $\qquad$ so $\qquad A_0 = 4.91$ in^2

$$\varepsilon_f = \frac{\Delta H}{H_0} = \frac{0.14}{6.98} = 0.020$$

$$A_f = \frac{A_0}{1 - \varepsilon_f} = \frac{4.91}{1 - 0.020} = 5.01 \text{ in}^2$$

$$q_u = \frac{P_f}{A_f} = \frac{105}{5.01} = 21.0 \text{ lb/in}^2$$

$$s_u = c = \frac{q_u}{2} = \frac{21.0}{2} = 10.5 \text{ lb/in}^2 = 1500 \text{ lb/ft}^2$$

As indicated in Fig. 9.41, $d = D/H = 20/40 = 0.5$. Enter chart (Fig. 9.41) at $\beta = 27°$, intersect line corresponding to $d = 0.5$, and therefore $N_0 = 6.3$.

$$F = \frac{N_0 c}{\gamma_t H} = \frac{6.3\ (1500)}{125\ (40)} = 1.89$$

Earthquake analysis: Using the fully softened location on the stress-strain curve yields

$P_f = 38$ lb $\qquad H_0 = 6.98$ in $\qquad \Delta H = 0.42$ in $\qquad D_0 = 2.5$ in $\qquad$ so $\qquad A_0 = 4.91$ in^2

$$\varepsilon_f = \frac{\Delta H}{H_0} = \frac{0.42}{6.98} = 0.060$$

$$A_f = \frac{A_0}{1 - \varepsilon_f} = \frac{4.91}{1 - 0.060} = 5.22 \text{ in}^2$$

$$q_u = \frac{P_f}{A_f} = \frac{38}{5.22} = 7.27 \text{ lb/in}^2$$

$$s_u = c = \frac{q_u}{2} = \frac{7.27}{2} = 3.6 \text{ lb/in}^2 = 520 \text{ lb/ft}^2$$

As indicated in Fig. 9.41, $d = D/H = 20/40 = 0.5$. Enter chart (Fig. 9.41) at $\beta = 27°$, intersect line corresponding to $d = 0.5$, and $N_0 = 6.3$.

$$F = \frac{N_0 c}{\gamma_t H} = \frac{6.3 \ (520)}{125 \ (40)} = 0.66$$

CHAPTER 10

10.1 *Static analysis:*

Equation (10.2):

$$k_A = \tan^2 (45° - \tfrac{1}{2}\phi) = \tan^2 [45° - \tfrac{1}{2} (32°)] = 0.307$$

Equation (10.4):

$$k_p = \tan^2 (45° + \tfrac{1}{2}\phi) = \tan^2 [45° + \tfrac{1}{2} (32°)] = 3.25$$

Equation (10.1):

$$P_A = \tfrac{1}{2} k_A \gamma_t H^2 = \tfrac{1}{2} (0.307) (20) (4)^2 = 49.2 \text{ kN/m}$$

Equation (10.3):

$$P_p = \tfrac{1}{2} k_p \gamma_t D^2 = \tfrac{1}{2} (3.25) (20) (0.5)^2 = 8.14 \text{ kN/m}$$

With reduction factor $= 2$, allowable $P_p = 4.07$ kN/m.

Resultant value of N and distance of N from the toe of footing:

Footing weight $= 3 \ (0.5) \ (32.5) = 35.3$ kN/m

Stem weight $= 0.4 \ (3.5) \ (23.5) = 32.9$ kN/m

$N =$ weight of concrete wall $= 35.3 + 32.9 = 68.2$ kN/m

Take moments about the toe of the wall to determine location of N:

$$Nx = -P_A \left(\frac{4}{3}\right) + (W) \text{ (moment arms)}$$

$$68.2x = -49.2 \left(\frac{4}{3}\right) + 35.5 \left(\frac{3}{2}\right) + 32.9 \ (2.8)$$

$$x = \frac{79.2}{68.2} = 1.165 \text{ m}$$

Maximum and minimum bearing pressure:

$$q' = \frac{N(B + 6e)}{B^2} \qquad q'' = \frac{N(B - 6e)}{B^2} \qquad \text{Eqs. (8.7a) and (8.7b)}$$

Note that Eqs. (8.7a) and (8.7b) are identical to the analysis presented in Fig. 10.5c.

Eccentricity $e = 1.5 - 1.165 = 0.335$ m

$$q' = 68.2 \ \frac{3 \mid 6 \ (0.335)}{3^2} = 37.9 \text{ kPa}$$

$$q'' = 68.2 \ \frac{3 - 6 \ (0.335)}{3^2} = 7.5 \text{ kPa}$$

Factor of safety for sliding:

$$FS = \frac{N \tan \delta + P_p}{P_A} \qquad \text{Eq. (10.11)}$$

$$= \frac{68.2 \tan 38° + 4.07}{49.2} = 1.17$$

Factor of safety for overturning:

$$FS = \frac{Wa}{\tfrac{1}{3}P_A H} \qquad \text{Eq. (10.12)}$$

$$= \frac{35.3 \ (3/2) + 32.9 \ (2.8)}{\tfrac{1}{3} \ (49.2) \ (4)} = 2.2$$

Earthquake analysis. To find the value of P_E:

$$\frac{a_{max}}{g} = 0.20$$

$$P_E = \tfrac{1}{2} k_A^{1/2} \left(\frac{a_{max}}{g} \right) H^2 \gamma_t \qquad \text{Eq. (10.7)}$$

$$= \tfrac{1}{2} \ (0.307)^{1/2} \ (0.20) \ (4^2) \ (20) = 17.7 \text{ kN/m}$$

So P_E is located at a distance of $\tfrac{2}{3}H$ above base of wall, or $\tfrac{2}{3}H = \tfrac{2}{3}(4) = 2.67$ m.

Resultant value of N and distance of N from the toe of footing: Take moments about the toe of the wall to determine location of N:

$$Nx = -P_A \left(\frac{4}{3} \right) - P_E \ (2.67) + W \ (\text{moment arms})$$

$$68.2x = -49.2 \left(\frac{4}{3} \right) - 17.7 \ (2.67) + 35.5 \left(\frac{3}{2} \right) + 32.9 \ (2.8)$$

$$x = \frac{32.5}{68.2} = 0.48 \text{ m}$$

Middle third of the foundation: $x = 1$ to 2 m, therefore N is not within the middle third of the foundation.

Factor of safety for sliding:

$$FS = \frac{N \tan \delta + P_p}{P_H + P_E} \qquad \text{Eq. (10.13)}$$

$$= \frac{68.2 \tan 38° + 4.07}{49.2 + 17.7} = 0.86$$

Factor of safety for overturning:

$$FS = \frac{Wa}{\tfrac{1}{3} P_H H + \tfrac{2}{3} H P_E} \qquad \text{Eq. (10.14) with } P_v = 0$$

$$= \frac{35.3 (3/2) + 32.9 (2.8)}{\tfrac{1}{3} (49.2) (4) + \tfrac{2}{3} (4) (17.7)} = 1.29$$

10.2 *Static analysis:* Static values will remain unchanged (see solution for Prob. 10.1). *Earthquake analysis:* To find the value of P_E:

$$\frac{a_{max}}{g} = 0.20$$

$$P_E = \tfrac{3}{8} \frac{a_{max}}{g} H^2 \gamma_t \qquad \text{Eq. (10.8)}$$

$$= \tfrac{3}{8} (0.20) (4^2) (20) = 24 \text{ kN/m}$$

So P_E is located at a distance of $0.6H$ above base of wall, or $0.6H = 0.6(4) = 2.4$ m.

Resultant value of N and distance of N from the toe of footing: Take moments about the toe of the wall to determine location of N:

$$Nx = -P_A \left(\frac{4}{3} \right) - P_E (2.67) + W \text{ (moment arms)}$$

$$68.2x = -49.2 \left(\frac{4}{3} \right) - 24 (2.4) + 35.5 \left(\frac{3}{2} \right) + 32.9 (2.8)$$

$$x = \frac{22.2}{68.2} = 0.33 \text{ m}$$

Middle third of the foundation: $x = 1$ to 2 m, therefore N is not within the middle third of foundation.

Factor of safety for sliding:

$$FS = \frac{N \tan \delta + P_p}{P_H + P_E} \quad \text{Eq. (10.13)}$$

$$= \frac{68.2 \tan 38° + 4.07}{49.2 + 24} = 0.78$$

Factor of safety for overturning:

$$FS = \frac{Wa}{\frac{1}{3}P_H H + 0.6HP_E} \quad \text{Eq. (10.15) with } P_v = 0$$

$$= \frac{35.3 \, (\frac{3}{2}) + 32.9 \, (2.8)}{\frac{1}{3} \, (49.2) \, (4) + 0.6 \, (4) \, (24)} = 1.18$$

10.3 *Static analysis:*

$$\delta = \phi_w = 24° \qquad \phi = 32° \qquad \theta = 0° \qquad \beta = 0°$$

Inserting the above values into Coulomb's equation (Fig. 10.3) gives

$$K_A = 0.275$$

From Eq. (10.1),

$$P_A = \frac{1}{2}k_A \gamma_t H^2$$

$$= \frac{1}{2} \, (0.275) \, (20) \, (4^2) = 43.9 \text{ kN/m}$$

$$P_H = P_A \cos 24° = 43.9 \cos 24° = 40.1 \text{ kN/m}$$

$$P_v = P_A \sin 24° = 43.9 \sin 24° = 17.9 \text{ kN/m}$$

Resultant location of N and distance of N from the toe of footing:

Footing weight = 3 (0.5) (23.5) = 35.3 kN/m

Stem weight = 0.4 (3.5) (23.5) = 32.9 kN/m

N = weight of concrete wall + P_v = 35.3 + 32.9 + 17.9 = 86.1 kN/m

Take moments about the toe of the wall:

$$Nx = -P_H \left(\frac{4}{3}\right) + W \text{ (moment arms)} + 3P_v$$

$$86.1x = -40.1 \left(\frac{4}{3}\right) + 35.3 \left(\frac{3}{2}\right) + 32.9 \, (2.8) + 17.9 \, (3)$$

$$x = \frac{145}{86.1} = 1.69 \text{ m}$$

Maximum and minimum bearing pressures:

$$q' = \frac{N(B + 6e)}{B^2} \qquad q'' = \frac{N(B - 6e)}{B^2} \qquad \text{Eqs. (8.7a) and (8.7b)}$$

Eccentricity $e = 1.5 - 1.69 = -0.19$ m

$$q' = \frac{86.1\,[3 + 6\,(0.19)]}{3^2} = 39.6 \text{ kPa}$$

$$q'' = \frac{86.1\,[3 - 6\,(0.19)]}{3^2} = 17.8 \text{ kPa}$$

Factor of safety for sliding:

$$FS = \frac{N \tan \delta + P_p}{P_H} \qquad \text{Eq. (10.11)}$$

where $N = W + P_v = 86.1$ kN/m

$$FS = \frac{86.1 \tan 38° + 4.07}{40.1} = 1.78$$

Factor of safety for overturning:

$$\text{Overturning moment} = P_H \, \tfrac{1}{3}H - 3P_v$$
$$= 40.1\,(\tfrac{1}{3})\,(4) - 17.9\,(3) = -0.23$$

Therefore

$$FS = \infty$$

Earthquake analysis: To find P_{AE}:

$$\frac{a_{max}}{g} = 0.20$$

$$\psi = \tan^{-1} k_h = \tan^{-1} \frac{a_{max}}{g} = \tan^{-1} 0.20 = 11.3° \qquad \text{Eq. (10.10)}$$

$$\delta = \phi_w = 24° \qquad \phi = 32° \qquad \theta = 0° \qquad \beta = 0°$$

Inserting the above values into the k_{AE} equation in Fig. 10.3 gives

$$K_{AE} = 0.428$$
$$P_{AE} = P_A + P_E = \tfrac{1}{2}k_{AE}H^2\gamma_t \qquad \text{Eq. (10.9)}$$
$$= \tfrac{1}{2}\,(0.428)\,(4^2)\,(20) = 68.5 \text{ kN/m}$$

Resultant location of N and distance of N from the toe of footing:

$$N = \text{weight of concrete wall} + P_{AE} \sin 24°$$

$$= 35.3 + 32.9 + 68.5 \sin 24° = 96.1 \text{ kN/m}$$

Take moments about the toe of the wall:

$$Nx = -(P_{AE} \cos \delta)\left(\frac{4}{3}\right) + (P_{AE} \sin \delta)(e) + W \text{ (moment arms)}$$

$$96.1x = -(68.5 \cos 24°)\left(\frac{4}{3}\right) + (68.5 \sin 24°)(3) + 35.3\left(\frac{3}{2}\right) + 32.9\,(2.8)$$

$$x = \frac{145.5}{96.1} = 1.51 \text{ m}$$

Maximum and minimum bearing pressures: Since N is essentially at the center of the footing,

$$q' = q'' = \frac{N}{B} = \frac{96.1}{3} = 32.0 \text{ kPa}$$

Factor of safety for sliding:

$$FS = \frac{N \tan \delta + P_p}{P_{AE} \cos \delta} \qquad \text{Eq. (10.16)}$$

where $N = W + P_{AE} \sin \delta = 96.1 \text{ kN/m}$

$$FS = \frac{96.1 \tan 38° + 4.07}{68.5 \cos 24°} = 1.26$$

Factor of safety for overturning:

$$\text{Overturning moment} = (P_{AE} \cos \delta)(\tfrac{1}{3}H) - (P_{AE} \sin \delta)(3)$$

$$= (68.5 \cos 24°)(\tfrac{1}{3})(4) - (68.5 \sin 24°)(3) = -0.15$$

therefore FS $= \infty$

Summary:

		Static analysis					Earthquake analysis			
Problem	N, kN/m	Location of N from toe, m	q', kPa	q'', kPa	FS sliding	FS over-turning	P_E or P_{AE}, kN/m	Location of N from toe, m	FS sliding	FS over-turning
10.1	68.2	1.16	37.9	7.5	1.17	2.2	17.7	0.48	0.86	1.29
10.2	68.2	1.16	37.9	7.5	1.17	2.2	24.0	0.33	0.78	1.18
10.3	86.1	1.69	39.6	17.8	1.78	∞	68.5	1.51	1.26	∞

10.4 *Static analysis:*

$$\delta = 32° \qquad \phi = 32° \qquad \theta = 0° \qquad \beta = 0°$$

Inserting the above values into Coulomb's equation (Fig. 10.3) gives $k_A = 0.277$.
Equation (10.4):

$$k_p = \tan^2 (45° + \tfrac{1}{2}\phi) = \tan^2 [45° + \tfrac{1}{2} (32°)] = 3.25$$

Using Eq. (10.1) gives

$$P_A = \tfrac{1}{2} k_A \gamma_t H^2$$

$$P_A = \tfrac{1}{2} (0.277) (20) (4^2) = 44.3 \text{ kN/m}$$

$$P_v = P_A \sin 32° = 44.3 \sin 32° = 23.5 \text{ kN/m}$$

$$P_H = P_A \cos 32° = 44.3 \cos 32° = 37.6 \text{ kN/m}$$

Equation (10.3):

$$P_p = \tfrac{1}{2} k_p \gamma_t D^2 = \tfrac{1}{2} (3.25) (20) (0.5)^2 = 8.14 \text{ kN/m}$$

with reduction factor $= 2$, allowable $P_p = 4.07 \text{ kN/m}$.
   Resultant value of N and distance of N from the toe of footing:

$$\text{Footing weight} = 2 (0.5) (23.5) = 23.5 \text{ kN/m}$$

$$\text{Stem weight} = 0.4 (3.5) (23.5) = 32.9 \text{ kN/m}$$

$$\text{Soil weight on top of footing} = 0.8 (3.5) (20) = 56.0 \text{ kN/m}$$

$$N = \text{weights} + P_v = 23.5 + 32.9 + 56.0 + 23.5 = 135.9 \text{ kN/m}$$

Take moments about the toe of the wall to determine x:

$$Nx = -P_H \left(\frac{4}{3}\right) + W \text{ (moment arms)} + 2P_v$$

$$135.9x = -37.6 \left(\frac{4}{3}\right) + 56.4 (1) + 56.0 (1.6) + 23.5 (2)$$

$$x = \frac{143}{135.9} = 1.05 \text{ m}$$

Maximum and minimum bearing pressures:

$$q' = \frac{N (B + 6e)}{B^2} \qquad q'' = \frac{N (B-6e)}{B^2} \qquad \text{Eqs. (8.7a) and (8.7b)}$$

Eccentricity $e = 1.0 - 1.05 = -0.05 \text{ m}$

$$q' = \frac{135.9 [2 + 6 (0.05)]}{2^2} = 78.1 \text{ kPa}$$

$$q'' = \frac{135.9 \, [2 - 6 \, (0.05)]}{2^2} = 57.8 \text{ kPa}$$

Factor of safety for sliding:

Friction angle between bottom of wall footing and underlying soil = 24°

$$\text{FS} = \frac{N \tan \delta + P_p}{P_H} \qquad \text{Eq. (10.11)}$$

$$N = W + P_v = 135.9 \text{ kN/m}$$

$$\text{FS} = \frac{135.9 \tan 24° + 4.07}{37.6} = 1.72$$

Factor of safety for overturning: Taking moments about the toe of the wall gives

$$\text{Overturning moment} = P_H \frac{H}{3} - 2P_v$$

$$= 37.6 \left(\frac{4}{3} \right) - 23.5 \, (2) = 3.1 \text{ kN} \cdot \text{m/m}$$

$$\text{Moment of weights} = 56.4 \, (1) + 56 \, (1.6) = 146 \text{ kN} \cdot \text{m/m}$$

$$\text{FS} = \frac{146}{3.1} = 47$$

Earthquake analysis: To find P_{AE},

$$\frac{a_{max}}{g} = 0.20$$

$$\psi = \tan^{-1} k_h = \tan^{-1} \frac{a_{max}}{g} = \tan^{-1} 0.20 = 11.3° \qquad \text{Eq. (10.10)}$$

$$\delta = 32° \qquad \phi = 32° \qquad \theta = 0° \qquad \beta = 0°$$

Inserting the above values into the k_{AE} equation in Fig. 10.3 yields

$$k_{AE} = 0.445$$

$$P_{AE} = P_A + P_E = \tfrac{1}{2} k_{AE} H^2 \gamma_t \qquad \text{Eq. (10.9)}$$

$$= \tfrac{1}{2} \, (0.445) \, (4^2) \, (20) = 71.2 \text{ kN/m}$$

Resultant location of N and distance of N from the toe of footing:

$$N = \text{weight of concrete wall} + P_{AE} \sin 32°$$

$$= 23.5 + 32.9 + 56.0 + 71.2 \sin 32° = 150.1 \text{ kN/m}$$

Take moments about the toe of the wall:

$$Nx = -(P_{AE} \cos \delta)\left(\frac{4}{3}\right) + (P_{AE} \sin \delta)(e) + W \text{ (moment arms)}$$

$$150.1x = -(71.2 \cos 32°)\left(\frac{4}{3}\right) + (71.2 \sin 32°)(2) + 56.4(1) + 56.0(1.6)$$

$$x = \frac{140.9}{150.1} = 0.94 \text{ m}$$

Maximum and minimum bearing pressures:

$$q' = \frac{N(B + 6e)}{B^2} \qquad q'' = \frac{N(B - 6e)}{B^2} \qquad \text{Eqs. (8.7a) and (8.7b)}$$

Eccentricity $e = 1.0 - 0.94 = -0.06$ m

$$q' = \frac{150.1 [2 + 6(0.06)]}{2^2} = 88.6 \text{ kPa}$$

$$q'' = \frac{150.1 [2 - 6(0.06)]}{2^2} = 61.5 \text{ kPa}$$

Factor of safety for sliding:

Friction angle between bottom of wall footing and underlying soil = 24°

$$FS = \frac{N \tan \delta + P_p}{P_{AE} \cos \delta} \qquad \text{Eq. (10.16)}$$

where

$$N = W + P_{AE} \sin \delta = 150.1 \text{ kN/m}$$

$$FS = \frac{150.1 \tan 24° + 4.07}{71.2 \cos 32°} = 1.17$$

Factor of safety for overturning:

Overturning moment $= (P_{AE} \cos \delta)(\frac{1}{3}H) - (P_{AE} \sin \delta)(2)$

$$= (71.2 \cos 32°)(\frac{1}{3})(4) - (71.2 \sin 32°)(2) = 5.1$$

Moment of weights $= 56.4(1) + 56(1.6) = 146 \text{ kN} \cdot \text{m/m}$

$$FS = \frac{146}{5.1} = 29$$

10.5 *Static analysis:*

$$P_Q = QHk_A = 200\,(20)\,(0.297) = 1190 \text{ lb/ft at 10 ft}$$

$$P_{QH} = 1190 \cos \phi_w = 1190 \cos 30° = 1030 \text{ lb/ft}$$

$$P_{Qv} = 1190 \sin \phi_w = 1190 \sin 30° = 595 \text{ lb/ft}$$

$$N = 15{,}270 + 595 = 15{,}870 \text{ lb/ft}$$

$$P_H = 5660 + 1030 = 6690 \text{ lb/ft}$$

Factor of safety for sliding:

$$FS = \frac{N \tan \delta + P_p}{P_H} \qquad \text{Eq. (10.11) where } \delta = \phi_{cv} = 30°$$

$$= \frac{15{,}870 \tan 30° + 750}{6690} = 1.48$$

Factor of safety for overturning:

Overturning moment $= 14{,}900 + 1030\,(10) - 595\,(7) = 21{,}030 \text{ ft} \cdot \text{lb/ft}$

Moment of weight $= 55{,}500 \text{ ft} \cdot \text{lb/ft}$

$$FS = \frac{55{,}500}{21{,}030} = 2.64$$

Location of *N:*

$$x = \frac{55{,}500 - 21{,}030}{15{,}870} = 2.17 \text{ ft}$$

Middle third of the foundation: $x = 2.33$ to 4.67 ft, therefore *N* is not within the middle third of the foundation.

Earthquake analysis: Calculate P_E for surcharge (neglect wall friction):

$$\frac{a_{max}}{g} = k_h = 0.20$$

The surcharge adds weight to the active wedge, where $L =$ top length of the active wedge, so

$$W_s = QL = (200 \text{ lb/ft}^2)\,(20 \tan 30°) = 2300 \text{ lb/ft}$$

$$P_E = k_h W = 0.20\,(2300) = 460 \text{ lb/ft} \qquad \text{Eq. (10.5)}$$

Total $P_E = 2540 + 460 = 3000 \text{ lb/ft}$

Factor of safety for sliding:

$$FS = \frac{N \tan \delta + P_p}{P_H + P_E} \qquad \text{Eq. (10.13) where } \delta = \phi_{cv} = 30°$$

$$= \frac{15{,}870 \tan 30° + 750}{6690 + 3000} = 1.02$$

Factor of safety for overturning:

Overturning moment $= 21{,}030 + \frac{2}{3}\,(20)\,(3{,}000) = 61{,}030$ ft · lb/ft

Moment of weight $= 55{,}500$ ft · lb/ft

$$FS = \frac{55{,}500}{61{,}030} = 0.91$$

Location of N:

$$x = \frac{55{,}500 - 61{,}030}{15{,}870} = -0.35 \text{ ft}$$

Middle third of the foundation: $x = 2.33$ to 4.67 ft, therefore N is not within the middle third of the foundation.

10.6 *Static analysis:*

$$\delta = \phi_w = 30° \qquad \phi = 30° \qquad \theta = 0° \qquad \beta = 18.4°$$

Inserting the above values into Coulomb's equation (Fig. 10.3) gives $k_A = 0.4065$. From Eq. (10.1),

$$P_A = \tfrac{1}{2} k_A \gamma_t H^2$$

$$= \tfrac{1}{2}\,(0.4065)\,(110)\,(20^2) = 8940 \text{ lb/ft}$$

$$P_v = P_A \sin 30° = 8940 \sin 30° = 4470 \text{ lb/ft}$$

$$P_H = P_A \cos 30° = 8940 \cos 30° = 7740 \text{ lb/ft}$$

$$N = 12{,}000 + 4470 = 16{,}470 \text{ lb/ft}$$

Factor of safety for sliding:

$$FS = \frac{N \tan \delta + P_p}{P_H} \qquad \text{Eq. (10.11) where } \delta = \phi_{cv} = 30°$$

$$= \frac{16{,}470 \tan 30° + 750}{7740} = 1.32$$

Factor of safety for overturning:

Overturning moment $= 7740 \left(\dfrac{20}{3} \right) - 4470\,(7) = 20{,}310$ ft · lb/ft

Moment of weight $= 55{,}500$ ft · lb/ft

$$FS = \frac{55{,}500}{20{,}310} = 2.73$$

Location of N:

$$x = \frac{55{,}500 - 20{,}310}{16{,}470} = 2.14 \text{ ft}$$

For the middle third of the retaining wall foundation, $x = 2.33$ to 4.67 ft. Therefore, N is not within the middle third of the retaining wall foundation.

Earthquake analysis: To find P_{AE},

$$\frac{a_{max}}{g} = 0.20$$

$$\psi = \tan^{-1} k_h = \tan^{-1} \frac{a_{max}}{g} = \tan^{-1} 0.20 = 11.3° \qquad \text{Eq. (10.10)}$$

$$\delta = \phi_w = 30° \qquad \phi = 30° \qquad \theta = 0° \qquad \beta = 18.4°$$

Inserting the above values into the k_{AE} equation in Fig. 10.3 gives

$$k_{AE} = 1.045$$

$$P_{AE} = P_A + P_E = \tfrac{1}{2} k_{AE} H^2 \gamma_t \qquad \text{Eq. (10.9)}$$

$$= \tfrac{1}{2} (1.045) (20^2) (110) = 23{,}000 \text{ lb/ft}$$

Factor of safety for sliding:

Friction angle between bottom of wall footing and underlying soil $= 30°$

$$\text{FS} = \frac{N \tan \delta + P_p}{P_{AE} \cos \delta} \qquad \text{Eq. (10.16)}$$

where

$$N = W + P_{AE} \sin \delta = 12{,}000 + 23{,}000 \sin 30° = 23{,}500 \text{ lb/ft}$$

$$\text{FS} = \frac{23{,}500 \tan 30° + 750}{23{,}000 \cos 30°} = 0.72$$

Factor of safety for overturning:

Overturning moment $= (P_{AE} \cos \delta)(\tfrac{1}{3}H) - (P_{AE} \sin \delta)(7)$

$$= (23{,}000 \cos 30°)(\tfrac{1}{3})(20) - (23{,}000 \sin 30°)(7)$$

$$= 52{,}300 \text{ ft} \cdot \text{lb/ft}$$

Moment of weights $= 55{,}500 \text{ ft} \cdot \text{lb/ft}$

$$\text{FS} = \frac{55{,}500}{52{,}300} = 1.06$$

Location of N:

$$x = \frac{55,500 - 52,300}{23,500} = 0.14 \text{ ft}$$

For the middle third of the retaining wall foundation, $x = 2.33$ to 4.67 ft. Therefore, N is not within the middle third of the retaining wall foundation.

10.7 *Static analysis:*

$$P_Q = P_2 = QHk_A = 200\,(20)\,(0.333) = 1330 \text{ lb/ft at 10 ft}$$

$$P_H = P_A + P_2 = 7330 + 1330 = 8660 \text{ lb/ft}$$

$$N = 33,600 \text{ lb/ft}$$

Factor of safety for sliding:

$$FS = \frac{N \tan \delta + P_P}{P_H} \qquad \text{Eq. (10.11)}$$

$$= \frac{33,600 \tan 23° + 740}{8660} = 1.73$$

Factor of safety for overturning:

$$\text{Overturning moment} = 7330 \left(\frac{20}{3}\right) + 1330\,(10) = 62,200$$

$$\text{Moment of weight} = 235,000$$

$$FS = \frac{235,000}{62,200} = 3.78$$

Maximum pressure exerted by the bottom of the wall:

$$x = \frac{235,000 - 62,200}{33,600} = 5.14$$

$$\text{Eccentricity } e = 7 - 5.14 = 1.86 \text{ ft}$$

$$q' = \frac{33,600\,[14 + 6\,(1.86)]}{(14)^2} = 4300 \text{ lb/ft}^2$$

Earthquake analysis: Calculate P_E for surcharge (neglect wall friction):

$$\frac{a_{max}}{g} = k_h = 0.20$$

The surcharge adds weight to the active wedge, where $L = $ top length of the active wedge, so

$$W_s = QL = (200 \text{ lb/ft}^2)(20 \tan 30°) = 2300 \text{ lb/ft}$$

$$P_E = k_h W = 0.20(2300) = 460 \text{ lb/ft} \qquad \text{Eq. (10.5)}$$

$$\text{Total } P_E = 2540 + 460 = 3000 \text{ lb/ft}$$

Factor of safety for sliding:

$$FS = \frac{N \tan \delta + P_p}{P_H + P_E} \qquad \text{Eq. (10.13) where } \delta = 23°$$

$$= \frac{33,600 \tan 23° + 740}{8660 + 3000} = 1.29$$

Factor of safety for overturning:

$$\text{Overturning moment} = 62,200 + \tfrac{2}{3}(20)(3000) = 102,200 \text{ ft} \cdot \text{lb/ft}$$

$$\text{Moment of weight} = 235,000 \text{ ft} \cdot \text{lb/ft}$$

$$FS = \frac{235,000}{102,200} = 2.30$$

Location of N:

$$x = \frac{235,000 - 102,200}{33,600} = 3.95 \text{ ft}$$

Middle third of the base of the wall: $x = 4.67$ to 9.33 ft. Therefore N is not within the middle third of the base of the wall.

10.8 *Static analysis:*

$$\delta = 0° \qquad \phi = 30° \qquad \theta = 0° \qquad \beta = 18.4°$$

Inserting the above values into Coulomb's equation (Fig. 10.3) gives $k_A = 0.427$. Using Eq. (10.1), we get

$$P_A = P_H = \tfrac{1}{2}k_A\gamma_t H^2 = \tfrac{1}{2}(0.427)(110)(20^2) = 9390 \text{ lb/ft}$$

Factor of safety for sliding:

$$FS = \frac{N \tan \delta + P_p}{P_H} = \frac{33,600 \tan 23° + 740}{9390} = 1.60$$

Factor of safety for overturning:

$$\text{Overturning moment} = 9390\left(\frac{20}{3}\right) = 62,600$$

$$\text{Moment of weight} = 235,200$$

$$FS = \frac{235,200}{62,600} = 3.76$$

Maximum pressure exerted by the wall foundation:

$$x = \frac{235,200 - 62,600}{33,600} = 5.14$$

Eccentricity $e = 7 - 5.14 = 1.86$ ft

$$q' = \frac{33,600 \, [14 + 6 \, (1.86)]}{(14)^2} = 4310 \text{ lb/ft}^2$$

Earthquake analysis: To find P_{AE},

$$\frac{a_{max}}{g} = 0.20$$

$$\psi = \tan^{-1} k_h = \tan^{-1} \frac{a_{max}}{g} = \tan^{-1} 0.20 = 11.3° \qquad \text{Eq. (10.10)}$$

$$\delta = \phi_w = 0° \qquad \phi = 30° \qquad \theta = 0° \qquad \beta = 18.4°$$

Inserting the above values into the k_{AE} equation in Fig. 10.3 yields

$$k_{AE} = 0.841$$

$$P_{AE} = P_A + P_E = \frac{1}{2} k_{AE} H^2 \, \gamma_t \qquad \text{Eq. (10.9)}$$

$$= \frac{1}{2} \, (0.841) \, (20^2) \, (110) = 18,500 \text{ lb/ft}$$

Since $\delta = 0°$, then $P_H = P_{AE} = 18,500$ lb/ft and $P_v = 0$.

Factor of safety for sliding:

$$FS = \frac{N \tan \delta + P_p}{P_H} = \frac{33,600 \tan 23° + 740}{18,500} = 0.81$$

Factor of safety for overturning:

$$\text{Overturning moment} = 18,500 \left(\frac{20}{3} \right) = 123,300 \text{ ft} \cdot \text{lb/ft}$$

$$\text{Moment of weight} = 235,200$$

$$FS = \frac{235,200}{123,300} = 1.91$$

Location of N:

$$x = \frac{235,200 - 123,300}{33,600} = 3.33 \text{ ft}$$

Middle third of the base of the wall: $x = 4.67$ to 9.33 ft. Therefore N is not within the middle third of the base of the wall.

10.9 *Static analysis:* For the failure wedge,

$$W = \tfrac{1}{2} (20) (11.1) (120) = 13{,}300 \text{ lb/ft} \qquad R = 12{,}000 \text{ lb/ft}$$

$$FS = \frac{(W \cos \alpha + R \sin \alpha) (\tan \phi)}{W \sin \alpha - R \cos \alpha}$$

$$= \frac{(13{,}300 \cos 61° + 12{,}000 \sin 61°) (\tan 32°)}{13{,}300 \sin 61° - 12{,}000 \cos 61°}$$

$$= 1.82$$

Earthquake analysis: Using Eq. (9.2a) with $c = 0$ and including R, we find

$$FS = \frac{(W \cos \alpha - k_h W \sin \alpha + R \sin \alpha) (\tan \phi)}{W \sin \alpha - R \cos \alpha + k_h W \cos \alpha}$$

where

$$k_h W = 0.20 (13{,}300) = 2660 \text{ lb/ft}$$

$$FS = \frac{(13{,}300 \cos 61° - 2660 \sin 61° + 12{,}000 \sin 61°) (\tan 32°)}{13{,}300 \sin 61° - 12{,}000 \cos 61° + 2660 \cos 61°}$$

$$= 1.29$$

10.10 *Static analysis:*

$$P_Q = QHk_A = 200 (50) (0.295) = 2950 \text{ lb/ft at 25 ft}$$

Moment due to active forces $= 7.8 \times 10^5 + 2950 (25 - 4) = 8.4 \times 10^5$

$$FS = \frac{\text{resisting moment}}{\text{destabilizing moment}} = \frac{1.71 \times 10^6}{8.4 \times 10^5} = 2.03$$

$$A_p = P_A + P_Q - \frac{P_p}{FS} = 27{,}500 + 2950 - \frac{43{,}400}{2.03} = 9070 \text{ lb/ft}$$

For a 10-ft spacing, therefore,

$$A_p = 10 (9070) = 90{,}700 \text{ lb} = 90.7 \text{ kips}$$

Earthquake analysis, pseudostatic method:

$$\frac{a_{max}}{g} = k_h = 0.20$$

The surcharge adds weight to the active wedge, where $L =$ top length of the active wedge, so

$$W_s = QL = (200 \text{ lb/ft}^2) (50 \tan 28.5°) = 5430 \text{ lb/ft}$$

$$P_E = k_h W = 0.20 (5430) = 1090 \text{ lb/ft} \qquad \text{Eq. (10.5)}$$

Total $P_E = 8690 + 1090 = 9780$ lb/ft

Now P_E acts at a distance of $\frac{2}{3}(H + D)$ above the bottom of the sheet pile wall. Thus

Moment due to $P_E = 9780 [\frac{1}{3} (50) - 4] = 1.24 \times 10^5$

Total destabilizing moment $= 8.40 \times 10^5 + 1.24 \times 10^5 = 9.64 \times 10^5$

Moment due to passive force $= 1.71 \times 10^6$

$$\text{FS} = \frac{\text{resisting moment}}{\text{destabilizing moment}} = \frac{1.71 \times 10^6}{9.64 \times 10^5}$$

$$= 1.77$$

$$A_p = P_A + P_E - \frac{P_p}{\text{FS}} = 30,450 + 9780 - \frac{43,400}{1.77} = 15,700 \text{ lb/ft}$$

For a 10-ft spacing, therefore,

$$A_p = 10 (15,700) = 157,000 \text{ lb} = 157 \text{ kips}$$

Earthquake analysis, liquefaction of passive wedge:

Total $P_E = 8690 + 1090 = 9780$ lb/ft

And P_E acts at a distance of $\frac{2}{3}(H + D)$ above the bottom of the sheet pile wall. So

Total destabilizing moment $= 8.40 \times 10^5 + 1.24 \times 10^5 = 9.64 \times 10^5$

Moment due to passive force $= 1.35 \times 10^6$

$$\text{FS} = \frac{\text{resisting moment}}{\text{destabilizing moment}} = \frac{1.35 \times 10^6}{9.64 \times 10^5}$$

$$= 1.40$$

$$A_p = P_A + P_E - \frac{P_p}{\text{FS}} = 30,450 + 9780 - \frac{32,600}{1.40} = 16,900 \text{ lb/ft}$$

For a 10-ft spacing, therefore,

$$A_p = 10 (16,900) = 169,000 \text{ lb} = 169 \text{ kips}$$

10.11   *Static analysis:*

$$\delta = 0° \qquad \phi = 33° \qquad \theta = 0° \qquad \beta = 18.4°$$

Inserting the above values into Coulomb's equation (Fig. 10.3) gives $k_A = 0.369$.

$$\text{Destabilizing moment} = \left(\frac{0.369}{0.295}\right)(7.8 \times 10^5) = 9.76 \times 10^5$$

$$FS = \frac{\text{resisting moment}}{\text{destabilizing moment}} = \frac{1.71 \times 10^6}{9.76 \times 10^5}$$

$$= 1.75$$

$$A_p = P_A - \frac{P_p}{FS} = 27,500 \left(\frac{0.369}{0.295} \right) - \frac{43,400}{1.75} = 9600 \text{ lb/ft}$$

For a 10-ft spacing, therefore,

$$A_p = 10 \, (9600) = 96,000 \text{ lb} = 96 \text{ kips}$$

Earthquake analysis, pseudostatic method:

$$\frac{a_{max}}{g} = 0.20$$

$$\psi = \tan^{-1} k_h = \tan^{-1} \frac{a_{max}}{g} = \tan^{-1} 0.20 = 11.3° \qquad \text{Eq. (10.10)}$$

$$\delta = \phi_w = 0° \qquad \phi = 33° \qquad \theta = 0° \qquad \beta = 18.4°$$

Inserting the above values into the k_{AE} equation in Fig. 10.3 gives

$$k_{AE} = 0.641$$

$$P_{AE} = P_A + P_E = \frac{1}{2} k_{AE} H^2 \gamma_t \qquad \text{Eq. (10.9)}$$

Since the effect of the water pressure tends to cancel out on both sides of the wall, use Eq. (10.9) and estimate P_E based on the buoyant unit weight ($\gamma_b = 64$ lb/ft^3).

$$P_{AE} = \frac{1}{2} \, (0.641) \, (50^2) \, (64) = 51,300 \text{ lb/ft}$$

Since $\delta = 0°$,

$$P_H = P_{AE} = 51,300 \text{ lb/ft} \qquad \text{and} \qquad P_v = 0$$

And P_{AE} acts at a distance of $1/3(H + D)$ above the bottom of the sheet pile wall.

Moment due to $P_{AE} = 51,300 \, [\frac{2}{3} \, (50) - 4] = 1.50 \times 10^6$

$$FS = \frac{\text{resisting moment}}{\text{destabilizing moment}} = \frac{1.71 \times 10^6}{1.50 \times 10^6}$$

$$= 1.14$$

$$A_p = P_{AE} - \frac{P_p}{FS} = 51,300 - \frac{43,400}{1.14} = 13,200 \text{ lb/ft}$$

For a 10-ft spacing, therefore,

$$A_p = 10 \, (13,200) = 132,000 \text{ lb} = 132 \text{ kips}$$

Earthquake analysis, liquefaction of passive wedge:

Moment due to passive force $= 1.35 \times 10^6$

$$FS = \frac{\text{resisting moment}}{\text{destabilizing moment}}$$

$$= \frac{1.35 \times 10^6}{1.50 \times 10^6}$$

$$= 0.90$$

Note that the substantial differences in the factor of safety for toe kick-out for the earth-quake analysis in Probs. 10.10 and 10.11 are due in large part to the assumed location of P_E and P_{AE} [that is, P_E is assumed to act at a distance of $\frac{2}{3}(H + D)$ while P_{AE} is assumed to act at a distance of $\frac{1}{3}(H + D)$ above the bottom of the sheet pile wall].

10.12   *Static analysis:*   Use the following equation (Day 1999):

$$k_p = \frac{\cos^2 \phi}{[1 - (\sin^2 \phi + \sin \phi \cos \phi \tan \beta)^{0.5}]^2}$$

where ϕ = friction angle of the soil in front of the retaining wall and β = slope inclination measured from a horizontal plane, where a descending slope in front of the retaining wall has a negative β value. Although not readily apparent, if $\beta = 0$, the above equation will give exactly the same values of k_p as Eq. (10.4). Inserting $\beta = -18.4°$ and $\phi' = 33°$ into the above equation, we get $k_p = 1.83$.

$$\text{Resisting moment} = \left(\frac{1.83}{3.39}\right)(1.71 \times 10^6) = 9.23 \times 10^5$$

$$FS = \frac{\text{resisting moment}}{\text{destabilizing moment}} = \frac{9.23 \times 10^5}{7.8 \times 10^5} = 1.18$$

Earthquake analysis:

$$P_E = \frac{1}{2}k_A^{\frac{1}{2}}\left(\frac{a_{max}}{g}\right)H^2\gamma_b = \frac{1}{2}\,(0.295)^{\frac{1}{2}}(0.20)(50^2)(64) = 8690 \text{ lb/ft}$$

And P_E acts at a distance of $\frac{2}{3}(H + D)$ above the bottom of the sheet pile wall.

Moment due to $P_E = 8690\,[\frac{1}{3}\,(50) - 4] = 1.10 \times 10^5$

Total destabilizing moment $= 7.80 \times 10^5 + 1.10 \times 10^5 = 8.90 \times 10^5$

$$\text{Moment due to passive force} = \left(\frac{1.83}{3.39}\right)(1.71 \times 10^6) = 9.23 \times 10^5$$

$$FS = \frac{\text{resisting moment}}{\text{destabilizing moment}} = \frac{9.23 \times 10^5}{8.90 \times 10^5}$$

$$= 1.04$$

10.13   *Static analysis:*

Weight of anchor block $W = (5^2)\,(2)\,(150 \text{ lb/ft}^3) = 7500 \text{ lb}$

Determine the frictional resistance at the top and bottom of the anchor block:

$$\text{Friction coefficient} = \tfrac{2}{3}\phi = \tfrac{2}{3}(33°) = 22°$$

$$\text{Friction at top of block} = N \tan 22°$$

where

$$N = (5^2)(3)(120 \text{ lb/ft}^3) = 9000 \text{ lb}$$

Therefore,

$$\text{Friction at top of block} = N \tan 22° = 9000 \tan 22° = 3640 \text{ lb}$$

$$\text{Friction at bottom of block} = (N + W)(\tan 22°) = (9000 + 7500)(\tan 22°) = 6670 \text{ lb}$$

For passive resistance along front side of block,

$$k_p = \tan^2(45° + \tfrac{1}{2}\phi) = \tan^2[45° + \tfrac{1}{2}(33°)] = 3.39 \quad \text{Eq. (10.4)}$$

Passive pressure at depth of 4 ft $= k_p \gamma_t D = 3.39 (120 \text{ lb/ft}^2)(4 \text{ ft}) = 1630 \text{ lb/ft}^2$

Assuming block slides out from soil (i.e., passive resistance only at a depth of 3 to 5 ft below ground surface), we find

$$\text{Passive resistance} = 2 (5)(1630 \text{ lb/ft}^2) = 16,300 \text{ lb}$$

Neglecting friction along sides of block, we therefore find

$$\text{Total lateral resistance} = \text{top friction} + \text{bottom friction} + \text{passive resistance}$$

$$= 3640 + 6670 + 16,300 = 26,610 \text{ lb} = 26.6 \text{ kips}$$

Earthquake analysis: If all the soil behind the sheet pile wall were to liquefy, then the total lateral resistance of the anchor block would equal zero (i.e., for liquefied soil, friction angle is 0°, hence top friction and bottom friction are zero and passive resistance is zero). The anchor block will also tend to sink into the liquefied soil.

10.14 *Static analysis:*

$$k_A = \tan^2(45° - \tfrac{1}{2}\phi) = \tan^2[45° - \tfrac{1}{2}(32°)] = 0.307$$

From Fig. 10.10,

$$\sigma_h = 0.65 k_A \gamma_t H = 0.65 (0.307)(120)(20) = 480 \text{ lb/ft}^2$$

$$\text{Resultant force} = \sigma_h H = 480 (20) = 9600 \text{ lb/ft}$$

Earthquake analysis:

$$\frac{a_{max}}{g} = 0.20$$

$$P_E = \tfrac{1}{2} k_A^{1/2} \left(\frac{a_{max}}{g} \right) H^2 \gamma_t \quad \text{Eq. (10.7)}$$

$$= \tfrac{1}{2}(0.307)^{1/2}(0.20)(20^2)(120) = 2700 \text{ lb/ft}$$

10.15 *Static analysis:*

$$N_0 = \frac{\gamma_t H}{c} = \frac{120\,(20)}{300} = 8$$

Therefore use case (b) in Fig. 10.10.

$$k_A = 1 - \frac{m\,(4c)}{\gamma_t H}$$

Use $m = 1$. Therefore,

$$k_A = 1 - \frac{4\,(300)}{120\,(20)} = 0.5$$

From Fig. 10.10, $\sigma_h = k_A \gamma_t H = 0.5(120)(20) = 1200$ lb/ft². Using the earth pressure distribution shown in Fig. 10.10, i.e., case (b), gives

Resultant force = ½ (1200) (0.25) (20) + 1200 (0.75) (20) = 21,000 lb/ft

Earthquake analysis:

$$\frac{a_{max}}{g} = 0.20$$

$$P_E = \frac{3}{8}\left(\frac{a_{max}}{g}\right)H^2\gamma_t \qquad \text{Eq. (10.8)}$$

$$= \tfrac{3}{8}\,(0.20)\,(20^2)\,(120) = 3600 \text{ lb/ft}$$

10.16 *Static analysis:*

$$N_0 = \frac{\gamma_t H}{c} = \frac{120\,(20)}{1200} = 2$$

Therefore use case (c) in Fig. 10.10. From Fig. 10.10,

$$\sigma_{h2} = 0.4\gamma_t H = 0.4\,(120)\,(20) = 960 \text{ lb/ft}^2$$

Using the earth pressure distribution shown in Fig. 10.10, i.e., case (c), yields

Resultant force = ½ (960) (0.5) (20) + 960 (0.5) (20) = 14,400 lb/ft

Earthquake analysis:

$$\frac{a_{max}}{g} = 0.20$$

$$P_E = \frac{3}{8}\left(\frac{a_{max}}{g}\right)H^2\gamma_t \qquad \text{Eq. (10.8)}$$

$$= \tfrac{3}{8}\,(0.20)\,(20^2)\,(120) = 3600 \text{ lb/ft}$$

10.17 *Static analysis:*

Equation (10.2):

$$k_A = \tan^2 (45° - \tfrac{1}{2}\phi) = \tan^2 [45° - \tfrac{1}{2} (30°)] = 0.333$$

Equation (10.4):

$$k_p = \tan^2 (45° + \tfrac{1}{2}\phi) = \tan^2 [45° + \tfrac{1}{2} (30°)] = 3.0$$

To simplify the calculations for P_A, use Eq. (10.1) and assume that the groundwater table is at the top of the wall. Using γ_b in place of γ_t, therefore, gives

Equation (10.1):

$$P_A = \tfrac{1}{2}k_A\gamma_b H^2 = \tfrac{1}{2} (0.333) (9.7) (8^2) = 103 \text{ kN/m}$$

Equation (10.3):

$$P_p = \tfrac{1}{2}k_p\gamma_t D^2$$

Using γ_b in place of γ_t then gives

$$P_p = \tfrac{1}{2} (3.0) (9.7) (1^2) = 14.6 \text{ kN/m}$$

With reduction factor $= 2$, allowable $P_p = 7.3$ kN/m.
Resultant value of N and distance of N from the toe of footing:

Total weight of concrete box structure $= 823$ kN/m

Water pressure at bottom of box structure $= (8 - 0.4) (9.81 \text{ kN/m}^3) = 74.6$ kPa

Water force $= 74.6 (5) = 373$ kN/m

$W = N =$ buoyant unit weight of wall
$= 823 - 373 = 450$ kN/m

Take moments about the toe of the wall to determine the location of N:

$$Nx = -P_A \left(\frac{8}{3}\right) + W \text{ (moment arm)}$$

$$450x = -103 \left(\frac{8}{3}\right) + 450 \left(\frac{5}{2}\right)$$

$$x = \frac{850}{450} = 1.89 \text{ m}$$

Maximum and minimum bearing pressures:

$$q' = \frac{N (B + 6e)}{B^2} \qquad q'' = \frac{N (B - 6e)}{B^2} \qquad \text{Eqs. (8.7}a\text{) and (8.7}b\text{)}$$

Eccentricity $e = \dfrac{5}{2} - 1.89 = 0.61$ m

$$q' = \frac{450\,[5 + 6\,(0.61)]}{5^2} = 156 \text{ kPa}$$

$$q'' = \frac{450\,[5 - 6\,(0.61)]}{5^2} = 24 \text{ kPa}$$

Factor of safety for sliding:

$$\delta = \tfrac{2}{3}\phi' = \tfrac{2}{3}\,(30°) = 20°$$

$$\text{FS} = \frac{N \tan \delta + P_p}{P_A} \qquad \text{Eq. (10.11)}$$

$$= \frac{450 \tan 20° + 7.3}{103} = 1.66$$

Factor of safety for overturning:

$$\text{FS} = \frac{Wa}{\tfrac{1}{3}\,P_A H} \qquad \text{Eq. (10.12)}$$

$$= \frac{450\,(5/2)}{\tfrac{1}{3}\,(103)\,(8)} = 4.1$$

Earthquake analysis: To find P_L, we do as follows: From Prob. 6.15, $a_{max}/g = 0.16$, and the soil will liquefy from a depth of 1.2 to 6.7 m behind the retaining wall. To simplify the calculations, assume that the entire active wedge will liquefy during the earthquake. For the passive wedge and the soil beneath the bottom of the wall, the factor of safety against liquefaction is very high up to a depth of 13 m.

As indicated in Sec. 10.3.2, when the water levels are approximately the same on both sides of the retaining wall, use Eq. (10.1) with $k_A = 1$ [i.e., for $\phi' = 0$, $k_A = 1$, see Eq. (10.2)] and use γ_b (buoyant unit weight) in place of γ_t. Using Eq. (10.1) with $k_A = 1$ and $\gamma_b = 9.7$ kN/m³ gives

$$P_L = \tfrac{1}{2}k_A\gamma_b H^2 = \tfrac{1}{2}\,(1.0)\,(9.7)\,(8^2) = 310 \text{ kN/m}$$

Resultant value of N and distance of N from the toe of footing: Take moments about the toe of the wall to determine location of N.

$$Nx = -P_L\left(\frac{8}{3}\right) + W \text{ (moment arm)}$$

$$450x = -310\left(\frac{8}{3}\right) + 450\left(\frac{5}{2}\right)$$

$$x = \frac{298}{450} = 0.66 \text{ m}$$

Middle third of the foundation, $x = 1.67$ to 3.33 m. Therefore, N is not within the middle third of the foundation.

Factor of safety for sliding:

$$FS = \frac{N \tan \delta + P_p}{P_L}$$

$$= \frac{450 \tan 20° + 7.3}{310} = 0.55$$

Factor of safety for overturning:

$$FS = \frac{Wa}{\frac{1}{3} P_L H}$$

$$= \frac{450 (5/2)}{\frac{1}{3} (310) (8)} = 1.36$$

10.18 *Static analysis:* For the initial active earth pressure resultant force, Equation (10.2):

$$k_A = \tan^2 (45° - \tfrac{1}{2}\phi) = \tan^2 [45° - \tfrac{1}{2} (30°)] = 0.333$$

Equation (10.1):

$$P_A = \tfrac{1}{2} k_A \gamma_t H^2 = \tfrac{1}{2} (0.333) (20) (3^2) = 30 \text{ kN/m}$$

Total force acting on wall due to rise in groundwater level:

$$\text{Water pressure} = \tfrac{1}{2} \gamma_w H^2 = \tfrac{1}{2} (9.81) (3^2) = 44.1 \text{ kN/m}$$

$$P_A = \tfrac{1}{2} (0.333) (20 - 9.81) (3^2) = 15.3 \text{ kN/m}$$

$$\text{Total force} = 44.1 + 15.3 = 59.4 \text{ kN/m}$$

Earthquake analysis: As indicated in Sec. 10.3.2, for a condition of a water level only behind the retaining wall and for liquefaction of the entire backfill soil, use Eq. (10.1) with

$k_A = 1$ [i.e., for $\phi' = 0$, $k_A = 1$, see Eq. (10.2)] and use $\gamma_t = \gamma_{sat} = 20$ kN/m³.

$$P_L = \tfrac{1}{2} k_A \gamma_t H^2 = \tfrac{1}{2} (1.0) (20) (3^2) = 90 \text{ kN/m}$$

CHAPTER 11

11.1 *Static analysis:* First we find the minimum design load. The pipeline has a diameter D of 24 in (2 ft) and a depth of overburden H of 20 ft, and the backfill soil has a total unit weight γ_t of 125 lb/ft³. Therefore, the minimum vertical load W_{min} acting on the pipeline is

$$W_{min} = (125 \text{ lb/ft}^3) (20 \text{ ft}) (2 \text{ ft}) = 5000 \text{ lb per linear foot of pipe length}$$

For the embankment condition, use $B = D = 2$ ft, $H = 20$ ft, and $\gamma_t = 125$ lb/ft³. Figure 11.9a is entered with $H = 20$ ft, the curve marked 24 in (2 ft) is intersected, and the value of W read from the vertical axis is about 7600 lb. Therefore,

$$W = 7600 \left(\frac{125}{100} \right) = 9500 \text{ lb per linear foot of pipeline length}$$

Note this value of 9500 lb is greater than the minimum dead load (5000 lb), and the above value (9500 lb) would be used for the embankment condition.

For the trench condition, use $D = 2$ ft, $H = 20$ ft, and $\gamma_t = 125$ lb/ft³. Also assume that the trench width at the top of the pipeline will be 4 ft ($B = 4$ ft), and the trench will be backfilled with sand. Figure 11.9a is entered with $H/B = 20/4 = 5$, the curve marked *sands* is intersected, and the value of C_W of about 2.3 is obtained from the vertical axis. Therefore,

$$W = C_w \gamma_t B^2 = 2.3 \, (125) \, (4^2) = 4600 \text{ lb per linear foot}$$

Note this value of 4600 lb is less than the minimum value (5000 lb), and thus 5000 lb would be used for the trench condition.

For the jacked or driven pipelines, use $D = B = 2$ ft, $H = 20$ ft, and $\gamma_t = 125$ lb/ft³, and the pipeline will be jacked through a sand deposit. Figure 11.9 is entered with $H/B = 20/2 = 10$, the curve marked *sand* is intersected, and the value of C_w of about 1.6 is obtained from the vertical axis. Therefore,

$$W = C_w \gamma_t B^2 = 1.6 \, (125) \, (2^2) = 800 \text{ lb per linear foot}$$

Note this value of 800 lb is less than the minimum load value (5000 lb), and thus the value of 5000 lb would be used for the jacked or driven pipe condition. Basic soil mechanics indicates that the long-term load for rigid pipelines will be at least equal to the overburden soil pressure (i.e., the minimum design load).

Earthquake conditions: Use $B = 2$ ft, $H = 20$ ft, $\gamma_t = 125$ lb/ft³; and assume that for the design earthquake, the peak ground acceleration $a_{max} = 0.30g$. Using $k_v = \frac{2}{3}k_h = \frac{2}{3}(0.30) = 0.20$, the pseudostatic forces are as follows:

Minimum pseudostatic force:

$$F_v = k_v W_{min} = 0.20 \, (5000) = 1000 \text{ lb per linear foot}$$

Embankment condition:

$$F_v = k_v W = 0.20 \, (9500) = 1900 \text{ lb per linear foot}$$

For the trench condition,

$$F_v = k_v W_{min} = 0.20 \, (5000) = 1000 \text{ lb per linear foot}$$

For jacked or driven pipeline,

$$F_v = k_v W_{min} = 0.20 \, (5000) = 1000 \text{ lb per linear foot}$$

In summary, for the example problem of a 2-ft-diameter pipeline having 20 ft of overburden soil with a total unit weight of 125 lb/ft³, the soil loads are as follows:

Pipeline design	Minimum design load, lb/ft	Embankment condition, lb/ft	Trench condition, lb/ft	Jacked or driven pipeline, lb/ft
Static load W	5000	9500	5000*	5000*
Pseudostatic load F_v	1000	1900	1000*	1000*

*Using minimum design values.

11.2 Use the following data: soil profile type = S_C and seismic zone = 1.

1. Seismic coefficient C_a (Table 11.2): Entering Table 11.2 with soil profile type S_C and zone 1, the value of $C_a = 0.09$.
2. Seismic coefficient C_v (Table 11.5): Entering Table 11.5 with soil profile type S_C and zone 1, the value of $C_v = 0.13$.
3. Values of T_s and T_0 [Eqs. (11.5) and (11.6)]: The values of T_s and T_0 can be calculated as follows:

$$T_S = \frac{C_v}{2.5C_a} = \frac{0.13}{2.5\,(0.09)} = 0.58 \text{ s}$$

$$T_0 = 0.2T_s = 0.20\,(0.58) = 0.12 \text{ s}$$

By using Fig. 11.10 and the values $C_a = 0.09$, $C_v = 0.13$, $T_s = 0.58$ s, and $T_0 = 0.12$ s, the response spectrum can be determined and is shown in Fig. 11.12.

CHAPTER 13

13.1 See next page.

13.2 Divide the soil into two layers, as follows:

Layer 1:

$$\text{Depth} = 0 \text{ to } 1.5 \text{ m}$$

The first layer is located above the groundwater table. Consider conditions at the average depth = $(0 + 1.5)/2 = 0.75$ m. Assume pore water pressures are equal to zero above the groundwater table.

$$\sigma_v' = \gamma_t z = (18.3 \text{ kN/m}^3)\,(0.75 \text{ m}) = 13.7 \text{ kPa}$$

$$\sigma_h' = k_0 \sigma_v' = 0.5\,(13.7) = 6.9 \text{ kPa}$$

$$\text{Down-drag load} = (\text{pile perimeter})\,(\text{layer thickness})\,(\sigma_h' \tan \phi')$$

$$= \pi\,(0.3 \text{ m})\,(1.5 \text{ m})\,(6.9 \tan 28°) = 5.2 \text{ kN}$$

Layer 2:

$$\text{Depth} = 1.5 \text{ to } 6 \text{ m}$$

The second layer is located below the groundwater table. Consider conditions at the average depth = $(1.5 + 6)/2 = 3.75$ m.

13.1

Depth, m		Figure 7.1						Figure 7.2			
	$(N_1)_{60}$	FS	ε_v, percent	H, m	Settlement, cm		$(N_1)_{60}$	CSR	ε_v, percent	H, m	Settlement, cm
2	6.7	0.39	4.4	1.5	6.6		6.7	0.18	3.1	1.5	4.7
4	6.3	0.32	4.5	3	13.5		6.3	0.22	3.2	3	9.6
					Total = 20 cm						Total = 14 cm

Notes: Data obtained from Prob. 9.13.

$$\sigma'_v = \gamma_t \, (1.5 \text{ m}) + \gamma_b \, (3.75 - 1.5)$$

$$= (18.3 \text{ kN/m}^3) \, (1.5 \text{ m}) + (9.7 \text{ kN/m}^3) \, (2.25 \text{ m}) = 49.3 \text{ kPa}$$

$$\sigma'_h = k_0 \sigma'_v = 0.5 \, (49.3) = 24.6 \text{ kPa}$$

$$\text{Down-drag load} = (\text{pile perimeter}) \, (\text{layer thickness}) \, (\sigma'_h \tan \phi')$$

$$= \pi \, (0.3 \text{ m}) \, (4.5 \text{ m}) \, (24.6 \tan 28°) = 55.5 \text{ kN}$$

Total down-drag load $= 5.2 + 55.5 = 61 \text{ kN}$

13.3 Soil will liquefy down to a depth of about 20 m. Thus the piles should be at least 20 m long.

APPENDIX F
REFERENCES

AASHTO (1996). *Standard Specifications for Highway Bridges,* 16th ed. Prepared by the American Association of State Highway and Transportation Officials (AASHTO), Washington.

ASCE (1982). *Gravity Sanitary Sewer Design and Construction.* Manuals and Reports on Engineering Practice, No. 60. Joint publication: American Society of Civil Engineering (ASCE) and Water Pollution Control Federation, New York.

ASTM (2000). *Annual Book of ASTM Standards,* vol. 04.08: *Soil and Rock (I).* Standard No. D 653-97, "Standard Terminology Relating to Soil, Rock, and Contained Fluids." Terms prepared jointly by ASCE and American Society for Testing and Materials (ASTM), West Conshohocken, PA, pp. 43–77.

ASTM (2000). *Annual Book of ASTM Standards,* vol. 04.08: *Soil and Rock (I).* Standard No. D 1143-94, "Standard Test Method for Piles Under Static Axial Compressive Load," West Conshohocken, PA, pp. 96–106.

ASTM (2000). *Annual Book of ASTM Standards,* vol. 04.08: *Soil and Rock (I).* Standard No. D 1586-99, "Standard Test Method for Penetration Test and Split-Barrel Sampling of Soils," West Conshohocken, PA, pp. 139–143.

ASTM (2000). *Annual Book of ASTM Standards,* vol. 04.08: *Soil and Rock (I).* Standard No. D 2166-98, "Standard Test Method for Unconfined Compressive Strength of Cohesive Soil," West Conshohocken, PA, pp. 191–196.

ASTM (2000). *Annual Book of ASTM Standards,* vol. 04.08: *Soil and Rock (I).* Standard No. D 2844-94, "Standard Test Method for Resistance R-Value and Expansion Pressure of Compacted Soils," West Conshohocken, PA, pp. 267–274.

ASTM (2000). *Annual Book of ASTM Standards,* vol. 04.08: *Soil and Rock (I).* Standard No. D 2850-95, "Standard Test Method for Unconsolidated-Undrained Triaxial Compression Test on Cohesive Soils," West Conshohocken, PA, pp. 281–285.

ASTM (2000). *Annual Book of ASTM Standards,* vol. 04.08: *Soil and Rock (I).* Standard No. D 3080-98, "Standard Test Method for Direct Shear Test of Soils Under Consolidated Drained Conditions," West Conshohocken, PA, pp. 324–329.

ASTM (2000). Annual Book of ASTM Standards, vol. 04.08: *Soil and Rock (I).* Standard No. D 3441-98, "Standard Test Method for Mechanical Cone Penetration Tests of Soil," West Conshohocken, PA, pp. 373–377.

ASTM (2000). *Annual Book of ASTM Standards,* vol. 04.08: *Soil and Rock (I).* Standard No. D 4253-96, "Standard Test Methods for Maximum Index Density and Unit Weight of Soils Using a Vibratory Table," West Conshohocken, PA, pp. 525–536.

ASTM (2000). *Annual Book of ASTM Standards,* vol. 04.08, *Soil and Rock (I).* Standard No. D 4254-96, "Standard Test Method for Minimum Index Density and Unit Weight of Soils and Calculation of Relative Density," West Conshohocken, PA, pp. 538–545.

ASTM (2000). *Annual Book of ASTM Standards,* vol. 04.08: *Soil and Rock (I).* Standard No. D 4767-95, "Standard Test Method for Consolidated Undrained Triaxial Compression Test for Cohesive Soils," West Conshohocken, PA, pp. 882–891.

ASTM (2000). *Annual Book of ASTM Standards,* vol. 04.08: *Soil and Rock (I).* Standard No. D 4945-96, "Standard Test Method for High-Strain Dynamic Testing of Piles," West Conshohocken, PA, pp. 933–939.

ASTM (2000). *Annual Book of ASTM Standards,* vol. 04.08: *Soil and Rock (I).* Standard No. D 5311-96, "Standard Test Method for Load Controlled Cyclic Triaxial Strength of Soil," West Conshohocken, PA, pp. 1104–1113.

ASTM (2000). *Annual Book of ASTM Standards,* vol. 04.09: *Soil and Rock (II), Geosynthetics.* Standard No. D 4439-99, "Standard Terminology for Geosynthetics," West Conshohocken, PA, pp. 852–855.

ASTM (2000). *Annual Book of ASTM Standards,* vol. 04.09: *Soil and Rock (II), Geosynthetics.* Standard No. D 6467-99, "Standard Test Method for Torsional Ring Shear Test to Determine Drained Residual Shear Strength of Cohesive Soils," West Conshohocken, PA, pp. 832–836.

Abbott, P. L. (1996). *Natural Disasters.* William C. Brown Publishing Co., Dubuque, IA.

Ambraseys, N. N. (1960). "On the Seismic Behavior of Earth Dams." *Proceedings of the Second World Conference on Earthquake Engineering,* vol. 1, Tokyo and Kyoto, Japan, pp. 331–358.

Ambraseys, N. N. (1988). "Engineering Seismology." *Earthquake Engineering and Structural Dynamics,* vol. 17, pp. 1–105.

Ambraseys, N. N., and Menu, J. M. (1988). "Earthquake-Induced Ground Displacements." *Earthquake Engineering and Structural Dynamics,* vol. 16, pp. 985–1006.

Andrus, R. D., and Stokoe, K. H. (1997). "Liquefaction Resistance Based on Shear Wave Velocity." *Proceedings, NCEER Workshop on Evaluation of Liquefaction Resistance of Soils,* Technical Report NCEER-97-0022, T. L. Youd and I. M. Idriss, eds., National Center for Earthquake Engineering Research, Buffalo, pp. 89–128.

Andrus, R. D., and Stokoe, K. H. (2000). "Liquefaction Resistance of Soils from Shear-Wave Velocity." *Journal of Geotechnical and Geoenvironmental Engineering,* vol. 126, no. 11, pp. 1015–1025.

Arango, I. (1996). "Magnitude Scaling Factors for Soil Liquefaction Evaluations." *Journal of Geotechnical Engineering,* ASCE, vol. 122, no. 11, pp. 929–936.

Arnold, C., and Reitherman, R. (1982). *Building Configuration and Seismic Design.* Wiley, New York.

Asphalt Institute (1984). *Thickness Design—Asphalt Pavements for Highways and Streets.* The Asphalt Institute, College Park, MD.

Atkins, H. N. (1983). *Highway Materials, Soils, and Concretes,* 2d ed. Reston Publishing, Reston, VA.

Bartlett, S. F., and Youd, T. L. (1995). "Empirical Prediction of Liquefaction-Induced Lateral Spread." *Journal of Geotechnical Engineering,* ASCE, vol. 121, no. 4, pp. 316–329.

Bishop, A. W. (1955). "The Use of the Slip Circle in the Stability Analysis of Slopes." *Geotechnique,* London, England, vol. 5, no. 1, pp. 7–17.

Bjerrum, L. (1967). "The Third Terzaghi Lecture: Progressive Failure in Slopes of Overconsolidated Plastic Clay and Clay Shales." *Journal of the Soil Mechanics and Foundations Division,* ASCE, vol. 93, no. SM5, Part 1, pp. 1–49.

Bjerrum, L. (1972). "Embankments on Soft Ground." *Proceedings of the ASCE Specialty Conference on Performance of Earth and Earth-Supported Structures,* vol. 2. Purdue University, West Lafayette, IN, pp. 1–54.

Blake, T. F. (2000a). EQFAULT Computer Program, Version 3.00. Computer Program for Deterministic Estimation of Peak Acceleration from Digitized Faults. Thousand Oaks, CA.

Blake, T. F. (2000b). EQSEARCH Computer Program, Version 3.00. Computer Program for Estimation of Peak Acceleration from California Earthquake Catalogs. Thousand Oaks, CA.

Blake, T. F. (2000c). FRISKSP Computer Program, Version 4.00. Computer Program to Perform Probabilistic Earthquake Hazard Analyses Using Multiple Forms of Ground-Motion-Attenuation Relations. Thousand Oaks, CA.

Bolt, B. A. (1988). *Earthquakes.* Freeman, New York.

Bonilla, M. G. (1970). "Surface Faulting and Related Effects." Chapter 3 of *Earthquake Engineering,* Robert L. Wiegel, coordinating ed. Prentice-Hall, Englewood Cliffs, NJ, pp. 47–74.

Boone, S. T. (1996). "Ground-Movement-Related Building Damage." *Journal of Geotechnical Engineering,* ASCE, vol. 122, no. 11, pp. 886–896.

Boscardin, M. D., and Cording, E. J. (1989). "Building Response to Excavation-Induced Settlement." *Journal of Geotechnical Engineering,* ASCE, vol. 115, no. 1, pp. 1–21.

Bowles, J. E. (1982). *Foundation Analysis and Design,* 3d ed. McGraw-Hill, New York.

Brown, D. R., and Warner, J. (1973). "Compaction Grouting." *Journal of the Soil Mechanics and Foundations Division,* ASCE, vol. 99, no. SM8, pp. 589–601.

Brown, R. W. (1992). *Foundation Behavior and Repair, Residential and Light Commercial.* McGraw-Hill, New York.

Bruneau, M. (1999). "Structural Damage: Kocaeli, Turkey Earthquake, August 17, 1999." MCEER Deputy Director and Professor, Department of Civil, Structural and Environmental Engineering, University at Buffalo. Posted on the Internet.

California Department of Water Resources (1967). "Earthquake Damage to Hydraulic Structures in California." Bulletin 116-3, California Department of Water Resources, Sacramento.

California Division of Highways (1973). *Flexible Pavement Structural Design Guide for California Cities and Counties.* Sacramento.

California Institute of Technology (1971). *Strong Motion Earthquake Accelerograms: Corrected Accelerograms and Integrated Ground Velocities and Displacements.* Pasadena.

Casagrande, A. (1975). "Liquefaction and Cyclic Deformation of Sands, A Critical Review." *Proceedings of the 5th Panamerican Conference on Soil Mechanics and Foundation Engineering,* vol. 5, Buenos Aires, Argentina, pp. 79–133.

Cashman, P. M., and Harris, E. T. (1970). *Control of Groundwater by Water Lowering.* Conference on Ground Engineering, Institute of Civil Engineers, London.

Castro, G. (1975). "Liquefaction and Cyclic Mobility of Saturated Sands." *Journal of the Geotechnical Engineering Division,* ASCE, vol. 101, no. GT6, pp. 551–569.

Castro, G., Seed, R. B., Keller, T. O., and Seed, H. B. (1992). "Steady-State Strength Analysis of Lower San Fernando Dam Slide." *Journal of Geotechnical Engineering,* ASCE, vol. 118, no. 3, pp. 406–427.

Caterpillar Performance Handbook (1997). 28th ed. Prepared by Caterpillar, Inc., Peoria, IL.

Cedergren, H. R. (1989). *Seepage, Drainage, and Flow Nets,* 3d ed. Wiley, New York.

Cernica, J. N. (1995a). Geotechnical Engineering: Foundation Design. Wiley, New York.

Cernica, J. N. (1995b). *Geotechnical Engineering: Soil Mechanics.* Wiley, New York.

Christensen, D. H. (2000). "The Great Prince William Sound Earthquake of March 28, 1964." Geophysical Institute, University of Alaska, Fairbanks. Report obtained from the Internet.

Christensen, D. H., and Ruff, L. J. (1988). "Seismic Coupling and Outer Rise Earthquakes." *Journal of Geophysical Research,* vol. 93, no. 13, pp. 421–444.

Close, U., and McCormick, E. (1922). "Where the Mountains Walked." *National Geographic Magazine,* vol. 41, no. 5, pp. 445–464.

Cluff, L. S. (1971). "Peru Earthquake of May 31, 1970; Engineering Geology Observations." *Bulletin of the Seismological Society of America,* vol. 61, pp. 511–533.

Coduto, D. P. (1994). *Foundation Design, Principles and Practices.* Prentice-Hall, Englewood Cliffs, NJ.

Dakoulas, P., and Gazetas, G. (1986). "Seismic Shear Strains and Seismic Coefficients in Dams and Embankments." *Soil Dynamics and Earthquake Engineering,* vol. 5, no. 2, pp. 75–83.

Day, R. W. (1999). *Geotechnical and Foundation Engineering: Design and Construction.* McGraw-Hill, New York.

Day, R. W. (2000). *Geotechnical Engineer's Portable Handbook.* McGraw-Hill, New York.

Day, R. W. (2001a). *Soil Testing Manual: Procedures, Classification Data, and Sampling Practices.* McGraw-Hill, New York.

Day, R. W. (2001b). "Soil Mechanics and Foundations." Section 6 of *Building Design and Construction Handbook,* 6th ed. Frederick S. Merritt and Jonathan T. Ricketts, eds. McGraw-Hill, New York, pp. 6.1 to 6.121.

Day, R. W., and Thoeny, S. (1998). "Reactivation of a Portion of an Ancient Landslide." *Journal of the Environmental and Engineering Geoscience,* joint publication, AEG and GSA, vol. 4, no. 2, pp. 261–269.

de Mello, V. F. B. (1971). "The Standard Penetration Test." State-of-the-Art Report, *Fourth Pan American Conference on Soil Mechanics and Foundation Engineering,* vol. 1, San Juan, Puerto Rico, pp. 1–86.

Division of Mines and Geology (1997). *Guidelines for Evaluating and Mitigating Seismic Hazards in California,* Special Publication 117. Department of Conservation, Division of Mines and Geology, California.

Duke, C. M. (1960). "Foundations and Earth Structures in Earthquakes." *Proceedings of the Second World Conference on Earthquake Engineering,* vol. 1, Tokyo and Kyoto, Japan, pp. 435–455.

Duncan, J. M. (1996). "State of the Art: Limit Equilibrium and Finite-Element Analysis of Slopes." *Journal of Geotechnical Engineering.* ASCE, vol. 122, no. 7, pp. 577–596.

Earthweek: A Diary of the Planet (2001). "Salty Shifts." *The Los Angeles Times,* Feb. 11, 2001, Los Angeles, CA.

EERC (1995). "Geotechnical Reconnaissance of the Effects of the January 17, 1995, Hyogoken-Nanbu Earthquake, Japan." Report No. UCB/EERC-95/01. Earthquake Engineering Research Center (EERC), College of Engineering, University of California, Berkeley.

EERC (2000). Photographs and Descriptions from the Steinbrugge Collection, EERC, University of California, Berkeley. Photographs and description obtained from the Internet.

Ehlig, P. L. (1992). "Evolution, Mechanics, and Migration of the Portuguese Bend Landslide, Palos Verdes Peninsula, California." *Engineering Geology Practice in Southern California.* B. W. Pipkin and R. J. Proctor, eds. Star Publishing Company, Association of Engineering Geologists, southern California section, Special Publication No. 4, pp. 531–553.

EQE Summary Report (1995). "The January 17, 1995 Kobe Earthquake." Report posted on the EQE Internet site.

Federal Emergency Management Agency (1994). "Phoenix Community Earthquake Hazard Evaluation Maricopa County Arizona." Prepared for State of Arizona, Department of Emergency and Military Affairs, Division of Emergency Management, FEMA/NEHRP, Federal Emergency Management Agency Cooperative Agreement No. AZ102EPSA.

Fellenius, W. (1936). "Calculation of the Stability of Earth Dams." *Proceedings of the Second Congress on Large Dams,* vol. 4, Washington, pp. 445–463.

Finn, W. D. L., Bransby, P. L., and Pickering, D. J. (1970). "Effect of Strain History on Liquefaction of Sands." *Journal of the Soil Mechanics and Foundations Division,* ASCE, vol. 96, no. SM6, pp. 1917–1934.

Florensov, N. A., and Solonenko, V. P., eds. (1963). "Gobi-Altayskoye Zemletryasenie." *Iz. Akad. Nauk SSSR.* Also *The Gobi-Altai Earthquake,* U.S. Department of Commerce (English translation), Washington, 1965.

Fowler, C. M. R. (1990). *The Solid Earth: An Introduction to Global Geophysics.* Cambridge University Press, Cambridge, England.

Geologist and Geophysicist Act (1986). Prepared by the Board of Registration for Geologists and Geophysicists, Department of Consumer Affairs, state of California, Sacramento.

Geo-Slope (1991). *User's Guide, SLOPE/W for Slope Stability Analysis. Version 2.* Geo-Slope International, Calgary, Canada.

Gere, J. M., and Shah, H. C. (1984). *Terra Non Firma.* Freeman, New York.

Graf, E. D. (1969). "Compaction Grouting Techniques." *Journal of the Soil Mechanics and Foundations Division,* ASCE, vol. 95, no. SM5, pp. 1151–1158.

Grantz, A., Plafker, G., and Kachadoorian, R. (1964). *Alaska's Good Friday Earthquake, March 27, 1964.* Department of the Interior, Geological Survey Circular 491, Washington.

Gutenberg, B., and Richter, C. F. (1956). "Earthquake Magnitude, Intensity, Energy and Acceleration (Second Paper)." *Bulletin of the Seismology Society of America,* vol. 46, no. 2, pp. 143–145.

Hands, S. (1999). "Soft Stories Teach Hard Lesson to Taiwan's Construction Engineers." *Taipei Times,* Saturday, Oct. 9, 1999.

Hanks, T. C., and Kanamori, H. (1979). "A Moment Magnitude Scale." *Journal of Geophysical Research,* vol. 84, pp. 2348–2350.

Hansen, W. R. (1965). *Effects of the Earthquake of March 27, 1964 at Anchorage, Alaska.* Geological Survey Professional Paper 542-A. U.S. Department of the Interior, Washington.

Hawkins, A. B., and Privett, K. D. (1985). "Measurement and Use of Residual Shear Strength of Cohesive Soils." *Ground Engineering,* vol. 18, no. 8, pp. 22–29.

Heaton, T. H., Tajima, F., and Mori, A. W. (1982). *Estimating Ground Motions Using Recorded Accelerograms.* Report by Dames & Moore to Exxon Production Research Company, Houston, TX.

Hofmann, B. A., Sego, D. C., and Robertson, P. K. (2000). "In Situ Ground Freezing to Obtain Undisturbed Samples of Loose Sand." *Journal of Geotechnical and Geoenvironmental Engineering.* ASCE, vol. 126, no. 11, pp. 979–989.

Holtz, R. D., and Kovacs, W. D. (1981). *An Introduction to Geotechnical Engineering,* Prentice-Hall, Englewood Cliffs, NJ.

Housner, G. W. (1970). "Strong Ground Motion." Chapter 4 of *Earthquake Engineering,* Robert L. Wiegel, coordinating ed. Prentice-Hall, Englewood Cliffs, NJ, pp. 75–91.

Housner, G., ed. (1963). "An Engineering Report on the Chilean Earthquakes of May 1960." *Bulletin of the Seismological Society of America,* vol. 53, no. 2.

Hudson, D. E. (1970). "Dynamic Tests of Full-Scale Structures." Chapter 7 of *Earthquake Engineering,* Robert L. Wiegel, coordinating ed. Prentice-Hall, Englewood Cliffs, NJ, pp. 127–149.

Hynes-Griffin, M. E., and Franklin, A. G. (1984). "Rationalizing the Seismic Coefficient Method." Miscellaneous Paper GL-84-13. U.S. Army Corps of Engineers Waterways Experiment Station, Vicksburg, MS.

Idriss, I. M. (1985). "Evaluating Seismic Risk in Engineering Practice." *Proceedings of the Eleventh International Conference on Soil Mechanics and Foundation Engineering,* vol. 1, San Francisco, pp. 255–320.

Idriss, I. M. (1999). "Presentation Notes: An Update of the Seed-Idriss Simplified Procedure for Evaluating Liquefaction Potential." *Proceedings, TRB Workshop on New Approaches to Liquefaction Analysis.* Publication no. FHWA-RD-99-165, Federal Highway Administration, Washington.

Iida, K., Cox, D. C., and Pararas-Carayannis, G. (1967). "Preliminary Catalog of Tsunamis Occurring in the Pacific Ocean." Report No. HIG-67-10. Hawaii Institute of Geophysics, University of Hawaii, Honolulu.

Ishihara, K. (1985). "Stability of Natural Deposits During Earthquakes." *Proceedings of the Eleventh International Conference on Soil Mechanics and Foundation Engineering,* vol. 1, San Francisco, pp. 321–376.

Ishihara, K. (1993). "Liquefaction and Flow Failure During Earthquakes." *Geotechnique,* vol. 43, no. 3, London, England, pp. 351–415.

Ishihara, K., Sodekawa, M., and Tanaka, Y. (1978). "Effects of Overconsolidation on Liquefaction Characteristics of Sands Containing Fines." *Dynamic Geotechnical Testing,* ASTM Special Technical Publication 654, ASTM, Philadelphia, pp. 246–264.

Ishihara, K., and Yoshimine, M. (1992). "Evaluation of Settlements in Sand Deposits Following Liquefaction During Earthquakes." *Soils and Foundations,* vol. 32, no. 1, pp. 173–188.

Janbu, N. (1957). "Earth Pressure and Bearing Capacity Calculation by Generalized Procedure of Slices." *Proceedings of the 4th International Conference on Soil Mechanics and Foundation Engineering,* vol. 2, London, England, pp. 207–212.

Janbu, N. (1968). "Slope Stability Computations." Soil Mechanics and Foundation Engineering Report. The Technical University of Norway, Trondheim.

Johansson, J. (2000). "1964 Niigata Earthquake, Japan." Liquefaction Internet site, University of Washington, Department of Civil Engineering.

Kanamori, H. (1977). "The Energy Release in Great Earthquakes." *Journal of Geophysical Research,* vol. 82, pp. 2981–2987.

Kayan, R. E., Mitchell, J. K., Seed, R. B., Lodge, A., Nishio, S., and Coutinho, R. (1992). "Evaluation of SPT-, CPT-, and Shear Wave-Based Methods for Liquefaction Potential Assessments Using Loma Prieta Data." *Proceedings, 4th Japan-U.S. Workshop on Earthquake Resistant Design of Lifeline Facilities and Countermeasures for Soil Liquefaction,* NCEER-92-0019. National Center for Earthquake Engineering, Buffalo, NY, pp. 177–192.

Keefer, D. K. (1984). "Landslides Caused by Earthquakes." *Geological Society of America Bulletin,* vol. 95, no. 2, pp. 406–421.

Kennedy, M. P. (1975). "Geology of the Western San Diego Metropolitan Area, California." Bulletin 200. California Division of Mines and Geology, Sacramento.

Kerwin, S. T., and Stone, J. J. (1997). "Liquefaction Failure and Remediation: King Harbor Redondo Beach, California." *Journal of Geotechnical and Geoenvironmental Engineering,* ASCE, vol. 123, no. 8, pp. 760–769.

Kramer, S. L. (1996). *Geotechnical Earthquake Engineering.* Prentice-Hall, Englewood Cliffs, NJ.

Krinitzsky, E. L., Gould, J. P., and Edinger, P. H. (1993). *Fundamentals of Earthquake-Resistant Construction.* Wiley, New York.

Ladd, C. C., Foote, R., Ishihara, K., Schlosser, F., and Poulos, H. G. (1977). "Stress-Deformation and Strength Characteristics." State-of-the-Art Report, *Proceedings, 9th International Conference on Soil Mechanics and Foundation Engineering,* vol. 2, Japanese Society of Soil Mechanics and Foundation Engineering, Tokyo, pp. 421–494.

Lambe, T. W., and Whitman, R. V. (1969). *Soil Mechanics.* Wiley, New York.

Lawson, A. C., et al. (1908). *The California Earthquake of April 18, 1906—Report of the State Earthquake Investigation Commission,* vol. 1, pt. 1, pp. 1–254; pt. 2, pp. 255–451. Publication 87, Carnegie Institution of Washington.

Maksimovic, M. (1989). "On the Residual Shearing Strength of Clays." *Geotechnique,* London, England, vol. 39, no. 2, pp. 347–351.

Marcuson, W. F. (1981). "Moderator's Report for Session on 'Earth Dams and Stability of Slopes Under Dynamic Loads.'" *Proceedings, International Conference on Recent Advances in Geotechnical Earthquake Engineering and Soil Dynamics,* vol. 3, St. Louis, Missouri, p. 1175.

Marcuson, W. F., and Hynes, M. E. (1990). "Stability of Slopes and Embankments During Earthquakes." *Proceedings,* Geotechnical Seminar Sponsored by ASCE and the Pennsylvania Department of Transportation, Hershey, PA.

Marston, A. (1930). "The Theory of External Loads on Closed Conduits in the Light of the Latest Experiments." Bulletin No. 96, Iowa Engineering Experiment Station.

McCarthy, D. F. (1977). *Essentials of Soil Mechanics and Foundations.* Reston Publishing, Reston, VA.

Meyerhof, G. G. (1951). "The Ultimate Bearing Capacity of Foundations." *Geotechnique,* vol. 2, no. 4, pp. 301–332.

Meyerhof, G. G. (1953). "Bearing Capacity of Foundations Under Eccentric and Inclined Loads." *Proceedings of the Third International Conference on Soil Mechanics and Foundation Engineering,* vol. 1, Zurich, pp. 440–445.

Meyerhof, G. G. (1961). Discussion of "Foundations Other Than Piled Foundations." *Proceedings of Fifth International Conference on Soil Mechanics and Foundation Engineering,* vol. 3, Paris, p. 193.

Meyerhof, G. G. (1965). "Shallow Foundations." *Journal of the Soil Mechanics and Foundations Division,* ASCE, vol. 91, no. SM2, pp. 21–31.

Mitchell, J. K. (1970). "In-Place Treatment of Foundation Soils." *Journal of the Soil Mechanics and Foundation Division,* ASCE, vol. 96, no. SM1, pp. 73–110.

Mononobe, N., and Matsuo, H. (1929). "On the Determination of Earth Pressures During Earthquakes." *Proceedings, World Engineering Congress.*

Morgenstern, N. R., and Price, V. E. (1965). "The Analysis of the Stability of General Slip Surfaces." *Geotechnique,* vol. 15, no. 1, London, England, pp. 79–93.

Mulilis, J. P., Townsend, F. C., and Horz, R. C. (1978). "Triaxial Testing Techniques and Sand Liquefaction." *Dynamic Geotechnical Testing,* ASTM Special Technical Publication 654. ASTM, Philadelphia, pp. 265–279.

Nadim, F. (1982). "A Numerical Model for Evaluation of Seismic Behavior of Gravity Retaining Walls." Research Report R82-33. Department of Civil Engineering, Massachusetts Institute of Technology, Cambridge.

Nadim, F., and Whitman, R. V. (1984). "Coupled Sliding and Tilting of Gravity Retaining Walls During Earthquakes." *Proceedings, 8th World Conference on Earthquake Engineering,* vol. 3, San Francisco, pp. 477–484.

Namson, J. S., and Davis, T. L. (1988). "Seismically Active Fold and Thrust Belt in the San Joaquin Valley, Central California." *Geological Society of America Bulletin,* vol. 100, pp. 257–273.

National Information Service for Earthquake Engineering (2000). "Building and Its Structure Should Have a Uniform and Continuous Distribution of Mass, Stiffness, Strength and Ductility." University of California, Berkeley. Report obtained from the Internet.

National Research Council (1985). *Liquefaction of Soils During Earthquakes.* National Academy Press, Washington.

NAVFAC DM-7.1 (1982). *Soil Mechanics, Design Manual 7.1.* Department of the Navy, Naval Facilities Engineering Command, Alexandria, VA.

NAVFAC DM-7.2 (1982). *Foundations and Earth Structures, Design Manual 7.2.* Department of the Navy, Naval Facilities Engineering Command, Alexandria, VA.

NAVFAC DM-7.3 (1983). *Soil Dynamics, Deep Stabilization, and Special Geotechnical Construction, Design Manual 7.3.* Department of the Navy, Naval Facilities Engineering Command, Alexandria, VA.

Nelson, N. A. (2000). "Slope Stability, Triggering Events, Mass Wasting Hazards." Tulane University, Natural Disasters Internet site.

Newmark, N. (1965). "Effects of Earthquakes on Dams and Embankments." *Geotechnique,* London, vol. 15, no. 2, pp. 139–160.

Ohsaki, Y. (1969). "The Effects of Local Soil Conditions upon Earthquake Damage." Paper presented at Session no. 2, Soil Dynamics, *Proceedings of the Seventh International Conference on Soil Mechanics and Foundation Engineering,* vol. 3, Mexico, pp. 421–422.

Ohta, Y., and Goto, N. (1976). "Estimation of S-Wave Velocity in Terms of Characteristic Indices of Soil." *Butsuri-Tanko,* vol. 29, no. 4, pp. 34–41, in Japanese.

Okabe, S. (1926). "General Theory of Earth Pressures." *Journal of the Japan Society of Civil Engineering,* vol. 12, no. 1.

Oldham, R. D. (1899). "Report on the Great Earthquake of 12th June, 1897." India Geologic Survey Memorial, Publication 29.

Olson, S. M., Stark, T. D., Walton, W. H., and Castro, G. (2000). "1907 Static Liquefaction Flow Failure of the North Dike of Wachusett Dam." *Journal of Geotechnical and Geoenvironmental Engineering,* ASCE, vol. 126, no. 12, pp. 1184–1193.

Orange County Grading Manual (1993). Part of the *Orange County Grading and Excavation Code,* prepared by Orange County, CA.

Pacheco, J. F., and Sykes, L. R. (1992). "Seismic Moment Catalog of Large Shallow Earthquakes, 1900 to 1989." *Bulletin of the Seismological Society of America,* vol. 82, pp. 1306–1349.

Peck, R. B., Hanson, W. E., and Thornburn, T. H. (1974). *Foundation Engineering.* Wiley, New York.

Peckover, F. L. (1975). "Treatment of Rock Falls on Railway Lines." Bulletin 653, American Railway Engineering Association, Chicago, IL, pp. 471–503.

Perloff, W. H., and Baron, W. (1976). *Soil Mechanics, Principles and Applications.* Wiley, New York.

Piteau, D. R., and Peckover, F. L. (1978). "Engineering of Rock Slopes." *Landslides, Analysis and Control.* Special Report 176, chap. 9, pp. 192–228. Transportation Research Board, National Academy of Sciences.

Plafker, G. (1972). "Alaskan Earthquake of 1964 and Chilean Earthquake of 1960: Implications for Arc Tectonics." *Journal of Geophysical Research,* vol. 77, pp. 901–925.

Plafker, G., Ericksen, G. E., and Fernandez, C. J. (1971). "Geological Aspects of the May 31, 1970, Peru Earthquake." *Bulletin of the Seismological Society of America,* vol. 61, pp. 543–578.

Post-Tensioning Institute (1996). "Design and Construction of Post-Tensioned Slabs-on-Ground," 2d ed. Report. Phoenix, AZ.

Poulos, S. J., Castro, G., and France, J. W. (1985). "Liquefaction Evaluation Procedure." *Journal of Geotechnical Engineering,* ASCE, vol. 111, no. 6, pp. 772–792.

Pyke, R., Seed, H. B., and Chan, C. K. (1975). "Settlement of Sands Under Multidirectional Shaking." *Journal of the Geotechnical Engineering Division,* ASCE, vol. 101, no. GT 4, pp. 379–398.

Richter, C. F. (1935). "An Instrumental Earthquake Magnitude Scale." *Bulletin of the Seismological Society of America,* vol. 25, pp. 1–32.

Richter, C. F. (1958). *Elementary Seismology.* Freeman, San Francisco.

Ritchie, A. M. (1963). "Evaluation of Rockfall and Its Control." *Highway Research Record 17*, Highway Research Board, Washington, pp. 13–28.

Robertson, P. K., and Campanella, R. G. (1983). "Interpretation of Cone Penetration Tests: Parts 1 and 2." *Canadian Geotechnical Journal*, vol. 20, pp. 718–745.

Robertson, P. K., Woeller, D. J., and Finn, W. D. L. (1992). "Seismic Cone Penetration Test for Evaluating Liquefaction Potential Under Cyclic Loading." *Canadian Geotechnical Journal*, Ottawa, vol. 29, pp. 686–695.

Rollings, M. P., and Rollings, R. S. (1996). *Geotechnical Materials in Construction*. McGraw-Hill, New York.

Savage, J. C., and Hastie, L. M. (1966). "Surface Deformation Associated with Dip-Slip Faulting." *Journal of Geophysical Research*, vol. 71, no. 20, pp. 4897–4904.

Seed, H. B. (1970). "Soil Problems and Soil Behavior." Chapter 10 of *Earthquake Engineering*, Robert L. Wiegel, coordinating ed. Prentice-Hall, Englewood Cliffs, NJ, pp. 227–251.

Seed, H. B. (1979a). "Soil Liquefaction and Cyclic Mobility Evaluation for Level Ground During Earthquakes." *Journal of the Geotechnical Engineering Division*, ASCE, vol. 105, no. GT2, pp. 201–255.

Seed, H. B. (1979b). "Considerations in the Earthquake-Resistant Design of Earth and Rockfill Dams." *Geotechnique*, vol. 29, no. 3, pp. 215–263.

Seed, H. B. (1987). "Design Problems in Soil Liquefaction." *Journal of Geotechnical Engineering*, ASCE, vol. 113, no. 8, pp. 827–845.

Seed, H. B., and De Alba, (1986). "Use of SPT and CPT Tests for Evaluating the Liquefaction Resistance of Sands." *Proceedings, In Situ 1986, ASCE Specialty Conference on Use of In Situ Testing in Geotechnical Engineering*. Special Publication no. 6, ASCE, New York.

Seed, H. B., and Idriss, I. M. (1971). "Simplified Procedure for Evaluating Soil Liquefaction Potential." *Journal of the Soil Mechanics and Foundations Division*, ASCE, vol. 97, no. SM9, 1249–1273.

Seed, H. B., and Idriss, I. M. (1982). *Ground Motions and Soil Liquefaction During Earthquakes*. Earthquake Engineering Research Institute, University of California, Berkeley.

Seed, H. B., Idriss, I. M., and Arango, I. (1983). "Evaluation of Liquefaction Potential Using Field Performance Data." *Journal of Geotechnical Engineering*, ASCE, vol. 109, no. 3, pp. 458–482.

Seed, H. B., and Lee, K. L. (1965). "Studies of Liquefaction of Sands Under Cyclic Loading Conditions." Report TE-65-65. Department of Civil Engineering, University of California, Berkeley.

Seed, H. B., and Martin, G. R. (1966). "The Seismic Coefficient in Earth Dam Design." *Journal of the Soil Mechanics and Foundations Division*, ASCE, vol. 92, no. SM3, pp. 25–58.

Seed, H. B., Mori, K., and Chan, C. K. (1975). "Influence of Seismic History on the Liquefaction Characteristics of Sands." Report EERC 75-25. Earthquake Engineering Research Center, University of California, Berkeley.

Seed, H. B., and Peacock, W. H. (1971). "Test Procedures for Measuring Soil Liquefaction Characteristics." *Journal of the Soil Mechanics and Foundations Division*, ASCE, vol. 97, no. SM8, pp. 1099–1119.

Seed, H. B., and Silver, M. L. (1972). "Settlement of Dry Sands During Earthquakes." *Journal of the Soil Mechanics and Foundations Division*, ASCE, vol. 98, no. SM4, pp. 381–397.

Seed, H. B., Tokimatsu, K., and Harder, L. F. (1984). "Moduli and Damping Factors for Dynamic Analyses of Cohesionless Soils." Report No. UCB/EERC-84/14. Earthquake Engineering Research Center, University of California, Berkeley.

Seed, H. B., Tokimatsu, K., Harder, L. F., and Chung, R. (1985). "Influence of SPT Procedures in Soil Liquefaction Resistance Evaluations." *Journal of Geotechnical Engineering*, ASCE, vol. 111, no. 12, pp. 1425–1445.

Seed, H. B., and Whitman, R. V. (1970). "Design of Earth Retaining Structures for Dynamic Loads." *Proceedings, ASCE Specialty Conference on Lateral Stresses in the Ground and Design of Earth Retaining Structures*, ASCE, pp. 103–147.

Seed, H. B., Wong, R. T., Idriss, I. M., and Tokimatsu, K. (1986). "Moduli and Damping Factors for Dynamic Analyses of Cohesionless Soils." *Journal of Geotechnical Engineering*, ASCE, vol. 112, no. 11, pp. 1016–1032.

Seed, R. B. (1991). *Liquefaction Manual.* Course Notes for CE 275: Geotechnical Earthquake Engineering, College of Engineering, University of California, Berkeley.

Seed, R. B., Dickenson, S. E., Reimer, M. F., Bray, J. D., Sitar, N., Mitchell, J. K., Idriss, I. M., Kayen, R. E., Kropp, A., Harder, L. F., and Power, M. S. (1990). "Preliminary Report on the Principal Geotechnical Aspects of the October 17, 1989 Loma Prieta Earthquake." Report UCB/EERC-90/05. Earthquake Engineering Research Center, University of California, Berkeley.

Seed, R. B., and Harder, L. F. (1990). "SPT-Based Analysis of Cyclic Pore Pressure Generation and Undrained Residual Strength." J. M. Duncan, ed. *Proceedings, H. Bolton Seed Memorial Symposium,* vol. 2. University of California, Berkeley, pp. 351–376.

Shannon and Wilson, Inc. (1964). *Report on Anchorage Area Soil Studies, Alaska, to U.S. Army Engineer District, Anchorage, Alaska.* Seattle, WA.

Sherif, M. A., and Fang, Y. S. (1984a). "Dynamic Earth Pressures on Wall Rotating About the Base." *Proceedings, 8th World Conference on Earthquake Engineering,* vol. 6, San Francisco, pp. 993–1000.

Sherif, M. A., and Fang, Y. S. (1984b). "Dynamic Earth Pressures on Wall Rotating About the Top." *Soils and Foundations,* vol. 24, no. 4, pp. 109–117.

Sherif, M. A., Ishibashi, I., and Lee, C. D. (1982)."Earth Pressures Against Rigid Retaining Walls." *Journal of the Geotechnical Engineering Division,* ASCE, vol. 108, no. GT5, pp. 679–695.

Siddharthan, R., Ara, S., and Norris, G. M. (1992). "Simple Rigid Plastic Model for Seismic Tilting of Rigid Walls." *Journal of Structural Engineering,* ASCE, vol. 118, no. 2, pp. 469–487.

Silver, M. L., and Seed, H. B. (1971). "Volume Changes in Sands During Cyclic Loading." *Journal of the Soil Mechanics and Foundations Division,* ASCE, vol. 97, no. SM9, pp. 1171–1182.

Skempton, A. W. (1964). "Long-Term Stability of Clay Slopes." *Geotechnique,* London, England, vol. 14, no. 2, pp. 75–101.

Skempton, A. W. (1985). "Residual Strength of Clays in Landslides, Folded Strata and the Laboratory." *Geotechnique,* London, England, vol. 35, no.1, pp. 3–18.

Skempton, A. W., (1986). "Standard Penetration Test Procedures and the Effects in Sands of Overburden Pressure, Relative Density, Particle Size, Aging and Overconsolidation." *Geotechnique,* vol. 36, no. 3, pp. 425–447.

Skempton, A. W. and Hutchinson, J. (1969). "State of-the-Art Report: Stability of Natural Slopes and Embankment Foundations." *Proceedings of the Seventh International Conference on Soil Mechanics and Foundation Engineering,* Mexico, pp. 291–340.

Sokolowski, T. J. (2000). "The Great Alaskan Earthquake & Tsunamis of 1964." West Coast & Alaska Tsunami Warning Center, Palmer, AK. Report obtained from the Internet.

Southern California Earthquake Data Center (2000). Internet site for the San Fernando earthquake.

Southern Nevada Building Code Amendments (1997). Adopted by Clark County, Boulder City, North Las Vegas, City of Las Vegas, City of Mesquite, and City of Henderson. Published December 15, 1997.

Sowers, G. B., and Sowers, G. F. (1970). *Introductory Soil Mechanics and Foundations.* 3d ed., Macmillan, New York.

Spencer, E. (1967). "A Method of Analysis of the Stability of Embankments Assuming Parallel Interslice Forces." *Geotechnique,* London, England, vol. 17, pp. 11–26.

Spencer, E. (1968). "Effect of Tension on Stability of Embankments." *Journal of the Soil Mechanics and Foundations Division,* ASCE, vol. 94, no. SM5, pp. 1159–1173.

Stark, T. D., and Eid, H. T. (1994). "Drained Residual Strength of Cohesive Soils." *Journal of Geotechnical Engineering,* ASCE, vol. 120, no. 5, pp. 856–871.

Stark, T. D., and Olson, S. M. (1995). "Liquefaction Resistance Using CPT and Field Case Histories." *Journal of Geotechnical Engineering,* ASCE, vol. 121, no. 12, pp. 856–869.

Stark, T. D., and Mesri, G. (1992). "Undrained Shear Strength of Liquefied Sands for Stability Analysis." *Journal of Geotechnical Engineering,* ASCE, vol. 118, no. 11, pp. 1727–1747.

State of California Special Studies Zones Maps (1982). Prepared by the state of California based on the Alquist-Priolo Special Studies Zones Act.

Stearns, R. G., and Wilson, C. W. (1972). *Relationships of Earthquakes and Geology in West Tennessee and Adjacent Areas.* Tennessee Valley Authority, Knoxville.

Steedman, R. S., and Zeng, X. (1990). "The Seismic Response of Waterfront Retaining Walls." *Proceedings, ASCE Speeialty Conference on Design and Performance of Earth Retaining Structures.* Special Technical Publication 25. Cornell University, Ithaca, NY, pp. 872–886.

Steinbrugge, K. V. (1970). "Earthquake Damage and Structural Performance in the United States." Chapter 9 of *Earthquake Engineering,* Robert L. Wiegel, coordinating ed. Prentice-Hall, Englewood Cliffs, NJ, pp. 167–226.

Stokes, W. L., and Varnes, D. J. (1955). "Glossary of Selected Geologic Terms with Special Reference to Their Use in Engineering." *Colorado Scientific Society Proceedings,* Denver, vol. 16.

Stone, W. C., Yokel, F. Y., Celebi, M., Hanks, T., and Leyendecker, E. V. (1987). "Engineering Aspects of the September 19, 1985 Mexico Earthquake." *NBS Building Science Series 165.* National Bureau of Standards, Washington.

Sykora, D. W. (1987). "Creation of a Data Base of Seismic Shear Wave Velocities for Correlation Analysis." Geotechnical Laboratory Miscellaneous Paper GL-87-26. U.S. Army Engineer Waterways Experiment Station, Vicksburg, MS.

Taniguchi, E., and Sasaki, Y. (1986). "Back Analysis of Landslide due to Naganoken Seibu Earthquake of September 14, 1984." *Proceedings, XI ISSMFE Conference, Session 7B, San Francisco, California.* University of Missouri, Rolla.

Terzaghi, K. (1943). *Theoretical Soil Mechanics.* Wiley, New York.

Terzaghi, K. (1950). "Mechanisms of Landslides." *Engineering Geology Volume.* Geological Society of America.

Terzaghi, K., and Peck, R. B. (1967). *Soil Mechanics in Engineering Practice,* 2d ed. Wiley, New York.

Tokimatsu, K., and Seed, H. B. (1984). *Simplified Procedures for the Evaluation of Settlements in Sands due to Earthquake Shaking.* Report no. UCB/EERC-84/16, sponsored by the National Science Foundation, Earthquake Engineering Research Center, College of Engineering, University of California, Berkeley.

Tokimatsu, K., and Seed, H. B. (1987). "Evaluation of Settlements in Sands due to Earthquake Shaking." *Journal of Geotechnical Engineering,* ASCE, vol. 113, no. 8, pp. 861–878.

Townsend, F. C. (1978). "A Review of Factors Affecting Cyclic Triaxial Tests." *Dynamic Geotechnical Testing,* ASTM Special Technical Publication 654. ASTM, Philadelphia, pp. 356–383.

Uniform Building Code (1997). International Conference of Building Officials, three volumes, Whittier, CA.

USCOLD (1985). "Guidelines for Selecting Seismic Parameters for Dam Projects," *Report of Committee on Earthquakes.* U.S. Committee on Large Dams.

USGS (1975). *Encinitas Quadrangle, San Diego, California.* Topographic map was mapped, edited, and published by the U.S. Geological Survey, Denver, CO.

USGS (1994). *USGS Response to an Urban Earthquake, Northridge.* Open-File Report 96-263. Report posted by U.S. Geological Survey on the Internet.

USGS (1996). *Peak Ground Acceleration Maps.* Maps prepared by the U.S. Geological Survey, National Seismic Hazards Mitigation Project. Maps obtained from the Internet.

USGS (1997). *Index of Publications of the Geological Survey.* Department of the Interior, Washington.

USGS (2000a). Earthquake Internet site. Prepared by U.S. Geological Survey, National Earthquake Information Center, Golden, CO.

USGS (2000b). *Glossary of Some Common Terms in Seismology.* United States Geological Survey Earthquake Hazards Program, National Earthquake Information Center, World Data Center for Seismology, Denver, CO. Glossary obtained from the Internet.

Vesic, A. S. (1963). "Bearing Capacity of Deep Foundations in Sand." *Highway Research Record, 39.* National Academy of Sciences, National Research Council, Washington, pp. 112–153.

Vesic, A. S. (1967). "Ultimate Loads and Settlements of Deep Foundations in Sand." *Proceedings of the Symposium on Bearing Capacity and Settlement of Foundations.* Duke University, Durham, NC, p. 53.

Vesic, A. S. (1975). "Bearing Capacity of Shallow Foundations." Chapter 3 of *Foundation Engineering Handbook*. Hans F. Winterkorn and Hsai-Yang Fang, eds. Van Nostrand Reinhold, New York, pp. 121–147.

Warner, J. (1978). "Compaction Grouting—A Significant Case History." *Journal of the Geotechnical Engineering Division*, ASCE, vol. 104, no. GT7, pp. 837–847.

Warner, J. (1982). "Compaction Grouting—The First Thirty Years." *Proceedings of the Conference on Grouting in Geotechnical Engineering*. W. H. Baker, ed. ASCE, New York, pp. 694–707.

Watry, S. M., and Ehlig, P. L. (1995). "Effect of Test Method and Procedure on Measurements of Residual Shear Strength of Bentonite from the Portuguese Bend Landslide." *Clay and Shale Slope Instability*, vol. 10. W. C. Haneberg and S. A. Anderson, eds. Geological Society of America, Reviews in Engineering Geology, Boulder, CO, pp. 13–38.

Wellington, A. M. (1888). "Formulae for Safe Loads of Bearing Piles." *Engineering News*, no. 20, pp. 509–512.

Whitman, R. V. (1990). "Seismic Design Behavior of Gravity Retaining Walls." *Proceedings, ASCE Specialty Conference on Design and Performance of Earth Retaining Structures*. Geotechnical Special Publication 25. ASCE, New York, pp. 817–842.

Whitman, R. V., and Bailey, W. A. (1967). "Use of Computers for Slope Stability Analysis." *Journal of the Soil Mechanics and Foundations Division*, ASCE, vol. 93, no. SM4, pp. 475–498.

Wilson, R. C., and Keefer, D. K. (1985). "Predicting Areal Limits of Earthquake-Induced Lansliding." In *Evaluating Earthquake Hazards in the Los Angeles Region*, J. I. Ziony, ed. Professional Paper 1360. U.S. Geological Survey, Reston, VA, pp. 317–345.

Woods, R. D., ed. (1994). *Geophysical Characterization of Sites*. Balkema, Rotterdam, The Netherlands.

Yeats, R. S., Sieh, K., and Allen, C. R. (1997). *The Geology of Earthquakes*. Oxford University Press, New York.

Yoshimi, Y., Tokimatsu, K., and Hasaka, Y. (1989). "Evaluation of Liquefaction Resistance of Clean Sands Based on High-Quality Undisturbed Samples." *Soils and Foundations*, vol. 29, no. 1, pp. 93–104.

Youd, T. L. (1978). "Major Cause of Earthquake Damage Is Ground Failure." *Civil Engineering Magazine*, ASCE, vol. 48, no. 4, pp. 47–51.

Youd, T. L. (1984). "Geologic Effects—Liquefaction and Associated Ground Failures." *Proceedings, Geologic and Hydrologic Hazards Training Program*. Open File Report 84-760. U.S. Geological Survey, Menlo Park, CA, pp. 210–232.

Youd, T. L., and Gilstrap, S. D. (1999). "Liquefaction and Deformation of Silty and Fine-Grained Soils." *Earthquake Geotechnical Engineering*, 2d 3d., Balkema, Rotterdam, pp.1013–1020.

Youd, T. L., and Hoose, S. N. (1978). "Historic Ground Failures in Northern California Triggered by Earthquakes." Professional Paper 933. U. S. Geological Survey, Washington.

Youd, T. L., and Idriss, I. M., eds. (1997). *Proceedings, NCEER Workshop Evaluation of Liquefaction Resistance of Soils*. National Center for Earthquake Engineering Research, State University of New York, Buffalo.

Youd, T. L., and Idriss, I. M. (2001). "Liquefaction Resistance of Soils: Summary Report from the 1996 NCEER and 1998 NCEER/NSF Workshops on Evaluation of Liquefaction Resistance of Soils." *Journal of Geotechnical and Geoenvironmental Engineering*, ASCE, vol. 127, no. 4, pp. 297–313.

Youd, T. L., and Noble, S. K. (1997). "Liquefaction Criteria Based on Statistical and Probabilistic Analyses." *Proceedings, NCEER Workshop on Evaluation of Liquefaction Resistance of Soils*, Technical Report NCEER-97-0022. T. L. Youd and I. M. Idriss, eds., National Center for Earthquake Engineering Research, Buffalo, NY, pp. 201–215.

Zhang, Z., and Lanmin, W. (1995). "Geological Disasters in Loess Areas During the 1920 Haiyuan Earthquake, China." *GeoJournal*, vol. 36, pp. 269–274.

INDEX

ABOUT THE AUTHOR

ROBERT W. DAY is a leading geotechnical engineer and the Chief Engineer at American Geotechnical in San Diego, California. The author of over 200 published technical papers and four textbooks (*Forensic Geotechnical and Foundation Engineering, Geotechnical and Foundation Engineering: Design and Construction, Geotechnical Engineer's Portable Handbook*, and *Soil Testing Manual*), he serves on advisory committees for several professional associations, including ASCE, ASTM, and NCEES. He holds four college degrees: two from Villanova University (bachelor's and master's degrees majoring in structural engineering), and two from the Massachusetts Institute of Technology [master's and the Civil Engineer degree (highest degree) majoring in geotechnical engineering]. He is also a registered civil engineer in several states and a registered geotechnical engineer in California.

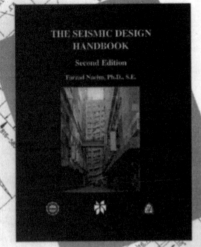